Electrical
Wiring
Residential

Based on the 1996
NATIONAL ELECTRICAL CODE®

Electrical Wiring

Residential
Twelfth Edition

Ray C. Mullin

Delmar Publishers

I(T)P An International Thomson Publishing Company

Albany • Bonn • Boston • Cincinnati • Detroit • London • Madrid • Melbourne • Mexico City
New York • Pacific Grove • Paris • San Francisco • Singapore • Tokyo • Toronto • Washington

NOTICE TO THE READER

Cover Image by Mick Brady

Delmar Staff
Publisher: Susan Simpfenderfer
Acquisitions Editor: Paul Shepardson
Developmental Editor: Jeanne M. Mesick
Project Editor: Eleanor Isenhart
Production Coordinator: Dianne Jensis
Art and Design Coordinator: Cheri Plasse

COPYRIGHT © 1996
By Delmar Publishers
a division of International Thomson Publishing Inc.

The ITP logo is a trademark under license.

Printed in the United States of America

For more information, contact:

Delmar Publishers
3 Columbia Circle, Box 15015
Albany, New York 12212-5015

International Thomson Publishing Europe
Berkshire House 168-173
High Holborn
London, WC1V 7AA
England

Thomas Nelson Australia
102 Dodds Street
South Melbourne, 3205
Victoria, Australia

Nelson Canada
1120 Birchmont Road
Scarborough, Ontario
Canada, M1K 5G4

International Thomson Editores
Campos Eliseos 385, Piso 7
Col Polanco
11560 Mexico D F Mexico

International Thomson Publishing GmbH
Konigswinterer Strasse 418
53227 Bonn
Germany

International Thomson Publishing Asia
221 Henderson Road
#05-10 Henderson Building
Singapore 0315

International Thomson Publishing—Japan
Hirakawacho Kyowa Building, 3F
2-2-1 Hirakawacho
Chiyoda-ku, Tokyo 102
Japan

1 2 3 4 5 6 7 8 9 10 XXX 01 00 99 98 97 96 95

Library of Congress Cataloging-in-Publication Data

Mullin, Ray C.
    Electrical wiring, residential / Ray C. Mullin. — 12th ed.
      p.   cm.
    "Based on the 1996 National Electrical Code®."
    Includes index.
    ISBN 0-8273-6842-9 (hc). — ISBN 0-8273-6841-0 (softcover)
    1. Electric wiring, Interior. I. Title.
TK3285.M84 1995
621.319'24'0218 — dc20
                         95-9817
                         CIP

CONTENTS

PLANS FOR SINGLE-FAMILY DWELLING
     Sheet   1 of 10   Basement Plan
     Sheet   2 of 10   Floor Plan
     Sheet   3 of 10   South Elevation; Window Schedule, Door Schedule
     Sheet   4 of 10   East Elevation; Interior Elevations
     Sheet   5 of 10   Plot Plan; North Elevation
     Sheet   6 of 10   West Elevation; Room Finish Schedule
     Sheet   7 of 10   Section A-A
     Sheet   8 of 10   Electrical Basement Plan
     Sheet   9 of 10   Electrical Floor Plan
     Sheet 10 of 10   Code Requirements for Swimming Pool Wiring

FOREWORD

THE ROLE OF THE ELECTRICAL INSPECTOR AND THE IMPORTANCE OF PROPER TRAINING

When one considers how different groups within the electrical industry affect the end product used by consumers, it is easy to see the importance of those groups working together in a coordinated manner. From the time electrical products are designed to the time when electrical installations are completed or when electrical equipment is purchased by the consumer, many steps must be taken to make it all possible. During this process, close attention to safety standards must be given to the design, installation, and use of electrical products. Product safety standards include testing guidelines which can be used to help ensure that products perform in a safe manner. Products so evaluated by appropriate third party testing laboratories may label the electrical equipment with authorized markings as a means of verifying that the product complies with the minimum requirements of the safety standard.

The *National Electrical Code*® is an electrical safety standard containing rules affecting products, installation procedures and use of electrical equipment. It is the most widely recognized safety standard in this country. The *NEC*® is usually adopted through a legislative process of a governmental agency and becomes enforceable as a law within that jurisdiction. The *NEC*® is amended every three years in response to such things as the development of new products, the recognition of different installation and use procedures, and the discovery of potential hazards to persons and property. Electrical inspectors have the responsibility of enforcing the safety requirements of the *National Electrical Code*.®

Electrical inspectors are an important part of the safety program. They are involved in the development of safety rules as well as in the application of those rules to electrical installations. The responsibility borne by electrical inspectors is significant. To be responsible for knowing safety rules, making proper interpretation of those rules, and properly applying them to a job site is a major endeavor. In order for the electrical inspector to properly do his or her job, knowledge and skill must be developed. This is achieved through adequate education and experience. It is vital that electrical inspectors be knowledgeable of the rules in the *National Electrical Code*® and know how to apply those rules.

The need for adequate training and experience is not limited to the electrical inspector. All who are a part of the electrical construction industry can benefit by it. In order for one to maintain an acceptable level of proficiency in a profession or trade, one must stay current with the ever changing electrical field. That proficiency is achieved through continuing education and training. There is no substitute for such training and experience. To emphasize this point, I refer to a philosophy taught by Ray Mullin: "The cost of education is small when compared to the price paid for ignorance."

Philip H. Cox
Executive Director
International Association
of Electrical Inspectors

PREFACE

INTRODUCTION

The twelfth edition of ELECTRICAL WIRING—RESIDENTIAL is based on the 1996 *National Electrical Code (NEC).*® The *NEC*® is used as the basic standard for the layout and construction of residential electrical systems. In this text, thorough explanations are provided of Code requirements as they relate to residential wiring. To gain the greatest benefit from this text, the student must use the *National Electrical Code*® on a continuing basis.

Why Is This Text Different Than Other House Wiring Books?

It is extremely difficult to learn the requirements of the Code by merely reading the *National Electrical Code.*® Try reading a dictionary from beginning to end. Do you remember what you read? Probably not. But, look up the definition of a word at the time you need to know about it, and you will remember its definition. In much the same manner, *Residential Wiring* does not cover the *National Electrical Code*® in a boring manner. Instead, it stresses specific Code requirements at the appropriate time. For example, the Code requirements for electric clothes dryers are discussed when studying about the electric clothes dryer hook-up.

This text is unique in that it includes a full set of an actual house plan. The format is for the student to simultaneously use the text, the plans, and the *National Electrical Code,*® taking the learner through a typical house, room by room, circuit by circuit. Also included are many recommendations that are above and beyond the basic *National Electrical Code*® requirements.

The *National Electrical Code*® (NFPA 70) becomes mandatory only after it has been adopted by a city, county, state, or other governing body. Until officially adopted, the *NEC*® is merely advisory in nature. State and local electrical codes may contain modifications of the *National Electrical Code*® to meet local requirements. In most cases, local codes will adopt certain more stringent regulations than those found in the *NEC.*® For example, the *NEC*® recognizes nonmetallic-sheathed cable as an acceptable wiring method for house wiring. Yet, the city of Chicago and surrounding counties do not permit nonmetallic-sheathed cable for house wiring. In these areas, all house wiring is done with electrical metallic tubing (thinwall).

There are also instances where a governing body has legislated action that waives specific *NEC*® requirements, feeling that the *NEC*® was too restrictive on that particular issue. Such instances are very rare. The Instructor is encouraged to furnish students with

---

National Electrical Code® and *NEC*® are registered trademarks of the National Fire Protection Association, Inc. Quincy, MA 02269.

any local variations from the *NEC*® that would affect this residential installation in a specific locality.

Electrical wiring is a skilled trade. Wiring should not be done by anyone not familiar with the hazards involved. It is a highly technical skill that requires much training. This text provides all of the electrical codes and standards information needed to approach house wiring in a safe manner. That is why the *National Electrical Code*® defines a **Qualified Person** as "one familiar with the construction and operation of the equipment and the hazards involved."

Do not work on live circuits! Always de-energize the system before working on it! There is no compromise when it comes to safety! Many injuries and deaths have occurred when individuals worked on live equipment. The question is always: "Would the injury or death have occurred had the power been shut off?" The answer is "No!"

All mandatory safety-related work practices are found in the Federal Regulation Occupational Safety and Health Act (OSHA), Title 29, Subpart S—Electrical, Sections 1910.331 through 1910.360.

ELECTRICAL WIRING—RESIDENTIAL is the most comprehensive text available anywhere. It covers residential wiring topics from the most advanced *National Electrical Code*® standpoint, to the fundamental requirements necessary for someone wanting to learn the basics. Every code rule is covered by text, illustrations, examples, and wiring diagrams.

Most electrical inspectors across the country are members of the International Association of Electrical Inspectors. This organization publishes one of the finest technical bimonthly magazines, devoted entirely to the *National Electrical Code*® and related topics. This organization is open to individuals that are not electrical inspectors. Electrical instructors, vo-tech students, apprentices, electricians, consulting engineers, contractors, and distributors are encouraged to join the IAEI so they can stay up to date on all *National Electrical Code*® issues, changes, and interpretations. An application form that explains the benefits of membership in the IAEI can be found in the **Appendix** of this text.

THE ELECTRICAL TRADE

Electricity is safe when it is handled properly. If electrical installations are made in an unsafe manner, great potential exists for damage to property as a result of fire. But more importantly, electrical installations that do not meet the minimum requirements of applicable codes and standards can and do result in personal injury and/or death.

Most communities, cities, counties, and states have regulations that adopt by reference the *National Electrical Code*® These regulations also specify any exceptions to the Code to cover local conditions.

Most building codes and standards contain definitions for the various levels of competency of workers in the electrical industry.

Apprentice shall mean a person who is required to be registered as such under Section XYZ, who is in compliance with the provisions of this article, and who is working at the trade in the employment of a registered electrical contractor and is under the direct supervision of a licensed master electrician, journeyman electrician, or residential wireman.

Residential Wireman shall mean a person having the necessary qualifications, training, experience, and technical knowledge to wire for and install electrical apparatus and equipment for wiring one-, two-, three-, and four-family dwellings. A residential wireman is sometimes referred to as a *Class B Electrician*.

Journeyman Electrician shall mean a person having the necessary qualifications, training, experience, and technical knowledge to wire for, install, and repair electrical apparatus and equipment for light, heat, power and other purposes, in accordance with standard rules and regulations governing such work.

Master Electrician means a person having the necessary qualifications, training, experience, and technical knowledge to properly plan, lay out, and supervise the installation and repair of wiring apparatus and equipment for electric light, heat, power and other purposes, in accordance with standard codes and regulations governing such work, such as the *National Electrical Code.*®

Electrical Contractor means any person, firm, copartnership, corporation, association, or combination thereof who undertakes or offers to undertake for another the planning, laying out, supervising and installing, or the making of additions, alterations, and repairs in the installation of wiring apparatus and equipment for electrical light, heat, and power.

THE TWELFTH EDITION

Continuing in the tradition of previous editions, this text thoroughly explains how Code changes affect wiring installations. New and revised illustrations supplement the explanations to ensure that electricians understand the new Code requirements. New photos reflect the latest wiring materials and components available on the market. Revised review questions test student understanding of the new content. New tables that summarize Code requirements offer a quick reference tool for students. Another reference aid is the tables reprinted directly from the 1996 edition of the *National Electrical Code.*® The extensive revisions for the twelfth edition make ELECTRICAL WIRING—RESIDENTIAL the most up-to-date and well-organized guide to house wiring.

A modern residence blueprint serves as the basis for the wiring schematics, cable layouts, and discussions provided in the text. Additional plans may be obtained by contacting Delmar Publishers Inc. Each unit dealing with a specific type of wiring is referenced to the appropriate plan sheet. All wiring systems are described in detail—lighting, appliance, heating, service entrance, and so on.

The house selected for this edition is scaled for current construction practices and costs. Note, however, that the wiring, lighting fixtures, appliances, number of outlets, number of circuits, and track lighting are not all commonly found in a home of this size. The wiring may incorporate more features than are absolutely necessary. This was done to present as many features and Code issues as possible to give the learner more experience in wiring a residence.

This text does *not* focus on such basic skills as drilling a hole, splicing two wires together, stripping insulations from wire, or fishing a wire through walls. These skills are certainly necessary for good workmanship, but it is assumed that the learner (electrical apprentice or journeyman electrician) has mastered the mechanical skills on the job.

This text *does* focus on the technical skills required to perform electrical installations. It covers such topics as calculating conductor sizes, calculating voltage drop, determining appliance circuit requirements, sizing service, connecting electric appliances, grounding service and equipment, installing recessed fixtures, and much more. These are critical skills that can make the difference between an installation that "meets Code" and one that does not. The electrician must understand the reasons for following Code regulations to achieve an installation that is essentially free from hazard to life and property.

CHANGES FOR THIS EDITION

The following list highlights changes and additions made in this twelfth edition. Many of these changes and additions are the result of suggestions sent to the author by users of the text.

- An introduction to the new National Fire Protection Association *Residential Electrical Maintenance Code for One- and Two-Family Dwellings*.

- More suggestions for wiring homes for the physically handicapped.

- A new comprehensive table of Conversion Multipliers for changing customary English terms to metrics.

- Current information on the background impact of Nationally Recognized Testing Agencies (NRTL).

- An introduction to the *harmonizing* of standards between Underwriters Laboratories (UL) and the Canadian Standards Association (CSA).

- Building Code regulations *now in effect* that are often overlooked, and that can prevent the spread of fire in residential construction. New materials are discussed and illustrated for use in preventing the spread of fire. Watch out when installing electrical wall boxes in the same partition between a residential garage and habitable rooms!

- New illustrations of electrical boxes for use in remodel work.

- New illustrations of electrical boxes for use with metal framing members.

- Clarification on where switch control of stairway lighting is required.

- The hazards of burying old knob-and-tube wiring in insulation.

- A new, simplified chart showing how to match conductor insulation temperature ratings and the terminals to which they will be connected, and "Meet Code."

- An update on the conductor temperature ratings in nonmetallic-sheathed cable and armored cable, and where they are permitted and/or required to be used.

- More wiring diagrams for three-way switches. Some are not so common.

- What a GFCI does? What a GFCI does not do?

- An update on the installation and wiring of recessed fixtures.

- A total rewrite and much additional text on incandescent and fluorescent lighting, dimming, color rendition, high-efficiency electronic ballasts, energy-saving lamps, lamp life and light output as affected by supply voltage.

- How many outlets are permitted on one branch-circuit?

- The latest technology for submersible water pumps and how to wire them.

- A complete rewrite on water heaters, required temperature settings, Shriner Burn Institute data on hot water hazards, corrosion problems, how pressure/temperature safety valves function, different types of electrical hook-ups showing *utility-controlled* and *customer-controlled* installations, time-of-use metering, and how to determine how long it will take an electric water heater to recover.

- New diagrams for the heat controls used on electric ranges.

- Latest wiring diagrams for bathroom *heat/vent/light* fixtures.

- Cord and plug connections not permitted for furnaces.

- Understanding the terms found on the nameplates of air conditioners and heat pumps, and what these mean to the electrician. Examples and calculations provided.

- Photos, diagrams, and the latest technical information on *digital satellite* systems that use a small 18-inch antenna instead of the large "dishes."

- The latest on telephone Standard Network Interfaces, the latest on color coding of telephone conductors and cables.

- How many telephones can be connected to one line before running into problems.

- "Cross-talk" on telephone lines, and how this nuisance can be avoided.

- Latest National Fire Protection Association Code requirements for household fire warning equipment. This is the new NFPA Standard 72.

- A complete rewrite of the low-voltage, remote-control systems chapter to show how the new relays, switches, master controls, and sequencers are used. New color coding of the cables and conductors is presented. Many new photos and diagrams.

- A complete rewrite and reorganization on the subject of grounding and bonding of service equipment for more clarity in the understanding of the subject. Many new diagrams.

- Added coverage on how to figure the cost of using electricity.

- A set of *blank* floor plans to use upon completion of studying the text. Students can design their own electrical layout.

- Extra *blank* forms for use in the proper sizing of residential services.

- A *Membership Application Form* for joining the International Association of Electrical Inspectors organization has been included in the Appendix of this text. Students and instructors are encouraged to become members of this educational association.

NEW CODE REQUIREMENTS FOR THE 1996 *NATIONAL ELECTRICAL CODE®*

Throughout this new edition of *Electrical Wiring—Residential*, 1996 Code changes are identified by the symbol ▶ before the change and ◀ after the change.

- New rule for "rounding up" overcurrent ratings for motor branch-circuits.

- New requirements for grounding electric ranges, dryers, ovens, and counter-mounted cooktops. Grounding to the neutral no longer permitted for new work.

- New and simplified requirements for locating receptacles on islands and peninsulas.

- New requirements that *all* 125-volt, single-phase 15- and 20-ampere receptacles in kitchens that serve countertops must be GFCI protected.

- New Code requirements for control circuit wiring.

- New rules that *only* service-drop conductors be supported by a service mast.

- New rule prohibiting the use of sheet metal screws for terminating equipment grounding conductors.

- Clarification on where to attach equipment grounding conductors and neutrals in a panel.

- New rules for reducing service neutrals on residences.

- New rules for waiving the bonding requirements of swimming pool double-insulated pump motors.
- New requirements for boxes that will support ceiling fans.
- New requirement that bathroom receptacles must be connected to a separate 20-ampere branch-circuit.
- New rules for outdoor receptacles.
- Clarification on terminating conductors based upon the conductor's insulation rating.
- New rules for supporting, securing, and protecting cables through framing members.
- New rules on using 3-wire and 4-wire cord- and plug-connection for electric ranges and dryers.
- New strict rules for temporary wiring.
- New requirements for hooking up the receptacle for refrigerators.
- New rules for low-voltage landscape lighting.
- New rules on when to install GFCI receptacles for appliances located in a dedicated space.
- Many changes have been made to Code section references because of editorial rearranging of numerous Code sections. These editorial changes have been picked up in this revision.

SUPPLEMENTS

An Instructor's Guide is available and consists of the following information: selected text references, answers to unit-end reviews, a final examination covering the content of the entire text, blank service-entrance calculation form, and answers to the final examination.

ABOUT THE AUTHOR

This text was written by Ray C. Mullin, former electrical instructor for the Wisconsin Schools of Vocational, Technical, and Adult Education. He is a former member of the International Brotherhood of Electrical Workers. He is a member of the International Association of Electrical Inspectors, the Institute of Electrical and Electronic Engineers, and the National Fire Protection Association, and has served on Code Making Panel 4 of the *National Electrical Code.*®

Mr. Mullin completed his apprenticeship training and worked as a journeyman and supervisor for residential, commercial, and industrial installations. He has taught both day and night electrical apprentice and journeyman courses, has conducted engineering seminars, and has conducted many technical Code workshops and seminars at International Association of Electrical Inspectors Chapter and Section meetings, and has served on their Code panels.

He has written many technical articles that have appeared in electrical trade publications. He has served as a consultant to electrical equipment manufacturers regarding conformance of their products to industry standards, and on legal issues relative to personal injury lawsuits resulting from the misuse of electricity and electrical equipment.

Mr. Mullin presents his knowledge and experience in this text in a clear-cut manner that is easy to understand. This presentation will help learners to fully understand the essentials required to pass the residential licensing examinations and to perform residential wiring that "meets Code."

Mr. Mullin is co-author of *Electrical Wiring—Commercial*, co-author of *Illustrated Electrical Calculations*, and co-author of *The Smart House*. He has contributed technical material for *Electrical Grounding*, and to the International Association of Electrical Inspectors text on *Grounding*.

He served on the Executive Board of the Western Section International Association of Electrical Inspectors, and on their *National Electrical Code*® Committee and Code Clearing Committee. He also serves on the Electrical Commission in his hometown.

Mr. Mullin is past Director, Technical Liaison, for a large electrical manufacturer. He has developed extensive technical literature for use by this company's field engineering personnel.

Mr. Mullin attended the University of Wisconsin, Colorado State University, and the Milwaukee School of Engineering.

Delmar Publishers Is Your Electrical Book Source!

Whether you're a beginning student or a master electrician, Delmar Publishers has the right book for you. Our complete selection of proven best-sellers and all-new titles is designed to bring you the most up-to-date, technically accurate information available.

DC/AC THEORY

Electricity 1-4/Kubala - Alerich/Keljik
Electricity 1 Order #0-8273-6574-8
Electricity 2 Order #0-8273-6575-6
Electricity 3 Order #0-8273-6594-2
Electricity 4 Order #0-8273-6593-4

Delmar's Standard Textbook of Electricity-Revised Edition/Herman

This full-color, highly illustrated book sets the new standard with its comprehensive and up-to-date content, plus a complete teaching/learning package, including: Lab-Volt and "generic" lab manuals and transparencies.
Order # 0-8273-6849-6

CODE & CODE-BASED

1996 National Electrical Code®/NFPA

The standard for all electrical installations, the 1996 NEC® is now available from Delmar Publishers!
Order # 0-87765-402-6

Understanding the National Electrical Code, 2E/Holt

This easy-to-use introduction to the NEC® helps you find your way around the NEC® and understand its very technical language. Based on the 1996 NEC®.
Order # 0-8273-6805-4

Illustrated Changes in the 1996 NEC®/O'Riley

Illustrated explanations of Code changes and how they affect a job help you learn and apply the changes in the 1996 NEC® more effectively and efficiently!
Order # 0-8273-6773-2

Interpreting the National Electrical Code®, 4E/Surbrook

This excellent book provides the more advanced students, journeyman electricians, and electrical inspectors with an understanding of NEC® provisions. Based on the 1996 NEC®.
Order # 0-8273-6803-8

Electrical Grounding, 4E/O'Riley

This illustrated and easy-to-understand book will help you understand the subject of electrical grounding and Article 250 of 1996 NEC®.
Order # 0-8273-6657-4

WIRING

Electrical Wiring—Residential, 12E/Mullin

This best-selling book takes you through the wiring of a residence in compliance with the 1996 NEC®. Complete with working drawings for a residence.
Order # 0-8273-6841-0 (SC)   0-8273-6842-9 (HC)

Smart House Wiring/Stauffer & Mullin

This unique book provides you with a complete explanation of the hardware and methods involved in wiring a house in accordance with the Smart House, L.P. system.
Order # 0-8273-5489-4

Electrical Wiring—Commercial, 9E/Mullin & Smith

Learn to apply the 1996 NEC® as you proceed step-by-step through the wiring of a commercial building. Complete with working drawings of a commercial building.
Order # 0-8273-6655-8

Electrical Wiring—Industrial, 9E/Smith & Herman

Learn industrial wiring essentials based on the 1996 NEC®. Complete with industrial building plans.
Order # 0-8273-6653-1

Cables and Wiring/AVO Multi-Amp

Your comprehensive practical guide to all types of electrical cables.
Order # 0-8273-5460-6

Raceways and Other Wiring Methods, 2E/Loyd

This excellent new edition provides you with complete information on metallic and nonmetallic raceways and other common wiring methods used by electricians and electrical designers.
Order # 8273-6659-0

Illustrated Electrical Calculations/Sanders

Your quick reference to all of the formulas and calculations electricians use.
Order # 8273-5462-2

Electrician's Formula and Reference Book/Holt

A concise and easy-to-use pocket reference that features up-to-date formulas for proper electrical calculations as they apply to any and all electrical installations. This handy pocket reference is a great companion to the NFPA's National Electrical Code® and any other electrical book.
Order # 0-8273-6961-1

MOTOR CONTROL

Electric Motor Control, 5E/Alerich

The standard for almost 30 years, this best-selling textbook explains how to connect electromagnetic and electric controllers.
Order # 0-8273-5250-6

Industrial Motor Control, 3E/Herman

This text combines a solid explanation of theory with practical instructions and information on controlling industrial motors with magnetic and solid-state controllers.
Order # 0-8273-5252-2

Electric Motors and Motor Controls/Keljik

This new book focuses on the hows and why behind all types of motors and controls. Special attention is given to why electrical systems work, thoroughly explaining the function and purpose behind all types of electric motors and controllers.
Order # 0-8273-6174-2

EXAM PREPARATION

Journeyman Electrician's Review, 2E/Loyd
Order # 0-8273-6680-9

Master Electrician's Review, 2E/Loyd
Order # 0-8273-6678-7

Master Electrician's Preparation Exam/Holt
Order #0-8273-7623-5

Journeyman's Exam Preparation/Holt
Order # 0-8273-7621-9

To request examination copies or a catalog of all of our titles, call or write to:

Delmar Publishers
3 Columbia Circle
P.O. Box 15015
Albany, NY 12212-5015

Phone: 1-800-347-7707 • 1-518-464-3500 • Fax: 1-518-464-0301

ACKNOWLEDGMENTS

Helen J. Mullin, my wife for her editing, typing, and encouragement.

Special thanks to everyone on the Delmar team who worked so hard at "pulling it all together" as we focused on specific target dates to get this text off the presses at the same time the 1996 *National Electrical Code*® became available. They did an outstanding job! Congratulations!

The author and Delmar Publishers would like to thank the following reviewers of the 1993 edition, who offered their technical expertise and suggestions for this revision. Our appreciation goes out to:

Robert S. Boiko, Vice President
Northbrook Heater and Parts Supply, Inc.
Northbrook, IL 60062

Samuel J. Geisler
Career Institute of Technology
Easton, PA 18042

Dallas Phillips
Kentucky Tech LaGrange Vocational
    Education Center
LaGrange, KY 40031

George Wilson
Mississippi Gulf Coast Community
    College
Long Beach, MS 39560

A. J. Pearson, Executive Director
National Joint Apprenticeship and
    Training Committee
Upper Marlboro, MD 20772

Special appreciation is expressed to Charles W. Talcott and Home Planners, Inc. for the basic plans upon which the residence plans found at the back of the text are based.

The author wishes to thank the following companies for their contributions of data, illustrations, and technical information:

Advance Transformer Company
AFC/A Nortek Co.
American Home Lighting Institute
Anchor Electric Division, Sola Basic Industries
Appleton Electric Co.
Arrow-Hart, Inc.
BRK Electronics, A Division of Pittway Corporation
Bussmann Division, Cooper Industries
Carlon, A Lamson & Sessions Company
Chromalox
Electri-Flex Co.
General Electric Co.
Halo Lighting Division, Cooper Industries
Heyco Molded Products, Inc.
Honeywell
Hubbell Incorporated, Wiring Devices Division

International Association of Electrical Inspectors
Juno
The Kohler Co.
Leviton Manufacturing Co., Inc.
LUTRON
Midwest Electric Products, Inc.
Milbank Manufacturing Co.
Moe Light Division, Thomas Industries
Motorola
NuTone Inc.
Pass & Seymour, Inc.
Progress Lighting
Reiker Enterprises Inc.
Rheem Manufacturing Co.
Seatek Co., Inc.
Sierra Electric Division, Sola Basic Industries
SMART HOUSE, L.P.
Square D Co.
Superior Electric Co.
THERM-O-DISC
Thomas & Betts Corporation
Winegard Company
Wiremold Co.

UNIT 1

General Information for Electrical Installations

OBJECTIVES

After studying this unit, the student will be able to

- understand the basic safety rules for working on electrical systems.
- explain how electrical wiring information is conveyed to the electrician at the construction or installation site.
- demonstrate how the specifications are used in estimating cost and in making electrical installations.
- explain why symbols and notations are used on electrical drawings.
- list the Nationally Recognized Testing agencies that are responsible for establishing electrical standards and insuring that materials meet the standards.
- develop a better understanding of the metric system of measurement.
- begin to refer to the latest edition of the *National Electrical Code®*.
- have a better understanding of the new *Residential Electrical Maintenance Code for One and Two-Family Dwellings*.

SAFETY IN THE WORKPLACE

Before venturing into the study of residential wiring, let's discuss safety. The electrician might be involved in new construction where the only electrical power is temporary power, or in remodel work where the home already has electrical power.

Serious injury can result if the power has not been properly turned off before beginning to work on an electrical problem. Electricity is dangerous! The voltage levels in a home are 120 volts between one "hot" conductor and the "neutral" conductor or a grounded surface. Between the two "hot" conductors, the voltage is 240 volts. These are the *line-to-ground* or *line-to-line* voltages. From basic electrical theory, *line voltage appears across an open in a series circuit*. Getting caught "in series" with a 120-volt circuit will give you a 120-volt shock. For

example, open-circuit voltage across the two wires of a single-pole switch on a lighting circuit is 120 volts, when the switch is in the OFF position and the lamp(s) are in place. Likewise, getting caught "in series" with a 240-volt circuit will give you a 240-volt shock. Working on electrical equipment with the power turned ON can result in death or serious injury either as a direct result of electricity (electrocution or burns), or from an indirect secondary reaction such as falling off a ladder or jerking away from the "hot" conductor into moving parts of equipment, such as into the turning blades of a furnace fan. Dropping a metal tool onto live parts, allowing metal shavings from a drilling operation to fall onto live parts of electrical equipment, or cutting into a "live" conductor and a "neutral" conductor at the same time, or touching the "live" wire

1

and the "neutral" wire or a grounded surface at the same time can cause injury. A short-circuit or ground fault can result in an *arc blast* that can cause serious injury or death. The heat of an electrical arc has been determined to be hotter than the sun. Tiny hot "balls" of copper can fly into your eye or onto your skin. Working on switches, receptacles, fixtures, or appliances with the power turned ON is dangerous.

Dirt, debris, and moisture can also set the stage for equipment failures and personal injury. Neatness and cleanliness in the workplace are a must.

It's the Law!

Not only is it a good idea to use proper safety measures as you work on and around electrical systems, it is *required* by law. As an electrician or electrical contractor, you need to be aware of these regulations.

The Federal Regulations in the Occupational Safety and Health Act (OSHA) Number 29, Subpart S, in Part 1910.332 discusses the training needed for those who face a risk of electrical injury. Proper training means trained in and familiar with the safety-related work practices required by paragraphs 1910.331 through 1910.335. Numerous texts are available that delve into the OSHA requirements in great detail.

The *National Electrical Code*® defines a *qualified person* as "one familiar with the construction and operation of the equipment and the hazards involved." Merely telling someone or being told "be careful" does not meet the definition of proper training, and does not make the person qualified.

Only qualified persons are permitted to work on or near exposed energized equipment. To become qualified, a person must have the skill and techniques necessary to distinguish exposed live parts from the other parts of electrical equipment, must be able to determine the voltage of exposed live parts, and must be trained in the use of special precautionary techniques such as personal protective equipment, insulations, shielding material, and insulated tools.

Subpart S, 1910.333 requires that safety related work practices shall be employed to prevent electric shock or other injuries resulting from either direct or indirect electrical contact. Live parts to which an employee may be exposed shall be de-energized before the employee works on or near them, unless the employer can demonstrate that de-energizing introduces additional or increased hazards.

Working on equipment "live" is acceptable only if there would be a greater hazard if the system was de-energized. Examples of this would be hospital life support systems, some alarm systems, certain ventilation systems in hazardous locations, and the power for critical illumination circuits. Working on energized equipment requires proper insulated tools, proper nonflammable clothing, rubber gloves, protective shields and goggles, and in some cases, rubber blankets. None of these reasons to work on electrical equipment "hot" are found in residential wiring.

OSHA regulations allow only qualified personnel to work on or near electrical circuits or equipment that have not been de-energized. The OSHA regulations provide rules regarding "lockout and tagging" to make sure that the electrical equipment being worked on will not inadvertently be turned ON while someone is working on the supposedly "dead" equipment. As the OSHA regulations state, "a lock and a tag shall be placed on each disconnecting means used to de-energize circuits and equipment. . . ."

Some electricians' contractual agreements require that as a safety measure, two or more qualified electricians must work together when working on energized circuits. They do not allow untrained apprentices to work on "live" equipment but do allow apprentices to stand way back and observe.

The National Fire Protection Association publications *Safety Related Work Practices NFPA 70E* and *Electrical Equipment Maintenance NFPA 70B* present much of the same material regarding electrical safety as does the OSHA regulation.

Safety cannot be compromised! The rule is **turn off and lock-off the power, then properly tag the disconnect with a description as to exactly what that particular disconnect controls**. With safety utmost in our minds, let us begin our course on the wiring of a typical residence.

THE WORKING DRAWINGS

The architect uses a set of working drawings or plans to make the necessary instructions available to the skilled crafts that are to build the structure shown in the plans. The sizes, quantities, and locations of the materials required and the construction features of the structural members are shown at a glance.

These details of construction must be studied and interpreted by each skilled construction craft— masons, carpenters, electricians, and others—before the actual work is started.

The electrician must be able to: (1) convert the two-dimensional plans into an actual electrical installation, and (2) visualize the many different views of the plans and coordinate them into a three-dimensional picture, as shown in figure 1-1.

The ability to visualize an accurate three-dimensional picture requires a thorough knowledge of blueprint reading. Because all of the skilled trades use a common set of plans, the electrician must be able to interpret the lines and symbols that refer to the electrical installation and also those used by the other construction trades. The electrician must know the structural makeup of the building and the construction materials to be used.

SPECIFICATIONS

Working drawings are usually complex because of the amount of information that must be included. To prevent confusing detail, it is standard practice to include with each set of plans a set of detailed written specifications prepared by the architect.

These specifications provide general information to be used by all trades involved in the construction. In addition, specialized information is given for the individual trades. The specifications include information on the sizes, the type, and the desired quality of the standard parts to be used in the structure.

Typical specifications include a section on "General Clauses and Conditions," which is applicable to all trades involved in the construction. This section is followed by detailed requirements for the various trades—excavating, masonry, carpentry, plumbing, heating, electrical work, painting, and others.

The plan drawings for the residence used as an example are included with this text. The specifications for the electrical work indicated on the plans are given in the Appendix.

In the electrical specifications, the listing of standard electrical parts and supplies frequently includes the manufacturers' names and the catalog numbers of the specified items. Such information insures that these items will be of the correct size, type, and electrical rating, and that the quality will

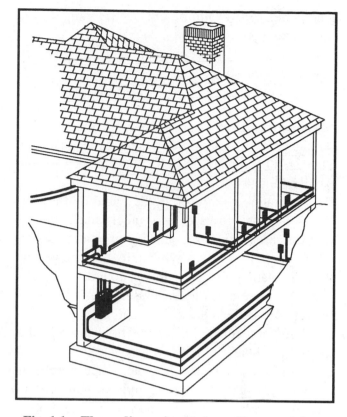

Fig. 1-1   Three-dimensional view of house wiring.

meet a certain standard. To allow for the possibility that the contractor will not always be able to obtain the specified item, the phrase "or equivalent" is usually added after the manufacturer's name and catalog number.

The specifications are also useful to the electrical contractor in that all of the items needed for a specific job are grouped together and the type or size of each item is indicated. This information allows the contractor to prepare an accurate cost estimate without having to find all of the data on the plans.

If there is a difference between the plans and specifications, the specifications will take preference. The electrical contractor should discuss the matter with the homeowner, architect, and engineer. The cost of the installation might vary considerably because of the difference(s), so obtain any changes to the plans and/or specifications in writing.

SYMBOLS AND NOTATIONS

The architect uses symbols and notations to simplify the drawing and presentation of information concerning electrical devices, appliances, and equipment. For example, an electric range outlet looks like this:

$$\equiv\!\ominus_R$$

Most symbols have a standard interpretation throughout the country as adopted by the American National Standards Institute (ANSI). Symbols are described in detail in unit 2.

A notation will generally be found on the plans (blueprints) next to a specific symbol calling attention to a variation, type, size, quantity, or other necessary information. In reality, a symbol might be considered to be a notation because symbols do represent a system that "represents words, phrases, numbers, quantities, etc." as defined in the dictionary.

Another method of using notations to avoid cluttering up a blueprint is to provide a system of symbols that refer to a specific table. For example, the written sentences on plans could be included in a table referred to by a notation. Figure 2-9 is an example of how this could be done. The special symbols that refer to the table would have been shown on the actual plan.

NATIONAL ELECTRICAL CODE® (NEC®)

The *National Electrical Code®* is published by the National Fire Protection Association, and is referred to as NFPA 70. Because of the ever-present danger of fire or shock hazard through some failure of the electrical system, the electrician and the electrical contractor must use listed materials and must perform all work in accordance with recognized standards. The *National Electrical Code®* is the basic standard that governs electrical work. The purpose of the Code is to provide information considered necessary for the safeguarding of people and property against electrical hazards. It states that the installation must be essentially hazard-free, but that such an installation is not necessarily efficient, convenient, or adequate, *Section 90-1(b)*. (Note: All *National Electrical Code®* section references used throughout this text are printed in italics.) It is the electrician's responsibility to insure that the installation meets these criteria. In addition to the *National Electrical Code®*, the electrician must also consider local and state codes. The purpose and scope of the *National Electrical Code®* are discussed in *Article*

90 of the Code book and should be studied by the student at this time.

Code Arrangement

It is important to understand the arrangement of the *NEC®*.

- Chapters 1, 2, 3, and 4 — These chapters contain general requirements that cover most situations.

- Chapters 5, 6, and 7 — These chapters pertain to special occupancies, special equipment, and special conditions. The requirements in these chapters are in addition to, or are changes to (modify or amend) the general rules found in Chapters 1, 2, 3, 4.

- Chapter 8 — Covers communication, telephone, fire and burglar alarm, radio, television, cable systems, etc.

- Chapter 9 — This chapter contains examples and tables.

Chapters are divided into **Articles**. **Articles** are divided into **Sections**. **Sections** are divided into **subsections** and **exceptions**.

The first *National Electrical Code®* was published in 1897. It is revised and updated every three years by changing, editing, adding technical material, adding sections, adding articles, adding tables, and so on, so as to be as up-to-date as possible relative to electrical installations. For example, the Code added requirements pertaining to circuits supplying computers and similar electronic equipment.

Another prime example is that we have included a chapter on a "SMART HOUSE," which relates to the requirements of *Article 780, Closed-Loop and Programmed Power Distribution*.

As material, equipment, and technologies change, the members of the Code-making panel solicit comments and proposals from individuals in the electrical industry or others interested in electrical safety. The panel members then meet to act upon the many proposals for changes to the Code. Final action (voting) on these proposals is taken at the National Fire Protection Association's annual meeting.

The Code-making panel consists of individuals representing over 65 organizations such as the International Association of Electrical Inspectors, the Institute of Electrical and Electronic Engineers, the National Electrical Contractors Association, the Edison Electric Institute, the Independent Electrical Contractors Association, Inc., Underwriters Laboratories, Inc., Consumer Product Safety Commission, International Brotherhood of Electrical Workers, and many others.

The *National Electrical Code®* does not become "law" until adopted by official action of the legislative body of a city, county, or state.

What Edition of the *National Electrical Code®* to Use?

This text is based on the 1996 edition of the *National Electrical Code.®* However, be aware that some municipalities, counties, and states have not yet adopted the 1996 edition, and might still be using the 1993 or earlier editions. Always check on this with the local authority having jurisdiction before starting a wiring project to make sure your installation complies with the proper electrical code.

Copies of the *National Electrical Code®* may be ordered from

National Fire Protection Association
1 Batterymarch Park
Quincy, Massachusetts 02269-9101

or

International Association of Electrical Inspectors
901 Waterfall Way
Suite 602
Richardson, TX 75080-7702

Residential Electrical Maintenance Code for One- and Two-Family Dwellings

This code is published by the National Fire Protection Association, and is referred to as NFPA 73. It is a rather brief seven-page code that provides requirements for evaluating installed electrical systems within and associated with *existing* one- and two-family dwellings to identify safety, fire, and shock hazards—such as improper installations,

overheating, physical deterioration, and abuse. This code lists most of the electrical things to check out in an existing dwelling that could result in a fire or shock hazard if not corrected. It only points out things to look for that are visible. It does not get into examining concealed wiring that would require removal of permanent parts of the structure. It also does not get into calculations, location requirements, and complex topics as does the *National Electrical Code.® NFPA 70.*

This code uses simple statements such as: *The service shall be adequate to serve the connected load; overcurrent protective devices shall be properly rated for conductor ampacities; conductors shall be properly terminated and supported at panelboards, boxes, and devices; raceways shall be securely fastened in place; cables and cable assemblies shall be properly secured and supported; fixture canopies shall be in place and properly secured; covers shall be in place and properly secured; and so forth.*

This code can be an extremely useful guide for electricians doing remodel work, and for electrical inspectors wanting to bring an existing dwelling to a reasonably safe condition. Many localities require that when a home changes ownership, the wiring must be brought up to some minimum standard, but not necessarily as extensively as would be the case for new construction. For example, a local code might require that a minimum 100-ampere, 120/240-volt service be installed, that the service ground be brought up to the requirements of the latest edition of the *National Electrical Code,®* that 20-ampere appliance circuits be installed in the kitchen and dining room, that GFCI receptacles be installed in bathrooms and other specified areas, and that a separate 20-ampere branch-circuit be installed for the laundry equipment (clothes washer).

Homes for Physically Challenged People

The *National Electrical Code®* does not address the wiring of homes for the physically or mentally challenged, or disabilities associated with the elderly.

Each installation for the physically challenged must be based upon the specific need(s) of the individual(s) who will occupy the home. Some physically challenged people are bed-ridden, require the mobility of a wheelchair, have trouble reaching, have trouble bending, etc. There are no hard and fast rules that MUST be followed—only many suggestions to consider.

Some of these are:

- Install more ceiling lighting fixtures instead of switching receptacles. Cords and lamps are obstacles.

- Install lighting fixtures having more than one bulb.

- Go "overboard" in the amount of lighting for all rooms, entrances, stairways, stairwells, closets, pantries, bathrooms, etc.

- Use higher wattage bulbs—not to exceed the wattage permitted in the specific fixture.

- Consider installing lighting fixtures in certain areas (such as bathrooms and hallways) to be controlled by motion detectors.

- Consider installing exhaust fans in certain areas (such as laundries, showers, and bathrooms) that turn on automatically when the humidity reaches a predetermined value.

- Consider height of switches and thermostats. Usually lower (42") instead of the standard 50–52".

- Consider rocker-type switches instead of toggle type.

- Install pilot light switches.

- Consider "jumbo" switches.

- Be sure stairways and stairwells are well lit.

- Consider stair tread lighting.

- Position lighting switches so as not to be over stairways or ramps.

- Locate switches and receptacles to be readily and easily accessible—not behind doors or other hard-to-reach places.

- Consider installing wall receptacle outlets higher (24–27") instead of normal 12" height.

- Install lighted doorbell buttons.

- Chimes: consider adding a strategically located "dedicated" lamp that will turn on when door bell buttons are pushed. The wiring diagram for this is found in figure 25-20A.

- Telephones: consider adding visible light(s) strategically located that will flash at the same time the telephone is ringing.

- Consider installing receptacle outlets and switches on the face of kitchen cabinets. Wall outlets and switches can be impossible for the physically challenged person to reach.

- Consider fire, smoke, and security systems that are directly connected to a central office for fast response to emergencies that do not depend upon the handicapped to initiate the call.

- Consider installing fuse box or breaker panel on first floor instead of in basement.

- Consider the SMART HOUSE concept of remote control of lighting, receptacles, appliances, television, telephones, etc. The control features can make life much easier for a handicapped person.

The American National Standards Institute (ANSI) publication *ANSI A117.1-1986* contains many suggestions and considerations for buildings and facilities for the physically challenged. The address is:

American National Standards Institute
1430 Broadway
New York, NY 10018

The National Association of Home Builders also has printed information and recommendations for homes for the physically challenged. Their address is:

NAHB National Research Center
400 Prince George's Boulevard
Upper Marlboro, MD 20772-8731

Building Codes

The majority of the building departments across the country have for the most part adopted the *National Electrical Code®* rather than attempting to develop their own electrical codes. As you study this text, you will note numerous references to the *National Electrical Code®*, the electrical inspector, or the Authority Having Jurisdiction.

The level of knowledge varies. The electrical inspector may be full-time or part-time and may also have responsibility for other trades, such as plumbing or heating. The heads of the building department in many communities are typically called the Building Commissioners or Directors of Development.

Regardless of title, they are responsible to make sure that the building codes in their communities are followed.

Communities, rather than writing their own codes, belong to one of the major building officials organizations, and have adopted the building codes of that particular organization. These organizations are:

- **BOCA** Building Official & Code Administrators International, Inc. 4051 W. Floosmoor Road Country Club Hills, Illinois 60478-5795 Phone: 708-799-2300

- **ICBO** International Conference of Building Officials 5360 South Workman Mill Road Whittier, California 90601 Phone: 213-699-0541

- **SBCCI** Southern Building Code Congress International, Inc. 900 Montclair Road Birmingham, Alabama 35212-1206 Phone: 205-591-1853

- **IAPMO** International Association of Plumbing and Mechanical Officials 2001 Walnut Drive South Walnut, California 91789-2825 Phone: 714-595-8449.

▶ An organization comprised of representatives from these organizations is called the Council of American Building Officials (**CABO**). The National Fire Protection Association (**NFPA**), developer and publisher of the *National Electrical Code,®* and **CABO** have recently been working together to come up with a uniform electrical code that is easily understood. This electrical code is one chapter in **CABO's** One- and Two-Family Dwelling Code. This code includes the basic electrical installation requirements for one- and two-family dwellings as found in the *National Electrical Code.®* ◀

As an electrician, you have the following electrical codes available:

National Electrical Code,® *NFPA 70*—needed for all types of electrical installations. *NFPA 70* is the basis of this text.

Electrical Code for One- and Two-Family Dwellings and Mobile Homes, NFPA 70A—

can be used when residential wiring only is involved.

CABO's One- and Two-Family Dwelling Code—contains residential electrical, plumbing, heating, and construction requirements.

Local electrical code requirements that may differ from those found in the *National Electrical Code.®*

Check with the local inspection authority to determine which electrical code is enforced, and what local amendments to this code might take precedence.

AMERICAN NATIONAL STANDARDS INSTITUTE

The American National Standards Institute is an organization that coordinates the efforts and results of the various standards-developing organizations, such as those mentioned in previous paragraphs. Through this process, ANSI approves standards that then become recognized as American national standards. One will find much similarity between the technical information found in ANSI standards, the Underwriters standards, the International Electronic and Electrical Engineers standards, and the *National Electrical Code.®*

Code Definitions

The electrical industry uses many words (terms) that are unique to the electrical trade. These terms need clear definitions to enable the electrician to understand completely the meaning intended by the Code.

Article 100 of the *National Electrical Code®* is a "dictionary" of these terms. *Article 90* also provides further clarification of terms used in the *NEC.®* A few of the terms defined in the Code follow.

Ampacity: The current in amperes a conductor can carry continuously under the conditions of use without exceeding its temperature rating.

Approved: Acceptable to the authority having jurisdiction.

Authority Having Jurisdiction: An organization, office, or individual responsible for "approving" equipment, an installation, or a procedure, *Section 90-4.*

Dwelling Unit: One or more rooms for the use of one or more persons as a housekeeping unit with space for eating, living, and sleeping, and permanent provisions for cooking and sanitation. (Note: The terms *dwelling* and *residence* are used interchangeably throughout this text.)

Fine Print Notes (FPN): Fine Print Notes (FPNs) are found throughout the Code. FPNs are "explanatory" in nature, in that they make reference to other Sections of the Code. FPNs also define things where further description is necessary. See *Section 110-1.*

Identified (as applied to equipment): Recognizable as suitable for a specific purpose, function, use, environment, or application, where described in a particular Code requirement. Suitability of use, marked on or provided with the equipment, may include labeling or listing.

Identified (as applied to conductors and terminals): The grounded circuit, feeder or service-entrance conductor is "identified" by its white or natural gray color. Terminals on wiring devices, panels, and disconnect switches have their terminals for the grounded circuit conductor "identified" with silver colored screws. In the electrical industry, you will hear the terms *grounded circuit conductor, neutral conductor, or the white wire.* For all practical purposes, these terms all mean the same thing. They refer to the grounded circuit conductor. An exception to this is that a neutral conductor is always a grounded conductor, but a grounded circuit conductor is not always a neutral as in the case of industrial wiring where a three-phase, grounded "B" phase system might be used. See *Section 250-5* of the *National Electrical Code.*®

Labeled: Equipment or materials to which has been attached a label, symbol, or other identifying mark of an organization acceptable to the authority having jurisdiction and concerned with product evaluation, that maintains periodic inspection of production of labeled equipment or materials and by whose labeling the manufacturer indicates compliance with appropriate standards or performance in a specified manner.

Listed: Equipment or materials included in a list published by an organization acceptable to the authority having jurisdiction and concerned with product evaluation, that maintains periodic inspection of production of listed equipment or materials and whose listing states either that the equipment or material meets appropriate standards or has been tested and found suitable for use in a specified manner.

Shall: Indicates a mandatory rule, S*ection 90-5.* As you study the *National Electrical Code,*® think of the word "shall" as meaning "must." Some examples found in the *NEC*® where the word "shall" is used in combination with other words are:

> **shall be, shall have, shall not,**
> **shall be permitted, shall not be permitted,**
> **and shall not be required.**

One of the most far-reaching *NEC*® rules is *Section 110-3(b).* This section states that the use and installation of listed or labeled equipment must conform to any instructions included in the listing or labeling. This means that an entire electrical system and all of the system's electrical equipment must be *installed* and *used* in accordance with the *National Electrical Code*® and the numerous standards against which the electrical equipment has been tested.

CODE USE OF METRIC (SI) MEASUREMENTS

The metric system is a "base-10" or "decimal" system in that values can be easily multiplied or divided by "ten" or "powers of ten." The metric system as we know it today is known as the International System of Units (SI) derived from the French term "le Systeme International d'Unites." The metric system is used by all industrial nations in the world but only to a limited extent in the United States.

As each edition of the *National Electrical Code*® comes out, more and more metric measurements appear. The electrician must become familiar with the metric system. In this country, we say "meter" (mee-tur) and we spell it "meter." Most other countries say "meter" but spell it "metre."

The 1981 *National Electrical Code*® introduced the use of metric measurements in addition to English measurements. Metric measurements appear in the Code as follows:

- in the Code paragraphs, the approximate metric measurement appears in parentheses following the English measurement.

Prefix	Symbol	Multiplier	Scientific Notation (Powers of Ten)	Value
tera	T	1 000 000 000 000	10^{12}	one trillion (1 000 000 000 000/1)
giga	G	1 000 000 000	10^{9}	one billion (1 000 000 000/1)
mega	M	1 000 000	10^{6}	one million (1 000 000/1)
kilo	k	1 000	10^{3}	one thousand (1 000/1)
hecto	h	100	10^{2}	one hundred (100/1)
deka	da	10	10^{1}	ten (10/1)
unit		1	—	one (1)
deci	d	0.1	10^{-1}	one tenth (1/10)
centi	c	0.01	10^{-2}	one hundredth (1/100)
milli	m	0.001	10^{-3}	one thousandth (1/1 000)
micro	μ	0.000 001	10^{-6}	one millionth (1/1 000 000)
nano	n	0.000 000 001	10^{-9}	one billionth (1/1 000 000 000)
pico	p	0.000 000 000 001	10^{-12}	one trillionth (1/1 000 000 000 000)

Fig. 1-2 Metric prefixes, symbols, multipliers, powers, and values.

- in the Code tables, a footnote shows the SI conversion factors.

A metric measurement is not shown for conduit size, box size, wire size, horsepower designation for motors, and other "trade sizes" that do not reflect actual measurements.

Guide to Metric Usage

In the metric system, the units increase or decrease in multiples of 10, 100, 1,000, and so on. For instance, one megawatt (1,000,000 watts) is 1 000 times greater than one kilowatt (1,000 watts).

By assigning a name to a measurement, such as a *watt*, the name becomes the unit. Adding a prefix to the unit, such as *kilo*, forms the new name *kilowatt*, meaning 1,000 watts. Refer to figure 1-2 for prefixes used in the metric system.

The prefixes used most commonly are *centi*, *kilo*, and *milli*. Consider that the basic unit is a meter (one). Therefore, a centimeter is 0.01 meter, a kilometer is 1,000 meters, and a millimeter is 0.001 meter.

Some common measurements of length and equivalents are shown in figure 1-3.

Electricians will find it useful to refer to the conversion factors and their abbreviations shown in figure 1-4.

Refer to the Appendix for a comprehensive metric conversion table. The table includes information on how to "round off" numbers for practical use on the job.

one inch	=	2.54	centimeters
	=	25.4	millimeters
	=	0.025 4	meter
one foot	=	12	inches
	=	0.304 8	meter
	=	30.48	centimeters
	=	304.8	millimeters
one yard	=	3	feet
	=	36	inches
	=	0.914 4	meter
	=	914.4	millimeters
one meter	=	100	centimeters
	=	1 000	millimeters
	=	1.093	yards
	=	3.281	feet
	=	39.370	inches

Fig. 1-3 Some common measurements of length and their equivalents.

inches (in) × 0.025 4	= meters (m)
inches (in) × 0.254	= decimeters (dm)
inches (in) × 2.54	= centimeters (cm)
centimeters (cm) × 0.393 7	= inches (in)
inches (in) × 25.4	= millimeters (mm)
millimeters (mm) × 0.039 37	= inches (in)
feet (ft) × 0.304 8	= meters (m)
meters (m) × 3.280 8	= feet (ft)
square inches (in²) × 6.452	= square centimeters (cm²)
square centimeters (cm²) × 0.155	= square inches (in²)
square feet (ft²) × 0.093	= square meters (m²)
square meters (m²) × 10.764	= square feet (ft²)
square yards (yd²) × 0.836 1	= square meters (m²)
square meters (m²) × 1.196	= square yards (yd²)
kilometers (km) × 1 000	= meters (m)
kilometers (km) × 0.621	= miles (mi)
miles (mi) × 1.609	= kilometers (km)

Fig. 1-4 Useful conversions (English/SI-SI/English) and their abbreviations.

NATIONALLY RECOGNIZED TESTING AGENCIES

How does one know if a product is safe to use? Manufacturers, consumers, regulatory authorities, and others recognize the importance of independent, "third-party" testing of products in an effort to reduce safety risks. Unless you have all of the necessary test equipment and knowledge of how to properly test a product for safety, the surest way is to accept the findings of a third-party testing agency. In other words, "look for the label." Nationally recognized testing laboratories (NRTL) have the knowledge and wherewithal to test and evaluate products for safety. Make sure the product has a listing marking on it from a nationally recognized testing laboratory. If the product is too small to have a listing mark on the product itself, then look for the marking on the carton the product came in. Of course, a listed product must also be used and installed properly to assure safety. The basic rule is found in Section 110-3(b) of the *National Electrical Code,*® which states that *"listed or labeled equipment shall be used or installed in accordance with any instructions included in the listing or labeling."*

There are a number of nationally recognized laboratories. The following laboratories do a considerable amount of testing and listing of electrical equipment.

Underwriters Laboratories Inc. (UL)

Underwriters Laboratories Inc. (UL), founded in 1894, is a highly qualified, nationally recognized testing laboratory. UL develops standards, and performs tests to these standards. Many reputable manufacturers of electrical equipment submit their products to UL where the equipment is subjected to numerous tests. These tests determine if the product can perform safely under normal and abnormal conditions to meet published standards. After UL tests, evaluates, and determines that a product complies with the specific standard, the manufacturer is then permitted to *label* its product with the UL Mark. The products are then *listed* in a UL Directory.

It should be noted that UL *does not approve* a product. Rather, UL *lists* those products that conform to a specific safety standard. A UL Listing Mark on a product means that representative samples of the product have been tested and evaluated to nationally recognized safety standards with regard to fire, electric shock, and related safety hazards.

Useful UL publications are:

Electrical Construction Materials Directory (Green Book)

Electrical Appliances and Utilization Equipment Directory (Orange Book)

Hazardous Location Equipment Directory (Red Book)

General Information for Electrical Construction, Hazardous Location, and Electric Heating and Air Conditioning Equipment (White Book)

It is extremely useful for an electrician, electrical contractor, and/or electrical inspector to refer to these directories when looking for specific requirements, permitted uses, limitations, and so on, for a certain product. These directories can be purchased from Underwriters Laboratories Inc. You may not be able to purchase all of the previous directories because of cost restraints. In this case, choose what is probably the best companion to the *National Electrical Code,*® which is the White Book.

Many times the answer to a product-related question cannot be found in the *National Electrical Code.*® It is quite possible that the answer might be found in the UL Directories.

The Green, Orange and Red Directories provide technical information regarding a particular product, plus they list the names and addresses of manufacturers and the manufacturers' identification numbers. The White Book provides the technical information by which a product is tested, but does not show manufacturers' names and addresses.

The previous directories can be obtained by writing or calling:

Underwriters Laboratories Inc.
333 Pfingsten Road
Northbrook, Illinois 60062
(708) 272-8800

Underwriters Laboratories Inc. and the Canadian Standards Association have worked out an agreement whereby either agency can test, evaluate, and list equipment for the other agency. For example, UL might test and list an air-conditioner unit to the requirements of UL Standard 1995 (Heating and Cooling Equipment) because the Canadian Standard C22.2 No. 236-M90 is a mirror image of UL Standard 1995. One by one, the UL and CSA Standards are becoming similar.

When UL tests and lists products that comply to the requirements of a particular CSA standard, the UL logo shown here will appear with a "C" outside of and to the left of the circle.

This means the product has been tested and evaluated for compliance *only* with Canadian requirements.

This harmonization of Codes and Standards is going on in the world today as a result of the North American Free Trade Act (NAFTA). Discussions are also going on with Mexico. When all of this is finalized, electrical equipment standards may be the same in the United States, Canada, and Mexico.

Canadian Standards Association (CSA)

The Canadian Standards Association is the Canadian counterpart of Underwriters Laboratories Inc. in the United States. CSA is the source of the *Canadian Electrical Code (CEC)* and of the Canadian Standards for the testing, evaluation, and listing of electrical equipment in Canada. The *Canadian Electrical Code* is quite different than the *National Electrical Code.*® A Canadian version of *Residential Wiring* is available in Canada.

The *Canadian Electrical Code* and CSA standards can be obtained by contacting:

Canadian Standards Association
178 Rexdale Blvd.
Rexdale, Ontario, Canada
M9W 1R3

Electrical Testing Laboratories (ETL)

Electrical Testing Laboratories is another nationally recognized testing laboratory. ETL provides a testing, evaluation, labeling, listing, and follow-up service for the safety testing of electrical products to nationally recognized safety standards or specifically designated requirements of jurisdictional authorities.

Information can be obtained by writing to:

ETL Testing Laboratories, Inc.
Main Office
Industrial Park
Cortland, New York 13045

REVIEW

Note: Refer to the *National Electrical Code*® or the plans where necessary.

1. What is the purpose of specifications? _____

_____

2. In what additional way are the specifications particularly useful to the electrical contractor? _____

_____

3. What is done to prevent a plan from becoming confusing because of too much detail?

_____

_____

4. Name three requirements contained in the specifications regarding material.

a. _____   c. _____

b. _____

5. The specifications state that all work shall be done _____

_____

6. What phrase is used when a substitution is permitted for a specific item? _____

_____

7. What is the purpose of an electrical symbol? _____

_____

8. What is a notation? _____

_____

9. Where are notations found? _____

10. List at least 12 electrical notations found on the plans for this residence. Refer to the plans at the back of the text. _____

_____

_____

_____

_____

_____

_____

11. What three parties must be satisfied with the completed electrical installation?

a. _____   b. _____   c. _____

12. What Code sets standards for electrical installation work? _____

_____

13. What authority enforces the standards set by the Code? _____

_____

14. Does the Code provide minimum or maximum standards? _____

15. What do the letters *UL* signify? _____

_____

16. What section of the Code states that all listed or labeled equipment shall be used or installed in accordance with any instructions included in the listing or labeling?

_____

17. When the words "shall be" appear in a Code reference, they mean that it (must)(may) be done. (Underline the correct word.)

18. What is the purpose of the *National Electrical Code®*?

_____

_____

_____

_____

_____

19. Does compliance with the Code always result in an electrical installation that is adequate, safe, and efficient? Why? _____
_____
_____
_____

20. Name two nationally recognized testing laboratories. _____
_____
_____

21. a. Do Underwriters Laboratories and the other recognized testing laboratories "approve" products? _____

    b. What do these testing laboratories do? _____
    _____
    _____

22. a. Has the *National Electrical Code*® been officially adopted by the community in which you live? _____

    b. By the state in which you live? _____

    c. If your answer is YES to a. or b., are there amendments to the *NEC*®? _____

    d. If your answer is YES to c., list some of the more important amendments.
    _____
    _____
    _____

23. Does the *National Electrical Code*® make suggestions about how to wire a house that will be occupied by handicapped persons? _____

24. A junction box on a piece of European equipment is marked 200 cubic centimeters. Convert this to square inches. _____
_____

25. Convert 4,500 watts to btu/hour. _____
_____

26. Residential lighting loads are based upon 3 volt-amperes per square foot. Determine the wattage required for an area of 186 square meters. _____
_____

UNIT 2

Electrical Symbols and Outlets

OBJECTIVES

After studying this unit, the student will be able to

- identify and explain the electrical outlet symbols used in the plans of the single-family dwelling.
- discuss the types of outlets, boxes, fixtures, and switches used in the residence.
- explain the methods of mounting the various electrical devices used in the residence.
- understand the meaning of the terms *receptacle outlet* and *lighting outlet*.
- understand the preferred way to position receptacles in wall boxes.
- understand issues involved in remodel work.
- know how to position wall boxes in relation to finished wall surfaces.
- know how to make surface extensions from concealed wiring methods.
- understand how to determine the number of wires permitted in a given size box.
- discuss interchangeable wiring devices.
- understand the concept of fire resistance rating of walls and ceilings.

ELECTRICAL SYMBOLS

Electrical symbols used on an architectural plan show the location and type of electrical device required. A typical electrical installation as taken from a plan is shown in figure 2-1.

The *National Electrical Code®* describes an *outlet* as "a point on a wiring system where current is taken to supply utilization equipment."

A receptacle outlet is "an outlet where one or more receptacles are installed," figure 2-2.

A lighting outlet is "an outlet intended for the direct connection of a lampholder, a lighting fixture, or a pendant cord terminating in a lampholder," figure 2-3.

A toggle switch is *not* an outlet.

The *NEC®* defines a *device* as "a unit of an electrical system which is intended to carry but not utilize electric energy."

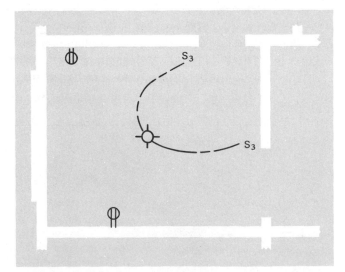

Fig. 2-1   Use of electrical symbols and notations on a floor plan.

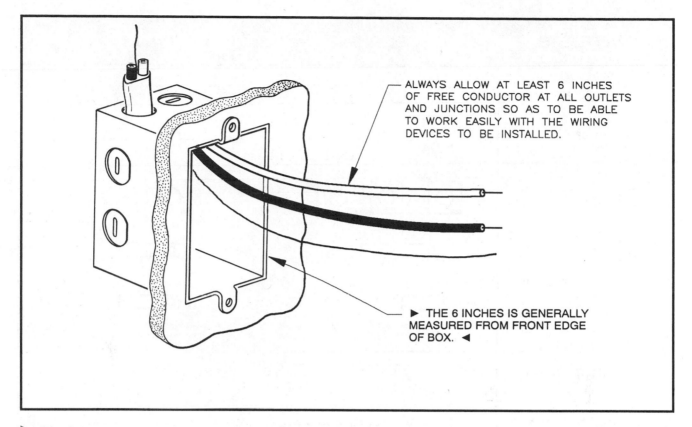

ALWAYS ALLOW AT LEAST 6 INCHES OF FREE CONDUCTOR AT ALL OUTLETS AND JUNCTIONS SO AS TO BE ABLE TO WORK EASILY WITH THE WIRING DEVICES TO BE INSTALLED.

▶ THE 6 INCHES IS GENERALLY MEASURED FROM FRONT EDGE OF BOX. ◀

▶ **Fig. 2-2**   When a receptacle is connected to the branch-circuit wires, the outlet is called a *receptacle outlet*. For ease in working with wiring devices and splicing conductors, the Code in *Section 300-14* requires that the branch-circuit conductors have at least 6 inches of free conductor length left at each outlet, junction, and switch point for splices or the connection of fixtures or devices. The 6-inch length is generally measured from the outer edge of the box. Do not leave too much wire length as this will result in the crowding of the wires into the box. ◀

The term *opening* is widely used by electricians and electrical contractors when estimating the cost of an installation. The term *opening* covers all lighting outlets, receptacle outlets, junction boxes, switches, etc. The electrician and/or electrical contractor will estimate a job at "X dollars per lighting outlet," "X dollars per switch," "X dollars per receptacle outlet," and so on. These estimates will include the "time" and "material" needed to complete the job. Each type of electrical *opening* is represented on the electrical plans as a symbol. In figure 2-1, the electrical openings are shown by the symbols:

Receptacle outlet:

Three-way switch: S_3

Ceiling lighting outlet:

Standard electrical symbols commonly found on architectural and electrical plans are detailed in figures 2-4 through 2-9.

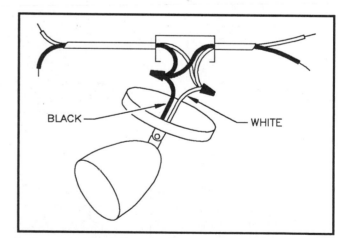

BLACK   WHITE

▶ **Fig. 2-3**   When a lighting fixture is connected to the branch-circuit wires, the outlet is called a *lighting outlet*. *Section 300-14* of the Code requires that the branch-circuit conductors have at least 6 inches of free conductor length left at each outlet, junction, and switch point for splices or the connection of fixtures or devices. The 6-inch length is generally measured from the outer edge of the box. Do not leave too much wire length as this will result in the crowding of the wires into the box. ◀

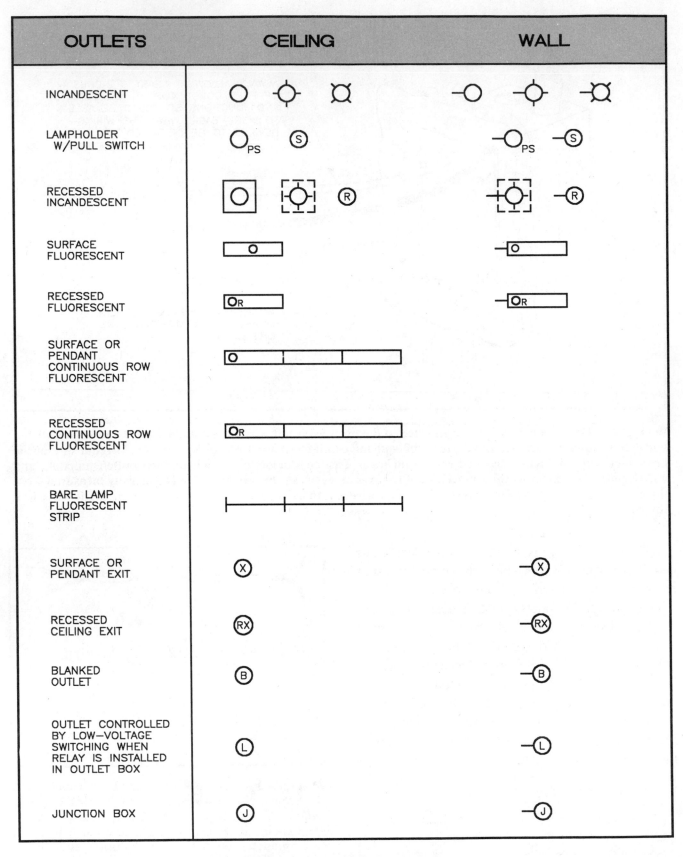

Fig. 2-4   Lighting outlet symbols.

RECEPTACLE OUTLETS

Symbol	Description
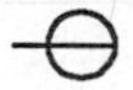 ⊝	SINGLE RECEPTACLE OUTLET
 ⊜	DUPLEX RECEPTACLE OUTLET
 ⊜IG	INSULATED (ISOLATED) GROUND RECEPTACLE OUTLET
 ⊕	TRIPLEX RECEPTACLE OUTLET
 ⊜	DUPLEX RECEPTACLE OUTLET, SPLIT–CIRCUIT
 ⊜	TRIPLEX RECEPTACLE OUTLET, SPLIT–CIRCUIT
 ⊜WP	WEATHERPROOF RECEPTACLE OUTLET
 ⊜GFCI	GROUND–FAULT CIRCUIT INTERRUPTER RECEPTACLE OUTLET
 ▲DW	SPECIAL–PURPOSE OUTLET (SUBSCRIPT LETTERS INDICATE SPECIAL VARIATIONS: DW = DISHWASWER. ALSO a, b, c, d, ETC. ARE LETTERS KEYED TO EXPLANATION ON DWGS. OR IN SPECIFICATIONS).
 ⊜R	RANGE OUTLET
 ⊜D	CLOTHES DRYER OUTLET
Ⓕ	FAN OUTLET
Ⓒ	CLOCK OUTLET
⊙	FLOOR OUTLET
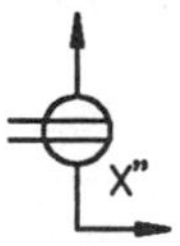 ↑⊜X"	MULTIOUTLET ASSEMBLY; ARROW SHOWS LIMIT OF INSTALLATION. APPROPRIATE SYMBOL INDICATES TYPE OF OUTLET. SPACING OF OUTLETS INDICATED BY "X" INCHES.
 ⊟	FLOOR SINGLE RECEPTACLE OUTLET
 ⊟	FLOOR DUPLEX RECEPTACLE OUTLET
 ▲	FLOOR SPECIAL–PURPOSE OUTLET

Fig. 2-5   Receptacle outlet symbols.

The dash lines in figure 2-1 run from the outlet to the switch or switches that control the outlet. These lines are usually curved so that they cannot be mistaken for invisible edge lines. Outlets shown on the plan without curved dash lines are independent outlets and have no switch control.

A study of the plans for the single-family dwelling shows that many different electrical symbols are used to represent the electrical devices and equipment used in the building.

In drawing electrical plans, most architects, designers, and electrical engineers use symbols approved by the American National Standards Institute (ANSI) wherever possible. However, plans may contain symbols that are not found in these standards. When such unlisted (nonstandard) symbols are used, the electrician must refer to a legend that interprets these symbols. The legend may be included on the plans or in the specifications. In many instances, a notation on the plan will clarify the meaning of the symbol.

Figures 2-4 through 2-9 list the standard, approved electrical symbols and their meanings. Many of these symbols can be found on the accompanying plans of the residence. Note in these figures that several symbols have the same shape. However, differences in the interior presentation indicate that the meanings of the symbols are different. For example, different meanings are shown in figure 2-10 for the outlet symbol. A good practice to follow in studying symbols is to learn the basic forms first and then add the supplemental information to obtain different meanings.

FIXTURES AND OUTLETS

The location of lighting outlets is determined by the amount and type of illumination required to provide the desired lighting effects. It is not the intent of this text to describe how proper and adequate lighting is determined. Rather, the text covers the proper methods of installing the circuits for such lighting. If the student is interested, standards have been developed to guide the design of adequate lighting. The

SWITCH SYMBOLS

S	SINGLE-POLE SWITCH
S_2	DOUBLE-POLE SWITCH
S_3	THREE-WAY SWITCH
S_4	FOUR-WAY SWITCH
S_D	DOOR SWITCH
S_{DS}	DIMMER SWITCH
S_K	KEY SWITCH
S_L	LOW-VOLTAGE SWITCH
S_{LM}	LOW-VOLTAGE MASTER SWITCH
S_P	SWITCH WITH PILOT LAMP
S_R	VARIABLE-SPEED SWITCH
S_T	TIME SWITCH
S_{WP}	WEATHERPROOF SWITCH

Fig. 2-6   Switch symbols.

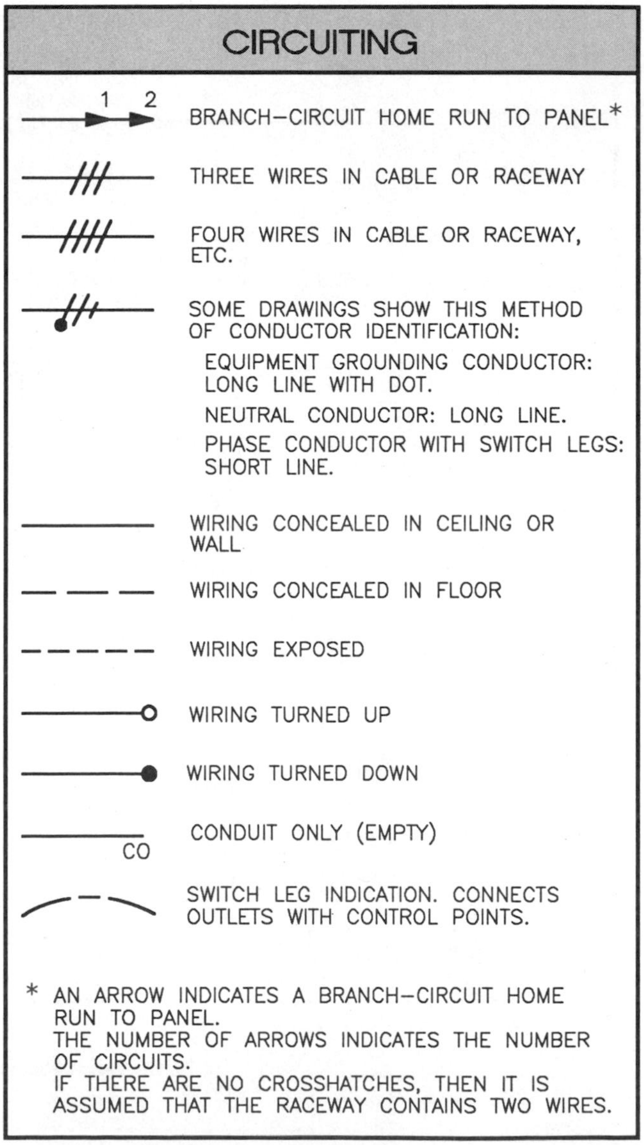

CIRCUITING

1 2 →	BRANCH-CIRCUIT HOME RUN TO PANEL*
—///—	THREE WIRES IN CABLE OR RACEWAY
—////—	FOUR WIRES IN CABLE OR RACEWAY, ETC.
—•//—	SOME DRAWINGS SHOW THIS METHOD OF CONDUCTOR IDENTIFICATION: EQUIPMENT GROUNDING CONDUCTOR: LONG LINE WITH DOT. NEUTRAL CONDUCTOR: LONG LINE. PHASE CONDUCTOR WITH SWITCH LEGS: SHORT LINE.
——	WIRING CONCEALED IN CEILING OR WALL
— — —	WIRING CONCEALED IN FLOOR
- - - -	WIRING EXPOSED
——o	WIRING TURNED UP
——●	WIRING TURNED DOWN
——— CO	CONDUIT ONLY (EMPTY)
⌒	SWITCH LEG INDICATION. CONNECTS OUTLETS WITH CONTROL POINTS.

* AN ARROW INDICATES A BRANCH-CIRCUIT HOME RUN TO PANEL.
   THE NUMBER OF ARROWS INDICATES THE NUMBER OF CIRCUITS.
   IF THERE ARE NO CROSSHATCHES, THEN IT IS ASSUMED THAT THE RACEWAY CONTAINS TWO WIRES.

Fig. 2-7   Circuiting symbols.

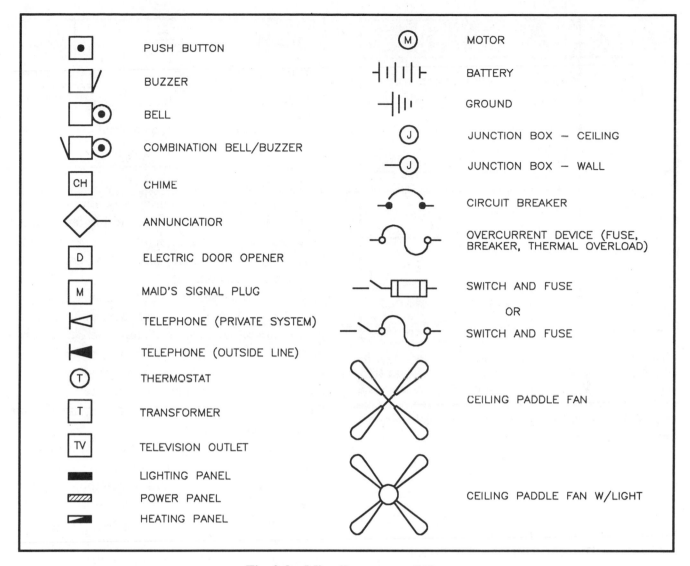

Fig. 2-8   Miscellaneous symbols.

local electric utility company can supply information on these standards. The Instructor's Guide lists excellent publications relating to proper residential lighting.

Architects often include in the specifications a certain amount of money for the purchase of lighting fixtures. The electrical contractor includes this amount in the bid, and the choice of fixtures is then left to the homeowner. If the owner selects fixtures whose total cost exceeds the fixture allowance, the owner is expected to pay the difference between the actual cost and the specification allowance. If the fixtures are not selected before the roughing-in stage of wiring the house, the electrician usually installs outlet boxes having standard fixture mounting studs.

Most modern surface-mount lighting fixtures can be fastened to a fixture stud in the box (figure 2-11, top row, picture A) or an outlet box or plaster ring using appropriate #8-32 metal screws and mounting strap furnished with the fixture (figure 2-11, third row, raised plaster cover; outlet boxes).

In conformance to *Section 300-15*, a box or fitting must be installed at every point in an electrical installation where there are splices, outlets, switches, junctions, or pull points. Some exceptions to this would be listed equipment that has an integral junction box as is typical on recessed lighting fixtures, or ▶ where the lighting fixtures are used as a raceway (i.e., end-to-end mounted recessed fluorescent lighting fixtures that are listed as O.K. to use as a raceway). ◀ For standard wiring methods, such as cable or conduit, a box is usually required. Outlet boxes must be accessible, *Section 370-29*.

SYMBOL	NOTATION
①	PLUGMOLD ENTIRE LENGTH OF WORKBENCH. OUTLETS 18" O.C. INSTALL 48" TO CENTER FROM FLOOR. GFCI PROTECTED.
②	TRACK LIGHTING. PROVIDE 5 LAMPHOLDERS.
③	TWO 40 WATT RAPID START FLUORESCENT LAMPS IN VALANCE. CONTROL WITH DIMMER SWITCH.

Fig. 2-9   Example of how certain *notations* might be added to a symbol when a symbol itself does not fully explain its meaning. The architect or engineer has a choice of explaining fully the meaning directly on the plan if there is sufficient room; if insufficient room, then a *notation* could be used.

Be careful when roughing in boxes for fixtures. UL states in its Green and White Books that it has not investigated hanging ceiling fixtures to:

- nonmetallic device (switch) boxes.
- nonmetallic device (switch) plaster rings.
- metallic device (switch) boxes.
- metallic device (switch) plaster rings.
- any nonmetallic box, unless specifically marked on the box or carton for use as a fixture support, or to support other equipment, or to accommodate heat-producing equipment.

Therefore, unless the box, device plaster ring, or carton is marked to indicate that it has been listed by UL for the support of fixtures, do not use where - fixtures are to be hung. Refer to figure 2-12A. Fixtures that weigh more than 50 pounds must be supported independently from the outlet box, *Section 410-16(a)*. *Section 370-17(c), Exception*, allows multiple cables to be run through a single knockout opening in a nonmetallic box. This permission is not given for metallic boxes. Refer to figure 2-12B.

Fig. 2-10   Variations in significance of outlet symbols.

Be careful when installing a ceiling outlet box for the purpose of supporting a ceiling fan. The box must be marked "Acceptable for Fan Support." Refer to unit 9 for detailed discussion of ceiling fans and their installation.

If the owner selects fixtures prior to construction, the architect can specify these fixtures in the plans and/or specifications. Thus, the electrician is provided with advance information on any special framing, recessing, or mounting requirements for the fixtures. This information *must* be provided in the case of recessed fixtures, which require a specific wall or ceiling opening.

Many types of lighting fixtures are presently available. Figure 2-11 shows several typical lighting fixtures that may be found in a dwelling unit. Also shown are the electrical symbols used on plans to designate these fixtures and the type of outlet boxes or switch boxes on which the lighting fixtures can be mounted. A standard receptacle outlet is shown as well. The switch boxes shown here are made of steel. Switch boxes may also be made of plastic, as shown in figure 2-12A. Other types of outlets are covered in later units.

FLUSH SWITCHES

Some of the standard symbols for various types of switches are shown in figure 2-13. Typical connection diagrams are also given. Any sectional switch box or 4-inch square box with a side mounting bracket and raised switch cover can be used to install these switches. Refer to figure 2-11.

JUNCTION BOXES AND SWITCH (DEVICE) BOXES *(ARTICLE 370)*

Article 370 of the *NEC®* contains much detail about outlet boxes, device boxes, pull boxes, junction

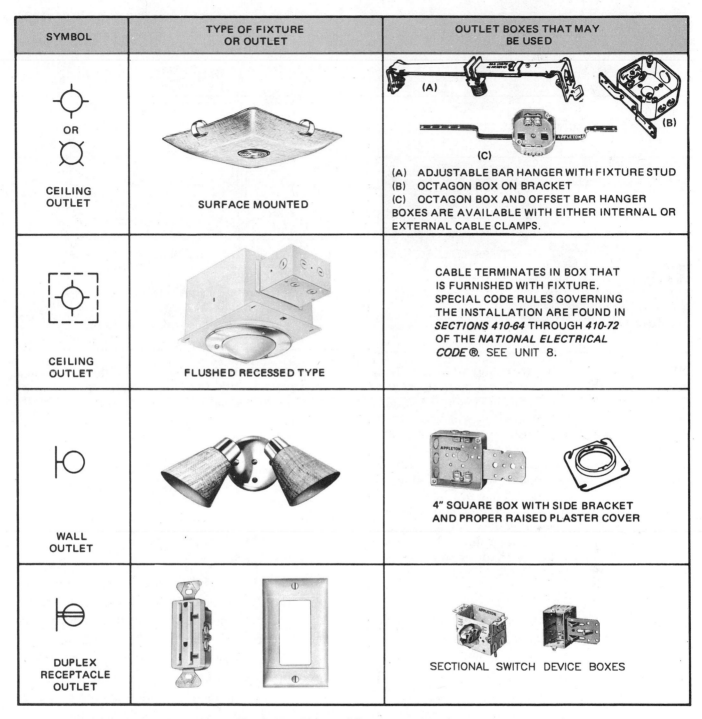

SYMBOL	TYPE OF FIXTURE OR OUTLET	OUTLET BOXES THAT MAY BE USED
OR CEILING OUTLET	SURFACE MOUNTED	(A) ADJUSTABLE BAR HANGER WITH FIXTURE STUD (B) OCTAGON BOX ON BRACKET (C) OCTAGON BOX AND OFFSET BAR HANGER BOXES ARE AVAILABLE WITH EITHER INTERNAL OR EXTERNAL CABLE CLAMPS.
CEILING OUTLET	FLUSHED RECESSED TYPE	CABLE TERMINATES IN BOX THAT IS FURNISHED WITH FIXTURE. SPECIAL CODE RULES GOVERNING THE INSTALLATION ARE FOUND IN *SECTIONS 410-64* THROUGH *410-72* OF THE *NATIONAL ELECTRICAL CODE®*. SEE UNIT 8.
WALL OUTLET		4" SQUARE BOX WITH SIDE BRACKET AND PROPER RAISED PLASTER COVER
DUPLEX RECEPTACLE OUTLET		SECTIONAL SWITCH DEVICE BOXES

Fig. 2-11 Types of fixtures and outlets.

boxes, and conduit bodies (LB, LR, LL, etc.). While studying this unit, you should also study *Article 370*, as there is too much material in the Article to repeat in this text.

Junction boxes are sometimes placed in a circuit for convenience in joining two or more cables or conduits. All conductors entering a junction box are joined to other conductors entering the same box to form proper hookups so that the circuit will operate in the manner intended.

All electrical installations must conform to the *National Electrical Code®* standards requiring that junction boxes be installed in such a manner that the wiring contained in them shall be accessible without removing any part of the building. In house wiring, this requirement limits the use of

Fig. 2-12A Supporting (hanging) a ceiling fixture from these types of boxes is not permitted unless specifically marked on the box or carton (UL requirement). Some electrical inspectors will accept metal device boxes and device plaster covers for small, lightweight, wall-mounted lighting fixtures, such as porch bracket fixtures and narrow, decorative bathroom strip lighting fixtures mounted above or along the sides of medicine cabinets and mirrors. Some of these fixtures are so narrow they will not completely cover the opening of a regular outlet box. Talk to the electrical inspector before roughing-in device boxes where the intent is to mount lighting fixtures on them.

junction boxes to unimproved basements, garages, and open attic spaces because flush blank covers exposed to view detract from the appearance of a room. Of course, an outlet box such as the one installed for the front hall ceiling fixture is really a junction box because it contains splices. Removing the fixture makes the box accessible, thereby meeting Code requirements. Refer to figures 2-14 and 2-15.

Section 300-15 requires that a box or fitting be installed wherever splices, switches, outlets, junction points, or pull points are required (figure 2-14). However, there are instances where a change is made from one wiring method to another, in which case a box is not required. This is permitted by *Exception No. 8* to *Section 300-15(b)*. Note that the

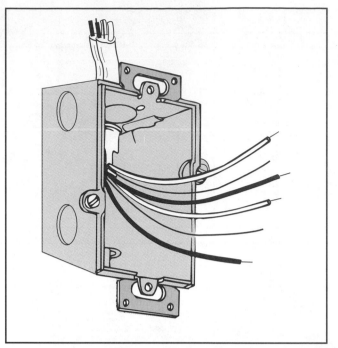

Fig. 2-12B Multiple cables may run through a single knockout opening in a nonmetallic box, *Section 370-17(c)*, *Exception*.

fitting where the change is made must be accessible after installation, figure 2-16.

Boxes shall be rigidly and securely mounted, *Section 370-23*.

NONMETALLIC OUTLET AND DEVICE BOXES

Section 370-3 of the Code permits nonmetallic outlet and device boxes to be installed where the wiring method is nonmetallic sheathed cable or nonmetallic raceway. The *NEC®* in *Section 370-3 (Exceptions)* allows metal raceways and metal-jacketed cables (BX) to be used with nonmetallic boxes provided all metal raceways or cables entering the box are bonded together to maintain the integrity of the grounding path to other equipment in the installation.

The house wiring system usually is formed by a number of specific circuits. Each circuit consists of a continuous run of cable from outlet to outlet or from box to box. The residence plans show many branch circuits for general lighting, appliances, electric heating, and other requirements. The specific Code rules for each of these circuits are covered in later units.

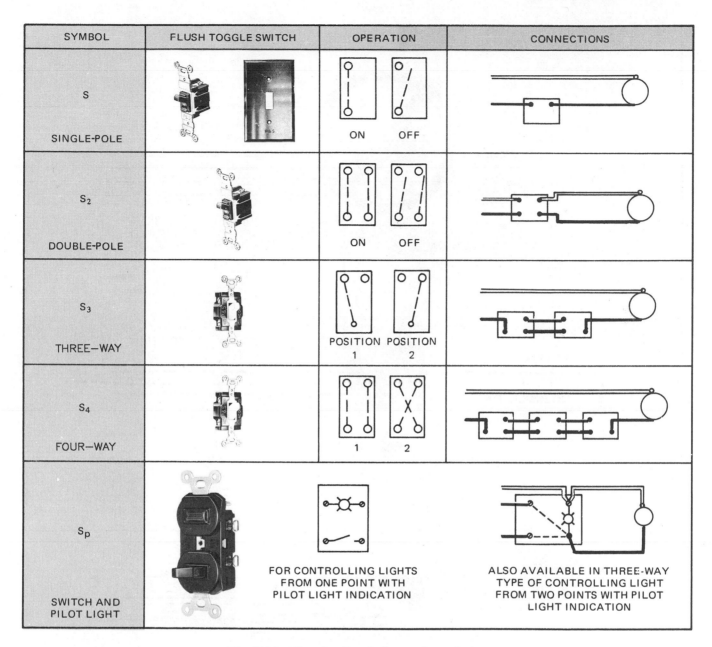

Fig. 2-13 **Standard switches and symbols.**

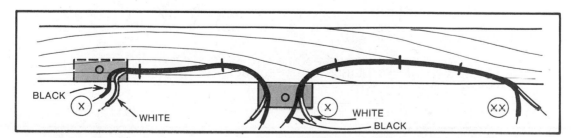

Fig. 2-14 **A box (or fitting) must be installed wherever there are splices, outlets, switches, or other junction points. Refer to the points marked X. A Code violation is shown at point XX,** *Section 300-15 (a)* **and** *(b)*.

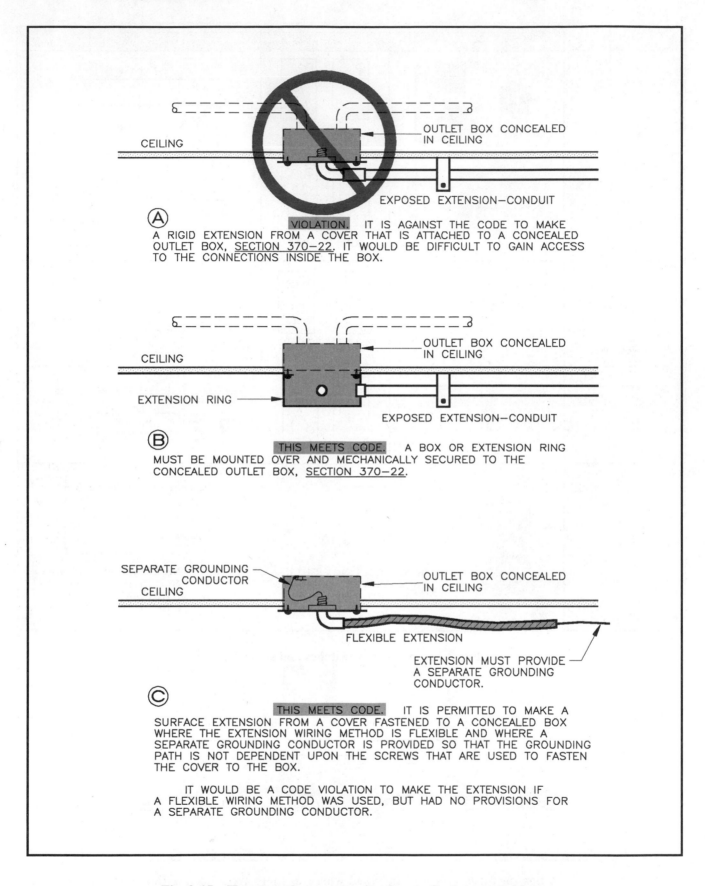

CEILING

OUTLET BOX CONCEALED
IN CEILING

EXPOSED EXTENSION—CONDUIT

(A) VIOLATION. IT IS AGAINST THE CODE TO MAKE
A RIGID EXTENSION FROM A COVER THAT IS ATTACHED TO A CONCEALED
OUTLET BOX, SECTION 370—22. IT WOULD BE DIFFICULT TO GAIN ACCESS
TO THE CONNECTIONS INSIDE THE BOX.

CEILING

OUTLET BOX CONCEALED
IN CEILING

EXTENSION RING

EXPOSED EXTENSION—CONDUIT

(B) THIS MEETS CODE. A BOX OR EXTENSION RING
MUST BE MOUNTED OVER AND MECHANICALLY SECURED TO THE
CONCEALED OUTLET BOX, SECTION 370—22.

SEPARATE GROUNDING
CONDUCTOR
CEILING

OUTLET BOX CONCEALED
IN CEILING

FLEXIBLE EXTENSION

EXTENSION MUST PROVIDE
A SEPARATE GROUNDING
CONDUCTOR.

(C) THIS MEETS CODE. IT IS PERMITTED TO MAKE A
SURFACE EXTENSION FROM A COVER FASTENED TO A CONCEALED BOX
WHERE THE EXTENSION WIRING METHOD IS FLEXIBLE AND WHERE A
SEPARATE GROUNDING CONDUCTOR IS PROVIDED SO THAT THE GROUNDING
PATH IS NOT DEPENDENT UPON THE SCREWS THAT ARE USED TO FASTEN
THE COVER TO THE BOX.

IT WOULD BE A CODE VIOLATION TO MAKE THE EXTENSION IF
A FLEXIBLE WIRING METHOD WAS USED, BUT HAD NO PROVISIONS FOR
A SEPARATE GROUNDING CONDUCTOR.

Fig. 2-15   How to make an extension from a flush-mounted box.

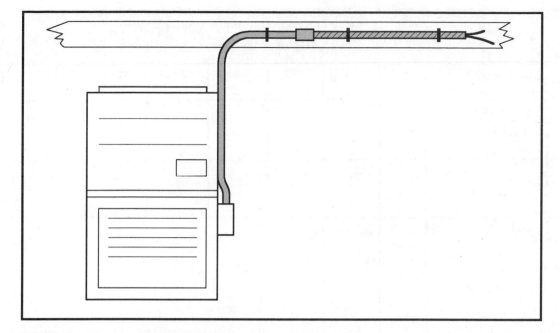

Fig. 2-16 *Section 300-15 (b), Exceptions No. 1* and *No. 8* permit a transition to be made from one wiring method to another wiring method. In this case, the armor of the Type AC cable is removed, allowing sufficient length of the conductors to be run through the conduit. A proper fitting must be used at the transition point, and the fitting must be accessible after installation.

GANGED SWITCH (DEVICE) BOXES

A flush switch or receptacle for residential use fits into a standard 2" × 3" sectional switch box (sometimes called a *device box*). When two or more switches or receptacles are located at the same point, the switch boxes are ganged or fastened together to provide the required mounts, figure 2-17.

Three switch boxes can be ganged together by removing and discarding one side from both the first and third switch boxes and both sides from the second (center) switch box. The boxes are then joined together as shown in figure 2-17. After the switches are installed, the gang is trimmed with a gang plate having the required number of switch

Fig. 2-17   Standard flush switches installed in ganged sectional device boxes.

QUIK-CHEK BOX SELECTION GUIDE
FOR BOXES GENERALLY USED FOR RESIDENTIAL WIRING

DEVICE BOXES	WIRE SIZE	3X2X1½ (7.5 in³)	3X2X2 (10 in³)	3X2X2¼ (10.5 in³)	3X2X2½ (12.5 in³)	3X2X2¾ (14 in³)	3X2X3 (16 in³)	3X2X3½ (18 in³)
	#14	3	5	5	6	7	7	9
	#12	3	4	4	5	6	7	8

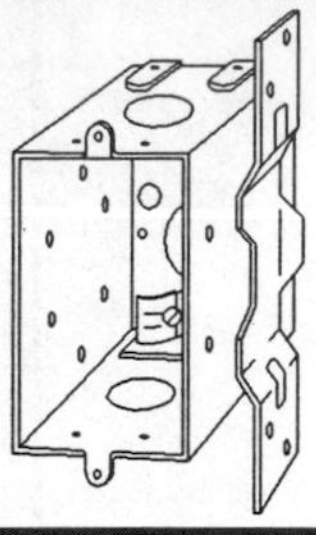

SQUARE BOXES	WIRE SIZE	4X4X1½ (21 in³)	4X4X2⅛ (30.3 in³)
	#14	10	15
	#12	9	13

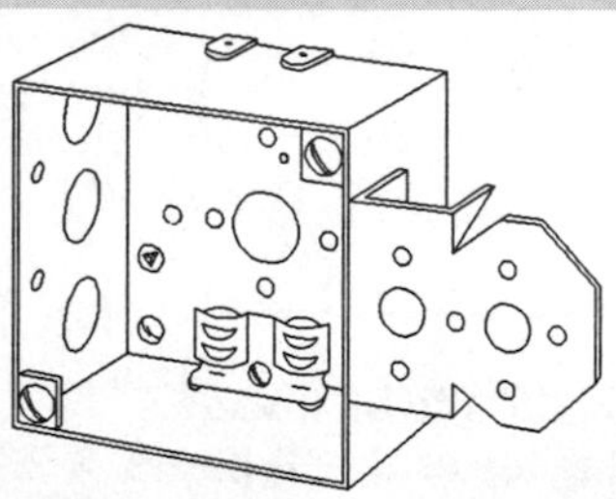

OCTAGON BOXES	WIRE SIZE	4X1½ (15.5 in³)	4X2⅛ (21.5 in³)
	#14	7	10
	#12	6	9

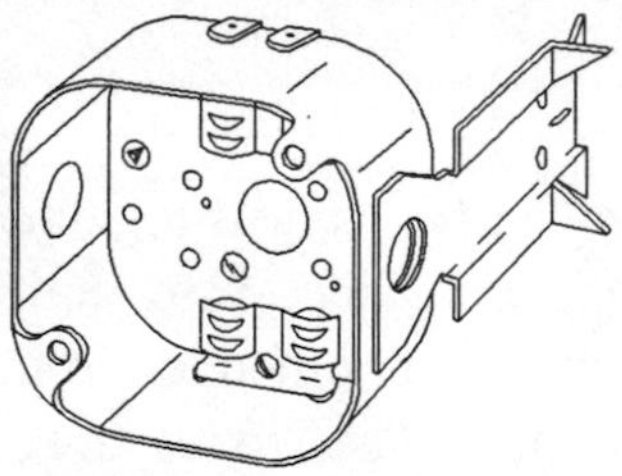

HANDY BOXES	WIRE SIZE	4X2⅛X1½ (10.3 in³)	4X2⅛X1⅞ (13 in³)	4X2X2⅛ (14.5 in³)
	#14	5	6	7
	#12	4	5	6

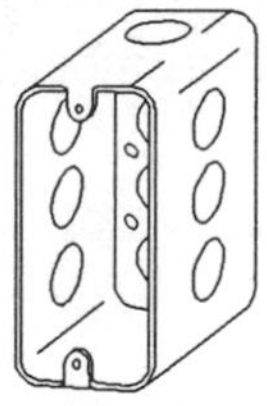

RAISED COVERS

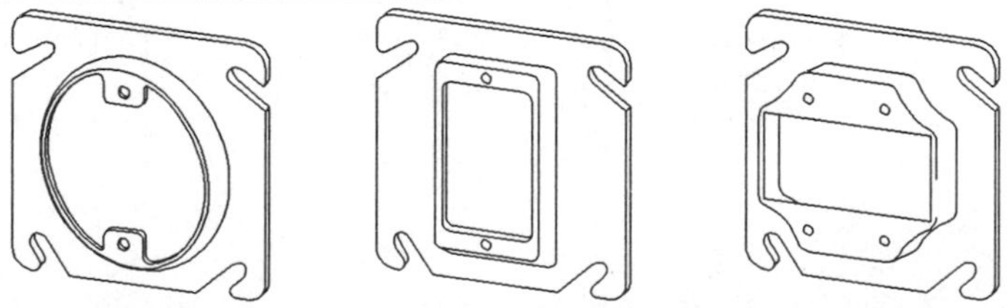

WHERE RAISED COVERS ARE MARKED WITH THEIR VOLUME IN CUBIC INCHES, THAT VOLUME MAY BE ADDED TO THE BOX VOLUME TO DETERMINE MAXIMUM NUMBER OF CONDUCTORS IN THE COMBINED BOX AND RAISED COVER.

NOTE: BE SURE TO MAKE DEDUCTIONS FROM THE ABOVE MAXIMUM NUMBER OF CONDUCTORS PERMITTED FOR WIRING DEVICES, CABLE CLAMPS, FIXTURE STUDS, AND GROUNDING CONDUCTORS. THE CUBIC INCH (IN³) VOLUME IS TAKEN DIRECTLY FROM TABLE 370—16(a) OF THE NEC ®.

Fig. 2-18   Quik-chek box selector guide.

handle or receptacle openings. These plates are called two-gang wall plates, three-gang wall plates, and so on, depending upon the number of openings.

The dimensions of a standard sectional switch box (2" × 3") are the dimensions of the opening of the box. The depth of the box may vary from 1½ inches to 3½ inches, depending upon the requirements of the building construction and the number of conductors and devices to be installed. *NEC® Article 370* covers outlet, switch, and junction boxes. See figure 2-18 for a complete listing of box dimensions.

BOX MOUNTING

Section 370-20 of the Code states that boxes must be mounted so that they will be set back not more than ¼ inch (6.35 mm) when the boxes are mounted in noncombustible walls or ceilings made of concrete, tile, or similar materials. When the wall or ceiling construction is of combustible material (wood), the box must be set flush with the surface, figure 2-19. These requirements are meant to prevent the spread of fire if a short circuit occurs within the box.

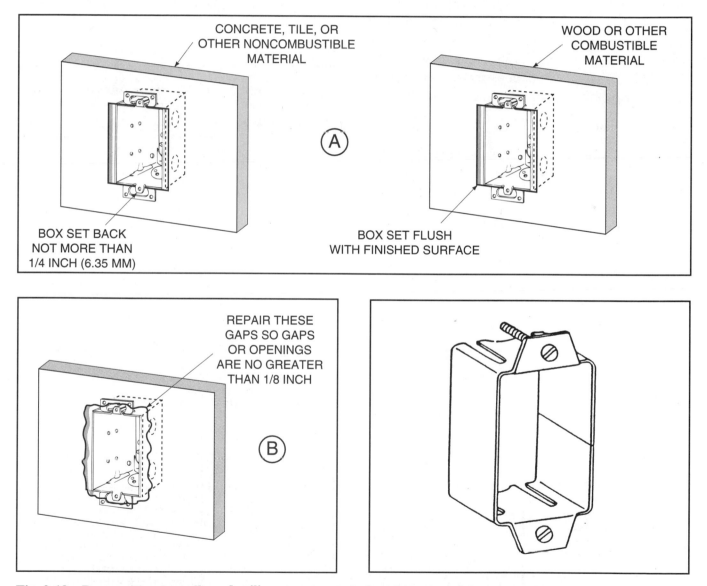

Fig. 2-19   Box position in walls and ceilings constructed of various materials. Where the box, for some reason is not flush with the finished wall, then install a switch box extension ring of the type shown in (C). See *Sections 370-20* and *370-21, NEC.®*

Ganged sectional switch (device) boxes can be installed using a pair of metal mounting strips. These strips are also used to install a switch box between wall studs, figure 2-20. The use of a bracket box in such an installation may result in an off-center receptacle outlet. When an outlet box is to be mounted at a specific location between joists, as for ceiling-mounted fixtures, an offset bar hanger is used (figure 2-11).

The Code states that when a switch box or outlet box is mounted to a stud or ceiling joist by nailing through the box, the nails must be not more than ¼ inch (6.35 mm) from the back or ends of the box, figure 2-21. This requirement ensures that when the nail passes through the box, it does not interfere with the wiring devices in the box.

Most residences are constructed with wood framing. However, metal studs and joists are being used more and more. Figure 2-21A shows types of boxes used for fast, easy attachment to metal framing members. These boxes "snap" onto the metal stud, and do not require drilling the metal studs and fastening the box with nuts and bolts. The first one is a 4-inch square box with ½- and ¾-inch knockouts. The second shows a 4-inch square box with nonmetallic sheathed cable (Romex) clamps. The third shows a 4-inch square box with armored cable (BX) clamps. See unit 4 for details about running cables through metal framing members.

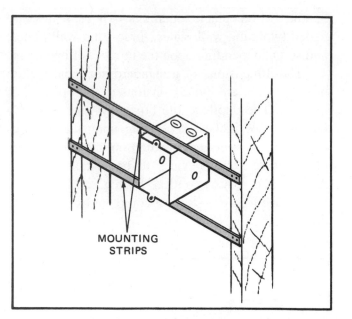

Fig. 2-20   Switch (device) boxes installed between studs using metal mounting strips, often referred to as "Kruse strips." If wood strips are used, they must have a cross-sectional dimension of not less than 1 inch × 2 inches, *Section 370-23(b)(2).*

Another type of device used for fastening electrical boxes to steel framing members is shown in figure 2-21B. The bracket is screwed to the steel framing members, then the box or boxes are attached to the bracket.

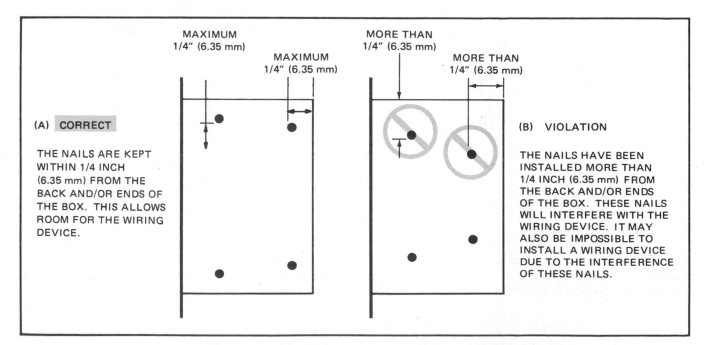

Fig. 2-21   Using nails to install a sectional switch box, *Section 370-23.*

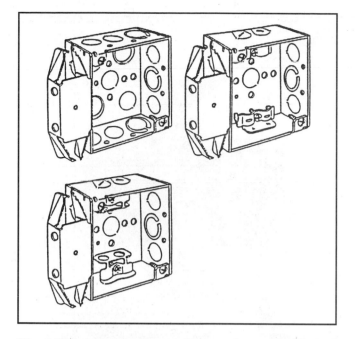

Fig. 2-21A   Types of boxes that can be used with steel framing construction. Note the side clamps that "snap" directly onto the steel framing members. *Courtesy* **of RACO.**

Remodel (Old Work)

When installing switches, receptacles, and lighting outlets in remodel work where the walls and ceilings (paneling, drywall, sheet rock and plaster, or lath and plaster) are already in place, first make sure you will be able to run cables to where the switches, receptacles, and lighting outlets are to be installed. After making sure you can get the cables through the concealed spaces in the walls and ceilings, you can then proceed to cut openings the size of the specific type of box to be installed at each outlet location. Cables are then "fished" through the stud and joist spaces behind or above the finished walls and ceilings to where you have cut the openings. Then, boxes having plaster ears and snap-in brackets (figure 2-21C) can be inserted from the front, through the holes already cut at the locations where the wall or ceiling boxes are to be installed. After the boxes are snapped into place through the hole from the front, they become securely locked into place.

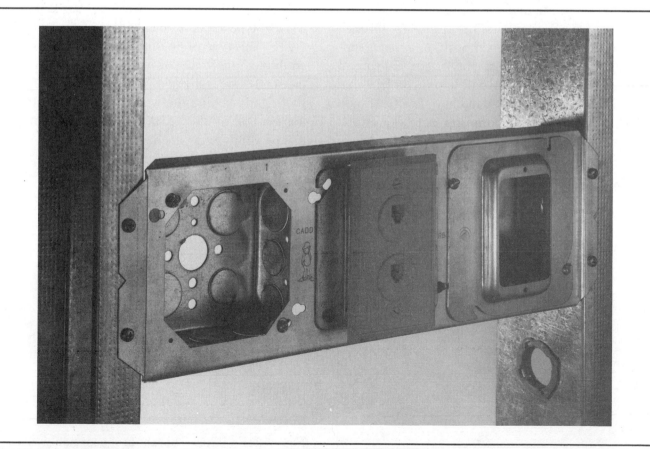

Fig. 2-21B   A bracket for use on steel framing members. The bracket spans the space between two joists. Note that more than one box can be mounted on this bracket. *Courtesy* **Erico® Inc.**

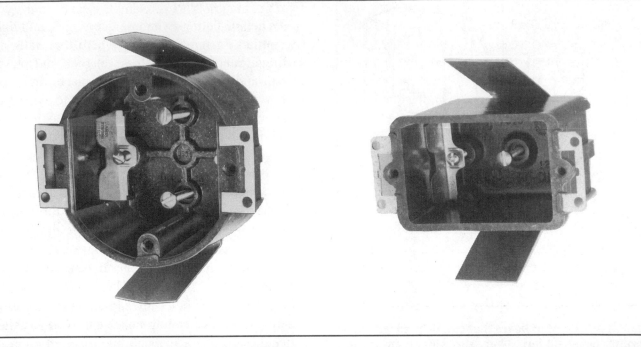

Fig. 2-21C Nonmetallic boxes are used for remodel work so that when they are inserted into an opening cut into a wall or ceiling, they will "snap" into place, thus conforming to the *National Electrical Code*® requirements for the adequate support of boxes. *Courtesy* **Challenger Electrical Equipment Corp.**

Another popular method of fastening wall and ceiling boxes in existing walls and ceilings is to use boxes that have plaster ears. Insert the box through the hole from the front. Then position a metal support into the hole on each side of the box. Be careful not to let the metal support fall into the hole. Next, bend the metal support over and into the box (figure 2-21D). Two metal supports are used. Over the years, these metal supports have become known as "Madison Holdits."

Also available for old work are boxes that have a screw-type support on each side of the box, figure 2-21E. This type of box has plaster ears. It is inserted through the hole cut in the wall or ceiling, then the screws are tightened. This pulls up a metal support tightly behind the wall or ceiling, firmly holding the box in place. The action is very similar to that of Molly screw anchors.

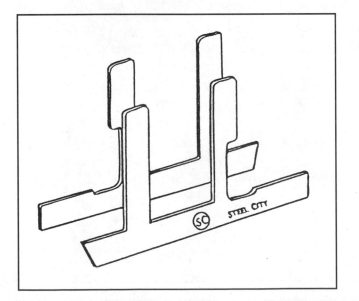

Fig. 2-21D Metal supports referred to as "Madison Holdits." *Courtesy* **of Thomas & Betts Corporation.**

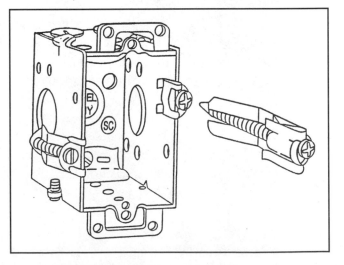

Fig. 2-21E These boxes are commonly used in remodel work. The box is installed from the front, then the screws are tightened. The box is adequately supported. *Courtesy* **of Steel City.**

Because these types of remodel boxes are supported by the drywall or plaster, as opposed to new work where the supporting of the boxes is to the framing members, be careful not to hang heavy lighting fixtures from them. Be sure that there is proper support if a ceiling fan is to be installed. This is covered in unit 9.

Spread of Fire

To protect lives and property, fire must be contained and not be allowed to spread.

This text covers one-family dwelling units. In walls, partitions, and ceilings that are *fire resistance rated*, special consideration must be given when installing wall and ceiling electrical boxes where you might want to mount the boxes "back-to-back" or in the same stud or joist space of common walls or ceilings, as might be found in multifamily buildings. Building codes such as the Council of American Building Officials (CABO) *One- and Two-Family Dwelling Code* require fire-resistance-rated walls and ceilings between occupancies in two-family dwellings and in townhouses. These building codes also require fire resistance rating between habitable areas and a garage, even in single-family dwellings. Ratings of fire-resistant materials are expressed in hours. Terms such as 1-hour fire resistance rating, 2-hour fire resistance rating, and so on, are used. Electrical wall boxes installed back-to-back or installed in the same stud or joist space defeat the fire resistance rating of the wall or ceiling.

Underwriters Standard 263 *Fire Tests of Building Construction and Materials* and NFPA 251 cover fire resistance ratings. In UL Standard 263, and in all building codes, you will find requirements such as: "the surface area of individual metallic outlet or switch boxes shall not exceed 16 square inches" (i.e., a 4-inch square box is $4 \times 4 = 16$ square inches), "the aggregate surface area of the boxes shall not exceed 100 square inches per 100 square feet of wall surface," "boxes located on opposite sides of walls or partitions shall be separated by a minimum horizontal distance of 24 inches," "the metallic outlet or switch boxes shall be securely fastened to the studs and the opening in the wallboard facing shall be cut so that the clearance between the box and the wallboard does not exceed ⅛ inch," and "the boxes shall be installed in compliance with the *National Electrical Code.*"

The UL *Fire Resistance Directory* covers fire resistant materials and assemblies, and contains a rather detailed listing of *Outlet Boxes and Fittings Classified for Fire Resistance (CEYY)*, showing all manufacturers' product part numbers for *nonmetallic boxes* to be installed in walls, partitions, and ceilings that meet the fire resistance standard. This category also covers special-purpose boxes for installation in floors.

When using nonmetallic boxes in fire-resistance-rated walls, the restrictions are more stringent than for metallic boxes. For example, the UL *Fire Resistance Directory* shows that one manufacturer of nonmetallic boxes limits the product to be installed so that no opening exceeds 10.0 square inches. Another manufacturer limits the product to be installed so that no opening exceeds 25.0 square inches.

For both metallic and nonmetallic electric outlet boxes, the *minimum* distance between boxes on opposite sides of a fire-rated wall or partition is 24 inches. There is an exception, which is discussed later.

For both metallic and nonmetallic electric outlet boxes, the maximum "gap" between the box and the wall material is ⅛ inch. This holds true for fire-resistance-rated walls and non-fire-resistance-rated walls.

Of interest to an electrician is an item listed in the UL *Fire Resistance Directory* that is an *intumescent* (expands when heated) fire-resistant material that comes in "pads." This moldable putty can be used to wrap electrical boxes that are installed in the same wall cavity. This moldable putty inhibits heat transfer from the fire-resistance-rated side of the wall to the non-fire-resistance rated side of the wall. It is pressed onto the metallic electrical outlet box, around electrical raceways that enter the box, and is pressed into the interface where the box, stud, and gypsum wallboard meet. Electricians commonly use the 6" × 7" pad, ⅛ inch thick. The ⅛ inch thickness provides a 1-hour fire resistance rating. A 2-hour rating is obtained with ¼ inch thickness. When this material is properly installed, the 24-inch separation between boxes is not required. However, the electrical outlet boxes **must not** be installed back-to-back. Using this material meets the requirements of *Section 300-21* of the *National Electrical Code®* and other building codes.

Rule 402.7 in the CABO *One- and Two-Family Dwelling Code* states that "firestopping be provided to cut off all concealed draft openings (both vertical and horizontal) and to form an effective fire barrier between stories, and between a top story and the roof space." This includes openings around vents, pipes, ducts, chimneys and fireplaces at ceiling and floor level. Oxygen supports fire. Firestopping cuts off or significantly reduces the flow of oxygen. Most communities that have adopted the CABO code require firestopping around electrical conduits. In residential work, this is generally done by the insulation installer, before the dry wall installer closes up the walls.

The previous discussion makes it clear that the electrician must become familiar with certain building codes, in addition to the *National Electrical Code.*® Knowing the *NEC*® is not enough!

It is extremely important to check with the electrical inspector and/or building official for complete clarification on the matter of fire resistance requirements before you find yourself cited with serious violations and lawsuits regarding noncompliance with fire codes. Changes can be costly, and will delay completion of the project. Improper wiring is hazardous. The *National Electrical Code*® makes us aware of this issue in *Section 300-21.*

Refer to unit 16 for further discussion and diagrams of installing electrical outlet boxes in fire-resistance-rated walls.

BOXES FOR CONDUIT WIRING

Some local electrical ordinances require conduit rather than cable wiring. Conduit wiring is discussed in unit 18. Examples of conduit fill (how many wires are permitted in a given size conduit) are presented.

When conduit is installed in a residence, it is quite common to use 4-inch square boxes trimmed with suitable plaster covers. The type of box shown in figure 2-22(A) is the most popular. There are sufficient knockouts in the top, bottom, sides, and back of the box to permit a number of conduits to run to the box. Plenty of room is available for the conductors and wiring devices. Note how easily these 4-inch square outlet boxes can be mounted back-to-back by installing a small fitting between the boxes. This is illustrated in figures 2-23(B) and (C).

(A)

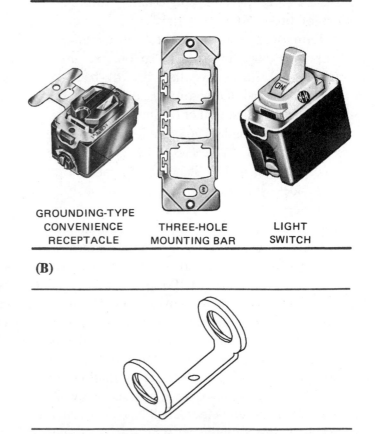

GROUNDING-TYPE
CONVENIENCE
RECEPTACLE

THREE-HOLE
MOUNTING BAR

LIGHT
SWITCH

(B)

(C)

Fig. 2-22   Interchangeable devices (A and B). A fixture hickey (C).

(A)

(B)

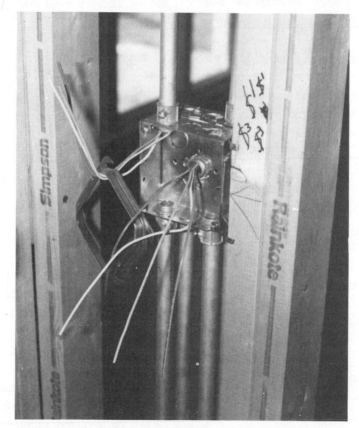

(C)

Examples of electrical metallic tubing showing how 4-inch square boxes are used for single outlets and switches. Note how the boxes are attached together for "back-to-back" installations. When mounting boxes back-to-back, be sure to consider possible transfer of sound between the rooms. Another thing to consider is the possible spread of fire. Some building and fire codes will not allow back-to-back installations in multifamily dwelling structures in those walls and partitions that separate the individual dwelling units.

Fig. 2-23   Photos of electrical metallic tubing (EMT) installation. Boxes are 4-inch square type.

Four-inch square outlet boxes can be trimmed with one-gang or two-gang plaster rings where wiring devices will be installed. Where lighting fixtures will be installed, a plaster ring having a round opening should be installed.

Any unused openings in outlet and device boxes must be closed per *Sections 110-12* and *370-18* of the Code, figure 2-24.

Close the Gap Around Boxes!

Section 370-21 of the *National Electrical Code®* requires that openings around electrical boxes shall be repaired so that "there will be no gaps or open space greater than ⅛ inch (3.18 mm) at the edge of the box or fitting." This is to minimize the spread of fire. The Underwriters Laboratories *Fire Resistance Directory* mirrors the *NEC®* rule by stating that "the outlet or switch boxes shall be securely fastened to the studs and the opening in the wallboard facing shall be cut so that the clearance between the box and the wallboard does not exceed ⅛ inch." See figure 2-19(B).

This is easier said than done. The electrician installs the wiring (rough-in), followed by an inspection by the electrical inspector. Next comes the dry wall/plaster/panel installer, who many times cuts out the box openings much larger than the outlet or switch box. In most cases, the walls and ceilings are painted or wall papered before the electrician returns to install the receptacles, switches, and lighting fixtures. For gaps or openings greater than ⅛ inch around the electrical boxes, should the electrician repair the gap with patching plaster, possibly damaging or marring the finished wall? Whose responsibility is it? Is it the electrician's? Is it the dry wall/plaster/panel installer's?

The electrician should check with the dry wall/plaster/panel installer to clarify who is to be responsible for seeing to it that gaps or openings around electrical outlet and switch boxes do not exceed ⅛ inch. *Section 370-21* puts the responsibility on the electrician.

INTERCHANGEABLE WIRING DEVICES

In the space of one standard wiring device, it is possible to install one, two, or three wiring devices by using an interchangeable line of devices. Up to three switches, pilot lights, receptacle outlets, or any

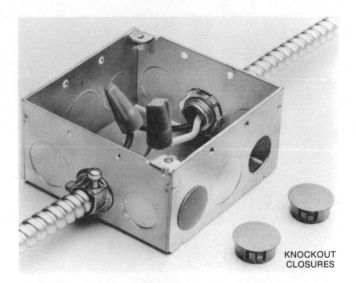

KNOCKOUT CLOSURES

Fig. 2-24   Unused openings in boxes must be closed according to the Code, *Sections 110-12* and *370-18*. This is done to contain electrical short-circuit problems inside the box or panel and to keep rodents out. Note that the conductors DO NOT meet the 6-inches of free conductor length requirement of *Section 300-14*. This is a good example of a Code violation. See figure 2-2 for proper installation regarding free conductor length at boxes.

combination of interchangeable wiring devices can be installed, one above the other, on a single strap in a standard 2" × 3" box opening. A total of six devices can be installed using two ganged boxes with standard openings.

Interchangeable devices are available in the same types and classifications of switches and receptacles as standard wiring devices. In other words, single-pole switches, three-way switches, four-way switches, pilot lights, and other devices are available in an interchangeable style. Both standard and interchangeable devices are available in silent style that makes little or no noise when the switch is actuated.

When interchangeable devices are to be installed, standard sectional switch boxes generally are not used because of the lack of wiring space in the box. Instead, 4-inch square and 4¹¹/₁₆-inch square boxes with raised plaster covers are used. This is not a requirement of the Code, however. Under certain conditions, 3-inch and 3½-inch deep sectional boxes are large enough for a group of interchangeable devices.

The Code requires that faceplates for switches and receptacles must completely cover the wall opening and must seat against the surface of the wall.

Figure 2-22 shows two interchangeable switches and one pilot light mounted on a single strap in a 4-inch square outlet box. The box has a plaster cover and a bracket. A single-gang, three-hole wall plate is required. This arrangement is useful where limited wall space, such as the space between a door casing and a window casing, prohibits the use of standard two-gang or three-gang wall plates.

SPECIAL-PURPOSE OUTLETS

Special-purpose outlets are usually indicated on the plans. These outlets are described by a notation and are usually detailed in the specifications. The plans indicate special-purpose outlets by a triangle inside a circle with subscript letters. In some cases, a subscript number is added to the letter.

When a special-purpose outlet is indicated on the plans or in the specifications, the electrician must check for special requirements. Such a requirement may be a separate circuit, a special 240-volt circuit, a special grounding or polarized receptacle, or other preparation.

A set of Plans and Specifications for this residence is found at the end of this text. The Specifications include a *Schedule of Special Purpose Outlets*.

NUMBER OF CONDUCTORS IN BOX

The Code *(Section 370-16)* dictates that outlet boxes, switch boxes, and device boxes should be large enough to provide ample room for the wires in that box, without having to jam or crowd the wires into the box. Jamming the wires into the box can not only damage the insulation on the wires, but can result in heat build-up in the box that can further damage the insulation on the wires. The Code specifies the maximum number of conductors allowed in standard metal outlet, device, and junction boxes, figure 2-25. Nonmetallic boxes are marked by the manufacturer with their cubic-inch capacity.

When conductors are the same size, the proper box size can be selected by referring to *Table 370-16(a)*. When conductors are of different sizes, refer to *Table 370-16(b)*, and use the cubic-inch volume for the particular size of wire being used.

Tables 370-16(a) and *(b)* do not take into con-sideration the space taken by fixture studs, cable clamps, hickeys (figure 2-22C), switches, pilot lights, or receptacles that may be in the box. These require additional space. Figure 2-26 shows the additional allowances that must be considered for different situations.

SELECTING THE CORRECT SIZE BOX

▶ **EXAMPLES:** Refer to figure 2-26.

When all conductors are the same size. A box contains one fixture stud and two internal cable clamps. Four No. 12 conductors enter the box.

four No. 12 conductors	4
one fixture stud	1
two cable clamps (only count one)	1
Total	6

Referring to *Table 370-16(a)*, a 4" × 1½" octagon box is suitable for this example.

Table 370-16(a). Metal Boxes								
Box Dimension, Inches Trade Size or Type	Min. Cu. In. Cap.	Maximum Number of Conductors						
		No. 18	No. 16	No. 14	No. 12	No. 10	No. 8	No. 6
4 x 1¼ Round or Octagonal	12.5	8	7	6	5	5	4	2
4 x 1½ Round or Octagonal	15.5	10	8	7	6	6	5	3
4 x 2⅛ Round or Octagonal	21.5	14	12	10	9	8	7	4
4 x 1¼ Square	18.0	12	10	9	8	7	6	3
4 x 1½ Square	21.0	14	12	10	9	8	7	4
4 x 2⅛ Square	30.3	20	17	15	13	12	10	6
4¹¹⁄₁₆ x 1¼ Square	25.5	17	14	12	11	10	8	5
4¹¹⁄₁₆ x 1½ Square	29.5	19	16	14	13	11	9	5
4¹¹⁄₁₆ x 2⅛ Square	42.0	28	24	21	18	16	14	8
3 x 2 x 1½ Device	7.5	5	4	3	3	3	2	1
3 x 2 x 2 Device	10.0	6	5	5	4	4	3	2
3 x 2 x 2¼ Device	10.5	7	6	5	4	4	3	2
3 x 2 x 2½ Device	12.5	8	7	6	5	5	4	2
3 x 2 x 2¾ Device	14.0	9	8	7	6	5	4	2
3 x 2 x 3½ Device	18.0	12	10	9	8	7	6	3
4 x 2⅛ x 1½ Device	10.3	6	5	5	4	4	3	2
4 x 2⅛ x 1⅞ Device	13.0	8	7	6	5	5	4	2
4 x 2⅛ x 2⅛ Device	14.5	9	8	7	6	5	4	2
3¾ x 2 x 2½ Masonry Box/Gang	14.0	9	8	7	6	5	4	2
3¾ x 2 x 3½ Masonry Box/Gang	21.0	14	12	10	9	8	7	4
FS—Minimum Internal Depth 1¾ Single Cover/Gang	13.5	9	7	6	6	5	4	2
FD—Minimum Internal Depth 2⅜ Single Cover/Gang	18.0	12	10	9	8	7	6	3
FS—Minimum Internal Depth 1¾ Multiple Cover/Gang	18.0	12	10	9	8	7	6	3
FD—Minimum Internal Depth 2⅜ Multiple Cover/Gang	24.0	16	13	12	10	9	8	4

Table 370-16(b). Volume Required per Conductor	
Size of Conductor	Free Space Within Box for Each Conductor
No. 18	1.5 cubic inches
No. 16	1.75 cubic inches
No. 14	2. cubic inches
No. 12	2.25 cubic inches
No. 10	2.5 cubic inches
No. 8	3. cubic inches
No. 6	5. cubic inches

Reprinted with permission from NFPA 70-1996, the *National Electrical Code.*® Copyright © 1995, National Fire Protection Association, Quincy, Massachusetts 02269. This reprinted material is not the complete and official position of the National Fire Protection Association on the referenced subject, which is repre-sented only by the standard in its entirety.

Fig. 2-25 Allowable number of conductors in boxes.

• If box contains no fittings, devices, fixture studs, cable clamps, hickeys, switches, receptacles, or equipment grounding conductors . . .	• refer directly to *Tables 370-16(a) or (b).*
• **Clamps.** If box contains one or more internal cable clamps . . .	• add a single-volume based on the largest conductor in the box.
• **Support Fittings.** If box contains one or more fixture studs or hickeys . . .	• add a single-volume for each type based on the largest conductor in the box.
• **Device or Equipment.** If box contains one or more wiring devices on a yoke . . .	• add a double-volume for each yoke based on the largest conductor connected to a device on that yoke.
• **Equipment Grounding Conductors.** If a box contains one or more equipment grounding conductors . . .	• add a single-volume based on the largest conductor in the box.
• **Isolated Equipment Grounding Conductor.** If a box contains one or more additional "isolated" (insulated) equipment grounding conductors as permitted by *Section 250-74, Exception No. 4* for "noise" reduction . . .	• add a single-volume based on the largest conductor in the box.
• For conductors running through the box without being spliced . . .	• add a single-volume for each conductor that runs through the box.
• For conductors that originated outside of the box and terminate inside the box . . .	• add a single-volume for each conductor that originates outside the box and terminates inside the box.
• If no part of the conductor leaves the box — for example, a "jumper" wire used to connect three wiring devices on one yoke, or pigtails as illustrated in Figure 8-9 . . .	• don't count this (these). No additional volume required.
• For an equipment grounding conductor or not more than 4 fixture wires smaller than No. 14 that originate from a fixture canopy or similar canopy (like a fan) and terminate in the box . . .	• don't count this (these). No additional volume required.
• For small fittings, such as lock-nuts and bushings . . .	• don't count this (these). No additional volume required.

▶ **Fig. 2-26   Quick checklist for possibilities to be considered when determining proper size boxes.** ◀

When the conductors are different sizes. What is the minimum cubic-inch volume required for a box that will contain one internal cable clamp, one switch, and one receptacle, all mounted on one yoke? Two No. 14 and two No. 12 wires enter the box. The wiring method is armored cable.

two No. 14 wires @ 2 cubic inches per wire	= 4.00 cubic inches
two No. 12 wires @ 2.25 cubic inches per wire	= 4.50 cubic inches
one cable clamp @ 2.25 cubic inches	= 2.25 cubic inches
one switch and one receptacle on one yoke @ 2.25 cubic inches × 2	= 4.50 cubic inches
Total	15.25 cubic inches

Select a box having a minimum volume of 15.25 cubic inches of space. The cubic-inch volume may be marked on the box, otherwise refer to the second column of *Table 370-16(a)* entitled "Min. Cu. In. Cap."

The Code in *Section 370-16(a)(2)* requires that all boxes *other* than those listed in *Table 370-16(a)* be durably and legibly marked by the manufacturer with their cubic-inch capacity. When sectional boxes are ganged together, the calculated volume is the total cubic-inch volume of the assembled boxes.

When installing a box that will have a raised cover attached to it: When marked with their cubic-inch volume, the *additional* space provided by plaster rings, domed covers, raised covers, and extension rings is permitted to be used when determining the overall volume. This is illustrated in figure 2-27.

How many No. 12 conductors are permitted in this box and raised plaster ring? Refer to *Section 370-16* and *Tables 370-16(a)* and *(b).*

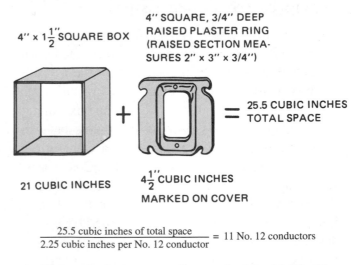

4″ × 1½″ SQUARE BOX

4″ SQUARE, 3/4″ DEEP RAISED PLASTER RING (RAISED SECTION MEASURES 2″ × 3″ × 3/4″)

= 25.5 CUBIC INCHES TOTAL SPACE

21 CUBIC INCHES

4½″ CUBIC INCHES MARKED ON COVER

$$\frac{25.5 \text{ cubic inches of total space}}{2.25 \text{ cubic inches per No. 12 conductor}} = 11 \text{ No. 12 conductors}$$

This calculation actually resulted in 11.33. The conductor fill numbers in *Table 370-16(a)* were created by dropping any excess above the whole number after the calculations were made. Following

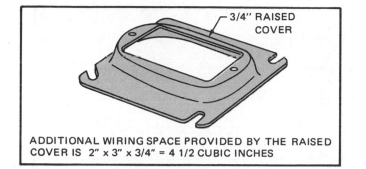

Fig. 2-27   Raised cover. Raised covers are sometimes called plaster rings.

the same procedure, this conductor fill calculation dropped off the excess. ◄

Electricians and electrical inspectors have become very aware of the fact that GFCI receptacles, dimmers, and timers take up a lot more space than regular wiring devices. Therefore, it is a good practice to install boxes that will provide "lots" of room for the wires, instead of pushing, jamming, and crowding the wires into the box.

Figure 2-28 shows that transformer leads No. 18 or larger must be counted when selecting a proper size box.

The minimum required size of equipment grounding conductors is shown in *Table 250-95.* The equipment grounding conductors are the same size as the circuit conductors in cables having No. 14, No. 12, and No. 10 circuit conductors. Thus, for normal house wiring, box sizes can be calculated using *Table 370-16(a).* Refer to figure 2-29.

Figure 2-18, a **quik-chek box selector guide,** shows some of the most popular types of boxes used in residential wiring. This guide also shows the box's cubic-inch volume.

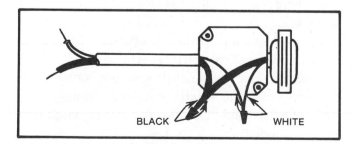

Fig. 2-28   Transformer leads No. 18 AWG or larger must be counted when selecting the proper size box, *Section 370-16(a)(1).* In the example shown, the box is considered to contain four conductors.

Figure 2-30 illustrates typical wiring using cable and conduit. Note in this figure how the conductor count is determined.

Using electrical metallic tubing (EMT) makes it possible to "loop" conductors through the box, which counts as one conductor only. Looping conductors through a box is not possible when using cable as the wiring method.

The basic rules for box fill are summarized in figure 2-26. Boxes that are intended to support ceiling fans are discussed in detail in unit 9.

Means for Connecting Equipment Grounding Conductors in a Box

► *Section 370-40(d)* requires that "a means shall be provided in each metal box for the connection of equipment grounding conductors." Generally, these are tapped $^{10}/_{32}$ holes in the box that are marked GR, GRN, GRND, or similar identification. The screws used for terminating the equipment grounding conductor must be hexagon shaped, and must be green. Electrical distributors sell green, hexagon shaped $^{10}/_{32}$ screws for this purpose. ◄

HEIGHT OF RECEPTACLE OUTLETS

There are no hard-and-fast rules for locating most outlets. A number of conditions determine the proper height for a switch box. For example, the

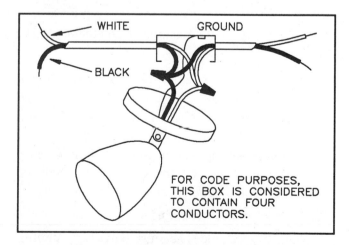

Fig. 2-29   Four or less fixture wires, smaller than No. 14 AWG, and/or one equipment grounding conductor that originates in the fixture and terminates in the box, need not be counted when calculating box fill, *Section 370-16(b)(1), Exception.*

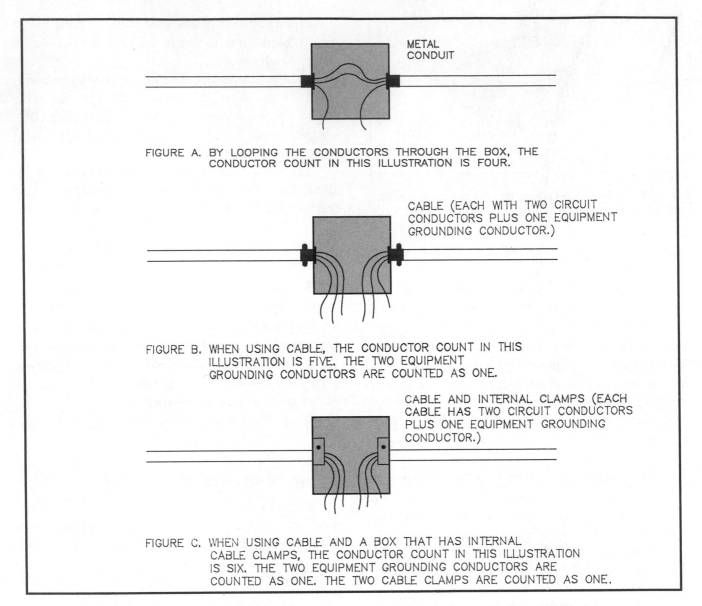

FIGURE A. BY LOOPING THE CONDUCTORS THROUGH THE BOX, THE CONDUCTOR COUNT IN THIS ILLUSTRATION IS FOUR.

FIGURE B. WHEN USING CABLE, THE CONDUCTOR COUNT IN THIS ILLUSTRATION IS FIVE. THE TWO EQUIPMENT GROUNDING CONDUCTORS ARE COUNTED AS ONE.

FIGURE C. WHEN USING CABLE AND A BOX THAT HAS INTERNAL CABLE CLAMPS, THE CONDUCTOR COUNT IN THIS ILLUSTRATION IS SIX. THE TWO EQUIPMENT GROUNDING CONDUCTORS ARE COUNTED AS ONE. THE TWO CABLE CLAMPS ARE COUNTED AS ONE.

Fig. 2-30   Example of conductor count for both metal conduit and cable installations.

height of the kitchen counter backsplash determines where the switches and receptacle outlets are located between the kitchen countertop and the cabinets.

The residence featured in this text is electrically heated and is discussed in unit 23. The type of electric heat could be:

• electric furnace (as in this text).

• electric resistance heating buried in ceiling plaster or "sandwiched" between two layers of drywall material.

• electric baseboard heaters.

Let us consider the electric baseboard heaters. In most cases, the height of these electric baseboard

units from the top of the unit to the finished floor seldom exceeds 6 inches (152 mm). The important issue here is that the manufacturer's receptacle accessories may have to be used to conform to the receptacle spacing requirements, as covered in *Section 210-52* of the Code.

Electrical receptacle outlets are not permitted to be located above an electric baseboard heating unit. Refer to the section "Location of Electric Baseboard Heaters," in unit 23.

It is common practice among electricians to consult the plans and specifications to determine the proper heights and clearances for the installation of electrical devices. The electrician then has these dimensions verified by the architect, electrical engineer, designer, or homeowner. This practice avoids unnecessary and costly changes in the locations of outlets and switches as the building progresses.

POSITIONING OF RECEPTACLES

Although no actual Code rules exist on positioning receptacles, there is a popular concept in the electrical industry that there is always the possibility of a metal wall plate coming loose and falling downward onto the blades of an attachment plug cap that is loosely plugged into the receptacle, thereby creating a potential shock and fire hazard. Here are the options:

Recommended

Grounding hole to the top. A loose metal plate could fall onto the grounding blade of the attachment plug cap, but no sparks would fly.

Not Recommended

Grounding hole to the bottom. A loose metal plate could fall onto both the grounded neutral and "hot" blades. Sparks would fly.

Recommended

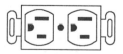

Grounded neutral blades on top. A loose metal plate could fall onto these grounded neutral blades. No sparks would fly.

Not Recommended

"Hot" terminal on top. A loose metal plate could fall onto these live blades. If this was a split-circuit receptacle fed by a 3-wire, 120/240-volt circuit, the short would be across the 240-volt line. Sparks would really fly!!

To insure uniform installation and safety, and in accordance with long-established custom, standard electrical outlets are located as shown in figure 2-31. These dimensions usually are satisfactory. However, the electrician must check the blueprints, specifications, and details for measurements that may affect the location of a particular outlet or switch. The cabinet spacing, available space between the countertop and the cabinet, and the tile height may influence the location of the outlet or switch. For example, if the top of wall tile is exactly 48 inches (1.22 m) from the finished floor line, a wall switch should not be mounted 48 inches (1.22 m) to center. This is considered poor workmanship. The switch should be located entirely within the tile area or entirely out of

SWITCHES	
	* Inches above floor
Regular	46 (1.17 m)
Between counter and kitchen cabinets (depends on backsplash)	44-46 (1.12-1.17 m)
RECEPTACLE OUTLETS	
	* Inches above floor
Regular . . . . (not permitted above electric baseboard heaters)	12 (305 mm)
Between counter and kitchen cabinets (depends on backsplash)	44-46 (1.12-1.17 m)
In garage	48 (1.22 m)
WALL LIGHTING OUTLETS	
	* Inches above floor
Outside	66 (1.68 m)
Inside	60 (1.52 m)
Side of medicine cabinet	60 (1.52 m)

*Note:
All dimensions given are from the finished floor to the center of the outlet box. Verify all dimensions before roughing in.

Fig. 2-31 Recommended heights for switches, receptacles, and wall lighting outlets.

the tile area, figure 2-32. This situation requires the full cooperation of all crafts involved in the construction job.

Faceplates

Faceplates for switches shall be installed so as to completely cover the wall opening and seat against the wall surface, *Section 380-9.*

Faceplates for receptacles shall be installed so as to completely cover the opening and seat against

the mounting surface. The mounting surface might be the wall, or it might be the gasket of a weatherproof box.

In either case, the intent of the Code is to prevent access to live parts that could cause electrical shock.

Faceplates should be level. In addition, a minor detail that can improve the appearance of the installation is to align the slots in all of the faceplate mounting screws in the same direction. See figure 2-33.

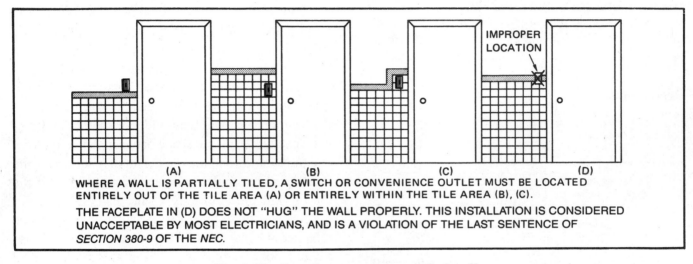

(A)   (B)   (C)   (D)

WHERE A WALL IS PARTIALLY TILED, A SWITCH OR CONVENIENCE OUTLET MUST BE LOCATED ENTIRELY OUT OF THE TILE AREA (A) OR ENTIRELY WITHIN THE TILE AREA (B), (C).

THE FACEPLATE IN (D) DOES NOT "HUG" THE WALL PROPERLY. THIS INSTALLATION IS CONSIDERED UNACCEPTABLE BY MOST ELECTRICIANS, AND IS A VIOLATION OF THE LAST SENTENCE OF *SECTION 380-9* OF THE *NEC.*

Fig. 2-32   Locating an outlet on a tiled wall.

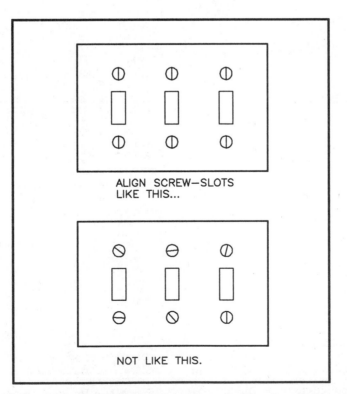

ALIGN SCREW—SLOTS LIKE THIS...

NOT LIKE THIS.

Fig. 2-33   Aligning the slots of the faceplate. Mounting screws in the same direction makes the installation look neater.

REVIEW

Note: Refer to the *National Electrical Code®* or the plans where necessary.

PART 1—ELECTRICAL FEATURES

1. What does a plan show about electrical outlets? _____

   _____

   _____

   _____

2. What is an outlet? _____

   _____

   _____

3. Match the following switch types with the proper symbol.

   a. single-pole S_p

   b. three-way S_4

   c. four-way S

   d. single-pole with pilot light S_3

4. The plans show dash lines running between switches and various outlets. What do these dash lines indicate? _____

   _____

5. Why are dash lines usually curved? _____

6. a. What are junction boxes used for? _____

   b. Are junction boxes normally used in wiring the first floor? Explain. _____

   _____

   _____

   _____

   c. Are junction boxes normally used to wire exposed portions of the basement? Explain. _____

   _____

7. How are standard sectional switch (device) boxes mounted? _____

   _____

   _____

   _____

8. a. What is an offset bar hanger? _____

   _____

   _____

   b. What types of boxes may be used with offset bar hangers? _____

   _____

   _____

9. What methods may be used to mount lighting fixtures to an outlet box fastened to an offset bar hanger? _____

_____

_____

_____

10. What advantage does a 4-inch octagon box have over a 3¼-inch octagon box?

_____

_____

11. What is the size of the opening of a switch (device) box for a single device? _____

_____

_____

12. The space between a door casing and a window casing is 3½ inches (88.9 mm). Two switches are to be installed at this location. What type of switches will be used?

_____

_____

13. Three switches are mounted in a three-gang switch (device) box. The wall plate for this assembly is called a _____ plate.

14. The mounting holes in a device (switch) box are tapped for ⁶/₃₂ screws. The mounting holes in an outlet box are tapped for ⁸/₃₂ screws. The mounting holes in metal boxes for attaching equipment grounding conductors are tapped for _____ screws.

15. a. How high above the finish floor are the switches located in the garage of this dwelling? _____

   b. In the living room of this dwelling? _____

16. How high above the finish floor are the receptacle outlets in the garage located? _____ In the living room? _____

17. Outdoor receptacle outlets in this dwelling are located _____ _____ inches above grade.

18. In the spaces provided, draw the correct symbol for each of the descriptions listed in (a) through (r).

   a. _____ Lighting panel

   b. _____ Clock outlet

   c. _____ Duplex outlet

   d. _____ Outside telephone

   e. _____ Single-pole switch

   f. _____ Motor

   g. _____ Duplex outlet, split-circuit

   h. _____ Lampholder with pull switch

   i. _____ Weatherproof outlet

   j. _____ Special-purpose outlet

   k. _____ Fan outlet

   l. _____ Range outlet

   m. _____ Power panel

   n. _____ Three-way switch

   o. _____ Push button

   p. _____ Thermostat

   q. _____ Electric door opener

   r. _____ Multioutlet assembly

19. The front edge of a box installed in a combustible wall must be _____ with the finished surface.

20. List the maximum number of No. 12 AWG conductors permitted in a

    a. 4" × 1½" octagon box. _____

    b. 4¹¹/₁₆" × 1½" square box. _____

    c. 3" × 2" × 3½" device box. _____

21. When a switch (device) box is nailed to a stud, and the nail runs through the box, the nail must not interfere with the wiring space. To accomplish this, keep the nail _____ (Select a, b, or c.)

    a. halfway between the front and rear of the box.

    b. a maximum of ¼ inch (6.35 mm) from the front edge of the box.

    c. a maximum of ¼ inch (6.35 mm) from the rear of the box.

22. Hanging a ceiling fixture directly from a plastic outlet box is permitted only if

    _____

    _____

    _____

    _____

23. It is necessary to count fixture wires when counting the permitted number of conductors in a box according to *Section 370-16*.

    (True)   (False)   Underline or circle the correct answer.

24. *Table 370-16(a)* allows a maximum of ten wires in a certain box. However, the box will have two cable clamps and one fixture stud in it. What is the maximum number of wires allowed in this box? _____

    _____

25. When laying out a job, the electrician will usually make a layout of the circuit, taking into consideration the best way to run the cables and/or conduits and how to make up the electrical connections. Doing this ahead of time, the electrician determines exactly how many conductors will be fed into each box. With experience, the electrician will probably select two or three sizes and types of boxes that will provide adequate space that will "meet Code." *Table 370-16(a)* of the Code shows the maximum number of conductors permitted in a given size box. In addition to counting the number of conductors that will be in the box, what is the additional volume that must be provided for the following items? Enter *single* or *double* volume allowance in the blank provided.

    a. one or more internal cable clamps: _____ volume allowance.

    b. for a fixture stud: _____ volume allowance.

    c. for one or more wiring devices on one yoke: _____ volume allowance.

    d. for one or more equipment grounding conductors: _____ volume allowance.

26. Is it permissible to install a receptacle outlet above an electric baseboard heater?

    _____

    _____

27. What is the maximum weight of a lighting fixture permitted to be hung directly to an outlet box in a ceiling? _____ pounds.

28. Two 12/2 and two 14/2 cables enter a box. Each cable has an equipment grounding conductor. The box will have one duplex receptacle and one toggle switch, each on a separate yoke. Calculate the minimum cubic inch area required for the box. The box contains two cable clamps.

29. Using the same number and size of conductors as in question 28, but using electrical metallic tubing, calculate the minimum cubic inch area required for the box. There will be no separate equipment grounding conductors, nor will there be any clamps in the box.

30. To allow for adequate conductor length to make up connections, *Section 300-14* requires that at least (4"), (6"), (8"), (10") of free conductor be provided. This distance is to be measured from the (raceway entry) (edge of box). Circle the correct answers.

31. When wiring a residence, what must be considered when installing wall boxes on both sides of a common partition that separates the garage and a habitable room? _____

_____

_____

32. Does the *NEC*® allow metal raceways to be used with nonmetallic boxes?

Yes _____ No _____ , Section _____

PART 2—STRUCTURAL FEATURES

1. To what scale is the basement plan drawn? _____

2. What is the size of the footing for the steel Lally columns in the basement? _____

_____

3. To what kind of material will the front porch lighting bracket fixture be attached?

_____

4. Give the size, spacing, and direction of the ceiling joists in the workshop. _____

_____

5. What is the size of the lot on which this residence is located? _____

_____

6. The front of the house is facing in what compass direction? _____

7. How far is the front garage wall from the curb? _____

8. How far is the side garage wall from the property lot line? _____

9. How many steel Lally columns are in the basement, and what size are they? _____

_____

10. What is the purpose of the I-beams that rest on top of the steel Lally columns? _____

_____

11. The garage walls are finished with _____.

12. Where is access to the attic provided? _____

13. Give the thickness of the outer basement walls. _____

14. What material is indicated for the foundation walls? _____

15. Where are the smoke detectors located in the basement? _____

_____

16. What is the ceiling height in the basement workshop from bottom of joists to floor?

_____

17. Give the size and type of the front door. _____

_____

18. What is the stud size for the partitions between the bathrooms in the bedroom area where substantial plumbing is to be installed? _____

19. Who is to furnish the range hood? _____

_____

20. Who is to install the range hood? _____

UNIT 3

Determining the Required Number and Location of Lighting and Small Appliance Circuits

OBJECTIVES

After studying this unit, the student will be able to

- understand the fundamental Code requirements for calculating branch-circuit sizing and loading.
- understand "volt-amperes per square foot" for arriving at total calculated general lighting loading.
- calculate the occupied floor area of a residence.
- determine the minimum number of lighting branch-circuits required.
- determine the minimum number of small appliance branch-circuits required.
- know where switched and nonswitched lighting outlets must be installed in homes.
- know where receptacle outlets must be installed in homes.
- know where GFCI receptacles must be installed in homes.

It is standard practice in the design and planning of dwelling units to permit the electrician to plan the circuits. Thus, the residence plans do not include layouts for the various branch-circuits. The electrician may follow the general guidelines established by the architect. However, any wiring systems designed and installed by the electrician must conform to the standards of the *National Electrical Code,*® as well as local and state code requirements.

This unit focuses on lighting branch-circuits and small appliance branch-circuits. The circuits supplying the electric range, oven, clothes dryer, and other specific circuitry not considered to be a lighting branch-circuit or a small appliance branch-circuit are covered later in this text. Refer to the index for the specific circuit being examined.

BASICS OF WIRE SIZING AND LOADING

The *National Electrical Code*® establishes some very important fundamentals that weave their way through the decision-making process for an electrical installation. They are presented here in brief form, and are covered in detail as required throughout this text.

The *NEC*® defines a *branch-circuit* as "the circuit conductors between the final overcurrent device protecting the circuit and the outlet(s)." See figure 3-1. In the residence discussed in this text, the wiring to wall outlets, the dryer, the range, etc., are all examples of a branch-circuit.

The *NEC*® defines a *feeder* as "All circuit conductors between the service equipment or the source of a separately derived system and the final branch-

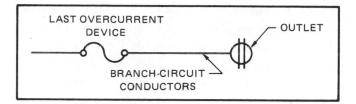

Fig. 3-1 The branch-circuit is that part of the wiring that runs from the final overcurrent device to the outlet. The rating of the overcurrent device, not the conductor size, determines the rating of the branch-circuit.

circuit overcurrent device." In the residence discussed in this text, the wiring between Main Panel A and Sub-Panel B is a feeder.

The ampacity (current-carrying capacity) of a conductor must not be less than the rating of the overcurrent device protecting that conductor, *Section 210-19*. A common exception to this is a motor branch-circuit, where it is quite common to have overcurrent devices (fuses or breakers) sized larger than the ampacity of the conductor. Motors and motor circuits are covered specifically in *Article 430* of the Code. The ampere rating of the branch-circuit overcurrent protective device (fuse or circuit breaker) determines the rating of the branch-circuit. For example, if a 20-ampere conductor is protected by a 15-ampere fuse, the circuit is considered to be a 15-ampere branch-circuit, *Sections 210-3, 210-19*.

Standard branch-circuits that serve more than one receptacle outlet or more than one lighting outlet are rated 15, 20, 30, 40, and 50 amperes. A branch-circuit that supplies an individual load can be of any ampere rating, *Section 210-3*.

The ampacity of branch-circuit conductors must not be less than the maximum load to be served. Where the circuit supplies receptacle outlets, as is typical in homes, the conductor's ampacity must not be less than the "rating" of the branch-circuit overcurrent device, *Sections 210-19 and 210-20*.

Where the ampacity of the conductor does not match up with a standard rating of a fuse or breaker, the next higher standard size overcurrent device may be used, provided the overcurrent device does not exceed 800 amperes, *Section 240-3*. This exception is not permitted, however, when the circuit supplies receptacles where "plug-connected" appliances, etc., could be used, because too many "plug-in" loads could result in an overload condition, *Section*

210-19(a). You may go to the next standard size overcurrent device only when the circuit supplies fixed lighting outlets.

For instance, when a conductor having an allowable ampacity of 25 amperes *(see Table 310-16)* No. 12 Type TW is derated to 70%:

$$25 \times 0.70 = 17.5 \text{ amperes}$$

It is permitted to use a 20-ampere overcurrent device if the circuit supplies *only* fixed lighting outlets or other fixed loads.

If the previous example were to supply receptacle outlets, then the rating of the overcurrent device would have to be dropped to 15 amperes; otherwise it is possible to overload the conductors by "plugging-in" more load than the conductors can safely carry.

The allowable ampacity of conductors commonly used in residential occupancies is found in *Table 310-16* of the Code. The ampacities in *Table 310-16* are subject to *correction factors* that must be applied where high ambient temperatures are encountered—for example, in attics. See *footnote to Table 310-16*.

Conductor ampacities are also *derated* when more than three current-carrying conductors are installed in a single raceway or cable. See *Note 8* to *Table 310-16*.

For loads that might be termed *continuous*, the load should not exceed 80% of the rating of the branch-circuit. The Code defines continuous as "a load where the maximum current is expected to continue for three hours or more." The lighting loads in residences are not considered to be "continuous" loads as would commercial lighting loads, such as store lighting. The Code permits 100% loading on an overcurrent device only if that overcurrent device is *listed* for 100% loading. At the time of this writing, Underwriters Laboratories has *no* listing of circuit breakers of the molded-case type used in residential installations that are suitable for 100% loading.

This is another judgment call on the part of the electrician and/or the electrical inspector. Will a branch-circuit supplying a heating cable in a driveway likely be on for 3 hours or more? The answer is "probably." Good design practice and experience tell us "never load a circuit conductor or overcurrent protective device to more than 80% of its rating."

The branch-circuit rating shall *not* be less than any noncontinuous load plus 125% of the continuous load. See *Sections 220-3(a), (b), (c),* and *(d)*.

The minimum lighting load per square foot of floor area for dwellings is 3 volt-amperes, *Section 220-3(b)* and *Table 220-3(b)*.

Most receptacle outlets in a residence are considered by the Code to be part of "general illumination," and additional load calculations are *not* necessary. See Footnote to *Table 220–3(b)*. Appliance receptacle outlets in the kitchen, dining room, laundry, and workshop are *not* to be considered part of general illumination. Additional loads are figured in for these receptacle outlets and are discussed in detail later in this text.

VOLTAGE

All calculations throughout this text use voltages of 120 volts and 240 volts. This is in conformance to *Section 220-2*, which states that "unless other voltages are specified, for purposes of computing branch-circuits and feeder loads, nominal system voltages of 120, 120/240, 208Y/120, 240, 480Y/277, 480, and 600 volts shall be used." This is repeated in the second paragraph of *Chapter 9, Part B*, which states that "for uniform application of Article 210, 215, and 220, a nominal voltage of 120, 120/240, 240, and 208Y/120 shall be used in computing the ampere load on the conductor."

The word *nominal* means **in name only, not a fact**. For example, in our calculations, we use 120 volts even though the actual voltage might be 110, 115, or 117 volts. This provides us with a uniformity that, without it, would lead to misleading and confusing results in our computations.

CALCULATING FLOOR AREA
First Floor Area

To estimate the total load for a dwelling, the occupied floor area of the dwelling must be calculated. Note in the residence plans that the first floor area has an irregular shape. In this case, the simplest method of calculating the occupied floor area is to determine the total floor area using the outside dimensions of the dwelling. Then, the areas of the

following spaces are subtracted from the total area: open porches, garages, or other unfinished or unused spaces if they are not adaptable for future use, *Section 220-3(b)*.

Many open porches, terraces, patios, and similar areas are commonly used as recreation and entertainment areas. Therefore, adequate lighting and receptacle outlets must be provided.

For practicality, the author has chosen to round up dimensions for the determining of total square footage and to round down dimensions for those areas (garage, porch, and portions of the inset at the front of the house) not to be included in the computation of the general lighting load. This results in a more "plus" answer as opposed to being on the conservative side.

Figure 3-2 shows the procedure for calculating the volt-amperes needed for the first floor of this residence.

Basement Area

The basement area of a home is not generally included in the calculating of general lighting loads because it could be classified as an "unfinished space not adaptable for future use," *Section 220-3(b)*. Yet in this residence, more than half of the total basement area is finished off as a recreation room, which certainly is considered living area. The workshop area also is intended to be used.

Therefore, the author again has taken the position that the basement in this residence will be fully utilized and has "invoked" *Section 90-4* of the Code, as do many electrical inspectors. In part, *Section 90-4* states that "the authority having jurisdiction of enforcement of the Code will have the responsibility for making interpretation of the rules."

By making this decision based upon *Section 90-4* of the Code, the basement square footage becomes the same as the area of the first floor, making it a simple addition problem to find total square footage of the residence.

The combined occupied area of the dwelling is found by adding the first floor and basement areas together:

First Floor	1616 ft^2
Basement	1616 ft^2
Total	3232 ft^2

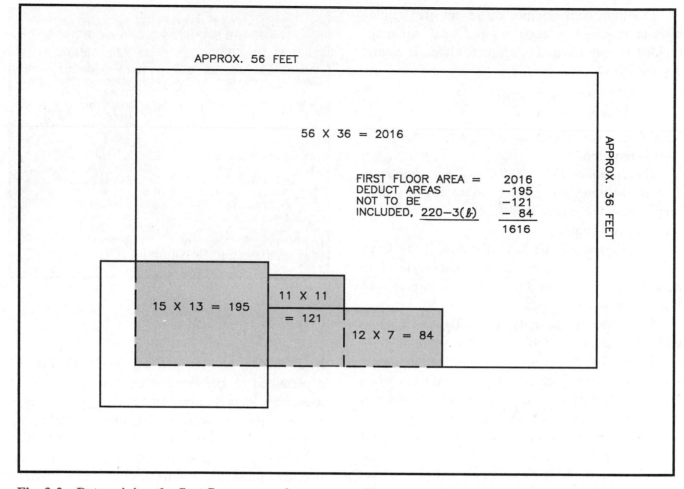

APPROX. 56 FEET

APPROX. 36 FEET

56 X 36 = 2016

FIRST FLOOR AREA =   2016
DEDUCT AREAS   −195
NOT TO BE   −121
INCLUDED, 220–3(*b*) − 84
                     1616

15 X 13 = 195

11 X 11
= 121

12 X 7 = 84

Fig. 3-2   Determining the first floor square footage area. The basement area is the same as the first floor. See text for explanation of how the dimensions of the residence were rounded off to make the calculations much simpler and much more practical for the electrician and electrical inspector to compute.

DETERMINING THE MINIMUM NUMBER OF LIGHTING BRANCH-CIRCUITS

To determine the minimum number of 15-ampere lighting branch-circuits required in a residence:

Step 1: $\dfrac{3 \text{ volt-amperes} \times \text{square feet}}{120 \text{ volts}} = \text{amperes}$

Step 2: $\dfrac{\text{amperes}}{15} = $ minimum number of 15-ampere circuits required

This equates to a *minimum* of:

- ONE 15-AMPERE BRANCH-CIRCUIT FOR EVERY 600 SQUARE FEET.

Although 20-ampere lighting branch-circuits are rarely used in residential installations, using the same procedure as indicated in Steps 1 and 2 would result in:

- ONE 20-AMPERE BRANCH-CIRCUIT FOR EVERY 800 SQUARE FEET.

Load calculations for electric water heaters, ranges, heat pumps, air-conditioning units, dryers, motors, electric heat, small appliance branch-circuits, and other specific loads are in addition to the general lighting load. These are discussed throughout this text as they appear.

Table 220-3(b) shows that the minimum load required for dwelling units is 3 volt-amperes (VA) per square foot (0.093 m^2) of occupied area. The calculated load for the total occupied area, 3232 ft^2, is:

$$1616 + 1616 = 3232 \text{ ft}^2$$

$$3232 \text{ (ft}^2) \times 3 \text{ (VA/ft}^2) = 9696 \text{ volt-amperes}$$

The total required amperage is determined as follows:

$$\text{Amperes} = \frac{\text{volt-amperes}}{\text{volts}}$$

$$= \frac{9696}{120} = 80.8 \text{ amperes}$$

The minimum number of 15-ampere lighting circuits required is equal to the total amperage divided by the maximum amperes of each circuit, *Section 220-4.*

$$\frac{80.8 \text{ amperes}}{15 \text{ amperes}} = 5.39 \text{ circuits}$$

Therefore, a minimum of six lighting branch-circuits is required.

The *National Electrical Code®* in *Chapter 9, Part B* illustrates a number of examples that show how to calculate circuit feeder, service-entrance, and motor circuit ampacities.

In *Chapter 9, Part B* of the *NEC®* the Code states that "except where the computations result in a major fraction of an ampere (0.5 or larger) such fractions may be dropped."

This residence actually has 13 lighting branch-circuits, as shown in Table 3-1.

Table 310-16 (footnote) shows that the maximum overcurrent protection is 15 amperes when No. 14 AWG copper wire is used, and 20 amperes when No. 12 AWG copper wire is used. It is significant to note that the lighting circuits are still rated as 15-ampere circuits even if No. 12 AWG copper wire is installed because the overcurrent protection for these lighting circuits is rated at 15 amperes, *Section 210-3.*

TRACK LIGHTING LOADS

Track lighting loads are considered to be part of the general lighting load for residences based upon the 3 volt-amperes per square foot as discussed previously. There is no need to add more wattage (volt-amperes) for track lighting. This is verified by the exception to *Section 410-102,* which exempts residential track lighting. ► However, for commercial installations the Code requires that every 2 feet (609.6 mm) of track lighting must be included in load calculations at 150 VA. ◄

DETERMINING THE NUMBER OF SMALL APPLIANCE BRANCH-CIRCUITS

Sections 220-16(a) and *(b)* state that an additional load of not less than 1500 watts shall be included for each two-wire small appliance circuit when calculating feeders and services.

SUMMARY OF 15 AMPERE LIGHTING AND 20 AMPERE SMALL APPLIANCE CIRCUITS			
15 AMPERE		**20 AMPERE**	
A15	Entry/Porch	A18	Workbench Receptacles
A17	Workshop Ltg.	A20	Workbench Receptacles on Window Wall
A19	Master Bedroom Ltg.	A22	Master Bath Receptacle
A21	Study/Bedroom Ltg.	A23	Hall Bath Receptacle
A14	Bathrooms, Hall Ltg.		
A16	Front Bedroom Ltg.		
B7	Kitchen/Nook Ltg.	B13	Kitchen
B9	Wet Bar Ltg. and Receptacles	B15	Kitchen Receptacles
		B16	Kitchen Receptacles
B11	Recreation Room Receptacles	B18	Washer Receptacle
		B20	Laundry Room Receptacles
B17	Living Room		
B10	Laundry, Rear Entry, Powder Room, Attic Ltg.	B21	Powder Room Receptacle
B12	Recreation Room Ltg.		
B14	Garage		

Table 3-1

► The receptacle for the refrigerator in the kitchen is permitted to be supplied by a separate 15- or 20-ampere branch-circuit rather than connecting it to one of the two required 20-ampere small appliance circuits. This permission is found in *Section 210-52(b)(1).* Because this tends to diversify the load, *Exception No. 1* to *Section 220-16(a)* permits this separate branch-circuit to be omitted from the calculations for services and feeders. Stated another way, an additional 1500 volt-amperes *does not* have to be included in the load calculations. However, there is nothing in the Code that prohibits adding another 1500 volt-amperes to the load calculations for this additional branch-circuit, as this would result in more capacity in the electrical system. ◄

Section 220-4(b) requires that two or more 20-ampere small appliance circuits must be provided for supplying receptacle outlets as specified in *Section 210-52.*

Section 210-52(b)(2) states that the 20-ampere appliance circuits mentioned in *Section 220-4(b)* shall feed only the receptacle outlets in the kitchen, pantry, dining room, breakfast room, or similar rooms of dwelling units. This section also requires that these 20-ampere small appliance circuits shall not supply other outlets.

► Exceptions in *Section 210-52(b)(2)* allow clock receptacles, and receptacles that are installed to plug-in the cords of gas-fired ranges, ovens, or

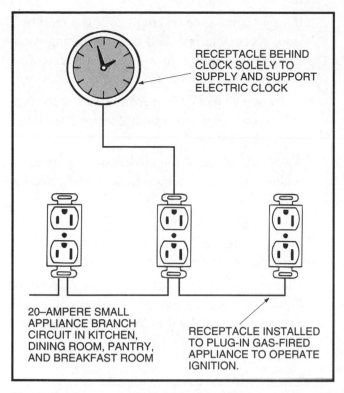

RECEPTACLE BEHIND CLOCK SOLELY TO SUPPLY AND SUPPORT ELECTRIC CLOCK

20–AMPERE SMALL APPLIANCE BRANCH CIRCUIT IN KITCHEN, DINING ROOM, PANTRY, AND BREAKFAST ROOM

RECEPTACLE INSTALLED TO PLUG-IN GAS-FIRED APPLIANCE TO OPERATE IGNITION.

Fig. 3-3   A receptacle installed solely to supply and support an electrical clock, or a receptacle that supplies power for clocks, clock timers, and electric ignition for gas ranges, ovens, and cook-tops may be connected to a 20-ampere small appliance circuit or to a general-purpose lighting circuit, *Section 210-52(b)(1) and (2)*.

counter-mounted cooking units in order to power up their timers and ignition systems, figure 3-3. ◀

Section 210-52(b)(3) requires that not less than two 20-ampere small appliance circuits must supply the receptacle outlets serving the countertop areas in kitchens.

Food waste disposers and/or dishwashers shall not be connected to these required 20-ampere small appliance circuits. They must be supplied by other circuits, as is discussed later. No. 12 AWG wire is used in small appliance circuits to minimize the voltage drop in the circuit. The use of this larger wire, rather than No. 14 AWG wire, helps improve appliance performance and lessens the danger of overloading the circuits.

Automatic washers draw a large amount of current during certain portions of their operating cycles. Thus, *Section 220-4(c)* requires that at least one additional 20-ampere branch-circuit must be provided to

supply the laundry receptacle outlet(s) as required by *Section 210-52(f)*.

A complete list of the branch-circuits for this residence is found in unit 28. The circuit directories for Main Panel A and Subpanel B show clearly the number of lighting, appliance, and special-purpose circuits provided.

It is recommended that the student now study the requirements of the following sections of the Code.

- *Section 210-7*—methods of connecting grounding-type receptacles.
- *Section 210-8*—ground-fault circuit interrupter (GFCI) protection for receptacle outlets.
- *Sections 210-50* and *210-52*—number of and spacing requirements for receptacle outlets in specific rooms and hallways.
- *Section 210-70*—switched lighting outlets.
- *Section 250-57*—methods of grounding fixed equipment.
- *Section 250-74*—connecting the receptacle grounding terminal to the box.
- *Section 250-114*—continuity and attachment of grounding conductors to boxes.
- *Section 305-4*—branch-circuits and receptacle requirements for temporary wiring, such as during construction.

RECEPTACLE OUTLET BRANCH-CIRCUIT RATINGS

▶ The receptacles in the kitchen, breakfast nook, workshop, powder room, bathrooms, and laundry room are connected to 20-ampere branch-circuits. All other receptacles are connected to 15-ampere branch-circuits. ◀

SUMMARY OF WHERE RECEPTACLE AND LIGHTING OUTLETS MUST BE INSTALLED IN RESIDENCES

The following is a brief explanation of where receptacle outlets and lighting outlets must be installed in residential occupancies. Throughout this text, each room is discussed in detail regarding outlet location. The following recaps all of the *National*

Electrical Code® rules that relate to receptacle and lighting outlet requirements. In the following paragraphs, the abbreviation GFCI stands for Ground Fault Circuit Interrupter.

Bear in mind that the Code provides *minimum* requirements for the placement of receptacles. A duplex receptacle located where there will be a concentration of such things as computers, printers, stereos, VCRs, CD players, television sets, radios, fax machines. answering machines, typewriters, adding machines, or lamps will probably not be adequate to serve these appliances without the use of a cord set having multiple receptacles. An electrician will do well to quiz the homeowner where this concentration might occur, then offer to install four receptacles at these locations as illustrated in figure 3-5(D). As an alternative, at least one manufacturer of wiring devices offers a quadplex receptacle (four receptacles) that can replace a standard duplex receptacle in a single-gang device box. This same quadplex receptacle can be mounted on a 4-inch square or octagon outlet box, on a 1- or 2-gang device box, or on a 4¹¹/₁₆-inch square box using a special adapter plate. See unit 6 for discussion of surge protection.

Receptacle Outlets
(125-Volt, Single-Phase, 15- or 20-Ampere)

General Lighting

- Wall receptacle outlets must be placed so that no point along the floor line is more than 6 feet (1.83 m) measured horizontally from an outlet. Fixed room dividers and railings are considered to be walls for the purpose of this requirement, *Section 210-52(a)*.

- Nonsliding fixed glass panels on exterior walls are considered to be wall space and are to be figured in when applying the rule that no point along the floor line is more than 6 feet (1.83 m) from a receptacle. Sliding panels on exterior walls are not considered to be a wall, *Section 210-52(a)*.

▶• Receptacle outlets located in the floor more than 18 inches (457 mm) from the wall are not to be counted as meeting the required number of wall receptacle outlets, *Section 210-52(a)*

third paragraph. An example of this would be a floor receptacle installed under a dining room table for plugging in warming trays, coffee pots, or similar appliances. ◀

- Hallways 10 feet or longer in homes must have at least one receptacle outlet, *Section 210-52(h)*. See figure 3-4.

- Any wall space 2 feet (610 mm) or more in width must have a receptacle outlet, *Section 210-52(a)*.

- Unfinished basement areas not generally "lived in," such as work areas and storage areas, are not included in the above 6-foot rule.

- Circuits supplying wall receptacle outlets for lighting loads are generally 15-ampere circuits, but are permitted by the Code to be 20 amperes.

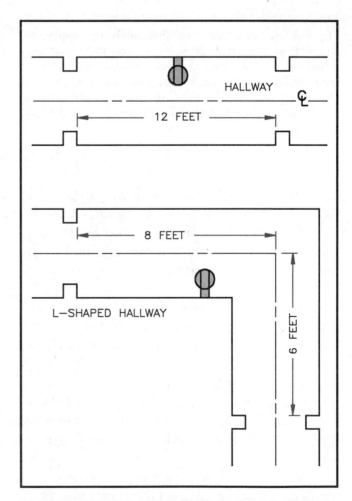

Fig. 3-4 Determination of hallway length. Hallways 10 feet or longer must have at least one receptacle outlet, *Section 210-52(h)* of the *NEC*.®

- In crawl spaces, and in attics, a receptacle outlet must be installed in an accessible location within 25 feet of heating, air-conditioning, or refrigeration equipment. Do *not* connect this receptacle outlet to the load side of the equipment disconnecting means, *Section 210-63*.

- For buildings containing more than two family dwellings (for example, a four-flat building), a GFCI protected receptacle outlet per *Section 210-8(b)(2)* must be installed on the roof if there is heating, air-conditioning, or refrigeration equipment that requires servicing on the roof. This receptacle outlet must be located in an accessible location within 25 feet of the equipment on the same level as the equipment, *Section 210-63*.

- A receptacle is *not* required on the rooftop of one- and two-family dwellings, *Section 210-63, Exception*.

- Circuits supplying electrical outlets in kitchens, dining rooms, and pantries are required to be 20 amperes where appliances will be plugged-in. Refer to *Section 210-52(b)* and *220-4(b)* of the Code.

Basements

- At least one receptacle outlet must be installed in addition to the laundry outlet, *Section 210-52(g)*.

- Connect to 15- or 20-ampere circuit.

- ▶ In unfinished basement areas that are not intended to be "lived in" such as storage areas or workshop areas, *all* 125-volt, single-phase, 15- or 20-ampere receptacles must have GFCI protection, *Section 210-8(a)(5)*. ◀ This includes receptacles that are an integral part of porcelain or plastic pull-chain or keyless lampholders. See figure 18-2.

- ▶ GFCI protection is *not* required for receptacles that are not readily accessible, or for a single receptacle or duplex receptacle for two appliances that are not easily moved, are cord- and plug-connected, and are located in a dedicated space. ◀ Examples of this include the receptacle for laundry equipment, a sump pump, a water softener, a freezer or a refrigerator. See figure 3-5.

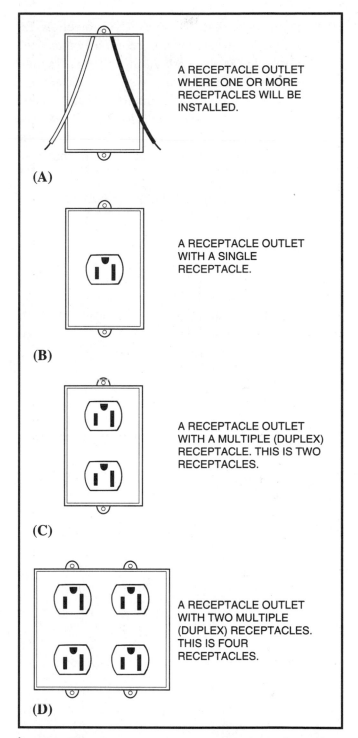

(A) A RECEPTACLE OUTLET WHERE ONE OR MORE RECEPTACLES WILL BE INSTALLED.

(B) A RECEPTACLE OUTLET WITH A SINGLE RECEPTACLE.

(C) A RECEPTACLE OUTLET WITH A MULTIPLE (DUPLEX) RECEPTACLE. THIS IS TWO RECEPTACLES.

(D) A RECEPTACLE OUTLET WITH TWO MULTIPLE (DUPLEX) RECEPTACLES. THIS IS FOUR RECEPTACLES.

▶ **Fig. 3-5   The branch-circuit wiring and the box where one or more receptacles will be installed is defined by the Code as a *receptacle outlet*. The receptacle itself, whether a single or multiple device (one, two, or three receptacles) on one strap (yoke) is defined as a *receptacle*. ◀ For house wiring, receptacle loads are considered to be part of the general lighting load calculations according to the *footnote* to *Table 220-3(b)*. For commercial installations, single or multiple receptacles (one, two, or three) on one strap (yoke) must be figured at 180 volt-amperes, *Section 220-3(c)*.**

▶• In finished basements, GFCI protection for receptacle outlets is *not* required unless the receptacles are located within 6 feet (1.83 mm) from the outer edge of a wet-bar sink and are intended to serve the countertop areas around the wet-bar sink, *Section 210-8(a)(7)*. ◀ The required number, spacing, and location of the receptacles would have to be in conformance to *Section 210-52(a)*.

• Refer to Sections *210-52(f)* and *210-52(g)* of the Code.

Appliance Receptacle Outlets

• For the receptacle outlets in the kitchen, dining areas, or pantry, at least two 20-ampere small appliance circuits must supply these receptacle outlets.

• See *Sections 220-4(b)*, *220-16*, and *210-52* of the Code.

Countertops *(See figure 3-6.)*

• In kitchens and dining rooms, receptacle outlets must be installed above all countertops 12 inches (305 mm) or wider.

• Receptacle outlets that serve countertops must be supplied by at least two 20-ampere small appliance circuits. These two 20-ampere circuits are permitted to supply other receptacle outlets in the kitchen, dining room, pantry, breakfast room, or similar location.

▶• Receptacle outlets that serve countertop surfaces in kitchens must be GFCI protected. This includes receptacles installed along the wall, and on peninsulas and freestanding islands. These receptacles *shall not* be installed more than 18 inches (458 mm) above the countertop. Where acceptable to the authority enforcing the Code, these receptacles are permitted to be mounted not more than 12 inches (305 mm)

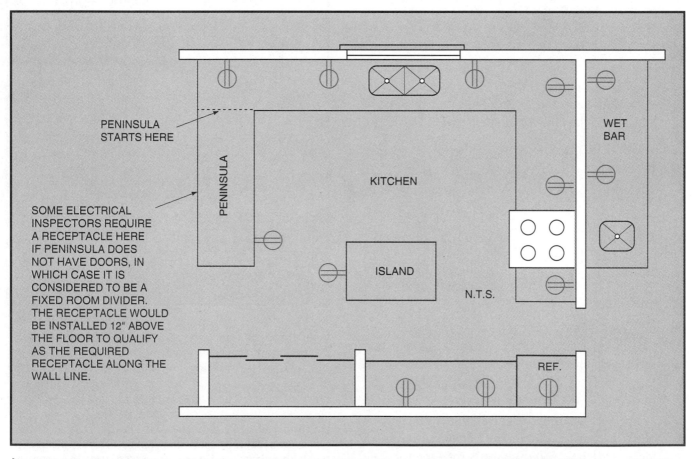

▶ Fig. 3-6   Floor plan showing placement of receptacles that serve countertop surfaces in kitchens and wet-bar areas in conformance to *Sections 210-8(a)* and *210-52(c)* of the Code. Drawing is not to scale. See text for detailed explanation. ◀

below the countertop. Receptacles mounted below the countertop must not be installed where the countertop extends more than 6 inches (153 mm) beyond the base cabinet. ◄ A receptacle located behind a refrigerator, or under the sink serving a food waste disposer and/or a dishwasher does not have to be GFCI protected because these receptacles do not serve countertop surfaces. See unit 6 for detailed coverage of GFCI protection.

►• Peninsulas and freestanding islands that have a long dimension of 24 inches (619 mm) or greater and a short dimension of 12 inches (305 mm) or greater must have at least one receptacle outlet. ◄ Measure the peninsula length from the edge where the peninsula connects to the wall base cabinet unit. The outlet(s) must be GFCI protected. See unit 6 for detailed coverage of GFCI protection.

►• Receptacle outlets installed within 6 feet (1.83 m) from the outside edge of a wet-bar sink must be GFCI protected. ◄

• Receptacle outlets in kitchen and dining areas located above countertops shall be positioned so that no point along the wall line is more than 24 inches (610 mm), measured horizontally, from a receptacle outlet in that space. See unit 6 for detailed coverage of GFCI protection. Example: A section of countertop measures 5'3" along the wall line. The minimum number of receptacles is: 5'3" ÷ 4 = 1 + (install two receptacles). Example: The countertop length from the edge of a sink to a range space measures 11'2" measured along the wall line. The minimum number of receptacles for this space is: 11'2" ÷ 4 = 2 + (install three receptacles).

Note: Dividing by 4 assures that there will be a receptacle outlet for every 4 linear feet or fraction thereof of counter length.

• Receptacles are not permitted to be installed "face up" on countertops.

• Receptacle outlets behind refrigerators or appliances fastened in place must not be thought of as meeting the previous requirements for receptacle outlets above countertops, because these outlets would be impossible to use once the refrigerator or appliance was fastened in place.

• See *Sections 210-8(a)* and *210-52* of the Code.

Laundry

• At least one receptacle outlet is required within 6 feet (1.83 m) of the intended location of the washing machine.

• The circuit must be a 20-ampere circuit.

• This receptacle *need not* be GFCI protected because acceptable levels of leakage current of some clothes washing machines, as determined by Underwriters Laboratories, can cause the GFCI to trip.

• This outlet and circuit are in addition to the basement receptacle outlet requirements already discussed, and must be in addition to the required two small appliance circuits that serve the receptacle outlets in the kitchen and dining areas. This circuit must *not* serve other outlets.

• See *Sections 210-8(a), 210-50(c), 210-52(c), 220-4(c)* and *220-16(b)* of the Code.

Bathrooms

Note: A *bathroom* is an area with a basin and one or more of the following: a toilet, a tub, or a shower.

• At least one receptacle outlet must be installed adjacent to each basin.

►• Connect the receptacle(s) to at least one separate 20-ampere branch-circuit. This separate circuit is permitted to supply other receptacles in the bathroom, and receptacles in other bathrooms. Other than the bathroom receptacles, no other outlets are permitted to be connected to this circuit. Although a minimum of one separate 20-ampere branch-circuit is required by the Code, more than one branch-circuit may be provided. For service and feeder load calculations, no additional load needs to be included for this separate circuit(s). ◄

• A receptacle that is an integral part of a fixture or medicine cabinet does *not* qualify for this required outlet.

• Must be GFCI protected.

• See *Sections 210-8(a)(1)* and *210-52(d)*.

Outdoors

▶• At least one receptacle outlet must be installed at grade level in the front and another in the back of all one-family dwellings and each unit of a two-family dwelling. These receptacles must be accessible and not more than 6½ feet (1.98 m) above grade level. See figure 3-7. ◀

• Connect to 15- or 20-ampere circuit.

• *All* outdoor receptacles must be GFCI protected. ▶An exception is that personnel GFCI protection is *not* required for an outdoor receptacle(s) installed for snow and ice-melting equipment (i.e., heating cables) that is not readily accessible, and is supplied by a separate branch-circuit. GFCI devices for *personnel protection* trip in the range of 4 to 6 milliamperes. ◀

▶• See *Section 426-28* for ground fault protection of *equipment*. GFPE protection must be provided for branch-circuits that supply "fixed" outdoor de-icing and snow melting equipment. Ground fault protection devices for equipment (GFPE) trip at approximately 30 milliamperes. GFPE is provided by the manufacturer of the equipment. An example of "fixed" outdoor de-icing and snow melting equipment is Type MI cable that is buried in the concrete of a driveway or sidewalk. ◀

• See *Sections 210-8(a)(3)* and *210-52(e)*.

Garages

• At least one receptacle outlet must be provided.

• Connect to 15- or 20-ampere circuit.

• Must be GFCI protected. An exception to this is for receptacles that are not readily accessible, such as a receptacle mounted on the ceiling for plugging in a garage door opener. Also exempt from the GFCI requirement are single receptacles or duplex receptacles intended for cord- and plug-connected appliances that are not easily moved, and are located in dedicated spaces. Examples are a freezer, refrigerator, or central vacuum equipment.

• If the garage is detached, the Code does not require electric power, but if electric power is

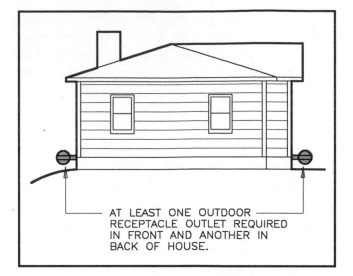

AT LEAST ONE OUTDOOR RECEPTACLE OUTLET REQUIRED IN FRONT AND ANOTHER IN BACK OF HOUSE.

▶ **Fig. 3-7 Outdoor receptacle requirements for residences, *Section 210-52(e)*, require that for one-family dwellings, and each unit of a two-family dwelling that is at grade level, at least one receptacle outlet must be installed in the front and another in the back. These receptacles must not be mounted more than 6 feet, 6 inches above grade. *Section 210-8(a)(3)* requires that *all* receptacle outlets installed outdoors must be GFCI protected, except those receptacles for snow and ice melting equipment that are not readily accessible, and are supplied from a dedicated branch-circuit. Additional receptacles may be installed higher, such as receptacles used for plugging in ice and snow melting heating cables and/or decorative outdoor lighting. ◀**

provided, then at least one GFCI receptacle outlet must be installed.

• See *Sections 210-8(a)(2)* and *210-52(g)*.

GFCI Exemption Recap

▶• Under certain conditions, GFCI protection is *not* required for receptacle outlets in garages, in crawl spaces at or below ground level, or in unfinished basements if the receptacles are:
  – not readily accessible.
  – single type receptacles or duplex type receptacles located in dedicated spaces to be used for cord- and plug-connected appliances that are not easily moved.

Examples of these would be such things as overhead door openers, refrigerators, freezers, water softeners, central vacuum equipment or sump pumps. These would be the receptacles that are dedicated to specific plug-in appliances. The homeowner would normally not unplug the appliance to plug-in an extension cord. These receptacles are exempt from the GFCI requirement because they are not intended to be used for plugging in extension cords for portable electric tools or portable appliances. These GFCI exempt receptacles are *in addition* to the receptacles required by *Section 210-52(g)* for garages and basements. See *Section 210-8(a)*. ◄

Specific Loads

• Appliances such as electric ranges and ovens, fans, air conditioners, electric heat, heat pumps,

electric water heaters, garage door openers, and similar loads are purchased as desired by the homeowner, *but* when included in the design of a residence the circuitry and installation must conform to the Code. See index for location of coverage for specific appliance or load.

Lighting Outlets *(Section 210-70)*

The *National Electrical Code*® does contain certain minimum requirements for providing lighting for dwellings. Per *Section 210-70*, lighting outlets in dwellings must be installed as follows:

Switched Lighting Outlets (See figure 3-8.)

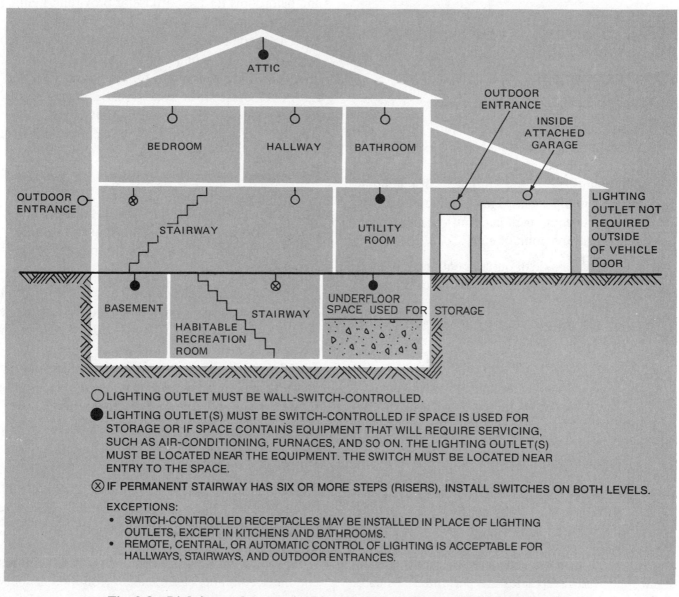

○ LIGHTING OUTLET MUST BE WALL-SWITCH-CONTROLLED.

● LIGHTING OUTLET(S) MUST BE SWITCH-CONTROLLED IF SPACE IS USED FOR STORAGE OR IF SPACE CONTAINS EQUIPMENT THAT WILL REQUIRE SERVICING, SUCH AS AIR-CONDITIONING, FURNACES, AND SO ON. THE LIGHTING OUTLET(S) MUST BE LOCATED NEAR THE EQUIPMENT. THE SWITCH MUST BE LOCATED NEAR ENTRY TO THE SPACE.

⊗ IF PERMANENT STAIRWAY HAS SIX OR MORE STEPS (RISERS), INSTALL SWITCHES ON BOTH LEVELS.

EXCEPTIONS:
• SWITCH-CONTROLLED RECEPTACLES MAY BE INSTALLED IN PLACE OF LIGHTING OUTLETS, EXCEPT IN KITCHENS AND BATHROOMS.
• REMOTE, CENTRAL, OR AUTOMATIC CONTROL OF LIGHTING IS ACCEPTABLE FOR HALLWAYS, STAIRWAYS, AND OUTDOOR ENTRANCES.

Fig. 3-8   Lighting outlets required in a typical dwelling unit, *Section 210-70*.

At least one switch-controlled lighting outlet shall be installed:

- in every habitable room. A habitable room is defined as "space in a structure for living, sleeping, eating, or cooking. Bathrooms, toilets, closets, halls, storage or utility space, and similar areas are not considered habitable space."

- in bathrooms.

- in hallways.

- in "interior" stairways. Where the permanent stairway has six or more steps (risers), switch control for the lighting outlet(s) that light up the stairway must be provided at each level. This includes permanent stairways to basements and attics, but does not include fold-down storable ladders commonly installed for access to residential attics.

- in attached garages.

- in detached garages if they have electric power.

- in attics, underfloor spaces, utility rooms, and basements where these areas are used for storage, or if these areas contain equipment that will require servicing (furnace, air-conditioning, heat pumps, etc.). This lighting outlet must be near the equipment and must be switch-controlled near the point of entry.

▶• at the exterior side of outdoor entrances or exits. A sliding glass door *is* considered to be an outdoor entrance. A vehicle door in an attached garage is *not* considered to be an outdoor entrance. ◀

- Clothes closets. The present edition of the *National Electrical Code*® does not require lighting outlet(s) in clothes closets. However, some Building Codes do require lighting outlet(s) in clothes closets. Be sure to check this out in your community.

Important! There are two exceptions to *Section 210-70*. The first exception recognizes the fact that in most residences, lighting is accomplished by plugging table and floor lamps into wall receptacle outlets. Other than kitchens and bathrooms, switch-controlled wall receptacles are permitted in lieu of ceiling and/or wall lighting outlets.

The Code requires that the receptacle outlets in kitchens, pantries, breakfast rooms (dinettes), and dining rooms be supplied by 20-ampere small appliance circuits, *Sections 210-52(b)* and *220-4(b)*.

Section 210-70 lists the rooms that require switched lighting outlets. If the homeowner wishes to install swag lighting, figure 3-9, in the dining room instead of a regular ceiling fixture, a switched receptacle connected to a general lighting circuit would have to be installed. The switched receptacle must be in addition to the 20-ampere small appliance receptacle outlets, *Section 210-52(b)(1)*, *Exception 1*, figure 3-10.

The Code, in *Section 210-70(a)*, *Exception 1*, prohibits the switching of receptacles in kitchens and bathrooms for the purpose of providing lighting rather than installing actual lighting fixtures.

The second exception is that a separate switch is not required for hallways, stairways, or outdoor entrances when automatic timers, photocells, central or remote control systems have been installed for the control of these lighting outlets. For example, a post light controlled by a photocell does not require a separate switch. See *Section 210-70(a)*, *Exception 2*.

How Many Branch-Circuits Are Required, and How Do You Determine the Number of Lighting and Receptacle Outlets to Be Connected to One Branch-Circuit?

In this unit, the *National Electrical Code*® rules on where receptacle and lighting outlets are to be installed was discussed. The question of *how many* receptacle outlets and lighting outlets should be connected to each branch-circuit is covered in detail in unit 8, where we begin the actual layout of the wiring for the residence discussed throughout this text.

Common Areas in Two-Family or Multifamily Dwellings

▶ *Section 210-25* of the *National Electrical Code*® contains specific requirements for circuits that serve common areas of two-family or multifamily dwellings. Be careful when wiring a two-family or multifamily dwelling. For areas such as a common entrance, common basement, or common laundry area, the branch-circuits for lighting, central fire/

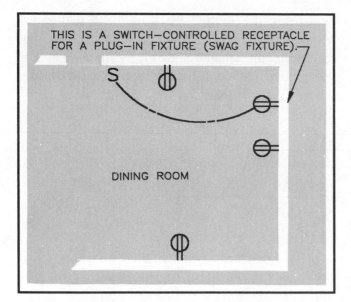

Fig. 3-9   When a switch-controlled receptacle outlet is installed in rooms where the receptacle outlets normally would be 20-ampere small appliance circuits (breakfast nook, dining room, etc.), it must be in addition to the required small appliance receptacle outlets, *Section 210-52(b)(1), Exception No. 1.*

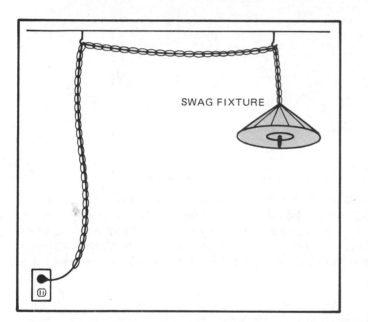

Fig. 3-10   A switch-controlled receptacle for a plug-in fixture (swag fixture) is supplied by a general lighting circuit and is *not* to be connected to the 20-ampere small appliance circuit that supplies the other three receptacles. The switch-controlled receptacle is in addition to the required small appliance outlets, *Section 210-52(b)(1), Exception 1.*

smoke/security alarms, communication systems, or any other electrical requirements that serve the common area *shall not* be served by circuits from any individual dwelling unit's power service. Common areas will require separate circuits, associated panels, and metering equipment. The reasoning for this requirement is that power could be lost in one occupant's electrical system, resulting in no lights or alarms for the other tenant(s) using the common shared area.

For example, a two-family dwelling would probably require two separate stairways to the basement.

If only one stairway is provided to serve the basement of a two-family dwelling, then lighting would have to be provided and served from a branch-circuit originating from each tenant's electrical system, so that each tenant would have control of the lighting. For a common entrance, two lighting fixtures would be installed and connected to each tenant's electrical system.

Likewise, fire/smoke/security systems in a two-family dwelling should be installed and connected to branch-circuits originating from each tenant's electrical panel. This becomes an architectural design issue. This is not generally a problem in two-family dwellings, but can become a problem in dwellings having more than two tenants. Rather than having to install a third watthour meter and associated electrical equipment on a two-family dwelling, the design of the structure can be such that there are no com-

mon areas. See figure 3-11. A four-plex will probably require a fifth watthour meter, four watthour meters for the tenants and one watthour meter to serve common area lighting and the fire/smoke/security systems. ◄

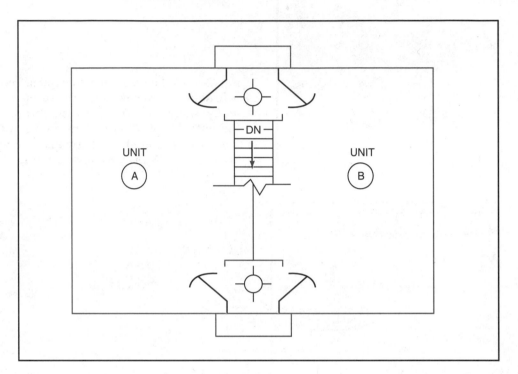

Fig. 3-11   This outline of a two-family dwelling shows a common rear entry and common stairway to the basement. It also shows a common front entry. Note that each entry has one lighting fixture. These common area lighting fixtures are not permitted to be connected to either of the tenant's electrical systems in the individual dwelling units, *Section 210-25*. The architect/designer must redo the layout of the building to eliminate the common entrances. Another possibility would be to provide two lighting fixtures in each entry, each lighting fixture to be connected to the respective individual tenant's electrical circuit.

REVIEW

Note: In the following questions the student may refer to the *National Electrical Code®* to obtain the answers. To help the student find these answers, references to Code sections are included after the questions. These references are not given in later reviews, where it is assumed that the student is familiar with the Code.

1. What is the meaning of computed load? _____

_____

2. How are branch-circuits rated? *Section 210-3* _____

_____

3. a. How is the rating of the branch-circuit protective device affected when the conductors used are of a larger size than called for by the Code? _____

_____

   b. How is the circuit classification affected? *Section 210-3* _____

4. What dimensions are used when measuring the area of a building? *Section 220-3(b)*

_____

5. What spaces are not included in the floor area when computing the load in volt-amperes per square foot? *Section 220-3(b)* _____

_____

6. What is the unit load per square foot for dwelling units? *Table 220-3(b)* _____

_____

7. According to *Section 210-50(c)*, a laundry equipment outlet must be placed within _____ feet (_____ meters) of the intended location of the laundry equipment.

8. How is the total load in volt-amperes for lighting purposes determined? *Section 220-3(b)* _____

_____

_____

_____

9. How is the total lighting load in amperes determined? _____

_____

10. How is the required number of branch-circuits determined? _____

_____

11. What is the minimum number of 15-ampere lighting circuits required if the dwelling has an occupied area of 4000 square feet? Show all calculations.

12. How many branch lighting circuits are provided in this dwelling? _____

13. a. What is the minimum load allowance for small appliance circuits for dwellings? *Section 220-16(a)* _____

▶b. A separate 15-ampere branch-circuit is run to the receptacle outlet behind the refrigerator instead of connecting it to one of the two 20-ampere small appliance branch-circuits that are required in kitchens. For this separate circuit, an additional 1500 volt-ampere (must) (does not have to) be added to the load calculations for dwellings. Circle the correct answer. ◀

14. What is the smallest size wire that can be used in a branch-circuit rated at 20 amperes?

_____

15. How is the load determined for outlets supplying the specific appliances? *Section 220-3(c)* _____

_____

16. What type of circuits must be provided for receptacle outlets in the kitchen, pantry, dining room, and breakfast room? *Section 220-4(b)* _____

_____

17. How is the minimum number of receptacle outlets determined for most occupied rooms?

_____

_____

_____

_____

_____

_____

_____

18. In a single-family dwelling, what types of overload protection for circuits are used?

_____

_____

19. Is a grounded circuit conductor included in this dwelling for

a. a 120-volt, 2-wire branch-circuit? _____

b. a 240-volt, 2-wire branch-circuit? _____

c. a 240-volt, 3-wire branch-circuit? _____

20. The minimum number of outdoor receptacles for a residence is _____,

Section _____ State the location. _____

21. The Code indicates the rooms in a dwelling that are required to have switched lighting outlets or switched receptacles. Write yes (switch required) or no (switch not required) for the following areas:

a. attic   _____

b. stairway   _____

c. crawl space (where
   used for storage)   _____

d. hallway   _____

e. bathroom   _____

22. Is a receptacle required in a bedroom on a 3-foot (914-mm) wall space behind the door? The door is normally left open._____

23. The Code requires that at least one 20-ampere circuit feeding a receptacle outlet must be provided for the laundry. May this circuit supply other outlets?_____

24. In a basement, at least one receptacle outlet must be installed in addition to the receptacle outlet installed for the laundry equipment. This additional receptacle outlet and any other receptacle outlets in unfinished areas must be _____ protected.

25. The Code in *Section 210-8(a)(2)* requires that ALL receptacles installed in garages must have GFCI protection, but there are certain exceptions. What are these exceptions?

_____

_____

_____

_____

_____

26. Define a branch-circuit. _____

_____

27. Although the Code contains many exceptions to the basic overcurrent protection requirements for conductors, in general, the rating of the branch-circuit overcurrent device must (not be less than) (not be more than) the ampacity of a conductor. Circle the correct answer.

28. The rating of a branch-circuit is based upon
    a. the rating of the overcurrent device
    b. the length of the circuit
    c. the branch-circuit wire size

29. a. A 25-ampere branch-circuit conductor is derated to 70%. It is important to provide proper overcurrent protection for these conductors. The derated conductor ampacity is _____

    _____

    b. If the connected load is a "fixed" nonmotor, noncontinuous load, the branch-circuit overcurrent device may be sized at (20) (25) amperes. Circle correct answer.

    c. If the above circuits supply receptacle outlets, the branch-circuit overcurrent device must be sized at not over (15) (20) (25) amperes. Circle correct answer.

30. Small appliance receptacle outlets are (included) (not included) in the 3 volt-amperes per square foot calculations. Circle correct answer.

31. If a homeowner wishes to have a switched receptacle outlet for a swag lamp in a dining room, may this switched receptacle outlet be considered to be one of the receptacle outlets required for the 20-ampere small appliance circuits?

32. How many receptacle outlets are required on a 13-foot wall space between two doors? Refer to *Section 210-52*. Draw the outlets in on the diagram.

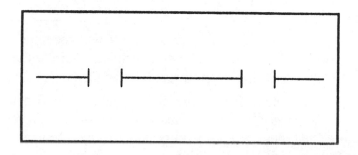

33. Is a receptacle required in a hallway in a home when the hallway is 8 feet long?

_____

34. Is a receptacle required in a hallway in a home when the hallway is 13 feet long?

_____

35. A sliding glass door is installed in a family room that leads to an outdoor deck. The sliding door has one 4-foot sliding section and one 4-foot permanently mounted glass section. Inside the recreation room, what does the Code say about wall receptacle outlets relative to the receptacle's position near the sliding glass door?

_____

_____

_____

_____

_____

_____

36. The Code permits certain other receptacles to be connected to a 20-ampere small appliance circuit. What are they? _____

_____

37. What is the minimum number of 20-ampere small appliance circuits required by the Code to supply the receptacle outlets that serve the countertop surfaces in the kitchen? In what section of the Code is this requirement found? _____

38. Does the Code allow a 120-volt receptacle located behind a gas range to be supplied by a small appliance circuit? (Yes) (No) Circle the correct answer. *Section* _____

39. When determining the location and number of receptacles required, fixed room dividers and railings (shall be) (need not be) considered to be wall space. Circle correct answer.

40. No point along the wall line shall be more than 6 feet from a receptacle outlet. This requirement (does apply) (does not apply) to unfinished residential basements. Circle correct answer.

41. Receptacle outlets that serve countertop surfaces in kitchens are required to be _____ protected, and shall be installed so that no point along the wall line above the base cabinets is more than (18 inches) (24 inches) (36 inches) from a receptacle. Circle the correct answer.

42. Does the Code permit an outdoor receptacle to be connected to one of the two required 20-ampere small appliance branch-circuits? (Yes) (No) Circle the correct answer. What *Section* supports your answer? *Section* _____

43. In kitchens, the Code requires that at least (one) (two) receptacle(s) be installed to serve the countertop surfaces on an island or peninsula. Circle the correct answer. In what *Section* of the Code is this requirement found? *Section* _____

44. In the past, it was common practice to connect the lighting and receptacle(s) in a bathroom to the same circuit. Because of overloads caused by high wattage grooming appliances, the Code now requires that these receptacle(s) be supplied by a separate 20-ampere branch-circuit. In what *Section* of the Code is this requirement found? *Section* _____

45. If a residence has two bathrooms, the Code states that:

    a. the receptacles in each bathroom must be connected to a separate 20-ampere branch-circuit. For two bathrooms, this would mean two circuits. (True) (False) Circle the correct answer.

    b. the receptacles in both bathrooms are permitted to be served by the same 20-ampere branch-circuit. (True) (False) Circle the correct answer.

46. In your own words, explain the GFCI exemptions for receptacle outlets installed in unfinished basements and garages. Give some examples.

    _____

    _____

    _____

    _____

47. *Section* _____ of the Code prohibits connecting lighting and/or fire/smoke/ security systems to individual tenant's electrical source where common (shared) areas are present, in which case the lighting provided for both tenants is connected on one tenant's electrical source.

48. When installing weatherproof outdoor receptacles high up under the eaves of a house intended to be used for plugging in de-icing and snow melting cable for the rain gutters, must these receptacles be GFCI protected for personnel protection? (Yes) (No) Circle the correct answer. *Section* _____

UNIT 4

Conductor Sizes and Types, Wiring Methods, Wire Connections, Voltage Drop, Neutral Sizing for Services

OBJECTIVES

After studying this unit, the student will be able to

- define the terms used to size and rate conductors.
- discuss the subject of aluminum conductors.
- describe the types of cables used in most dwelling unit installations.
- list the installation requirements for each type of cable.
- describe the uses and installation requirements for electrical conduit systems.
- describe the requirements for service grounding conductors.
- describe the use and installation requirements for flexible metal conduit.
- make voltage drop calculations.
- understand the fundamentals of the markings found on terminals of wiring devices and wire connectors.
- understand why the ampacity of high-temperature insulated conductors may not necessarily be suitable at that same rating on a particular terminal on a wiring device, breaker, or switch.
- understand why knob and tube wiring must not be buried in insulation.
- use a reduced neutral on residential services if certain criteria are met.

CONDUCTORS

Throughout this text, all references to conductors are for copper conductors, unless otherwise stated.

Wire Size

The copper wire used in electrical installations is graded for size according to the American Wire Gauge (AWG) Standard. The wire diameter in the AWG standard is expressed as a whole number. AWG sizes vary from fine, hairlike wire used in coils and small transformers to very large diameter wire required in industrial wiring to handle heavy loads.

The wire may be a single strand (solid conductor), or it may consist of many strands. Each strand of wire acts as a separate conducting unit. Conductors that are No. 8 AWG and larger generally are stranded. The wire size used for a circuit depends upon the maximum current to be carried. The *National Electrical Code®* states that the minimum conductor size permitted in house wiring is No. 14 AWG. Exceptions to this rule are covered in *Section 210-19(c)* and *Article 725* for the wires used in lighting fixtures, bell wiring, and remote-control low-energy circuits. See conductor application chart that follows.

CONDUCTOR APPLICATIONS CHART	
CONDUCTOR SIZE	**APPLICATIONS**
No. 16 and No. 18 AWG	cords, low-voltage control circuits, bell and chime wiring
No. 14 and No. 12 AWG	normal lighting circuits, and circuits supplying receptacle outlets
No. 10, No. 8, No. 6, and No. 4 AWG	clothes dryers, ovens, ranges, cooktops, water heaters, heat pumps, central air conditioners, furnaces, feeders to sub-panels
No. 3, No. 2, No. 1 (and larger) AWG	main service entrances, feeders to sub-panels

Ampacity

Ampacity means "the current in amperes a conductor can carry continuously under the condition of use without exceeding its temperature rating." This value depends upon the conductor's cross-sectional area, whether the conductor is copper or aluminum, and the type of insulation around the conductor. Ampacity values, also referred to as current-carrying capacity, are found in *Article 310*. The most commonly used conductor ampacities are found in *Table 310-16*.

The ampacity values in the tables are "derated" when more than three current-carrying conductors are installed in the same raceway or cable. The ampacity values are further "corrected" when the conductors are installed in high-temperature applications.

More on conductor ampacities, correction factors, and derating factors is discussed in unit 15.

Conductors must have an ampacity not less than the maximum load that they are supplying, figure 4-1. All conductors of a specific branch circuit must have an ampacity of the branch circuit's rating, figure 4-2. There are exceptions to this rule, such as taps for electric ranges, unit 20.

Conductor Sizing

The diameter of wire is usually given in a unit called a "mil." A *mil* is defined as one-thousandth of an inch (0.001 inch). Mils squared are known as "circular mils."

Table 8 in *Chapter 9* of the *NEC®* clearly shows that conductors are expressed in *AWG* numbers from No. 18 (1620 circular mils) through No. 4/0 (211,600 circular mils). These letters are the first letters of each word *American Wire Gauge*. The higher the AWG number, the smaller the wire.

Wire sizes larger than No. 4/0 are expressed in circular mils.

The letter *k* designates 1,000 in the metric system.

Large conductors, such as 500,000 circular mils, are generally expressed as 500 kcmils. The term kcmils means "thousand circular mils." This is much easier to express in both written and verbal terms.

Older texts used the term MCM, which also means "thousand circular mils." The first letter *M* refers to the Roman numeral that represents 1,000. Thus, 500 MCM means the same as 500

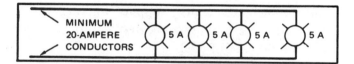

Fig. 4-1 Branch-circuit conductors shall have an ampacity not less than the maximum load to be served, *Section 210-19(a).* **See** *Section 210-22* **for maximum loads. See** *Section 210-23* **for permissible loads.**

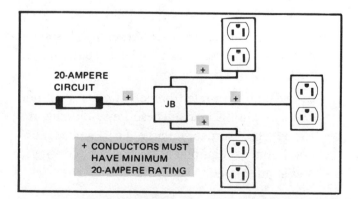

Fig. 4-2 All conductors in this circuit supplying readily accessible receptacles shall have an ampacity of not less than the rating of the branch-circuit. In this example, 20-ampere conductors must be used, *Section 210-19(a).*

kcmils. Roman numerals are no longer used in the electrical industry for expressing conductor sizes.

Circuit Rating

The rating of the overcurrent device (OCD) determines the rating of a circuit, figure 4-3.

PERMISSIBLE LOADS ON BRANCH-CIRCUITS *(SECTION 210-23)*

- the load shall not exceed the branch-circuit rating.

- the branch-circuit must be rated 15, 20, 30, 40, or 50 amperes when serving two or more outlets.

- an individual branch-circuit can supply any size load.

- 15 and 20 ampere branch-circuits:
  a. can supply lighting, other equipment, or both types of loads.
  b. cord- and plug-connected equipment shall not exceed 80% of the branch-circuit rating.
  c. equipment fastened in place shall not exceed 50% of the branch-circuit rating if the branch-circuit also supplies lighting, other cord- and plug-connected equipment, or both types of loads.

- the 20-ampere small appliance circuits in homes shall not supply other loads, *Section 220-4(b)*.

- 30 ampere branch-circuits can supply equipment such as dryers, cook-tops, etc. Cord- and plug-connected equipment shall not exceed 80% of the branch-circuit rating.

- 40 and 50 ampere branch-circuits can supply cooking equipment that is fastened in place, such as an electric range.

- over 50-ampere-rated branch-circuits shall not supply lighting loads. These large circuits are for electric furnaces, large heat pumps, air-conditioning equipment, large double-ovens, and similar large loads.

Aluminum Conductors

The conductivity of aluminum wire is not as great as that of copper wire for a given size. For example, checking *Table 310-16*, it is found that

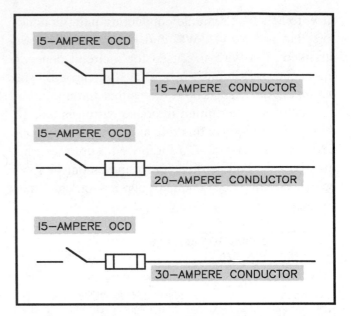

Fig. 4-3   All three of the above circuits are rated as 15-ampere circuits, even though larger conductors were used for some other reason, such as solving a voltage drop problem. The rating of the overcurrent device (OCD) determines the rating of a circuit, *Section 210-3*.

a No. 8 THHN copper wire has an allowable ampacity of 55 amperes, whereas a No. 8 THHN aluminum or copper-clad aluminum wire has an ampacity of only 45 amperes. These ampacities are further reduced if the conductors are installed in high-ambient-temperature locations. (See ampacity correction factors below the table.) Ampacities are also reduced when there are more than three current-carrying conductors in a raceway or cable. See *Note 8* to *Table 310-16*.

Another example is found in the *footnotes* to *Table 310-16* where the maximum overcurrent protection for No. 12 AWG copper wire is 20 amperes, but only 15 amperes for No. 12 AWG aluminum or copper-clad aluminum wire.

Resistance is an important consideration when installing aluminum conductors. An aluminum conductor has a higher resistance compared to a copper conductor for a given wire size, which therefore causes a greater voltage drop.

$$\text{Voltage Drop}(E_d) = \text{Amperes}(I) \times \text{Resistance}(R)$$

Common Connection Problems

Some common problems associated with aluminum conductors when not properly connected may be summarized as follows:

- a corrosive action is set up when dissimilar wires come in contact with one another when moisture is present.

- the surface of aluminum oxidizes as soon as it is exposed to air. If this oxidized surface is not broken through, a poor connection results. When installing aluminum conductors, particularly in large sizes, an inhibitor is brushed onto the aluminum conductor, then the conductor is scraped with a stiff brush where the connection is to be made. The process of scraping the conductor breaks through the oxidation, and the inhibitor keeps the air from coming into contact with the conductor. Thus, further oxidation is prevented. Aluminum connectors of the compression type usually have an inhibitor paste already factory installed inside of the connector.

- aluminum wire expands and contracts to a greater degree than does copper wire for an equal load. This factor is another possible cause of a poor connection. Crimp connectors for aluminum conductors are usually longer than those for comparable copper conductors, thus resulting in greater contact surface of the conductor in the connector.

Proper Installation Procedures

Proper, trouble-free connections for aluminum conductors require terminals, lugs, and/or connectors that are suitable for the type of conductor being installed.

Terminals on receptacles and switches must be suitable for the conductors being attached. Table 4-1 shows how the electrician can identify these terminals. Listed connectors provide proper connection when properly installed.

See also *Sections 110-14, 380-14(c)*, and *410-56(b)*.

Wire Connections

When splicing wires or connecting a wire to a switch, fixture, circuit breaker, panelboard, meter socket, or other electrical equipment, the wires may be twisted together; soldered; then taped. Usually, however, some type of wire connector is required.

Wire connectors are known in the trade by such names as *screw terminal, pressure terminal connec-*

tor, wire connector, Wing nut®, Wire-nut®, Scotchlok®, split-bolt connector, pressure cable connector, solderless lug, soldering lug, solder lug, and others. Solder-type lugs are not often used today. In fact, connections that depend upon solder are *not* permitted for connecting service-entrance conductors to service equipment. The labor costs and time spent make the use of solder-type connections prohibitive.

Solderless connectors, designed to establish connections by means of mechanical pressure, are quite common. Examples of some types of wire connectors, and their uses, are shown in figure 4-4.

As with the terminals on wiring devices (switches and receptacles), wire connectors must be marked *AL* when they are to be used with aluminum conductors. This marking is found on the connector itself, or appears on or in the shipping carton.

Connectors marked *AL/CU* are suitable for use with aluminum, copper, or copper-clad aluminum conductors. This marking is found on the connector itself, or it appears on or in the shipping carton.

Connectors not marked *AL* or *AL/CU* are for use with copper conductors only.

TYPE OF DEVICE	MARKING ON TERMINAL OR CONNECTOR	CONDUCTOR PERMITTED
15- or 20-ampere receptacles and switches	CO/ALR	aluminum, copper, copper-clad aluminum
15- and 20-ampere receptacles and switches	NONE	copper, copper-clad aluminum
30-ampere and greater receptacles and switches	AL/CU	aluminum, copper, copper-clad aluminum
30-ampere and greater receptacles and switches	NONE	copper only
Screwless pressure terminal connectors of the push-in type	NONE	copper or copper-clad aluminum
Wire connectors	AL	aluminum
Wire connectors	AL/CU	aluminum, copper, copper-clad aluminum
Wire connectors	CC	copper-clad aluminum only
Wire connectors	CC/CU	copper or copper-clad aluminum
Wire connectors	CU	copper only
Any of the above devices	COPPER OR CU ONLY	copper only

Table 4-1   **Terminal identification markings and the acceptable types of conductors permitted to be connected to a specific type of terminal.**

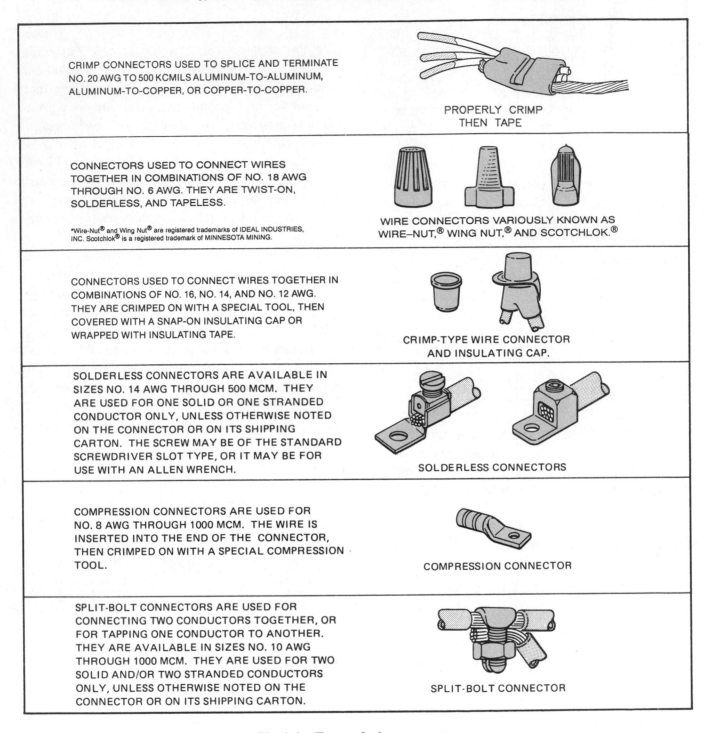

CRIMP CONNECTORS USED TO SPLICE AND TERMINATE NO. 20 AWG TO 500 KCMILS ALUMINUM-TO-ALUMINUM, ALUMINUM-TO-COPPER, OR COPPER-TO-COPPER.

PROPERLY CRIMP THEN TAPE

CONNECTORS USED TO CONNECT WIRES TOGETHER IN COMBINATIONS OF NO. 18 AWG THROUGH NO. 6 AWG. THEY ARE TWIST-ON, SOLDERLESS, AND TAPELESS.

*Wire-Nut® and Wing Nut® are registered trademarks of IDEAL INDUSTRIES, INC. Scotchlok® is a registered trademark of MINNESOTA MINING.

WIRE CONNECTORS VARIOUSLY KNOWN AS WIRE–NUT,® WING NUT,® AND SCOTCHLOK.®

CONNECTORS USED TO CONNECT WIRES TOGETHER IN COMBINATIONS OF NO. 16, NO. 14, AND NO. 12 AWG. THEY ARE CRIMPED ON WITH A SPECIAL TOOL, THEN COVERED WITH A SNAP-ON INSULATING CAP OR WRAPPED WITH INSULATING TAPE.

CRIMP-TYPE WIRE CONNECTOR AND INSULATING CAP.

SOLDERLESS CONNECTORS ARE AVAILABLE IN SIZES NO. 14 AWG THROUGH 500 MCM. THEY ARE USED FOR ONE SOLID OR ONE STRANDED CONDUCTOR ONLY, UNLESS OTHERWISE NOTED ON THE CONNECTOR OR ON ITS SHIPPING CARTON. THE SCREW MAY BE OF THE STANDARD SCREWDRIVER SLOT TYPE, OR IT MAY BE FOR USE WITH AN ALLEN WRENCH.

SOLDERLESS CONNECTORS

COMPRESSION CONNECTORS ARE USED FOR NO. 8 AWG THROUGH 1000 MCM. THE WIRE IS INSERTED INTO THE END OF THE CONNECTOR, THEN CRIMPED ON WITH A SPECIAL COMPRESSION TOOL.

COMPRESSION CONNECTOR

SPLIT-BOLT CONNECTORS ARE USED FOR CONNECTING TWO CONDUCTORS TOGETHER, OR FOR TAPPING ONE CONDUCTOR TO ANOTHER. THEY ARE AVAILABLE IN SIZES NO. 10 AWG THROUGH 1000 MCM. THEY ARE USED FOR TWO SOLID AND/OR TWO STRANDED CONDUCTORS ONLY, UNLESS OTHERWISE NOTED ON THE CONNECTOR OR ON ITS SHIPPING CARTON.

SPLIT-BOLT CONNECTOR

Fig 4-4   Types of wire connectors.

Unless specially stated on or in the shipping carton, or on the connector itself, conductors made of copper, aluminum, or copper-clad aluminum may not be used in combination in the same connector. When combinations are permitted, the connector will be identified for the purpose and for the conditions when and where they may be used. The conditions usually are limited to dry locations only. There are some "twist-on" wire connectors that have recently been listed for use with combinations of copper and aluminum conductors.

The preceding data is found in *Section 110-14* of the *National Electrical Code,*® and in the U.L. Standards.

Insulation of Wires

The Code requires that all wires used in electrical installations be insulated, *Section 310-2.* Exceptions to this rule are clearly indicated in the Code.

The most common type of insulation used on wires is a thermoplastic material, although rubber compound conductor insulations are still recognized by the Code. *Table 310-13* lists the various conductor insulations and applications. The insulation completely surrounds the metal conductor, has a uniform thickness, and runs the entire length of the wire or cable.

The allowable ampacities of copper conductors are given in *Table 310-16* for various types of insulation. ▶ *Note 3* to *Table 310-16* contains a table of conductor ampacities that are greater than the ampacities found in *Table 310-16*. This table is permitted to be used *only* for dwelling unit 120/240-volt, 3-wire, single-phase service-entrance conductors, service lateral conductors, and feeder conductors that serve as the main power feeder to a dwelling unit. This table is based upon the tremendous known load diversity found in homes. These special ampacities must not be used for feeder conductors that do not serve as the main power feeder to a dwelling. An example of this would be the feeder from the Main Panel A to Panel B in the residence discussed in this text.

Table 4-2 shows the special ampacities for types RH, RHH, RHW, THHW, THW, THWN, THHN, XHHW, and USE conductors. Table 4-2 is similar to the table found in *Note 3* to *Table 310-16*. ◀

The insulation covering wires and cables used in house wiring is usually rated at 600 volts or less. Exceptions to this statement are low-voltage bell wiring and fixture wiring.

Conductor Temperature Ratings

Conductors are also rated as to the temperature they can withstand. For example:

	Celsius	Fahrenheit
Type TW	60°	140°
Type THW	75°	167°
Type THHN	90°	194°

Without question, conductors with Type THHN insulation are the most popular and most commonly used, particularly in the smaller sizes, because of their small diameter (easy to handle; more conductors of a given size permitted in a given size raceway) and their ability to be installed where high temperatures are encountered, such as attics, buried in insulation, and supplying recessed fixtures.

The temperature rating of the conductors found in nonmetallic-sheathed cable, most commonly used for house wiring, is 90°C. However, the allowable ampacity of the conductors in nonmetallic-sheathed cable is that of the 60°C column of *Table 310-16*. See *Section 336-26*.

Knob-and-Tube Wiring

Older homes were wired with open, individual, insulated conductors supported by *knobs* and *tubes*. Knobs and tubes are made of porcelain. This text does not cover knob-and-tube wiring because this method is no longer used in new installations, but it is necessary to make some modifications to knob-and-tube wiring when doing remodeling work. The *National Electrical Code®* covers knob-and-tube

▶ SPECIAL AMPACITY RATINGS FOR RESIDENTIAL 120/240-VOLT, 3-WIRE, SINGLE-PHASE SERVICE-ENTRANCE CONDUCTORS, SERVICE LATERAL CONDUCTORS, AND FEEDER CONDUCTORS THAT SERVE AS THE MAIN POWER FEEDER TO A DWELLING UNIT. ◀

Copper Conductor AWG for Insulation of RH-RHH-RHW-THW-THWN-THHW-THHN-XHHW-USE	Aluminum or Copper-Clad Aluminum Conductors (AWG)	Allowable Special Ampacity
4	2	100
3	1	110
2	1/0	125
1	2/0	150
1/0	3/0	175
2/0	4/0	200
3/0	250 kcmil	225
4/0	300 kcmil	250
250 kcmil	350 kcmil	300
350 kcmil	500 kcmil	350
400 kcmil	600 kcmil	400

Table 4-2

wiring in *Article 324.* Many fires have occurred because insulation has been blown in attics of older homes, completely burying the open conductors. The conductors become overheated! Note the requirements of *Section 324-4,* which states in part that "concealed knob-and-tube wiring shall not be used in hollow spaces of walls, ceilings, and attics where such spaces are insulated by loose, rolled, or foamed-in-place insulating material that envelops the conductors." Keep a close watch on the companies that blow in insulation in older homes!

Table 310-16 shows the many types of insulations available on conductors for building wire of the types used in most electrical installations. Electricians refer to this table most often when selecting wire sizes for specific loads. These are the types of conductors found in nonmetallic-sheathed cable (Romex), armored cable (BX), and the type commonly installed in conduit.

Note that in the common building wire categories, we find 60°C, 75°C, and 90°C temperature ratings.

Section 110-14(c) and UL standards require that unless otherwise identified, terminals on wiring devices, switches, breakers, motor controllers, and so forth are based upon the ampacity of 60°C conductors where the circuit is rated 100 amperes or less, and the conductor sizes are No. 14 through No. 1 AWG. It is acceptable to install conductors having a higher temperature rating, such as 90°C THHN, but you must use the 60°C ampacity values.

For circuits rated over 100 amperes, or marked for use with conductors larger than No. 1 AWG, the conductors installed must have a minimum 75°C rating. The ampacity of the conductors would be based upon the 75°C values. It is acceptable to install conductors having a higher temperature rating, such as 90°C THHN, but you must use the 75°C ampacity values.

► Conductors having a temperature rating higher than the rating of the termination may be used. When using high temperature insulated conductors where adjustment factors are to be applied, such as *derating* for more than three current-carrying conductors in a raceway or cable, or *correcting* where high temperatures are encountered, the final results of these adjustments (in amperes) must comply with the requirements of *Section110-14(c).* ◄ Examples

of applying *derating factors* and *correcting factors* are found in unit 18.

Where separately installed "listed" pressure connectors or lugs are used, conductor ampacity is based upon the temperature rating of the connector or lug.

Therefore, we cannot arbitrarily check *Table 310-16* to find the allowable ampacity of a given conductor merely by selecting a conductor that has a high temperature rating, such as found in the 90°C column. We can use the high temperature ratings when high temperatures are encountered, such as in attics, or when we must derate because of the number of conductors in a conduit, but we must always check the temperature ratings of terminals to make sure we are not creating a "hot spot."

Terminals and/or the label in the equipment might be marked "75°C only" or "60°/75°C."

A conductor has two ends! It is important to check and verify the temperature limitations of the terminations at both ends of the conductor. For example, a panelboard may be listed for 75°C, a wiring device (switch or receptacle) might be marked 60°C, and a solderless wire connector used as a splicing means along the circuit might be marked 90°C. Always use the lowest temperature rating "in the chain."

See the following chart for correct application of conductor insulation data and termination data.

Termination Rating	Conductor Insulation Rating		
	60°C	75°C	90°C
60°C	OK	OK (at 60°C ampacity)	OK (at 60°C ampacity)
75°C	NO	OK	OK (at 75°C ampacity)
60/75°C	OK	OK (at 60°C or 75°C ampacity)	OK (at 60°C or 75°C ampacity)
90°C	NO	NO	OK (Only if *equipment* has a 90°C rating.)

It is recommended that the reader study *NEC®* *Article 310, Conductors for General Wiring.*

Neutral Size

▶ Figure 4-5 shows how *line-to-line* loads and *line-to-neutral* loads are connected. The grounded neutral (white) conductor Ⓝ for residential services and feeders is permitted to be smaller than the "hot" ungrounded phase (black, red) conductors Ⓓ only when the neutral conductor is properly and adequately sized to carry the maximum unbalance loads computed according to *Sections 215-2, 220-22, and 230-42*. ◀

Section 215-2 relates to feeders and refers us to *Article 220* for computation requirements.

Note that Section *230-42* makes direct reference to *Article 220*, where branch-circuit and feeder calculation requirements are found.

Focusing on the neutral conductor, *Section 220-22* states that "the feeder neutral load shall be the maximum unbalance of the load determined by this Article."

Section 220-22 further states that "the maximum unbalanced load shall be the maximum net computed load between the neutral and any one ungrounded conductor."

In this residence we find a number of loads that carry little or no neutral currents, such as the electric water heater, electric clothes dryer, electric oven and range, electric furnace, and air conditioner. These appliances for the most part are draw-

ing their biggest current through the "hot" conductors because they are fed with 240-volt Ⓒ or 120/240-volt circuits Ⓔ. Loads Ⓑ are connected line to neutral only.

Thus we find logic in the Code, *Note 3, Table 310-16*, which permits reducing the neutral size on services, feeders, and branch circuits where the computations prove that the neutral will be carrying less current than the "hot" phase conductors.

VOLTAGE DROP

Low voltage can cause lights to dim, television pictures to "shrink," motors to run hot, electric heaters to not produce their rated heat output, and appliances to not operate properly.

Low voltage in a home can be caused by:

• wire that is too small for the load being served,

• a circuit that is too long,

• poor connections at the terminals,

• conductors operating at high temperatures having higher resistance than when operating at lower temperatures.

A simple formula for calculating voltage drop on single-phase systems considers only the dc resistance of the conductors and the temperature of the conductor. The more accurate formulas consider

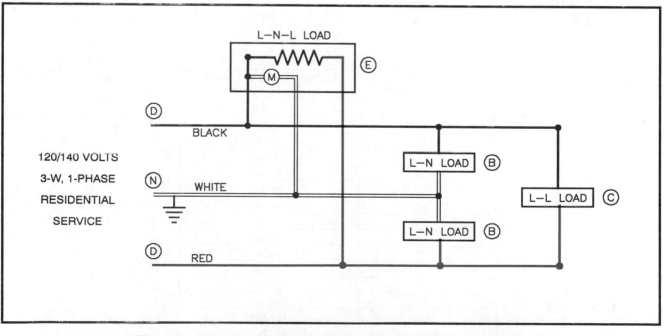

Fig. 4-5   Diagram showing line-to-line and line-to-neutral loads.

ac resistance, reactance, temperature, spacing, in metal conduit, and in nonmetallic conduit, *NEC®* *Table 9, Chapter 9*. Voltage drop is covered in great detail in *Electrical Wiring—Commercial*. The simple voltage drop formula is more accurate with smaller conductors, and gets less and less accurate as conductor size increases. It is sufficiently accurate for voltage drop calculations necessary for residential wiring.

To find voltage drop:

$$E_d = \frac{K \times I \times L \times 2}{CSA}$$

where: E_d = allowable voltage drop in volts.

K = resistance in ohms per circular-mil foot at approximately 75°C.
- For copper, approximately 12 ohms.
- For aluminum, approximately 19 ohms.

I = the current in amperes flowing through the conductors.

L = the length from the beginning of the circuit to the load.

CSA = cross-sectional area of the conductor in circular mils. Refer to Table 4-3.

To find conductor size:

$$CSA = \frac{K \times I \times L \times 2}{E_d}$$

Code Reference to Voltage Drop

Voltage drop percentages mentioned in the Fine Print Notes to *Sections 210-19* for branch-circuits, and *215-2* for feeders are RECOMMENDATIONS, NOT REQUIREMENTS. However, some local authorities make these Fine Print Notes actual requirements. The recommended voltage drop percentages are good numbers to follow. See figures 4-6 and 4-7.

The Code recommends that the voltage drop on a branch-circuit should not exceed 3%. Refer to figure 4-6.

When the voltage drop issue involves both a feeder and a branch-circuit, figure 4-7 explains the Code requirements.

Caution Using High-Temperature Wire

When trying to use smaller size conduits, be careful about selecting conductors on the ability of their insulation to withstand high temperatures.

A quick check of *Table 310-16* of the Code will confirm that the allowable ampacity of high-temperature conductors is greater than the ampacity of lower temperature conductors. For instance:

- No. 8 THHN copper has an ampacity of 55 amperes.

- No. 8 TW copper has an ampacity of 40 amperes.

CONDUCTOR SIZE	CROSS-SECTIONAL AREA IN CIRCULAR MILS
18	1,620
16	2,580
14	4,110
12	6,530
10	10,380
8	16,510
6	26,240
4	41,740
3	52,620
2	66,360
1	83,690
0	105,600
00	133,100
000	167,800
0000	211,600

Table 4-3

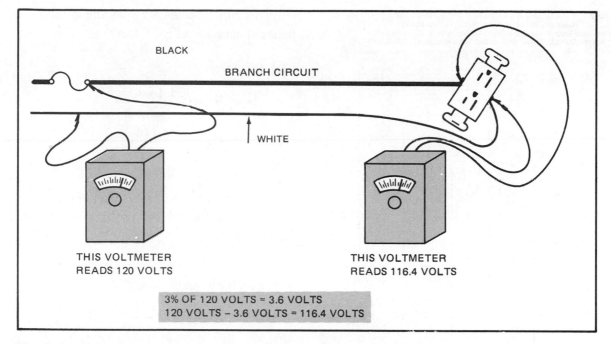

THIS VOLTMETER
READS 120 VOLTS

THIS VOLTMETER
READS 116.4 VOLTS

3% OF 120 VOLTS = 3.6 VOLTS
120 VOLTS – 3.6 VOLTS = 116.4 VOLTS

Fig. 4-6   Maximum recommended voltage drop on a branch-circuit is 3%, *Section 210-19, Fine Print Notes (FPN).*

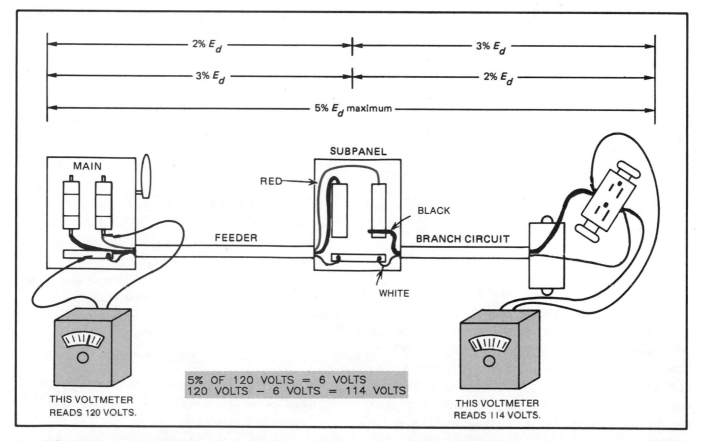

THIS VOLTMETER
READS 120 VOLTS.

5% OF 120 VOLTS = 6 VOLTS
120 VOLTS – 6 VOLTS = 114 VOLTS

THIS VOLTMETER
READS 114 VOLTS.

Fig. 4-7 *Sections 210-19 FPN No. 4* **and** *215-2 FPN No. 2* **of the Code state that the total voltage drop from the beginning of a feeder to the farthest outlet on a branch-circuit should not exceed 5%. In this figure, if the voltage drop in the feeder is 3%, then do not exceed 2% voltage drop in the branch-circuit. If the voltage drop in the feeder is 2%, then do not exceed 3% voltage drop in the branch-circuit.**

Table 310-16.   Allowable Ampacities of Insulated Conductors Rated 0–2000 Volts, 60° to 90°C (140° to 194°F) Not More Than Three Conductors in Raceway or Cable or Earth (Directly Buried), Based on Ambient Temperature of 30°C (96°F)

Size AWG kcmil	Temperature Rating of Conductor. See Table 310-13.						Size AWG kcmil
	60°C (140°F) TYPES TW†, UF†	75°C (167°F) TYPES FEPW†, RH†, RHW†, THHW†, THW†, THWN†, XHHW†, USE†, ZW†	90°C (194°F) TYPES TBS, SA, SIS, FEP†, FEPB†, MI RHH†, RHW-2, THHN†, THHW†, THW-2, THWN-2, USE-2, XHH, XHHW†, XHHW-2, ZW-2	60°C (140°F) TYPES TW†, UF†	75°C (167°F) TYPES RH†, RHW†, THHW†, THW†, THWN†, XHHW†, USE†	90°C (194°F) TYPES TBS, SA, SIS, THHN†, THHW†, THW-2, THWN-2, RHH†, RHW-2, USE-2, XHH, XHHW†, XHHW-2, ZW-2	
	COPPER			ALUMINUM OR COPPER-CLAD ALUMINUM			
18	....	....	14	....	....	....	....
16	....	....	18	....	....	....	....
14	20†	20†	25†	....	....	....	....
12	25†	25†	30†	20†	20†	25†	12
10	30	35†	40†	25	30†	35†	10
8	40	50	55	30	40	45	8
6	55	65	75	40	50	60	6
4	70	85	95	55	65	75	4
3	85	100	110	65	75	85	3
2	95	115	130	75	90	100	2
1	110	130	150	85	100	115	1
1/0	125	150	170	100	120	135	1/0
2/0	145	175	195	115	135	150	2/0
3/0	165	200	225	130	155	175	3/0
4/0	195	230	260	150	180	205	4/0
250	215	255	290	170	205	230	250
300	240	285	320	190	230	255	300
350	260	310	350	210	250	280	350
400	280	335	380	225	270	305	400
500	320	380	430	260	310	350	500
600	355	420	475	285	340	385	600
700	385	460	520	310	375	420	700
750	400	475	535	320	385	435	750
800	410	490	555	330	395	450	800
900	435	520	585	355	425	480	900
1000	455	545	615	375	445	500	1000
1250	495	590	665	405	485	545	1250
1500	520	625	705	435	520	585	1500
1750	545	650	735	455	545	615	1750
2000	560	665	750	470	560	630	2000

CORRECTION FACTORS							
Ambient Temp. °C	For ambient temperatures other than 30°C (86°F), multiply the allowable ampacities shown above by the appropriate factor shown below.						Ambient Temp. °C
21–25	1.06	1.05	1.04	1.06	1.05	1.04	70–77
26–30	1.00	1.00	2.00	1.00	1.00	1.00	78–86
31–35	.91	.94	.96	.91	.94	.96	87–95
36–40	.82	.88	.91	.82	.88	.91	96–104
41–45	.71	.82	.87	.71	.82	.87	105–113
46–50	.58	.75	.82	.58	.75	.82	114–122
51–55	.41	.67	.76	.41	.67	.76	123–131
56–60	....	.58	.75	....	.58	.71	132–140
61–70	....	.33	.58	....	.33	.58	141–158
71–80	....	....	.41	....	....	.41	159–176

†Unless otherwise specifically permitted elsewhere in this Code, the overcurrent protection for conductor types marked with an obelisk (†) shall not exceed 15 amperes for No. 14, 20 amperes for No. 12, and 30 amperes for No. 10 copper; or 15 amperes for No. 12 and 25 amperes for No. 10 aluminum and copper-clad aluminum after any correction factors for ambient temperature and number of conductors have been applied.

Therefore, always seriously consider doing a voltage-drop calculation, particularly when the decision to use the higher-temperature-rated conductors was based on the fact that the use of smaller size conduit would be possible. Select the proper size conductor to use for the job according to the load requirements, then run a voltage-drop calculation to see that the maximum permitted voltage drop is not exceeded.

EXAMPLE 1: What is the approximate voltage drop on a 120-volt, single-phase circuit consisting of No. 14 AWG copper conductors where the load is 11 amperes and the distance of the circuit from the panel to actual load is 85 feet?

SOLUTION:

$$E_d = \frac{K \times I \times L \times 2}{CSA}$$

$$E_d = \frac{12 \times 11 \times 85 \times 2}{4110}$$

$$E_d = 5.46 \text{ volts drop}$$

Note: Refer to Table 4-3 for the CSA value.

This exceeds the voltage drop permitted by the Code, which is:

$$3\% \text{ of } 120 \text{ volts} = 3.6 \text{ volts}$$

Let's try it again using No. 12 AWG:

$$E_d = \frac{12 \times 11 \times 85 \times 2}{6530}$$

$$E_d = 3.44 \text{ volts drop}$$

This "Meets Code."

EXAMPLE 2: Find the wire size needed to keep the voltage drop to no more than 3% on a single-phase, 240-volt, air-conditioner circuit. The nameplate reads: MINIMUM CIRCUIT AMPACITY 40 AMPERES. The circuit originates at the main panel located approximately 65 feet from the air-conditioner unit. No neutral is required.

SOLUTION: Checking *Table 310-16*, the conductors could be No. 6 Type TW copper or No. 8 Type THHN copper.

Remember that even though THHN conductors have a temperature rating of 90°C, their current-carrying ability must be based on the 60°C column or the 75°C column. It all depends upon the temperature rating of the disconnect switch and appliance. The labels on the equipment will give us this information.

The permitted voltage drop is:

$$E_d = 240 \times 0.03 = 7.2 \text{ volts}$$

Let's see what voltage drop we might experience if we installed the No. 8 Type THHN conductors.

$$E_d = \frac{12 \times 40 \times 65 \times 2}{16510}$$

$$= 3.78 \text{ volts drop}$$

This is well below the permissible 7.2 volts drop and would make an acceptable installation.

APPROXIMATE CONDUCTOR SIZE RELATIONSHIP

Rule One: For wire sizes up through 0000, every third size doubles or halves in circular mil area.

Thus, a No. 1 AWG conductor is 2X larger than a No. 4 AWG conductor (83 690 versus 41 470). Thus, a "0" wire is one-half the size of "0000" wire (105 600 versus 211 600).

Rule Two: For wire sizes up through 0000, every consecutive wire size is approximately 1.26 larger or smaller than the preceding wire size.

Thus, a No. 3 AWG conductor is approximately 1.26 larger than a No. 4 AWG conductor (41 740 × 1.26 = 52 592). Thus, a No. 2 AWG wire is approximately 1.26 smaller than a No. 1 AWG wire (83 690 ÷ 1.26 = 66 420).

Try to fix in your mind that a No. 10 AWG conductor has a cross-sectional area of 10,380 CM, and that it has a resistance of 1.2 ohms per 1000 feet. The resistance of aluminum wire is approximately 2 ohms per 1000 feet. By remembering these numbers, you will be able to perform voltage drop calculations without having the wire tables readily available.

EXAMPLE: What is the approximate cross-sectional area in circular mils, and resistance of a No. 6 AWG copper conductor?

SOLUTION:

WIRE SIZE	CSA (In Circular Mils)	OHMS PER 1000 FEET
10	10 380	1.2
9		
8		
7	20 760	0.6
6	26 158	0.0476

Note in this chart that when the CSA of a wire is doubled, then its resistance is cut in half. Inversely, when a given wire size is reduced to one-half, its resistance doubles.

NONMETALLIC-SHEATHED CABLE
(ARTICLE 336)

Description

Nonmetallic-sheathed cable is defined as a factory assembly of two or more insulated conductors having an outer sheath of moisture-resistant, flame-retardant, nonmetallic material. This cable is available with two or three current-carrying conductors. The conductors range in size from No. 14 through No. 2 for copper conductors, and from No. 12 through No. 2 for aluminum or copper-clad aluminum conductors. Two-wire cables contain one black conductor, one white conductor, and one bare equipment grounding conductor. Three-wire cables contain one black, one white, one red, and one bare equipment grounding conductor. Equipment grounding conductors are permitted to have green insulation, but bare equipment grounding conductors are the most common.

Figure 4-8 clearly shows a bare "wrapped" equipment grounding conductor. This wrap (sometimes paper, sometimes fiberglass) is a "filler" needed for the assembly of the cable in the manufacturing process, and is not considered to be conductor insulation. The equipment grounding conductor is *not* permitted to be used as a current-carrying conductor.

Underwriters Laboratories Inc. lists two types of nonmetallic-sheathed cables:

- Type NM-B cable has a flame-retardant, moisture-resistant, nonmetallic outer jacket covering over the conductors. The conductors are rated 90°C (194°F). The ampacity is based on 60°C (140°F).

- Type NMC-B cable has a flame-retardant, moisture-resistant, fungus-resistant, and corrosion-resistant nonmetallic outer jacket over the conductors. The conductors are rated 90°C (194°F). The ampacity is based on 60°C (140°F).

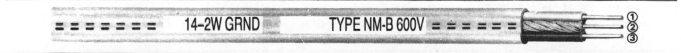

Fig. 4-8   Nonmetallic-sheathed cable with an uninsulated copper conductor. Note: ① "grounded" conductor; ② bare "wrapped" equipment "grounding" conductor; ③ black "ungrounded" (hot) conductor.

Although somewhat confusing, the *National Electrical Code®* in *Article 336* still refers to non-metallic-sheathed cable as Type NM and Type NMC. In *Section 336-26*, we find that the conductors in these cables must be rated 90°C (194°F).

Over the years, there have been many reported problems of conductor insulation becoming brittle, and breaking off of the conductors due to extreme high temperatures associated with recessed and surface-mounted lighting fixtures. Add to this the heat problem found in attics where the temperatures reach such high values, that when correctly applying the *Correction Factors* for the 60°C column of *Table 310-16* in the *NEC*,® we find that the allowable ampacity for the conductors in nonmetallic-sheathed cable in many instances is zero.

To solve this problem, Underwriters Laboratories Inc. required that all nonmetallic-sheathed cable manufactured after December 17, 1984, have 90°C insulated conductors, and required the manufacturers to mark the outer jacket with the suffix **B**. This was for ease in identifying the new 90°C cable from the older 60°C cable.

In the field today, you would find nonmetallic-sheathed cable marked Type NM-B or Type NMC-B (See figure 4-8). Although the conductors found in nonmetallic-sheathed cable are insulated for 90°C, the actual ampacity of the conductors must be based on the ampacity of 60°C conductors. Refer to *Section 336-26*. ▶ According to *Section 336-26, Exception*

Table 250-95. Minimum Size Equipment Grounding Conductors for Grounding Raceway and Equipment

Rating or Setting of Automatic Overcurrent Device in Circuit Ahead of Equipment, Conduit, etc., Not Exceeding (Amperes)	Size	
	Copper Wire No.	Aluminum or Copper-Clad Aluminum Wire No.*
15	14	12
20	12	10
30	10	8
40	10	8
60	10	8
100	8	6
200	6	4
300	4	2
400	3	1
500	2	1/0
600	1	2/0
800	1/0	3/0
1000	2/0	4/0
1200	3/0	250 kcmil
1600	4/0	350 "
2000	250 kcmil	400 "
2500	350 "	600 "
3000	400 "	600 "
4000	500 "	800 "
5000	700 "	1200 "
6000	800 "	1200 "

* See installation restrictions in Section 250-92(a).
Note: Equipment grounding conductors may need to be sized larger than specified in this table in order to comply with Section 250-51.

in the *National Electrical Code*,® the 90°C rating ampacity is permitted to be used for derating purposes such as would be necessary in attics where the temperatures can reach extremely high values. The final derated ampacity must not exceed the ampacity for 60°C rated conductors. *Table 310-16* shows the assigned ampacities for conductors. ◀

Most electricians still refer to nonmetallic-sheathed cable as *Romex*, named years ago after the Rome Wire and Cable Company.

Equipment grounding requirements are specified in *Sections 250-42, 250-43, 250-44,* and

FOR TYPICAL RESIDENTIAL WIRING TYPE NM AND NMC CABLE	TYPE NM	TYPE NMC
• May be used on circuits of 600 volts or less	Yes	Yes
• Has flame-retardant and moisture-resistant outer covering	Yes	Yes
• Has fungus-resistant and corrosion-resistant outer covering	No	Yes
• May be used to wire one- and two-family dwellings, or multifamily dwellings that do not exceed three floors above grade	Yes	Yes
• May be installed exposed or concealed in damp location	No	Yes
• May be embedded in masonry, concrete, plaster, adobe, fill	No	No
• May be exposed to corrosive fumes	No	Yes
• May be installed in dry, hollow voids in masonry blocks and similar locations	Yes	Yes
• May be installed in moist, damp, hollow voids in masonry blocks and similar locations	No	Yes
• May be used as service-entrance cable	No	No
• Must be protected against damage	Yes	Yes
• May be run in shallow chase of masonry, concrete, or adobe if protected by a steel plate at least 1/16 inch thick, then covered with plaster, adobe, or similar finish	No	Yes

Table 4-4 Uses permitted in typical residential wiring for Type NM and Type NMC nonmetallic-sheathed cable.

250-45. It is for the reasons stated in these sections that all boxes and fixtures in the residence are to be grounded.

Table 250-95 of the Code lists the sizes of the equipment grounding conductors used in cable assemblies. Note that the copper equipment grounding conductor shown in the portion of *Table 250-95* is the same size as the circuit conductors for 15-, 20-, and 30-ampere ratings.

Table 4-4 shows the uses permitted for Type NM and Type NMC cable.

Installation

Nonmetallic-sheathed cable is probably the least expensive of the various wiring methods. It is relatively light in weight and easy to install. It is widely used for dwelling unit installations on circuits of 600 volts or less. Figure 4-9 shows an example of a stripper used for stripping nonmetallic-sheathed cable. The installation of both types of nonmetallic-sheathed cable must conform to the requirements of *NEC® Article 336*. Refer to figures 4-10 through 4-14A.

- The cable must be strapped or stapled not more than 12 inches (305 mm) from a box or fitting.

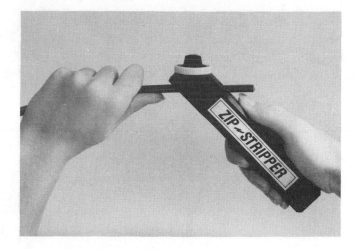

Fig. 4-9   This is an example of a stripper from one manufacturer. It is used to strip the outer jacket from nonmetallic-sheathed cable. *Courtesy* **of Seatek Co., Inc.**

- Do not staple 2-conductor cables on edge. See figure 4-13A.

- The intervals between straps or staples must not exceed 4½ feet (1.37 m).

- The cable must be protected against physical damage where necessary.

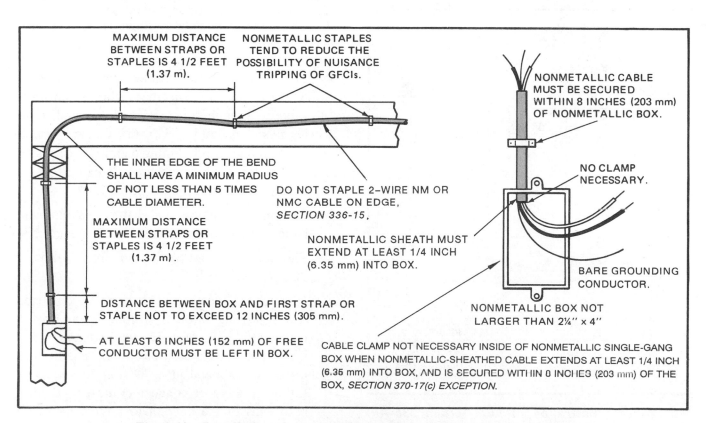

Fig. 4-10   Installation of nonmetallic-sheathed cable and armored cable.

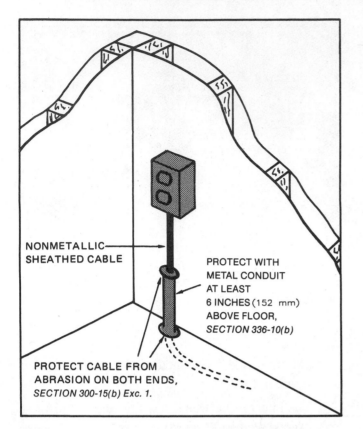

NONMETALLIC
SHEATHED CABLE

PROTECT WITH
METAL CONDUIT
AT LEAST
6 INCHES (152 mm)
ABOVE FLOOR,
SECTION 336-10(b)

PROTECT CABLE FROM
ABRASION ON BOTH ENDS,
SECTION 300-15(b) Exc. 1.

Fig. 4-11   Installation of exposed nonmetallic-sheathed cable where passing through a floor.

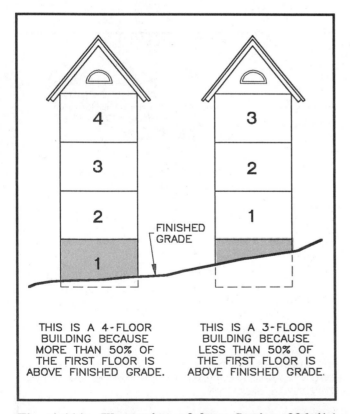

4

3

3

2

2

1

1

FINISHED
GRADE

THIS IS A 4-FLOOR
BUILDING BECAUSE
MORE THAN 50% OF
THE FIRST FLOOR IS
ABOVE FINISHED GRADE.

THIS IS A 3-FLOOR
BUILDING BECAUSE
LESS THAN 50% OF
THE FIRST FLOOR IS
ABOVE FINISHED GRADE.

Fig. 4-11A   Illustration of how *Section 336-4(a)* defines the first floor of a building.

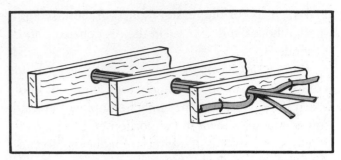

Fig. 4-12   When cables are "bundled" for distances more than 24 inches, their ampacities must be reduced (derated) according to the percentages found in *Note 8, Table 310-16* of the Code. See unit 18 for additional information.

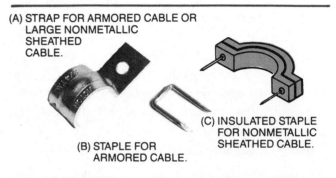

(A) STRAP FOR ARMORED CABLE OR
LARGE NONMETALLIC
SHEATHED
CABLE.

(C) INSULATED STAPLE
FOR NONMETALLIC
SHEATHED CABLE.

(B) STAPLE FOR
ARMORED CABLE.

Fig. 4-13   (A) One hole strap used for conduit, EMT, large nonmetallic-sheathed cable and armored cable. (B) Metal staple for armored cable and nonmetallic-sheathed cable. (C) Insulated staple for nonmetallic-sheathed cable.

▶• The 4½ foot securing requirement is not needed where nonmetallic-sheathed cable is run *horizontally* through holes in wood or metal framing members (studs, joists, rafters, etc.), or laid in notches in wood framing members. The cable is considered to be adequately supported by the framing members, *Section 336-15.* ◀

▶• The inner edge of the bend shall have a minimum radius not less than 5 times the cable diameter. ◀

• The cable must not be used in circuits of more than 600 volts.

• Nonmetallic-sheathed cable must be protected where passing through a floor by at least 6 inches (152 mm) of rigid metal conduit, intermediate metal conduit, electrical metallic tubing, or other metal pipe, *Section 336-10(b)*. A fitting (bushing or connector) must be used at both ends of the conduit to protect the cable from abrasion. See figure 4-11.

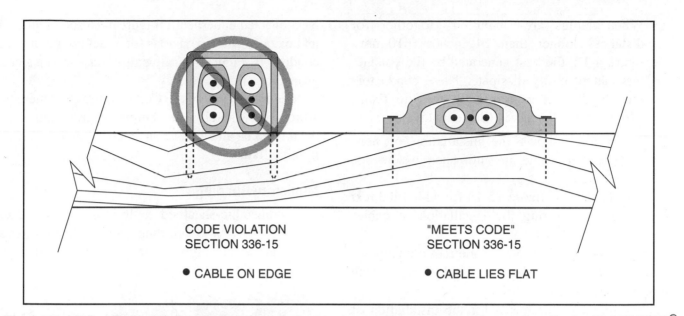

CODE VIOLATION
SECTION 336-15

● CABLE ON EDGE

"MEETS CODE"
SECTION 336-15

● CABLE LIES FLAT

Fig. 4-13A   It is a Code violation to staple two conductor nonmetallic-sheathed cables on edge, *Section 336-15, NEC.*®

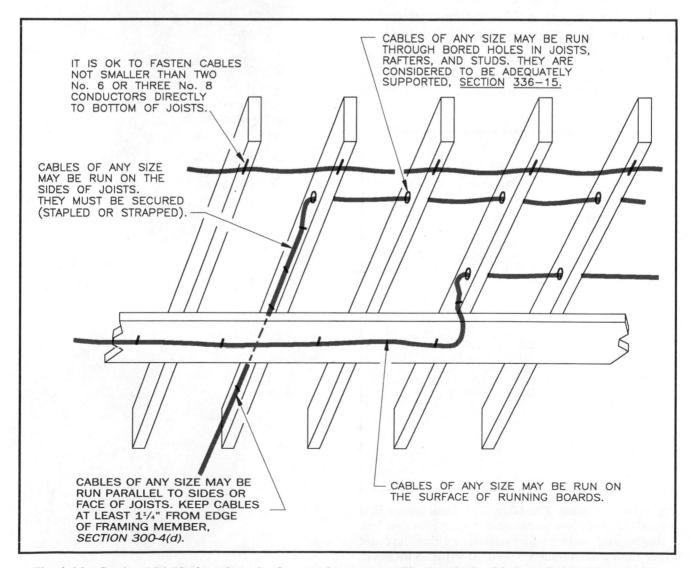

IT IS OK TO FASTEN CABLES
NOT SMALLER THAN TWO
No. 6 OR THREE No. 8
CONDUCTORS DIRECTLY
TO BOTTOM OF JOISTS.

CABLES OF ANY SIZE MAY BE RUN
THROUGH BORED HOLES IN JOISTS,
RAFTERS, AND STUDS. THEY ARE
CONSIDERED TO BE ADEQUATELY
SUPPORTED, <u>SECTION 336–15</u>.

CABLES OF ANY SIZE
MAY BE RUN ON THE
SIDES OF JOISTS.
THEY MUST BE SECURED
(STAPLED OR STRAPPED).

CABLES OF ANY SIZE MAY BE
RUN PARALLEL TO SIDES OR
FACE OF JOISTS. KEEP CABLES
AT LEAST 1¼" FROM EDGE
OF FRAMING MEMBER,
SECTION 300-4(d).

CABLES OF ANY SIZE MAY BE RUN ON
THE SURFACE OF RUNNING BOARDS.

Fig. 4-14 *Section 336-12* **gives the rules for running nonmetallic-sheathed cable in unfinished basements.**

• When cables are "bundled" together for distances longer than 24 inches (610 mm), figure 4-12, the heat generated by the conductors cannot easily dissipate. These conductors must be derated according to *Note 8* to *Tables 310-16* through *310-19.* (See unit 15.)

Figure 4-14 shows the installation of non-metallic-sheathed cable in unfinished basements, *Section 336-12.*

See unit 15 and figure 15-13 for additional text and diagrams covering the installation of cables in attics.

Figure 4-14A shows the Code requirements for securing nonmetallic-sheathed cable when using nonmetallic boxes.

Additional requirements for the installation of nonmetallic-sheathed cable are given in *Articles 200, 210, 220, 240, 250, 300,* and *310.*

► The Code in *Section 370-40(d)* requires that all metal boxes have provisions for the attachment of an equipment grounding conductor. ◄ Figure 4-15 shows a gang-type switch (device) box that is tapped for a screw by which the grounding conductor may be connected underneath. Figure 4-16 shows an outlet box, also with provisions for attaching grounding conductors. Figure 4-17 illustrates the use of a small grounding clip.

Section 410-20 of the Code requires that there be a provision whereby the equipment grounding conductor can be attached to the exposed metal parts of lighting fixtures.

Multistory Buildings

Nonmetallic-sheathed cable is not permitted to be used where the building exceeds three floors above grade, as might be the case in multifamily

Fig. 4-15   Gang-type switch (device) box.

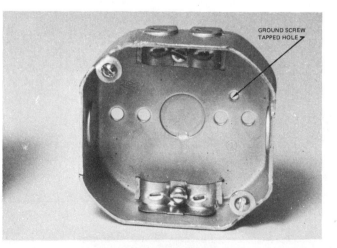

Fig. 4-16   Outlet box.

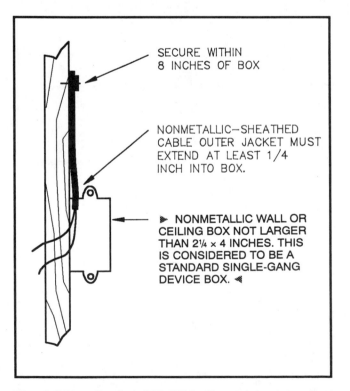

SECURE WITHIN 8 INCHES OF BOX

NONMETALLIC—SHEATHED CABLE OUTER JACKET MUST EXTEND AT LEAST 1/4 INCH INTO BOX.

► NONMETALLIC WALL OR CEILING BOX NOT LARGER THAN 2¼ × 4 INCHES. THIS IS CONSIDERED TO BE A STANDARD SINGLE-GANG DEVICE BOX. ◄

Fig. 4-14A   *Section 370-17(c), Exception* states that **nonmetallic-sheathed cable** *need not* **be clamped into a single-gang nonmetallic wall or ceiling box not larger than 2¼ × 4 inches if secured within 8 inches of the box.**

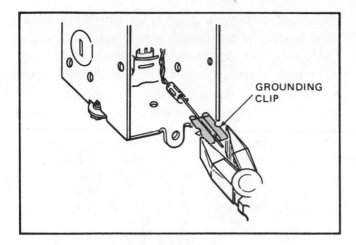

Fig. 4-17   Method of attaching ground clip to switch (device) box. See *Section 250-114*.

dwellings such as condominiums and apartments. *Section 336-4(a)* defines what constitutes the first floor of a building. The first floor of a building is that floor that has 50% or more of the exterior wall surface area level with or above finished grade. The Code allows one additional level for parking, storage, or similar use when such level is not used for human habitation. Walk around the entire building to determine the 50% criteria. Refer to figure 4-11A.

▶ For an *existing* one-family dwelling that already has three floors of habitable living area, an attic space, a storage space, or a parking space below the house might be remodeled into a bedroom or other living space. This additional "fourth floor" is permitted to be wired with nonmetallic-sheathed cable, *Section 336-4(a), Exception.* ◀

ARMORED CABLE *(ARTICLE 333)*

The *National Electrical Code®* covers the installation requirements for armored cable in *Article 333*.

Underwriters Laboratories Inc. tests armored cable per their standard No. 4. Because UL standards have evolved in numerical order, it is apparent that the armored cable standard is one of the first and oldest standards to be developed.

Description

The *National Electrical Code®* describes Type AC cable as a "fabricated assembly of insulated conductors in a flexible metallic enclosure." See figure 4-18. The armor can be steel or aluminum. If

aluminum, the words ALUMINUM ARMOR will appear on the armored cables marking tape *and* on the tag found on each coil.

Many electricians refer to armored cable as BX, supposedly derived from an abbreviation of the Bronx in New York City, where armored cable was once manufactured. At one time BX was a trademark owned by the General Electric Company. It has become a generic term.

Type AC have conductors insulated with thermosetting material. Type ACT have conductors insulated with thermoplastic material. Adding a suffix further describes Type AC or Type ACT:

- No suffix: Conductors are rated 60°C (140°F). Example: Type ACT.

- Suffix **B**: Conductors are rated 90°C (194°F). Use ampacity of 60°C (140°F) conductors. Example: Type ACTB.

- Suffix **H**. Use ampacity of 75°C (167°F) conductors. Example: Type ACTH.

- Suffix **HH**: Conductors are rated 90°C (194°F). Example: Type ACTHH. Use ampacity of 90°C (194°F) conductors for derating and temperature correction factor purposes. Use the ampacity of 60°C or 75°C conductors for maximum allowable current rating when the terminations are marked 60°C or 75°C. Currently, there are no equipment and device terminals rated more than 75°C. Also refer to *Section 110-14(c)*. This was covered earlier in unit 4.

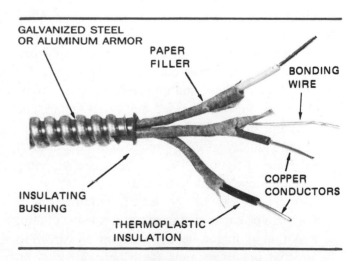

Fig. 4-18   Flexible armored cable.

Important! When armored cable is buried in or covered with insulation, the conductors must have a 90°C rating, but the ampacity must be based upon that of 60°C (140°F) conductors. See *Section 333-20, Exception.*

The ampacity of conductors is found in *Table 310-16* of the *NEC.®*

Today, most manufacturers of armored cable are producing and shipping armored cable with 90°C (194°F) insulated conductors. This is Type ACTHH armored cable.

Warning: Never use the internal bonding strip as a grounded neutral conductor, or as a separate equipment grounding conductor. It is an internal bonding strip—nothing more!

Connectors

Listed connectors for use with armored cable are permitted to be used on both steel and aluminum armored cable with one exception: Set-screw connectors are *not* permitted to be used with aluminum armored cable.

Listed connectors are suitable for grounding purposes.

For residential wiring, armored cable with two or three insulated conductors is most commonly used. Four-conductor, and even five-conductor, armored cable is available but is not commonly used for house wiring. In addition, armored cable could contain a separate green insulated equipment grounding conductor. This would be used if an isolated ground is required. Again, this is not too common in house wiring.

The current-carrying insulated conductors in armored cable can be No. 14 through No. 1 AWG copper, or No. 12 through No. 1 AWG aluminum. Two-conductor cable contains one black and one white conductor. Three-conductor cable contains one black, one white, and one red conductor. Four-conductor cable contains one black, one white, one red, and one blue conductor. Each insulated conductor is individually wrapped with a light brown kraft paper. Note in figure 4-19 that the paper wrapping has been removed back to the cut.

Section 250-51 of the *NEC®* requires that a ground path be capable of carrying *any* value of fault current that it might be called upon to carry. To assure that the impedance (resistance) of a given length of armored cable is low enough to meet the

demands of *Section 250-51,* an internal bonding strip (copper or aluminum) is built into the armored cable. See *Section 333-21* of the *NEC.®* This bonding strip is in direct contact with metal armor for the entire length of the cable. It is the combination of the metal armor *plus* the internal bonding wire that

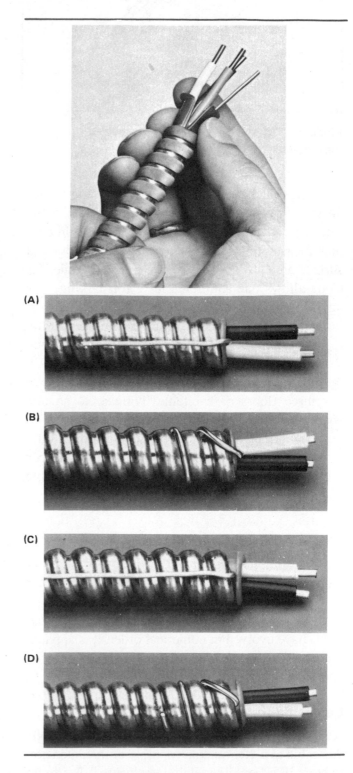

Fig. 4-19 **Antishort bushing prevents cutting of the conductor insulation by the sharp metal armor.**

guarantees that the impedance does not exceed a certain value established for that particular size and type of armored cable. The metal armor also provides excellent mechanical protection for the conductors.

Use and Installation

Armored cable can be used in more applications than nonmetallic-sheathed cable. Certain conditions govern the use of Types AC and ACT armored cable in dwelling unit installations. AC and ACT armored cable:

- may be used on circuits and feeders for applications of 600 volts or less.

- may be used for open and concealed work in dry locations.

- may be run through walls and partitions.

- may be embedded in the plaster finish on masonry walls or run through the hollow spaces of such walls if these locations are not considered damp or wet (see *Locations, Article 100*).

- must be secured within 12 inches (305 mm) from every outlet box or fitting and at intervals not exceeding 4½ feet (1.37 m). This securing requirement is not necessary when the armored cable is fished through the walls and/or ceilings, *Section 333-7, Exception No. 1.*

- must not be bent to a radius of less than five times the diameter of the cable (see figure 4-10).

▶• the 4½ foot securing requirement is *not* needed where AC cable is run *horizontally* through holes in wood or metal framing members (studs, joists, rafters, etc.), or laid in notches in wood framing members. The cable is considered to be adequately supported by the framing members, *Section 333-7, Exception No. 5.* ◀

- may be used in unsupported lengths of not over 6 feet (1.83 m) to run between an outlet box and a lighting fixture (or other equipment) above an accessible ceiling. The 12 inch (305 mm) and 4½ feet (1.37 m) securing requirement mentioned above is waived when the armored cable is used as a fixture "whip" to connect lighting fixtures or equipment, *Section 333-7, Exception No. 3.*

- must have an approved plastic insulating bushing (antishort) at the cable ends to protect the conductor insulation, figure 4-19.

- may be used where the dwelling or structure exceeds three floors above grade. Nonmetallic-sheathed cable is not permitted under this condition, figure 4-11A.

- when exposed, must closely follow the surface of the building or of running boards if used. You do not have to closely follow the surface when (1) hooking up equipment where flexibility is needed and the length of the cable is not more than 24 inches (610 mm), or (2) ▶ when the cable is supported on the underside of each joist and is not subject to physical damage, ◀ or (3) when using the cable as a fixture "whip" not longer than 6 feet (1.83 m).

- in accessible attics or roof spaces, must be protected by guard strips at least as high as the cable when run across the top of floor joists, or within 7 feet (2.13 m) of the floor or floor joists when the cable is run across the face of studs or rafters. If there is no permanent stairway or ladder, this protection is needed only within 6 feet (1.83 m) from the edge of the scuttle hole or entrance to the attic. The cable is considered to be adequately protected from physical damage when run parallel to the sides of joists, rafters, studs, in which case no guard strips or running boards are necessary.

- See unit 15 for additional discussion and diagrams regarding physical protection of cable in attics.

Armored cable (Type AC and ACT) is *not* suitable for use in the following situations:

- underground installations.

- burying in masonry, concrete, or fill of building during construction.

- installation in any location that is exposed to weather.

- installation in any location exposed to oil, gasoline, or other materials that have a destructive effect on the insulation.

To remove the outer metal cable armor, it is recommended that a tool similar to the one shown in figure 4-20 be used. Otherwise, a hacksaw can be used, figure 4-21, to cut through one of the raised convolutions of the cable armor. Be very careful not to cut too deep or you will cut into the conductor insulation. Then bend the cable armor at the cut. It will snap off easily, exposing the desired 8 to 10 inches of conductor.

To prevent cutting of the conductor insulation by the sharp metal armor, an antishort bushing is inserted at the cable ends, as shown in figure 4-19. Neither the Code nor UL specifies exactly how to insert the antishort bushing. The figure shows four ways to insert the bushings. Photos (A) and (B) show that the bonding wire holds the antishort bushing in place.

Nonmetallic-sheathed cable and armored cable each has advantages that make it suitable for particular types of installations. However, the type of cable to be used in a specific situation depends largely on the wiring method permitted or required by the local building code.

SERVICE-ENTRANCE CABLE

Service-entrance cable is covered in *Article 338* of the *National Electrical Code®.* ▶ Type SE service-entrance cable is primarily used for services, but is permitted to be used for interior branch-circuit

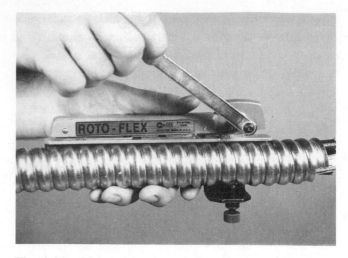

Fig. 4-20   This is an example of a tool from one manufacturer. It precisely cuts the outer armor of armored cable, making it easy to remove the armor with a few turns of the handle. *Courtesy* of Seatek Co., Inc.

and feeder wiring provided *all* of the circuit conductors are insulated. The equipment grounding conductor in these cables is permitted to be bare.

If the service-entrance cable has an uninsulated grounded conductor, a final outer nonmetallic covering, and no conductor in the cable operates over 150 volts to ground, its use is restricted to services or to supply other buildings on the same property.

The actual installation requirements for service-

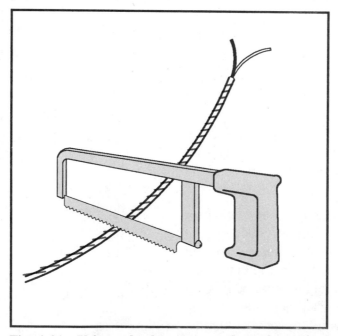

Fig. 4-21   Using a hacksaw to cut through a raised convolution of the cable armor.

entrance is the same as for nonmetallic-sheathed cable, *Article 336*. Units 15 and 20 have more information on the use of service-entrance cables for connecting electric clothes dryers and electric ranges, wall-mounted ovens, and counter-mounted cooking units. ◄ Type USE service-entrance cable is suitable for direct burial in the ground.

INSTALLING CABLES THROUGH WOOD AND METAL FRAMING MEMBERS
(SECTION 300-4)

To complete the wiring of a residence, cables must be run through studs, joists, and rafters. One method of installing cables in these locations is to run the cables through holes drilled at the approximate centers of wood building members, or at least 1¼ inches (31.8 mm) from the nearest edge. Holes bored through the center of standard 2 × 4s meet the requirements of *Section 300-4*. Refer to figure 4-22.

If the 1¼-inch distance cannot be maintained, or if the cables are to be laid in a notch, a metal plate at least ¹⁄₁₆ inch (1.59 mm) thick must cover the notch to protect the cable from nails, *Section 300-4*. Refer to figure 4-22.

To protect nonmetallic-sheathed cable from damage (being cut) when the cable is run through holes in metal joists and studs, the cable must be protected by a bushing securely fastened in place prior to installing the cable. Bushings that snap into a hole sized for the specific size bushing (½-inch, ¾-inch, etc.), or a two-piece snap in bushing that fits just about any size precut hole in the metal framing members is OK to use. See figure 4-23. These types of bushings (insulators) are not required in metal framing members when the wiring method is armored cable, electrical metallic tubing (EMT), flexible metal conduit, or electrical nonmetallic tubing.

To reduce the possibility of nails being driven into electrical cables and certain types of raceways, figure 4-23A illustrates some devices that can be used to *stand off* the cables from a framing member. Use of this type of device meets the requirements of *Section 300-4(d)*.

When running nonmetallic-sheathed cable or electrical nonmetallic tubing through metal framing members where there is a likelihood that nails or screws could be driven into the cables or tubing, a steel plate, steel sleeve, or steel clip at least ¹⁄₁₆-inch thick must be installed to protect the cables and/or tubing. This is spelled out in *Section 300-4 (b)(2)* of the Code. See figures 4-22 and 4-23B.

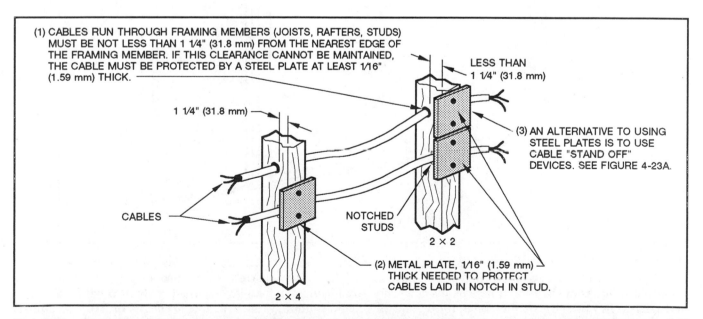

Fig. 4-22   Methods of protecting nonmetallic and armored cables so nails and screws that "miss" the studs will not damage the cables, *Section 300-4*. Because of their strength, intermediate and rigid metal conduit, rigid nonmetallic conduit, and electrical metallic tubing are exempt from this rule per the Exceptions found in *Section 300-4*. See figure 4-23A for alternative method of protecting cables.

Where cables and certain types of raceways are run concealed or exposed, parallel to a stud, joist, or rafter (any building framing member), the cables and raceways must be supported and installed in such a manner that there is not less than 1¼ inches (31.8 mm) from the edge of the stud or joist (framing member) to the cable or raceway. If the 1¼-inch clearance cannot be maintained, then a steel plate, sleeve, or equivalent that is at least ¹⁄₁₆-inch thick must be provided to protect the cable or raceway from damage that can occur when nails or screws are driven into the wall or ceiling, *Section 300-4(d)*, figures 4-22 and 4-23B.

▶ *Section 300-4(e)* addresses the wiring problems associated with cutting grooves into building material because there is no hollow space for the wiring.

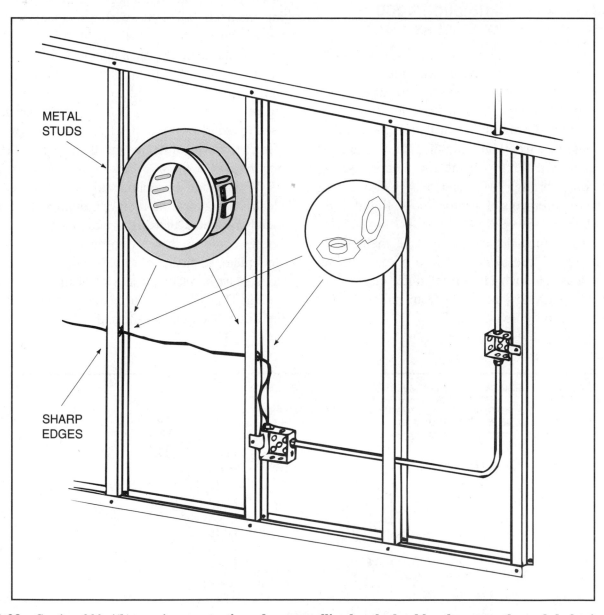

METAL
STUDS

SHARP
EDGES

Fig. 4-23 *Section 300-4(b)* **requires protection of nonmetallic-sheathed cable where run through holes in metal framing members. Shown are two types of bushings (grommets) that can be installed to protect nonmetallic-sheathed cable from abrasion. Electricians generally avoid using nonmetallic-sheathed cable through metal studs and joists. They use electrical metallic tubing, flexible metal conduit, armored cable (Type AC or MC), or electrical nonmetallic tubing. Where there is a likelihood that nails or screws might penetrate the nonmetallic-sheathed cable, ¹⁄₁₆-inch (1.59 mm) thick steel plates, sleeves, or clips must be installed to protect the cable or electrical nonmetallic tubing.**

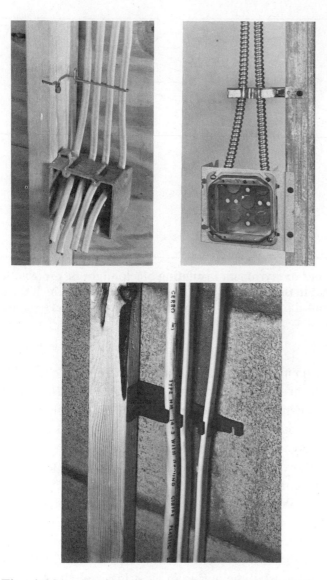

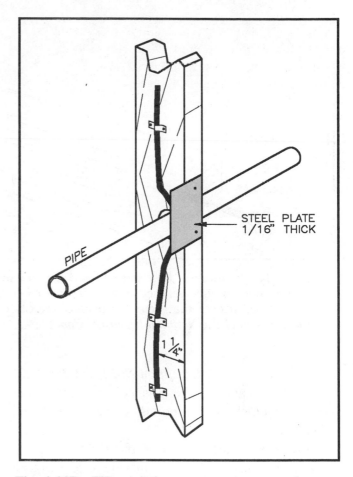

Fig. 4-23B   Where it is necessary to cross pipes or other similar obstructions, making it impossible to maintain at least 1¼-inch clearance from the framing member to the cable, install a steel plate at least ¹⁄₁₆-inch thick to protect the cable from possible damage from nails or screws. See *Section 300-4(d), NEC.*®

Fig. 4-23A   Devices that can be used to meet the Code requirement to maintain a 1¼-inch clearance from framing members.

Nonmetallic-sheathed cable and some raceways must be protected when laid in a groove. Protection must be provided with a ¹⁄₁₆ inch steel plate, sleeve, or equivalent, or the groove must be deep enough to allow not less than 1¼ inch free space for the entire length of the groove. This situation can be encountered where styrofoam insulation building blocks are grooved to receive the electrical cables, then covered with wallboard, wood paneling, or other finished wall material. Another typical example is where solid wood planking is installed on top of wood beams. The top side of the planking is covered with

roofing material, and the bottom side is exposed and serves as the finished ceiling. The only way to install wiring for ceiling fixtures and fans is to groove the planking in some manner. See figure 4-24. ◄

This additional protection is not required when the raceway is intermediate metal conduit *(Article 345)*, rigid metal conduit *(Article 346)*, rigid non-metallic conduit *(Article 347)*, or electrical metallic tubing *(Article 348)*. See figure 4-23.

Additional protection obviously cannot be installed inside the walls or ceilings of an existing building where the walls and ceilings are already closed in.

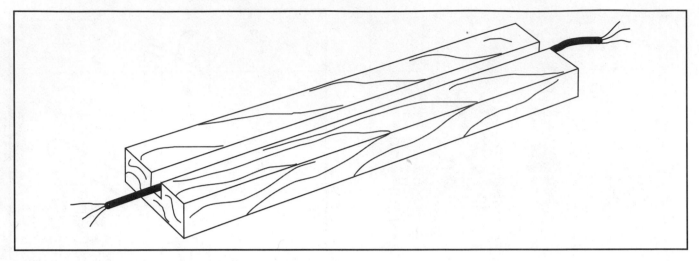

▶ **Fig. 4-24   Example of how a solid wood framing member or styrofoam insulating block might be grooved to receive the nonmetallic-sheathed cable. After the cable is laid in the groove, a steel plate not less than ¹/₁₆ inch thick must be installed over the groove to protect the cable from damage, such as from driven nails or screws. See *Section 300-4(e)* of the *National Electrical Code®*** ◀

Sections 336-15, Exception No. 1 and *300-4(d), Ex. 2* allow cables to be "fished" between outlet boxes or other access points without additional protection.

In the recreation room of the residence plans precaution must be taken so that there is adequate protection for the wiring. The walls in the recreation room are to be paneled. Therefore, if the carpenter uses 1- × 2-inch or 2- × 2-inch furring strips, figure 4-24A, nonmetallic-sheathed cable would require the additional ¹/₁₆-inch steel plate protection or the use of *stand off* devices similar to that illustrated in figure 4-23A.

Some building contractors attach 1-inch insula-

tion to the walls, then construct a 2- × 4-inch wall in front of the basement structural foundation wall, leaving 1-inch spacing between the insulation and the back side of the 2- × 4-inch studs. This makes it easy to run cables or conduits behind the 2- × 4-inch studs, and eliminates the need for the additional mechanical protection. See figure 4-24B.

Watch out for any wall partitions where 2- × 4-inch studs are installed "flat," see figure 4-24C. In this case, the choice is to provide the required mechanical protection as required by *Section 300-4*, or to install the wiring in electrical metallic tubing.

See unit 15 for the methods of cable installation and protection in attic areas. *Sections 333-10* and

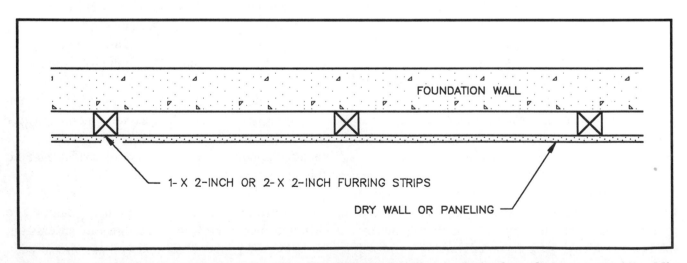

1- X 2-INCH OR 2- X 2-INCH FURRING STRIPS

DRY WALL OR PANELING —

Fig. 4-24A   When 1- × 2-inch or 2- × 2-inch furring strips are installed on the surface of a basement wall, additional protection as shown in figures 4-22, 4-23, 4-23A and 4-24 must be provided.

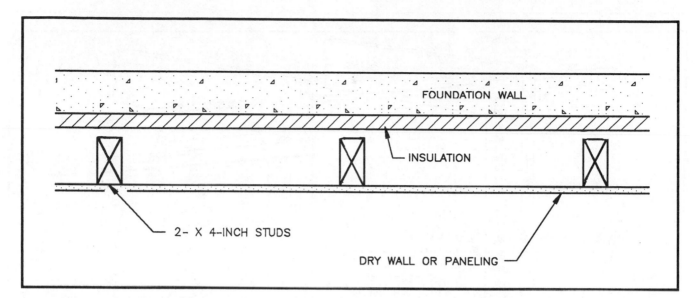

Fig. 4-24B   This sketch shows how to insulate a foundation wall and how to construct a wall that provides space to run cables and/or raceways that will not require the additional protection as shown in figures 4-22, 4-23, 4-23A and 4-24.

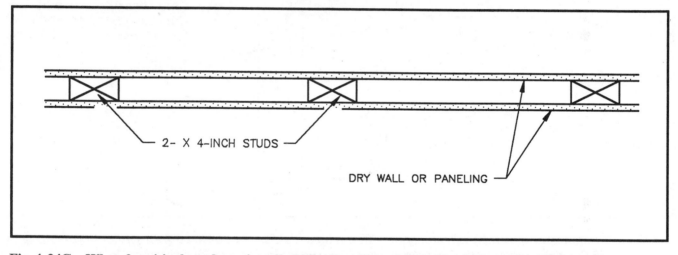

Fig. 4-24C   When 2- × 4-inch studs are installed "flat," such as might be found in nonbearing partitions, cables running parallel to or through the studs must be protected against the possibility of having nails or screws driven through the cables. This means protecting the cable with 1/16-inch steel plates or equivalent for the entire length of the cable, or installing the wiring using intermediate metal conduit, rigid metal conduit, rigid non-metallic conduit, or electrical metallic tubing. See figures 4-22, 4-23, 4-23A and 4-23B.

336-11 of the Code refer to cables run through or parallel to studs, joists, and rafters. These sections, in turn, refer to *Section 300-4* for more detailed information.

INSTALLATION OF CABLES THROUGH DUCTS

Section 300-22 of the Code is extremely strict as to what types of wiring methods are permitted for installation of cables through ducts or plenum chambers. These stringent rules are for fire safety.

The Code requirements have been relaxed slightly, however, to permit Types NM and NMC cable to be installed in joist and stud spaces (i.e., cold air returns) in dwellings, *Section 300-22, Exception No. 5.* This exception permits types NM and NMC to pass through such spaces only if they are run perpendicular to the long dimensions of such spaces, as illustrated in figure 4-25.

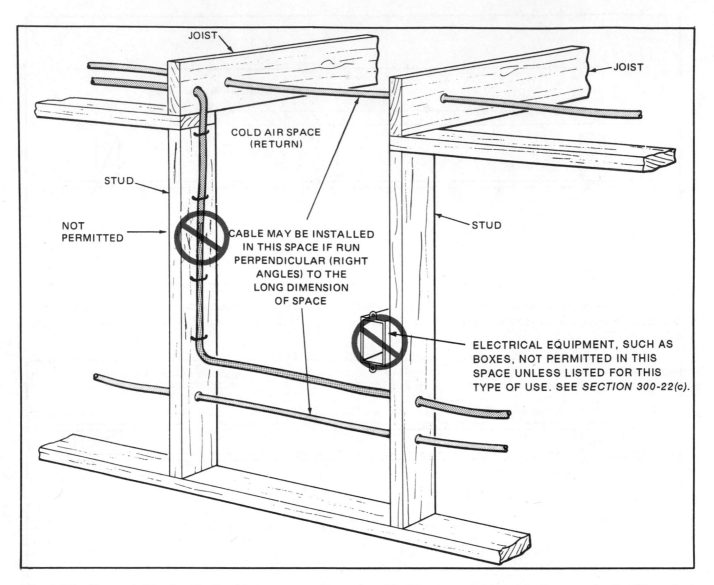

Fig. 4-25 Nonmetallic-sheathed cable may pass through cold air return, joist, or stud spaces in dwellings, but only if it is run at right angles to the long dimension of the space, *Section 300-22, Exception No. 5.*

CONNECTORS FOR INSTALLING NONMETALLIC-SHEATHED AND ARMORED CABLE

The connectors shown in figure 4-26 are used to fasten nonmetallic-sheathed cable and armored cable to the boxes and panels in which they terminate. These connectors clamp the cable securely to each outlet box. Many boxes have built-in clamps and do not require separate connectors.

The question continues to be asked: May more than one cable be inserted in one connector? Unless the UL listing of a specific cable connector indicates that the connector has been tested for use with more than one cable, the rule is *one cable, one connector.*

ELECTRICAL METALLIC TUBING *(ARTICLE 348)*, INTERMEDIATE METAL CONDUIT *(ARTICLE 345)*, RIGID METAL CONDUIT *(ARTICLE 346)*, AND RIGID NONMETALLIC CONDUIT *(ARTICLE 347)*

Some communities do not permit the installation of cable of any type in residential buildings. These communities require the installation of a raceway system of wiring, such as electrical metallic tubing (thinwall), intermediate metal conduit, rigid metal conduit, or rigid nonmetallic conduit.

Where the building construction is cement block, cinder block, or poured concrete, it will be necessary to make the electrical installation in conduit.

Sufficient data is given in this text for both cable and conduit wiring methods. Thus, the student will be able to complete the type of installation permitted or required by local codes.

According to installation requirements for a raceway system of wiring, electrical metallic tubing (EMT), intermediate metal conduit, and rigid metal conduit:

- may be buried in concrete or masonry and may be used for open or concealed work.

- may *not* be installed in cinder, concrete, or fill unless (1) protected on all sides by a layer of noncinder concrete at least 2 inches (50.8 mm)

thick, or (2) the conduit is at least 18 inches (457 mm) below the fill.

In general, conduit must be supported within 3 feet (914 mm) of each box or fitting and at 10-foot (3.05 m) intervals along runs, figure 4-27.

The number of conductors permitted in conduit and tubing is found in *Chapter 9* of the Code. Heavy-wall rigid conduit provides greater protection against mechanical injury to conductors than does either EMT or intermediate metal conduit, which have much thinner walls.

The residence specifications indicate that a meter pedestal is to be furnished by the utility and installed by the electrical contractor. Main Panel A

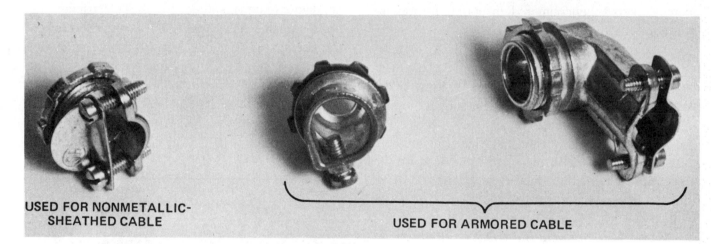

USED FOR NONMETALLIC-
SHEATHED CABLE

USED FOR ARMORED CABLE

Fig. 4-26   Cable connectors.

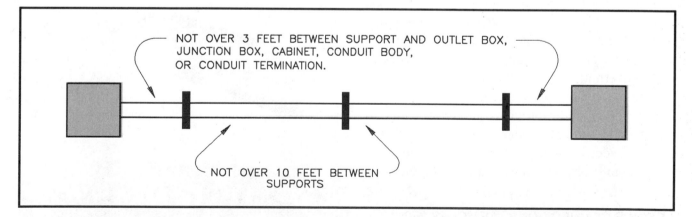

NOT OVER 3 FEET BETWEEN SUPPORT AND OUTLET BOX, JUNCTION BOX, CABINET, CONDUIT BODY, OR CONDUIT TERMINATION.

NOT OVER 10 FEET BETWEEN SUPPORTS

Fig. 4-27   Code requirements for support of intermediate metal conduit *(Section 345-12)*, rigid metal conduit *(Section 346-12)*, electrical metallic tubing *(Section 348-12)*. For supporting rigid nonmetallic conduit, refer to *Section 347-8*. ▶ Horizontal runs through bored holes or laid in notches in framing members (studs, joists, rafters) are considered to be adequately supported. The same as for flexible metal conduit, armored cable, and nonmetallic-sheathed cable, *Section 348-12, Exception No. 2* permits electrical metallic tubing to be "fished" through finished or concealed areas where it would be impractical to secure the tubing such as circumstances found in remodel work. ◀

is located in the workshop. The meter pedestal is located on the outside of the house near the air-conditioning unit. The electrical contractor must furnish and install a 1½-inch electrical metallic tubing between the meter pedestal and Main Panel A.

▶ *Sections 300-4(f)* and *373-6(c)* require that smoothly rounded insulating fittings or equivalent be used where No. 4 or larger conductors enter a raceway, box, cabinet, auxiliary gutter, or other enclosure. Examples of this are fittings that have an insulated throat, insulating bushings (figure 28-21), or an insulating sleeve that slides into the end of the raceway, separating the conductors from the metal fitting on the raceway. Where smoothly rounded or flared threaded hubs and bosses such as found on meter sockets are encountered, insulating fittings are not required, figure 28-14. ◀

In addition to these insulating fitting requirements, a bonding type bushing (figure 28-20) must be used where the service-entrance conduit enters Main Panel A. Unit 28 covers grounding and bonding of service equipment.

The plans also indicate that electrical metallic tubing is to be used in the workshop. Because the recreation room and basement stairwell have finished ceilings and walls, they will be wired with cable.

Many of the *National Electrical Code®* rules are applicable to both cable and conduit installations. Some Code requirements are applicable only to cable installations. For example, there are special requirements for the attachment of the grounding conductors and when protection from physical damage must be provided. Many of the Code requirements governing the installation of cables do not apply to a raceway system. Properly installed electrical metallic tubing provides good continuity for equipment grounding, a means of withdrawing or pulling in additional conductors, and excellent protection against physical damage to the conductors.

FLEXIBLE CONNECTIONS (ARTICLES 350 AND 351)

The installation of certain equipment requires flexible connections, both to simplify the installation and to stop the transfer of vibrations. In residential wiring, flexible connections are used to wire attic fans, food waste disposers, dishwashers, air conditioners, heat pumps, recessed fixtures, and similar equipment.

The three types of flexible conduit used for these connections are: flexible metal conduit, flexible liquidtight metal conduit, and flexible liquidtight nonmetallic conduit, figure 4-28. Figure 4-29 shows many of the types of connectors used with flexible metal conduit, flexible liquidtight metal conduit, and flexible liquidtight nonmetallic conduit.

FLEXIBLE METAL CONDUIT (ARTICLE 350)

Article 350 of the Code covers the use and installation of flexible metal conduit. This wiring method is similar to armored cable, except that the conductors are installed by the electrician. For armored cable, the cable armor is wrapped around the conductors at the factory to form a complete cable assembly. Flexible metal conduit is often referred to as *Greenfield*, named after Harry Greenfield, the individual who submitted the product for listing back in 1902.

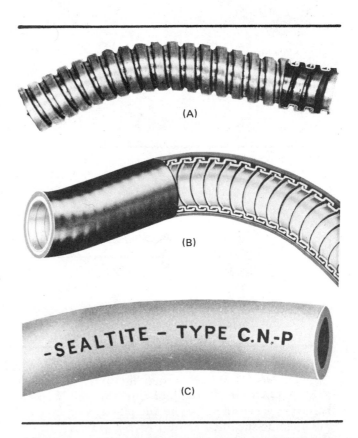

Fig. 4-28   (A) Flexible metal conduit. (B) Flexible liquid-tight metal conduit. (C) Flexible liquidtight non-metallic conduit.

Common installations using flexible metal conduit are shown in figure 4-30. Note that the flexibility required to make the installation is provided by the flexible metal conduit. The figure calls attention to the *National Electrical Code*® and Underwriters Laboratories restrictions on the use of flexible metal conduit with regard to relying on the metal armor as a grounding means.

The following are some of the common uses, limitations, and applications for flexible metal conduit, combining the material found in *Article 350* of the Code, and in the UL White Book:

• do not install in wet locations unless the conductors are suitable for use in wet locations (TW, THW, THWN), and the installation is made so water will not enter the enclosure or other raceways to which the flex is connected.

• do not bury in concrete.

• do not bury underground.

• do not use in locations subject to corrosive conditions.

▶• is acceptable as an equipment grounding conductor if:

– the flex is listed for grounding.

– the fittings are listed for grounding.

– the overcurrent protection does not exceed 20 amperes.

– the total length of the return ground path through the flex does not exceed 6 feet (1.83 m). See figure 4-30A. ◀

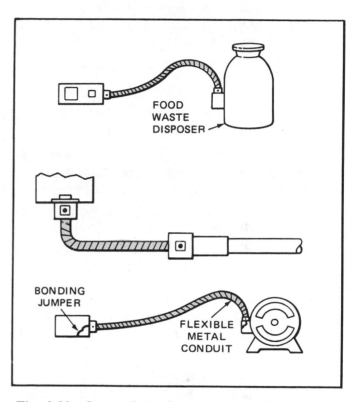

Fig. 4-30   Some of the more common places where flexible metal conduit may be used.

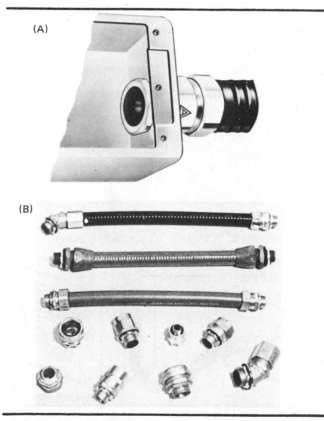

Fig. 4-29   (A) Liquidtight fitting. (B) Various types of connectors.

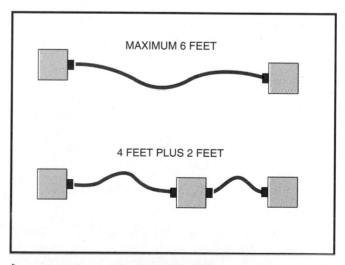

▶ Fig. 4-30A   If a flexible metal conduit is serving as an equipment ground return path, the total length shall not exceed 6 feet (1.83 m). ◀

- if longer than 6 feet (1.83 m), regardless of trade size, is not acceptable as grounding means. You must install a separate equipment grounding conductor.

- fittings larger than ¾ inch that have been listed as OK for grounding means will be marked "GRND" or some equivalent marking.

- ⅜-inch trade size not permitted longer than 6 feet (1.83 m). Fixture "whips" are a good example. See recreation room unit for a discussion of fixture whips.

- must be supported every 4½ feet (1.37 m).

- must be supported within 12 inches (305 mm) of box, cabinet, or fitting.

▶ • horizontal runs through holes and supported by framing members not more than 4½ feet (1.37 m) apart and secured within 12 inches (305 mm) of a box, cabinet, fitting, or enclosure are considered to be adequately supported. ◀

- if flexibility is needed, support within 36 inches (914 mm) of termination.

- if the flexible metal conduit is not acceptable as a grounding means, then install a bonding jumper inside or outside the flexible metal conduit.

- a bonding jumper must be installed inside the flexible metal conduit if the conduit is over 6 feet (1.83 m) long.

- do not have more than four quarter bends (360°) between boxes.

- do not conceal angle-type fittings because of the difficulty that would present itself when pulling wires in.

- if flexibility is needed, always install a separate equipment grounding conductor regardless of trade size.

- may be used for exposed or concealed installations.

- conductor fill same as regular conduit. See *Chapter 9, NEC.*®

▶ • conductor fill for ⅜-inch size. See *Table 350-12, NEC.*® ◀

- fittings used with flexible metal conduit must be listed.

- be sure to remove rough edges at "end-cuts" unless fittings are used that thread into the convolutions of the flex.

Section 250-91(b), Exception 1 also discusses flexible conduit installations and repeats the requirements already mentioned.

LIQUIDTIGHT FLEXIBLE METAL CONDUIT
(ARTICLE 351, PART A)

The use and installation of liquidtight flexible metal conduit are described in *Article 351* of the Code. Liquidtight flexible metal conduit has a "tighter" fit of its spiral turns as compared to standard flexible metal conduit. Liquidtight conduit also has a thermoplastic outer jacket that is liquidtight. Liquidtight flexible metal conduit is commonly used as the flexible connection to a central air-conditioning unit located outdoors, figure 4-31.

Figure 4-32 shows the Code rules for the use of liquidtight flexible metal conduit as a grounding means. These limitations are given in the Underwriters Laboratories Standards.

Listed here are some of the common uses, limitations, and applications for liquidtight flexible metal conduit, combining *National Electrical Code*® rules and UL White Book requirements:

- may be used for exposed and concealed installations.

- may be buried directly in the ground if so listed and marked.

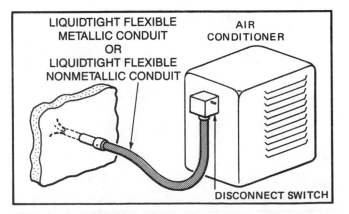

Fig. 4-31   Use of liquidtight flexible metal conduit or flexible liquidtight nonmetallic conduit.

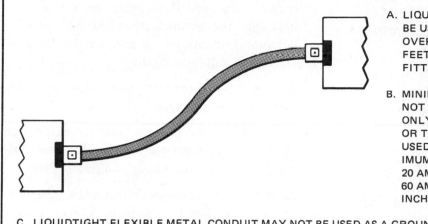

A. LIQUIDTIGHT FLEXIBLE METAL CONDUIT MAY BE USED AS A GROUNDING MEANS IF IT IS NOT OVER 1 1/4 INCH TRADE SIZE, IS NOT OVER 6 FEET (1.83 m) LONG, AND IS CONNECTED BY FITTINGS LISTED FOR GROUNDING PURPOSES.

B. MINIMUM SIZE 1/2 INCH, EXCEPT 3/8 INCH WHEN NOT LONGER THAN 6 FEET (1.83 m) PERMITTED ONLY AS PART OF AN APPROVED ASSEMBLY OR TO CONNECT LIGHTING FIXTURES. WHEN USED AS THE GROUNDING MEANS, THE MAXIMUM RATED OVERCURRENT DEVICE IS 20 AMPERES FOR 3/8- AND 1/2-INCH SIZES, AND 60 AMPERES FOR 3/4 INCH, 1 INCH, AND 1 1/4 INCHES.

C. LIQUIDTIGHT FLEXIBLE METAL CONDUIT MAY NOT BE USED AS A GROUNDING MEANS IN ANY SIZE WHEN LONGER THAN 6 FEET (1.83 m), OR IN SIZES 1 1/2 INCH AND LARGER OF ANY LENGTH. IN SUCH CASES, A BONDING JUMPER IS REQUIRED. SEE FIGURE 4–30.

Fig. 4-32   Liquidtight flexible metal conduit Code rules, *Article 351.*

- may not be used where subject to physical abuse.
- may not be used if ambient temperature and heat from conductors will exceed the temperature limitation of the nonmetallic outer jacket. UL listing will indicate maximum temperature.
- must not be used in sizes smaller than ½ inch except that ⅜-inch trade size is OK for enclosing the leads to a motor, or as a fixture "whip."
- conductor fill same as for regular conduit. See *Table 1, Chapter 9, NEC®*
- ▶• conductor fill for ⅜-inch size. See *Table 350-12.* ◀
- ▶• the flex must be "listed." ◀
- all fittings must be "listed."
- must be supported every 4½ feet (1.37 m).
- must be supported within 12 inches (305 mm) of box, cabinet, fitting, or enclosure.
- ▶• horizontal runs through holes and supported by framing members not more than 4½ feet (1.37 m) apart and secured within 12 inches (305 mm) of a box, cabinet, fitting, or enclosure are considered to be adequately supported. ◀
- ▶• the 4½ foot (1.37 m) and 12-inch (305 mm) securing requirements are waived when the flex is used as a fixture "whip" whose length is not over 6 feet (1.83 m). ◀

- the 12-inch (305 mm) securing requirement is waived when the flex is used for a flexible connection when length is not over 3 feet (914 mm).
- the 4½ foot (1.37 m) and 12-inch (305 mm) securing requirements are waived when the flex is "fished."
- if flexibility is needed, support within 3 feet (914 mm) of termination.
- ▶• ⅜-inch and ½-inch trade sizes are OK for grounding purposes if the flex and the fittings are "listed" for grounding, the total length of the return ground path through the flex does not exceed 6 feet (1.83 m), and the overcurrent device is not over 20 amperes. ◀
- ▶• ¾-inch, 1-inch, and 1¼-inch trade sizes are OK for grounding purposes if the flex and the fittings are "listed" for grounding, the total length of the return ground path through the flex does not exceed 6 feet (1.83 m), and the overcurrent device is not over 60 amperes. ◀
- is not suitable for grounding if 1½-inch trade size and larger.
- when the flex is not acceptable as a grounding means, then a separate equipment grounding conductor must be installed, sized per *Table 250-95, NEC®*
- if flexibility is needed, always install a separate equipment grounding conductor regardless of the trade size of the flex.

- do not conceal angle-type fittings.
- do not have more than 4 quarter bends (360°) between boxes.

Section 250-91(b), Exception 2 also discusses liquidtight flexible metal conduit installations, and repeats the listed requirements. See figure 4-33.

LIQUIDTIGHT FLEXIBLE NONMETALLIC CONDUIT *(ARTICLE 351, PART B)*

The following are some of the important Code and UL rules pertaining to liquidtight flexible nonmetallic conduit:

- do not use in direct sunlight unless specifically marked for use in direct sunlight.
- may be used in exposed or concealed installations.
- can become brittle in extreme cold applications.
- may be used outdoors when listed and marked as suitable for this application.
- may be buried directly in the earth when listed and marked as suitable for this application.
- may not be used where subject to physical abuse.

- may not be used if ambient temperature and heat from the conductors will exceed the temperature limitation of the nonmetallic material. UL listing will indicate this.
- do not have more than four quarter bends (360°) between boxes.
- must not be used in sizes smaller than ½ inch except that ⅜-inch trade size is OK for enclosing the leads to a motor, or as a fixture "whip."
- ►• must be secured every 3 feet (914 mm) and within 12 inches (305 mm) of a box, cabinet, fitting, or enclosure. ◄
- ►• horizontal runs through holes and supported by framing members not more than 3 feet (914 mm) apart and secured within 12 inches (305 mm) of a box, cabinet, fitting, or enclosure are considered to be adequately supported. ◄
- ►• the 3-foot (914 mm) and 12-inch (305 mm) securing requirements are waived when the flex is used as a fixture "whip" whose length is not over 6 feet (1.83 m). ◄
- ►• the 12-inch (305 mm) securing requirement is waived when the flex is used for a flexible connection whose length is not over 3 feet (914 mm). ◄

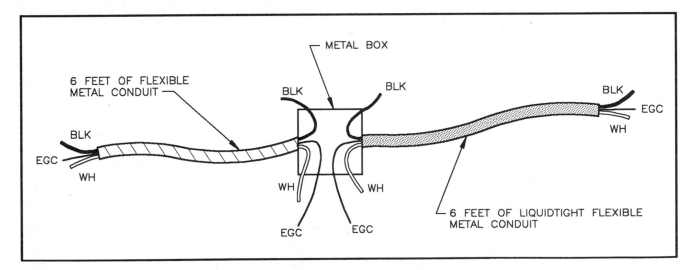

Fig. 4-33 The "combined" length of flexible metal conduit, liquidtight flexible metal conduit, and flexible metal tubing shall not exceed 6 feet (1.83 m) if it is to serve the same ground return path, *Section 250-91(b)*. In the above illustration, a separate equipment grounding conductor (EGC) would have to be provided.

▶• the 3-foot (914 mm) and 12-inch (305 mm) securing requirements are waived when the flex is "fished." ◀

• conductor fill same as for regular conduit. See *Table 1, Chapter 9, NEC.*®

▶• conductor fill for ⅜-inch size flex. See *Table 350-12, NEC.*® ◀

• do not conceal angle-type fittings.

▶• the flex must be "listed." ◀

• must be used with listed fittings.

• if an equipment grounding conductor is needed, the equipment grounding conductor may be run inside or outside of the flex. The equipment grounding conductor shall not be longer than 6 feet (1.83 m) if run on the outside of the flex.

• must not be used in lengths longer than 6 feet (1.83 m) unless absolutely necessary for flexibility, such as when connecting up a motor.

▶• must not be used in lengths longer than 6 feet (1.83 m) unless it is of the integral reinforcement type and is secured at intervals not over 3 feet (914 mm) and within 12 inches (305 mm) from a box or other similar enclosure. ◀

GROUNDING ELECTRODE CONDUCTOR

The sizing requirements for the grounding electrode conductor, which is the conductor between the main service-entrance equipment and the grounding electrode (street side of water meter, ground rods, concrete encased ground, etc., as mentioned in *Section 250-91*) is covered in unit 28. The grounding electrode conductor in this residence is a No. 4 AWG copper conductor contained in a flexible steel sheath, figure 4-34.

The Code specifies, in *Section 250-92(a)*, that a No. 4 or larger grounding conductor may be attached to the surface on which it is carried without the use of knobs, tubes, or insulators. Protection is not required unless the ground wire is expected to be exposed to severe physical damage.

If it is free from exposure to physical damage, a No. 6 grounding conductor may be run along the surface of the building construction without additional protection. If subject to physical damage, the conductor must be installed in rigid metal conduit, intermediate metal conduit, rigid non-metallic conduit, electrical metallic tubing, or cable armor. Grounding conductors smaller than No. 6 must be enclosed in rigid conduit, electrical metallic tubing, rigid nonmetallic conduit, or cable armor, *Section 250-92*.

In this residence, a No. 4 AWG armored grounding electrode conductor is run out of the top of the main service panel, across the ceiling of the workshop, over to the water pipe that comes through the basement wall in the front outside corner of the workshop area. Here the grounding electrode conductor is connected to the water pipe that leads to the well with a ground clamp. Good workmanship dictates that this armored ground wire closely follow the structure of the building. If possible, staple this armored ground wire up and between the joists, out of sight and not subject to physical damage.

The grounding electrode conductor must be connected to a metal underground water system when the water pipe is in direct contact with the earth for at least 10 feet (3.05 m). The well casing must also be bonded to the water pipe. A supplemental (additional) grounding electrode must be installed. In this residence, a driven ground rod has been installed.

Service-entrance grounding and bonding are discussed in detail in unit 28.

See unit 19 for additional reference material relating to submersible pump grounding and bonding.

See unit 18 for additional discussion on grounding.

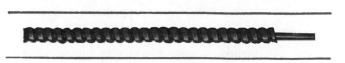

Fig. 4-34   Armored ground wire.

REVIEW

Note: Refer to the Code or the plans where necessary.

1. What is the largest size of solid building wire cable that is generally used rather than the stranded variety? _____

2. What is the minimum branch-circuit wire size that may be installed in a dwelling?

    _____

3. What exceptions, if any, are there to the answer for question 2? _____

    _____

    _____

4. What determines the ampacity of a wire? _____

    _____

    _____

5. What unit of measurement is used for the diameter of wires? _____

6. What unit of measurement is used for the cross-sectional area of wires? _____

    _____

7. What is the voltage rating of the conductors in Type NMC cable? _____

8. Indicate the allowable ampacity of the following Type THHN (copper) conductors. Refer to *Table 310-16*.

    a. 14 AWG _____ amperes   d. 8 AWG _____ amperes

    b. 12 AWG _____ amperes   e. 6 AWG _____ amperes

    c. 10 AWG _____ amperes   f. 4 AWG _____ amperes

9. What is the maximum operating temperature of the following conductors? Give the answer in degrees Fahrenheit and degrees Celsius.

    a. Type XHHW _____   c. Type THHN _____

    b. Type RH   _____   d. Type TW   _____

10. What are the colors of the conductors in nonmetallic-sheathed cable for

    a. two-wire cable? _____ , _____

    b. three-wire cable? _____ , _____ , _____

11. For nonmetallic-sheathed cable, can the uninsulated conductor be used for purposes other than grounding? _____

12. What size ground wire is used with the following sizes of nonmetallic-sheathed cable?

    a. 14 AWG _____   c. 10 AWG _____

    b. 12 AWG _____   d.   8 AWG _____

13. Under what condition may nonmetallic-sheathed cable (Type NM) be fished in the hollow voids of masonry block walls? _____

    _____

14. a. What is the maximum distance permitted between straps on a cable installation?

    _____

b. What is the maximum distance permitted between a box and the first strap in a cable installation? _____

c. Does the *NEC®* permit 2-wire Romex stapled on edge? _____

d. When Type NMC cable is run through holes in studs and joists, must additional support be provided? _____

15. What is the difference between Type AC and Type ACT cable? _____

_____

_____

_____

_____

16. Type ACT cable may be bent to a radius of not less than _____ times the diameter of the cable.

17. When armored cable is used, what protection is provided at the cable ends?

_____

_____

18. What protection must be provided when installing a cable in a notched stud or joist, or when a cable is run through bored holes in a stud or joist where the distance is less than 1¼ inches from the edge of the framing member to the cable, or where the cable is run parallel to a stud or joist and the distance is less than 1¼ inches from the edge of the framing member to the cable? _____

_____

_____

_____

19. For installing directly in a concrete slab, (armored cable, nonmetallic-sheathed cable, conduit) may be used. Circle the correct method of installation.

20. Describe the Code requirements for the mechanical protection of service grounding electrode conductors. _____

_____

_____

_____

_____

_____

21. The edge of a bored hole in a stud shall not be less than _____ inches from the edge of the stud.

22. Where is the main service entrance panel located in this residence? _____

_____

23. a. Is nonmetallic-sheathed cable permitted in your area for residential wiring?

_____

b. From what source is this information obtained? _____

24. Is it permitted to use flexible metal conduit over 6 feet (1.83 m) in length as a grounding means? (Yes) (No)   Circle one.

25. Liquidtight flexible metal conduit may serve as a grounding means in sizes up to and including _____ inches where used with approved fittings.

26. The allowable current-carrying capacity (ampacity) of aluminum wire, or the maximum overcurrent protection in the case of No. 14, No. 12, and No. 10 AWG conductors, is less than that of copper wire for a given size, insulation, and temperature of 86°F. Refer to *Table 310-16* and footnotes and complete the following table. Important: where the ampacity of the conductor does not match the rating of a standard fuse or circuit breaker as listed in *Section 240-6. Section 240-3* permits the selection of the next "higher" standard rating of fuse or circuit breaker if the next higher standard rating does not exceed 800 amperes. For ratings above 800 amperes, select the next "lower" rated over-current device.

WIRE	COPPER		ALUMINUM	
	Ampacity	Overcurrent Protection	Ampacity	Overcurrent Protection
No. 12 THHN				
No. 10 THHN				
No. 3 THW				
0000 THWN				
500 KCMIL THWN				

* ENTER BOTH AMPACITY AND MAXIMUM OVERCURRENT PROTECTION VALUES.

27. It is permissible for an electrician to connect aluminum, copper, or copper-clad aluminum conductors together in the same connector. True or False. Circle one.

28. Terminals of switches and receptacles marked CO/ALR are suitable for use with _____ , _____ , and _____ conductors.

29. Wire connectors marked AL/CU are suitable for use with _____ , _____ , and _____ conductors.

30. A wire connector bearing no marking or reference to AL, CU, or ALR is suitable for use with (copper) (aluminum) conductors only. Underline the correct answer.

31. When Type NM or NMC cable is run through a floor, it must be protected by at least _____ inches (_____ mm) of _____

_____

_____

32. When nonmetallic-sheathed cables are "bunched" or "bundled" together for distances longer than 24 inches (610 mm), what happens to their current-carrying ability?

_____

33. In diagrams A and B, nonmetallic-sheathed cable is run through the cold air return. Which diagram "Meets Code"? A _____ B _____ Check one.

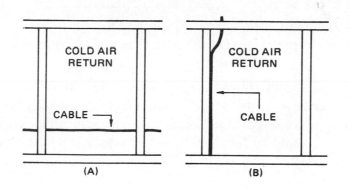

34. The marking on the outer jacket of a nonmetallic-sheathed cable indicates the letters NMC-B. What does the letter **B** signify?_____

_____

_____

35. A 120-volt branch-circuit supplies a resistive heating load of 10 amperes. The distance from the panel to the heater is approximately 140 feet. Calculate the voltage drop using (a) No. 14, (b) No. 12, (c) No. 10, (d) No. 8 AWG copper conductors.

36. In question 35, it is desired to keep the voltage drop to 3% maximum. What minimum size wire would be installed to accomplish this 3% maximum voltage drop?

_____

_____

_____

_____

_____

37. *Section 310-16, Note 3* states that the neutral conductor of a residential service or feeder can be smaller than the phase conductors, but only

_____

_____

_____

38. The allowable ampacity of a No. 4 THHN from *Table 310-16* is 95 amperes. What is this conductor's ampacity if connected to a terminal listed for use with 60°C wire?

_____

39. If, because of some obstruction in a wall space, it is impossible to keep an NMC cable at least 1¼ inches from the edge of the stud, then it shall be protected by a metal plate at least _____ inch thick.

40. The recessed fluorescent fixtures installed in the ceiling of the recreation room of this residence are connected with ⅜-inch flexible metal conduit. These flexible connections are commonly referred to as "fixture whips." Does the flexible metal conduit provide adequate grounding for the fixtures, or must a separate equipment grounding conductor be installed? _____

_____

_____

_____

_____

41. Flexible liquidtight metal conduit will be used to connect the air-conditioner unit. It will be ¾-inch trade size. Must a separate equipment grounding conductor be installed in this flex to ground the air conditioner properly?_____

_____

_____

_____

_____

42. What size overcurrent device protects the air-conditioning unit?_____

_____

43. May the 20-ampere small appliance circuits in the kitchen of a residence also supply the lighting above the kitchen sink and under-cabinet lighting? Check the correct answer. YES _____ NO _____

44. A 30-ampere branch-circuit is installed for an electric clothes dryer. What is the maximum ampere rating of a dryer permitted by the *NEC*® to be cord and plug connected to this circuit? Check the correct answer.

    a._____ 30 amperes

    b._____ 24 amperes

    c._____ 15 amperes

45. In many areas, metal framing members are being used in residential construction. When using nonmetallic-sheathed cable, what must be used where the cable is run through the metal framing members?_____

46. Are set-screw-type connectors permitted to be used with armored cable that has an aluminum armor? _____

47. Most armored cable today has 90°C conductors. What is the correct designation for this type of armored cable?_____

UNIT 5

Switch Control of Lighting Circuits, Receptacle Bonding, and Induction Heating Resulting from Unusual Switch Connections

OBJECTIVES

After studying this unit, the student will be able to

- identify the grounded and ungrounded conductors in cable or conduit (color coding).
- identify the various types of toggle switches for lighting circuit control.
- select a switch with the proper rating for the specific installation conditions.
- describe the operation that each type of toggle switch performs in typical lighting circuit installations.
- demonstrate the correct wiring connections for each type of switch per Code requirements.
- analyze some unusual "not-so-common" three-way switch connections.
- understand the various ways to bond wiring devices to the outlet box.
- understand how to design circuits to avoid heating by induction.

The electrician installs and connects various types of lighting switches. To do this, both the operation and method of connection of each type of switch must be known. In addition, the electrician must understand the meanings of the current and voltage ratings marked on lighting switches, as well as the *National Electrical Code®* requirements for the installation of these switches.

CONDUCTOR IDENTIFICATION *(ARTICLES 200 AND 210)*

Before making any wiring connections to devices, the electrician must be familiar with the ways in which conductors are identified. For alternating-current circuits, the Code requires that the grounded (identified) circuit conductor have an outer covering that is either white or a natural gray. In multiwire circuits, the grounded circuit conductor is also called a *neutral* conductor, *Sections 200-6(a) and 210-5.*

The ungrounded (unidentified) conductor of a circuit must be marked in a color other than green, white, or gray. This conductor generally is called the *hot* conductor. A shock is felt if this conductor and the grounded conductor are held at the same time, or if this conductor and a grounded surface such as a water pipe are touched at the same time.

The hazards associated with electric shock are discussed in detail in unit 6.

Neutral Conductor

In all residential, commercial, and industrial wiring, the grounded neutral conductor's insulation is white or natural gray. The neutral conductor in services is permitted to be bare, *Section 230-41*. Beyond the main service, the neutral conductor must be insulated, white or natural gray. As mentioned in the previous section on "Conductor Identification," the grounded circuit conductor is also called a neutral conductor when it is part of a multiwire branch-circuit. A *multiwire branch-circuit* is defined in the Code as a branch-circuit consisting of two or more ungrounded conductors having a potential difference between them, and a grounded conductor having equal potential difference between it and each ungrounded conductor of the circuit and that is connected to the neutral or grounded conductor of the system.

The requirements for establishing a grounded conductor are covered in *Section 250-5(b)* of the *National Electrical Code.* *Section 250-25(2)* tells us that for 3-wire, single-phase systems such as in residential wiring, the neutral conductor is to be grounded.

In most electrical systems today, the neutral conductor is always a grounded conductor (such as in 120/240-volt house wiring), but a grounded conductor is not always a neutral conductor (such as in 3-phase Delta corner grounded systems and 3-phase, 4-wire Delta mid-point grounded systems found in commercial and industrial wiring). Three-phase systems are covered in the *Electrical Wiring, Commercial* text © Delmar Publishers. Electricians often use the term "neutral" whenever it is the white grounded circuit conductor. Over the years, the *National Electrical Code®* has replaced the word "neutral" with the term "grounded circuit conductor" where confusion existed. And still, there is a gray area of the true meaning. Even electrical engi-

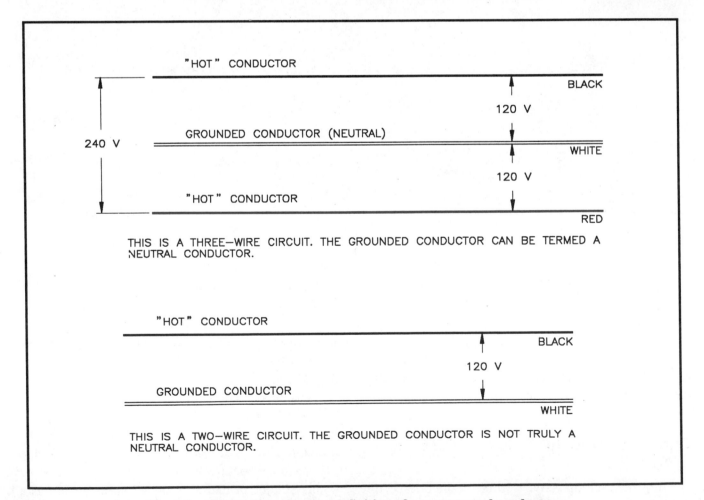

Fig. 5-1 Illustrations showing definition of a true neutral conductor.

neering textbooks are evasive as to whether the neutral definition refers to the actual source (transformer or main distribution equipment) or down into the system to a 2-wire branch-circuit.

Technically speaking, when a two-wire circuit is used, the white grounded circuit conductor is not truly a neutral conductor. To be called a neutral conductor correctly:

- it should be the conductor that carries only the unbalanced current from the other conductors, as in the case of multiwire circuits of *three or more* conductors, *Note 8* to *Table 310-16*.

- it should be the conductor where the voltage from every other conductor to it is equal under normal operating conditions.

Therefore, we can see that the white grounded circuit conductor of a two-wire circuit is not really a "neutral," even though many electricians refer to it as such. See figure 5-1.

For additional information relative to the hazards involved with multiwire circuits, refer to figures 17-5, 17-6, 17-7, 17-8, and 17-9. See unit 17.

Color Coding (Cable Wiring)

Nonmetallic-sheathed cable (Romex) and armored cable (BX) are color-coded as follows:

Two-Wire: one black ("hot" phase conductor)

              one white (grounded "identified" conductor)

              one bare (equipment grounding conductor)

Three-Wire: one black ("hot" phase conductor)

              one white (grounded "identified" conductor)

              one red ("hot" phase conductor)

              one bare (equipment grounding conductor)

Four-Wire: one black ("hot" phase conductor)

              one white (grounded "identified" conductor)

              one red ("hot" phase conductor)

              one blue ("hot" phase conductor)

              one bare (equipment grounding conductor)

Color Coding (Conduit Wiring)

When the installation is conduit, the electrician is permitted *(Section 210-5)* to use any color for the "hot" phase conductor except:

Green: reserved for use as a grounding conductor only

White or Gray: reserved for use as the grounded identified circuit conductor

Changing Colors

Should it become necessary to change the actual color of the conductor to meet Code requirements, the electrician may change the colors as follows:

FROM	TO	DO
red, black, blue, etc.	grounding conductor	for conductors larger than No. 6 AWG, strip off insulation to make it bare—or paint it green where exposed in the box—or mark exposed insulation with green tape *(Section 250-57)*
red, black, blue, etc.	grounded identified conductor	remark with colored tape or paint white or gray
white, gray or green	red, black, blue, etc.	reidentify with colored tape or paint, figure 5-2

For cable installations only, *Section 200-7* of the Code, *Exception No. 2*, permits the white conductor to be used as a switch "loop" for single-pole, three-way, or four-way switches, but further states that the white conductor must feed the switch. Therefore, the

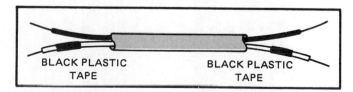

BLACK PLASTIC TAPE   BLACK PLASTIC TAPE

Fig. 5-2   A white wire may be wrapped with a piece of black plastic tape so that everyone will know that this wire is *not* a grounded conductor. Use red tape if it is desired to mark the conductor red, *Section 200-7, Exception 1 and 2*.

connection must be made up so that the return conductor from the switch is a colored conductor. When the connections are done in this manner, the Code does not require re-identification, but instead, the Code accepts the actual splices and attachment of the conductors to the terminals of the switch as adequate identification.

A good example of where this re-identification is applicable is in figure 5-6, where the white wire in the octagon box and the white wire in the switch box are in reality hot phase conductors.

Although *Section 200-7, Exception No. 2* specifically states that re-identification is *not* required, some electrical inspectors demand that these white wires be reidentified with black plastic tape to avoid any confusion. The electrical inspector has the right to require this according to the authority given him by *Section 90-4* of the Code, which states in part: "The authority having jurisdiction of enforcement of the Code will have responsibility for making interpretations of the rules, for deciding upon the approval of equipment and materials, and for granting the special permission contemplated in a number of the rules."

TOGGLE SWITCHES *(ARTICLE 380)*

The most frequently used switch in lighting circuits is the toggle flush switch, figure 5-3. When mounted in a flush switch box, the switch is concealed in the wall with only the insulated handle or toggle protruding.

Figure 5-4 shows how switches and receptacles are weatherproofed.

Toggle Switch Ratings

Underwriters Laboratories lists toggle switches used for lighting circuits as *general-use snap switches*. The UL requirements are a mirror image of *Section 380-14(a)* and *(b)* of the *National Electrical Code®.*

AC-DC General-Use Snap Switches

* alternating-current (ac) or direct-current (dc) circuits.
* resistive loads not to exceed the ampere rating of the switch at rated voltage.
* inductive loads not to exceed one-half the ampere rating of the switch at rated voltage unless otherwise marked.
* tungsten filament lamp loads not to exceed the ampere rating of the switch at 125 volts when marked with the letter **T**.
* for switches marked with a horsepower rating, a motor load shall not exceed the rating of the switch at rated voltage.

A tungsten filament lamp draws a very high momentary inrush current at the instant the circuit is energized. This is because the *cold resistance* of tungsten is very low. For instance, the cold resistance of a typical 100-watt lamp is approximately 9.5 ohms. This same lamp has a *hot resistance* of 144 ohms when operating at 100% of its rated voltage.

Normal operating current would be

$$I = \frac{E}{R} = \frac{120}{144} = 0.83 \text{ ampere}$$

But, *maximum* instantaneous inrush current could be as high as

$$I = \frac{E}{R} = \frac{170 \text{ (peak voltage)}}{9.5} = 17.9 \text{ amperes}$$

This instantaneous inrush current drops off to normal operating current in about 6 cycles (0.10

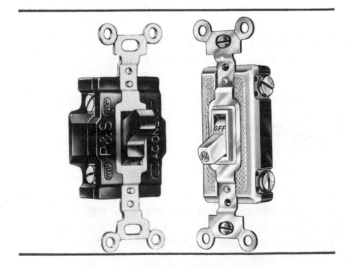

Fig. 5-3   Toggle flush switches.

Fig. 5-4   Receptacle outlet and toggle switch are protected by a weatherproof cover.

second). The contacts of T-rated switches are designed to handle these momentary high inrush currents. See unit 13 for more information pertaining to inrush currents.

The ac/dc general-use snap switch normally is not marked ac/dc. However, it is always marked with the current and voltage rating, such as 10A-125V or 5A-250V-T.

AC General-Use Snap Switches

- alternating-current (ac) only.
- resistive and inductive loads not to exceed the ampere rating of the switch at the voltage involved.
- tungsten-filament lamp loads not to exceed the ampere rating of the switch at 120 volts.
- motor loads not to exceed 80% of the ampere rating of the switch at rated voltage. A U.L. requirement is that the load shall not exceed 2 horsepower.

Ac general-use snap switches are marked "ac only," in addition to identifying their current and voltage ratings. A typical switch marking is 15A, 120-277V ac. The 277-volt rating is required on 277/480-volt systems.

Terminals of switches rated at 15 or 20 amperes, when marked *CO/ALR*, are suitable for use with aluminum, copper, and copper-clad aluminum conductors. Switches not marked *CO/ALR* are suitable for use with copper and copper-clad conductors only.

Screwless pressure terminals of the conductor push-in type may be used with copper and copper-clad aluminum conductors only. These push-in-type terminals are not suitable for use with ordinary aluminum conductors.

CONNECTING WIRING DEVICES

Wiring devices (switches and receptacles) most commonly have screw terminals under which the conductors are terminated.

Some wiring devices have back-wiring holes only. The conductor insulation is stripped off for the desired length, inserted into the hole, and held in place by the device's internal spring pressure.

Other wiring devices have back-wiring holes *and* screw terminals. Here, the conductor insulation is stripped off for the desired length, inserted into the hole, then the screw terminal is properly tightened. The conductors may also be terminated under the screw terminal in which case the back-wiring holes are not used. See figure 5-24.

The problem is that where a number of conductors are to be spliced, one might be tempted to use both back-wired holes *and* the screw terminals. On a receptacle, this could mean that as many as four wires would be connected to say, the white terminal. Although using the wiring device as a splicing means is permitted, replacing a receptacle wired in this manner is extremely difficult because of the number of wires connected to it. A better way is shown in figure 5-4A where all of the necessary splicing is done independent of the receptacle. Pigtails (short lengths of wire) are included in the splice, and these pigtails in turn connect to the receptacle.

Further information on switch ratings is given in *NEC® Section 380-14* and in the Underwriters Laboratories *Electrical Construction Materials List*.

Refer to Table 4-1 for more information regarding the markings found on wiring devices indicating the types of conductors they are listed for.

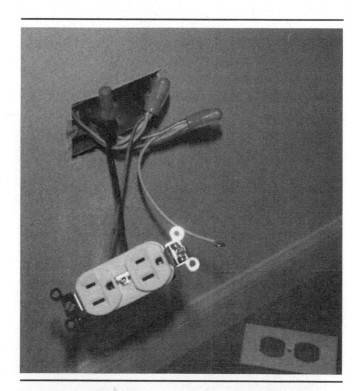

Fig. 5-4A   Photo of a receptacle connected with pigtails. This particular installation has one outlet switched, and one outlet on continuously. Six conductors enter the box because the wiring runs from receptacle to receptacle around the room. A 4-inch square outlet box trimmed with a single-gang plaster (drywall) cover is used. Removal of the receptacle does not affect the circuit to the other receptacles.

Toggle Switch Types

Toggle switches are available in four types: single-pole switch, three-way switch, four-way switch, and double-pole switch.

Single-Pole Switch.   A single-pole switch is used when a light or group of lights or other load is to be controlled from one switching point, figure 5-5. The

switch is identified by its two terminals and the toggle that is marked ON/OFF. The single-pole switch is connected in series with the ungrounded or hot wire feeding the load.

Figure 5-5 shows a single-pole switch controlling a light from one switching point. Note that the 120-volt source feeds directly through the switch location. In addition, the identified white wire goes

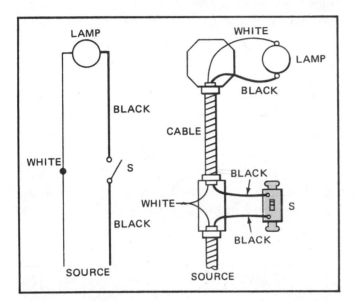

Fig. 5-5   Single-pole switch in circuit with feed at switch.

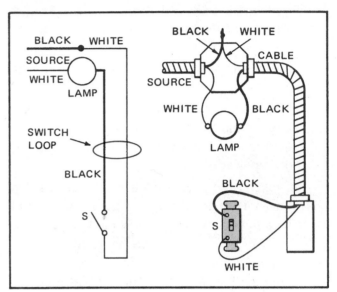

Fig. 5-6   Single-pole switch in circuit with feed at light.

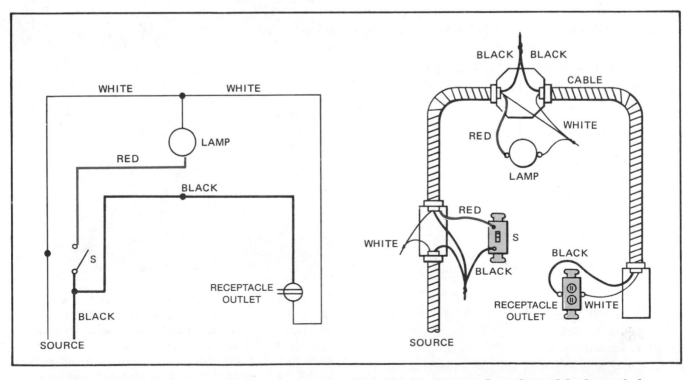

Fig. 5-7   Ceiling outlet controlled by single-pole switch with live receptacle outlet and feed at switch.

directly to the load and the unidentified black wire is broken at the single-pole switch.

In figure 5-6, the 120-volt source feeds the light outlet directly so that a two-wire cable with black and white wires is used as a switch loop between the light outlet and the single-pole switch. The use of a white wire in a single-pole switch loop is permitted in *Section 200-7, Exception No. 2*. The unidentified or black conductor must connect between the switch and the load as in figure 5-6.

Figure 5-7 shows another application of a single-pole switch. The feed is at the switch that controls the light outlet. The receptacle outlet is independent of the switch.

Three-Way Switch.   The name "three-way" is somewhat misleading because a three-way switch is used to control a light from two locations. A three-way switch has one terminal to which the internal switch mechanism is always connected. This terminal is called the *common* terminal. The two other terminals are called *traveler wire terminals*. The switching mechanism alternately switches between the common terminal and either one of the traveler terminals. Figure 5-8 shows the two positions of the three-way switch. Note that a three-way switch is actually a single-pole, double-throw switch.

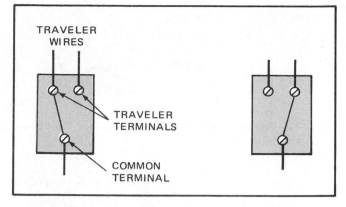

Fig. 5-8   Two positions of a three-way switch.

A three-way switch differs from a single-pole switch in that the three-way switch does not have an ON or OFF position. Thus, the switch handle does not have ON/OFF markings, figure 5-9. The three-way switch can be identified further by its three terminals. The common terminal is darker in color than the two traveler wire terminals, which are usually natural brass in color.

Three-Way Switch Control with Feed at Switch. Three-way switches are used when a load (or loads) is to be controlled from two different switching points. As shown in figure 5-10, two three-way switches are used. Note that the feed is at the first switch control point.

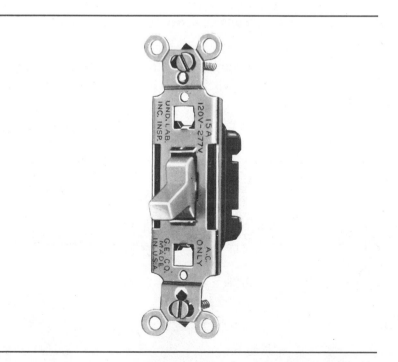

Fig. 5-9   Toggle switch: three-way flush switch.

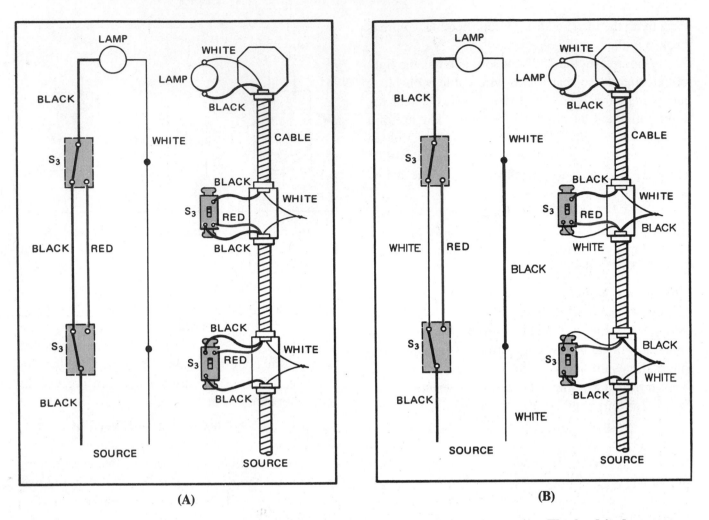

(A)   (B)

Fig. 5-10 Circuit with three-way switch control. Feed is at first switch control point. The load is downstream from the second switch control point. Diagram (A) uses black and red for the travelers. Diagram (B) uses red and white for the travelers. Either way is permitted by the Code. Be sure that the white conductor at the lampholder is connected to the lampholder's white (silver) terminal. See *Section 200-10, NEC*.®

Three-Way Switch Control with Feed at Light.
The circuit in figure 5-11 uses three-way switch control with the feed at the light. For this circuit, the white wire in the cable must be used as part of the three-way switch loop. The unidentified or black wire is used as the return wire to the light outlet, in compliance with *NEC® Section 200-7, Exception 2.* This exception makes it unnecessary to paint, tape,

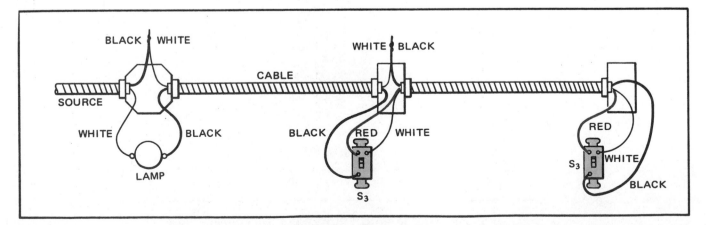

Fig. 5-11 Circuit with three-way switch control and feed at light.

or otherwise reidentify the white wire at the switch location when wiring single-pole, three-way, or four-way switch loops.

Alternate Configuration for Three-Way Switch Control with Feed at Light. Figure 5-12 shows another arrangement of components in a three-way switch circuit. The feed is at the light with cable runs from the ceiling outlet to each of the three-way switch control points located on either side of the light outlet.

The need for three-way switches occurs in rooms that have entry from more than one location. In this residence the garage lighting, entry hall lighting, living room receptacle outlets, rear outdoor lighting located between the living room sliding doors and the kitchen sliding doors, and the receptacle outlets in the study are excellent examples of the installation of three-way switches.

Conductor Color Coding for Switch Connections

The color coding for the travelers (dummies) for three-way and four-way switch connections when using cable can vary. The Code, in *Article 200*, requires that the white (identified) conductor always be connected to the white (silver) terminal of a lampholder or receptacle. Because of the many ways that 3-way and 4-way switches can be connected, the choice of colors to use for the travelers is left to the electrician. Most electrical contractors and electricians will establish some sort of color coding that works well for them.

One way is to establish white and red for the travelers, then make up the electrical splices accordingly. Refer to figure 5-11. This is not a Code requirement, but rather, a matter of choice.

Another way is illustrated in figure 5-10(A). The white conductor is spliced straight through to the lampholder, while using black and red for the travelers. Figure 5-10(B) illustrates still another way to connect the switches, where red and white are used for the travelers.

This issue does not arise when the wiring method is conduit, because there is an unlimited choice of conductor insulation colors to use for the travelers, switch returns, and so forth.

Four-Way Switch. The four-way switch is constructed so that the switching contacts can alternate their positions, as shown in figure 5-13. The four-way switch has two positions, but neither position is

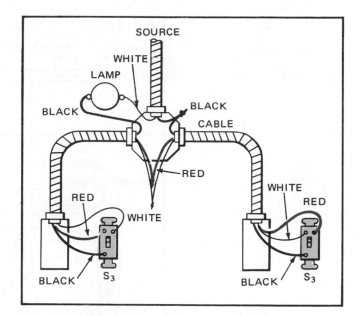

Fig. 5-12 Alternative circuit with three-way switch control and feed at light.

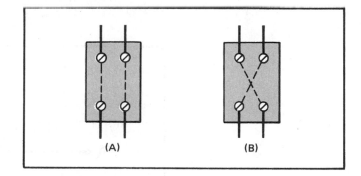

Fig. 5-13 Two positions of a four-way switch.

ON or OFF. The four-way switch can be identified readily by its four terminals and by the fact that the toggle does not have ON or OFF markings.

Four-way switches are used when a load must be controlled from more than two switching points. To accomplish this, three-way switches are connected to the source and to the load. The switches at all other control points, however, must be four-way switches.

Figure 5-14 shows how a lamp can be controlled from any one of three switching points. Care must be used in connecting the traveler wires to the proper terminals of the four-way switch: the two traveler wires from one three-way switch are connected to the two terminals on one side of the four-way switch, and the other two traveler wires from the second three-way switch are connected to the two terminals on the other side of the four-way switch.

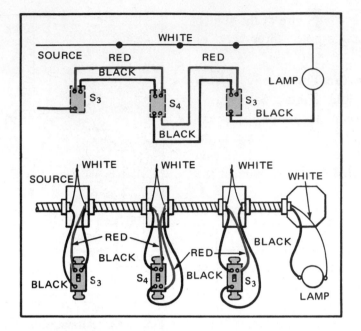

Fig. 5-14   Circuit with switch control at three different locations.

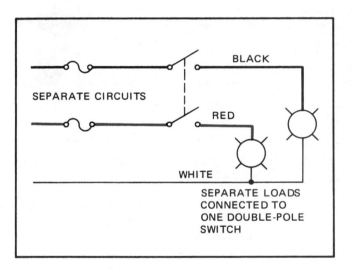

Fig. 5-15   Application of a double-pole switch.

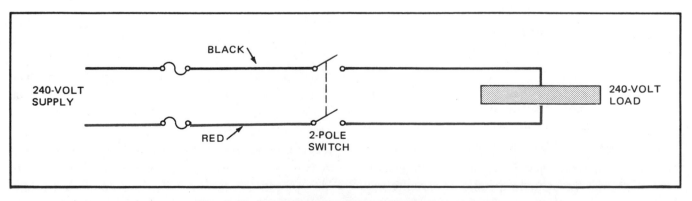

Fig. 5-16   Double-pole (2-pole) disconnect switch.

Double-Pole Switch. A double-pole (2-pole) switch may be used when two separate circuits must be controlled with one switch, figure 5-15. A 2-pole switch may also be used to provide 2-pole disconnecting means for a 240-volt load, figure 5-16.

Double-pole toggle switches are not commonly used in residential work. Double-pole disconnect switches, however, are used quite often in residences for the furnace, water pump motors, and other 240-volt feeders.

Switches with Pilot Lights. There are instances when a pilot light is desired at the switch location, as is the case for the attic lighting in this residence.

Pilot light switches are available in several styles. See figures 5-17, 5-18, and 5-19.

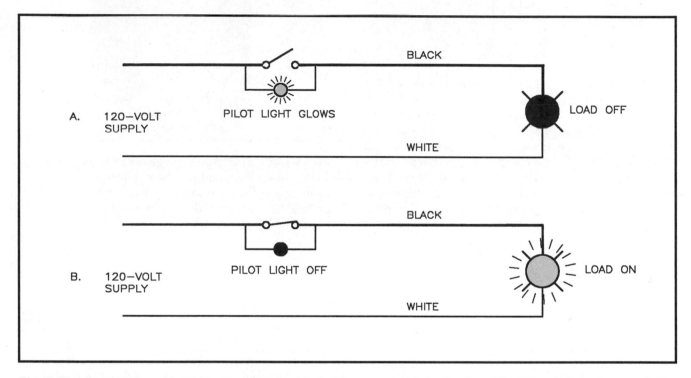

Fig. 5-17 In (A) the switch is in the OFF position. The neon lamp in the handle of the switch glows and the load is off. This is a series circuit. The neon lamp has extremely high resistance. In a series circuit voltage divides proportionately to the resistance. Therefore, the neon lamp "sees" line voltage (120 volts) for all practical purposes, and the load "sees" zero voltage. The neon lamp glows.

In (B) the switch is in the ON position. The neon lamp is shunted out and therefore has zero voltage across it. Thus, the neon lamp does not glow. Full voltage is supplied to the load. This type of switch might be referred to as a *locator* because it glows when the load is *off* and does not glow when the load is *on*.

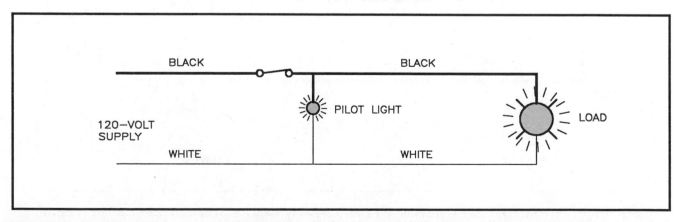

Fig. 5-18 This illustrates a true pilot light. The pilot lamp is an integral part of the switch. When the load is turned on, the pilot light is also on. When the load is turned off, the pilot light is also off. This switch has three terminals because it requires a grounded neutral circuit conductor.

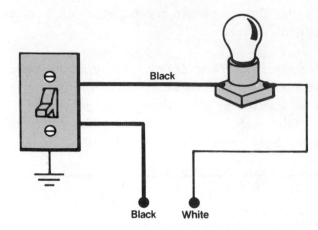

Fig. 5-19   This pilot light switch is a true pilot light. The pilot light is on when the load is on, and is off when the load is off. This is possible through its internal electronic circuit. This type of switch can be used on any 120-volt, ac grounded electrical circuit. It makes use of the system's equipment ground as its reference to permit the pilot light to glow. It does *not* require a grounded neutral conductor. The pilot light "pulsates" at approximately 60 times per minute. *Courtesy* of Hubbell Incorporated.

Fig. 5-19A   Single-pole toggle switch. Note that this switch has a terminal for the connection of an equipment grounding conductor. Thus, when a metal face plate is installed, it will be grounded per the requirement of *Section 380-12*.

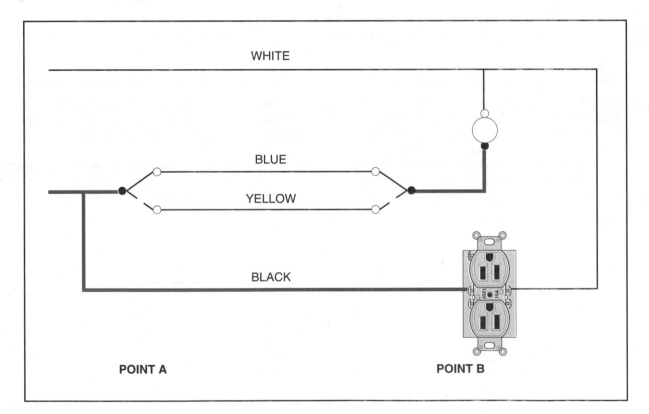

Fig. 5-19B   A lighting outlet controlled by two three-way switches. The receptacle is "live" at all times. Four conductors are run between Point A and Point B. A typical example of this might be between a house and a detached garage.

Section 380-12 requires that metal faceplates for toggle switches be grounded.

Metal faceplates for switches and receptacles are considered to be effectively grounded through the ⁶⁄₃₂ screws that attach the faceplate to the grounded yoke of the switch or receptacle.

When metal faceplates are used with non-metallic device boxes, toggle switches that have their yokes provided with a grounding screw can be installed, as illustrated in figure 5-19A.

Miscellaneous Connections. Figures 5-19B and 5-19C are other hook-ups that somewhat combine parts of the previous wiring diagrams. These hook-ups show lighting outlets controlled by two three-way switches, plus a receptacle outlet. These connections become rather complicated using cable as the wiring method because of the limited number of insulated conductors available in nonmetallic-sheathed cable or armored cable. These connections become more useful and cost effective when using conduit as the wiring method because there are many different colors of insulated conductors available.

Unusual Switch Connections. Over the years, "creative" electricians have come up with some very unusual ways to connect three-way switches. These are shown here as a challenge for the student. They were probably developed as a sort of game, to see if someone could hook up a circuit using less wires than normally required. Figures 5-19D, 5-19E, and 5-19F show these unusual connections. Cable layouts are not shown because these connections are not recommended for actual installations. They are confusing, impossible to troubleshoot, and in some cases, are not in conformance with the *National Electrical Code.*®

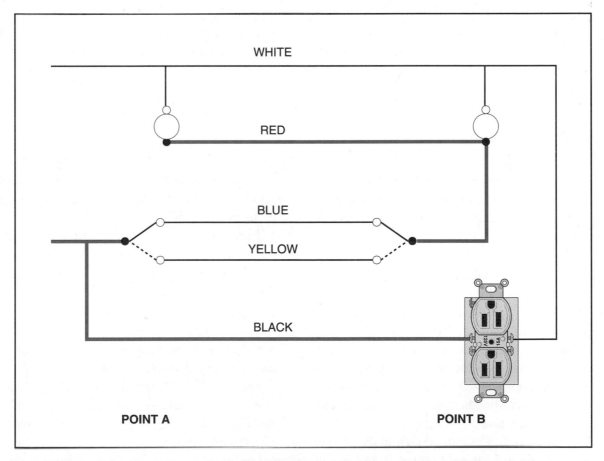

Fig. 5-19C Two three-way switches controlling two lighting outlets. The receptacle outlet is "live" at all times. Five conductors are needed between Point A and Point B. An example of this hook-up might be between a house and a detached garage, where one lighting outlet is on the house, and another lighting outlet is on the garage.

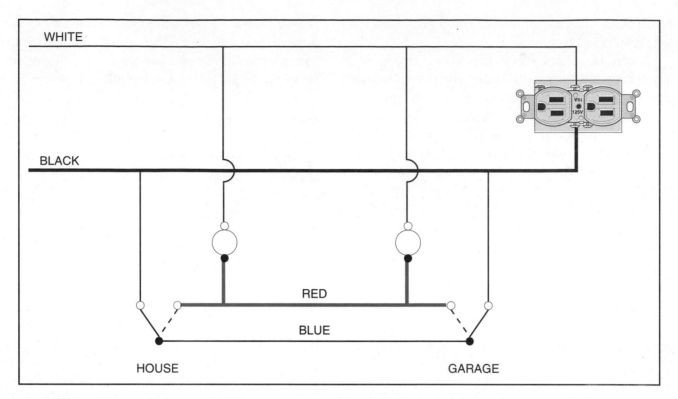

WHITE

BLACK

RED

BLUE

HOUSE   GARAGE

Fig. 5-19D Here we have a set of three-way switches controlling one light on the side of the house, and another light on the side of the garage. The receptacle is "hot" at all times when compared to figure 5-19C. The lights can be connected anywhere between the white and red wire. Normally, five wires are needed, whereas this hook-up gets the job done with only four wires. Insofar as the Code is concerned, there are no Code violations. This is a very confusing hook-up, and is not a recommended way to make the connections.

BONDING AT RECEPTACLES

Ground-fault protection is covered in detail in unit 6. However, the subject of how to provide bonding of the equipment grounding of conductor to a metal box and to the receptacles ground*ing* terminal is important.

To wire grounding-type receptacles with armored cable, the metal armor and the integral bonding strip under the armor are adequate to ground the metal outlet box and the grounding terminal of the receptacle. If nonmetallic-sheathed cable is used, the cable must contain a separate equipment grounding conductor, *Sections 210-7(c)* and *250-57(b)*. Figure 5-20 shows the grounding conductor of nonmetallic-sheathed cable attached to the receptacle per *Section 250-74*.

Almost all of the grounding terminals of the type shown in figure 5-20 are bonded to the metal yoke of the receptacle.

An exception would be the special "isolated ground" receptacles used to reduce "noise" that causes problems with computers. This is discussed in unit 6.

A metal box is considered to be adequately grounded when it is connected to grounded cable armor or metal raceway, *Section 250-57(a)*, or connected to a separate equipment grounding conductor, *Section 250-57(b)*.

Reference is made throughout this text to "equipment grounding conductors." An equipment grounding conductor is "the conductor used to connect the noncurrent-carrying metal parts of equipment, raceways, and other enclosures to the system grounded conductor and/or the grounding electrode conductor at the service equipment or at the source of a separately derived system."

Section 250-74 requires the connection of a separate equipment grounding conductor to the grounding screw of the grounding-type receptacle. This requirement holds regardless of the wiring method used—metal conduit, armored cable, or nonmetallic-sheathed cable with an equipment grounding conductor.

A bonding jumper, figure 5-20, is not necessary when the receptacle is fastened to a metal surface-mounted box such as the one in figure 18-7.

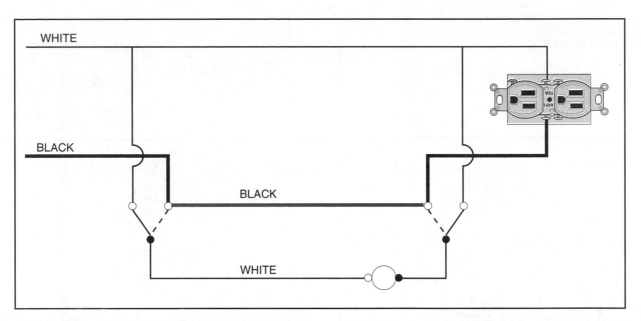

Fig. 5-19E This hook-up is NOT in conformance to the Code. Here we have a set of three-way switches controlling one light. The receptacle is "hot" at all times. The switches as shown have the light turned on. If you were to snap any one of the three-way switches to its other position, the light would go off. But note that the reason the light went off is that you are now connecting both terminals of the light with the same circuit conductor. Snapping the other switch to its other position turns the light on again. Each time a three-way switch is snapped, the wires on the light socket reverse. Sometimes the "shell" of the lamp-holder is connected to the grounded conductor as it should be, and other times the shell is connected to the "hot" circuit conductor. This is a violation of *Section 200-10(c)* of the Code. This is a very unsafe hook-up. Here are the four possible three-way switch positions and the respective polarity of the lampholder:

	Shell	Center Contact	Lamp
POSITION 1	grounded	"live"	on
POSITION 2	"live"	grounded	on
POSITION 3	grounded	grounded	off
POSITION 4	"live"	"live"	off

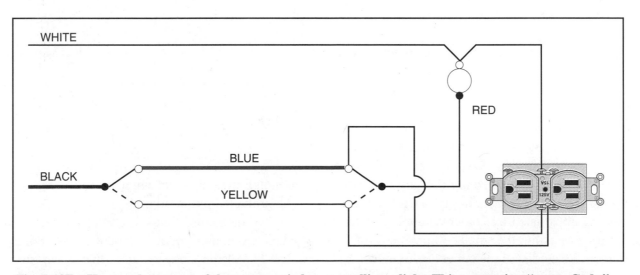

Fig. 5-19F Here we have a set of three-way switches controlling a light. This connection "meets Code." At a later date, a receptacle was installed. Note that the receptacle is of the "split circuit" type, so the white grounded conductor is common to both outlets on the receptacle, but a different "hot" conductor connects to each of the "hot" terminals on the receptacle outlet. Depending upon the position of the three-way switches, one or the other outlet is "hot." But both outlets cannot be "live" at the same time. In the diagram, the left outlet on the receptacle is "live." Snap any one of the three-way switches to another position, and the right outlet on the receptacle becomes "live." This is a very unusual "creative" hook-up.

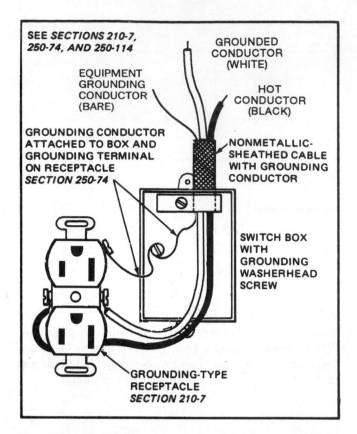

Fig. 5-20   Detail showing connections for grounding-type receptacles. See *Sections 210-7*, *250-74*, and *250-114*.

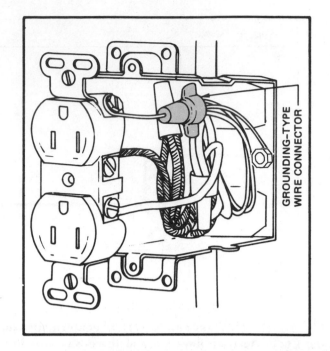

Fig. 5-21   An approved method of connecting the grounding conductor using a special grounding-type wire connector. See *Sections 250-114* and *250-74*.

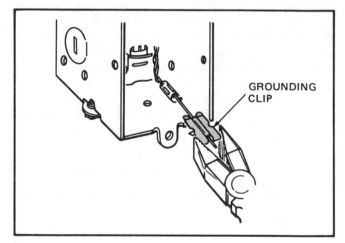

Fig. 5-22   Method of attaching grounding clip to switch (device) box. See *Sections 250-114* and *250-74*.

In this case, there is direct metal-to-metal contact between the receptacle yoke and the box *(Section 250-74, Exception No. 1)*. The No. 6-32 mounting screws of most receptacles and switches are held captive in the yoke by a small fiber or cardboard washer. Removing this washer gives metal-to-metal contact.

Some outlet and switch box manufacturers use washer head screws so that an equipment grounding conductor can be placed under them. This method is useful when installing armored cable or non-metallic-sheathed cable. That is, a means is provided to terminate the equipment grounding conductor in the metal outlet or switch box. Figure 5-21 shows an approved method of connecting the grounding conductor using a special grounding-type wire connector. Grounding clips listed by Underwriters Laboratories can be used to clamp the small equipment grounding conductor to the edge of the outlet box or switch box, figure 5-22. Grounding clips are convenient because special tools are not required to fasten them.

Exception No. 2 to *Section 250-74* states that the bonding jumper need not be used when the receptacle has a special self-grounding strap, figures 5-23

and 5-24. When such a receptacle is to be installed, a bonding jumper is not wired from the receptacle to the grounded outlet box.

To make sure that the continuity of the equipment grounding conductor path is insured, *Section 250-114* of the Code requires that when more than one equipment grounding conductor enters a box, they shall be spliced with devices "suitable for the use." Splices shall not depend on solder. The splicing must be done so that, if a receptacle or other wiring device is removed, the continuity of the equipment

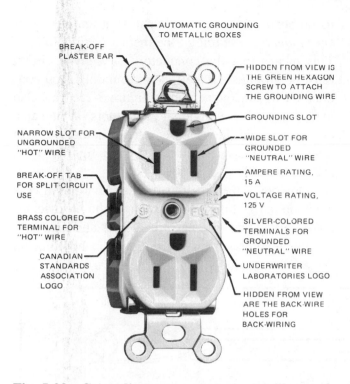

AUTOMATIC GROUNDING TO METALLIC BOXES

BREAK-OFF PLASTER EAR

HIDDEN FROM VIEW IS THE GREEN HEXAGON SCREW TO ATTACH THE GROUNDING WIRE

GROUNDING SLOT

NARROW SLOT FOR UNGROUNDED "HOT" WIRE

WIDE SLOT FOR GROUNDED "NEUTRAL" WIRE

AMPERE RATING, 15 A

BREAK-OFF TAB FOR SPLIT-CIRCUIT USE

VOLTAGE RATING, 125 V

BRASS COLORED TERMINAL FOR "HOT" WIRE

SILVER-COLORED TERMINALS FOR GROUNDED "NEUTRAL" WIRE

CANADIAN STANDARDS ASSOCIATION LOGO

UNDERWRITER LABORATORIES LOGO

HIDDEN FROM VIEW ARE THE BACK-WIRE HOLES FOR BACK-WIRING

Fig. 5-23 Grounding-type receptacle detailing various parts of the receptacle. *Courtesy* of Pass & Seymour, Inc.

grounding path shall not be interrupted. This is clearly shown in figure 5-21. When wiring with non-metallic-sheathed cable, the equipment grounding conductor is a noninsulated, bare conductor. When wiring with conduit, an equipment grounding conductor, when required, must be a green insulated conductor, sized per *Table 250-95* of the Code.

Splices must be made in accordance with *Section 110-14(b)*.

PUSH-IN TERMINALS

▶ "Push-in" terminals on receptacles are no longer permitted for use with No. 12 AWG conductors. They are "listed" by Underwriters Laboratories for use *only* with No. 14 AWG *solid copper* conductors. By design, the holes are only large enough to take a No. 14 AWG solid conductor. Push-in terminals are *not* to be used with stranded conductors. The screw terminals on these receptacles can be used for terminating a No. 12 AWG conductor.

Manufacturers had to comply with the above requirement effective January 1, 1995. Electricians,

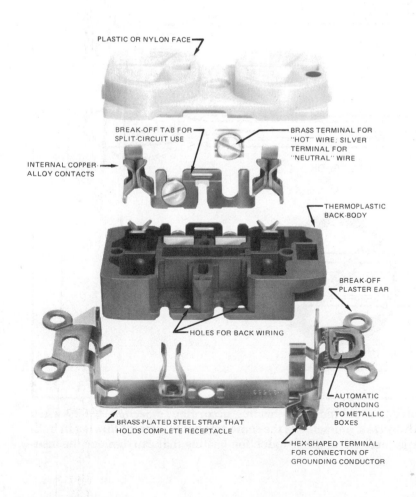

PLASTIC OR NYLON FACE

BREAK-OFF TAB FOR SPLIT-CIRCUIT USE

BRASS TERMINAL FOR "HOT" WIRE; SILVER TERMINAL FOR "NEUTRAL" WIRE

INTERNAL COPPER ALLOY CONTACTS

THERMOPLASTIC BACK-BODY

BREAK-OFF PLASTER EAR

HOLES FOR BACK WIRING

AUTOMATIC GROUNDING TO METALLIC BOXES

BRASS-PLATED STEEL STRAP THAT HOLDS COMPLETE RECEPTACLE

HEX-SHAPED TERMINAL FOR CONNECTION OF GROUNDING CONDUCTOR

▶ **Fig. 5-24 Exploded view of a grounding-type receptacle, showing all internal parts.** *Section 200-10(b)* **requires that terminals on receptacles, plugs, and connectors that are intended for the ground circuit conductor be substantially white (usually silver colored), or marked with the word "white" or the letter "W" adjacent to the terminal. The terminal for ungrounded "hot" conductor is usually brass colored.** ◀

contractors, electrical distributors, and other users are permitted to use up any inventory they might have of the older style "push-in" receptacles that would take No. 14 AWG and No. 12 AWG conductors. Until this old inventory is used up, it is O.K. to use the push-in terminal for No. 14 AWG solid copper conductors, but recommended to use the screw terminals for terminating No. 12 AWG conductors.

"Push-in" terminals for No. 12 AWG copper conductors are still acceptable on switches. ◄

INDUCTION HEATING

When alternating-current circuits are run in metal raceways, trenches, and through openings in metal boxes, the circuiting must be arranged to prevent *induction heating* of the metal, *Section 300-20.* This means that the conductors of a given circuit must be arranged so the flux surrounding each conductor will cancel each other, figure 5-25. This also means that when individual conductors of the same circuit are run in trenches, they should be kept together, not spaced far apart. See figures 5-26, 5-27, and 5-28.

Obviously, the previous situation is rarely if ever encountered in house wiring, *but* some electricians have learned that they can cut corners and save substantial lengths of three-wire cable in hooking up three-way and four-way switches in unusual ways.

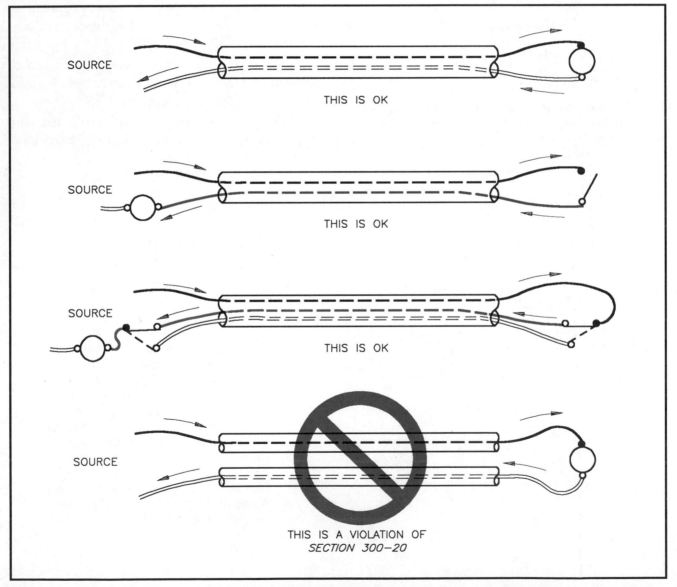

Fig. 5-25   A few examples of arranging circuitry to avoid induction heating according to *Section 300-20* when metal raceway or metal cables (BX) are used. Always ask yourself, "Is the same amount of current flowing in both directions in the metal raceway?" If the answer is "no," there will be induction heating that can damage the insulation on the conductors.

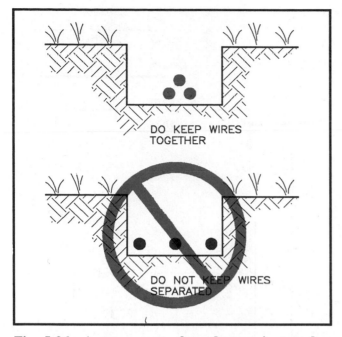

Fig. 5-26   Arrangement of conductors in trenches. Refer to unit 16 for complete information pertaining to underground wiring.

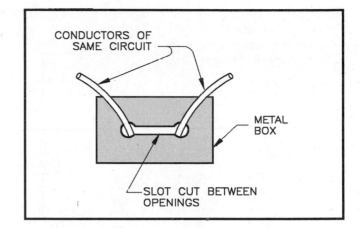

Fig. 5-28   To reduce the inductive effects the foregoing example can be minimized by cutting slots between the openings.

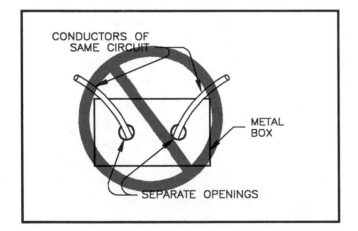

Fig. 5-27   Another typical example of how induction heating can occur is to run conductors of the same circuit through different openings of a metal box.

They use two-wire cable and make connections as shown in figure 5-29. This unusual hook-up, however, is permitted when nonmetallic boxes are used, but it violates *Section 300-20* when metal boxes are installed. The question is, is this unorthodox connection worth the confusion during installation and later on when troubleshooting the circuit might become necessary? Check with the electrical inspection authority before attempting to use this type of circuitry.

COMMON CODE VIOLATION—MAKING "TAPS"

Sometimes confusion exists in explaining "what is branch-circuit wiring, and what is a tap?" *Section 210-3* states that branch-circuits are rated according to the ampere rating of the overcurrent device protecting the branch-circuit. See figure 3-1. *Section 210-20* states that the branch-circuit conductors must be protected by fuses or circuit breakers that do not exceed the ratings established by *Section 210-3*.

There are some exceptions to this basic rule. One example is the overcurrent protection and conductor sizing for motor circuits discussed elsewhere in this text. For motor circuits, conductor overload protection is provided by the motor's thermal overload protection. Another example is where the conductors of a lighting fixture are smaller than the branch-circuit conductor size as permitted by *Section 210-19(c)*. These smaller fixture wires are considered to be "taps." Another example is where a "tap" is made for an electric surface-mounted cooking unit or wall-mounted oven as discussed in unit 20.

Table 210-24 is a summary of branch-circuit requirements. This table lists branch-circuit ratings, branch-circuit conductor sizes, tap conductor sizes, overcurrent protection, outlet (lampholder and receptacle ratings), maximum loads, and permissible loads.

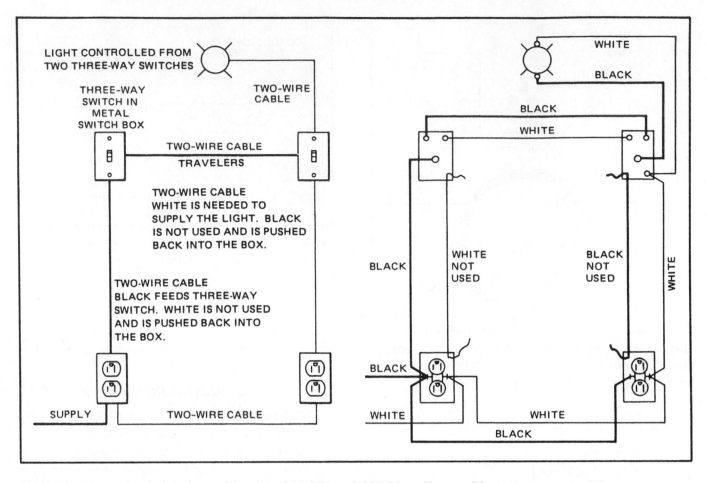

Fig. 5-29   Example of violations of *Sections 300-3(b)* and *300-20* as discussed in text.

Figure 5-30 illustrates a *code violation*. The switch leg (loop) conductors are part of the branch-circuit wiring and are *not* to be considered a tap! An argument could be made that the load on the switch leg is very small. So why does the switch leg have to be the same size as the branch-circuit conductors? The answer is that the actual connected load is not the problem, but the fact that the conductors of the switch leg must be capable of handling short-circuits and ground-faults. The switch leg conductors shown in figure 5-30 must be the same size as the branch-circuit wiring. In house wiring, most lighting branch-circuits are wired with No. 14 AWG conductors connected to a 15-ampere circuit breaker. Sometimes, to minimize voltage drop, No. 12 AWG conductors are used for long "home runs" in residential lighting branch-circuits, but the branch-circuit overcurrent device

remains at 15 amperes. This is permitted by the last sentence of *Section 210-3*, which states that "where conductors of higher ampacity are used for any reason, the ampere rating or setting of the specified overcurrent device shall determine the circuit rating."

In most commercial and industrial wiring, lighting branch-circuits are generally wired with No. 12 AWG conductors connected to a 20-ampere circuit breaker. Many times the specifications for commercial and industrial installations will call for a "home run" of 50 feet or more for a lighting branch-circuit to be wired with No. 10 AWG conductors to minimize voltage drop. The branch circuit breaker is rated 20 amperes. Again, this is permitted by *Section 210-3*. A *home run* is the branch-circuit wiring from the panel to the first outlet or junction box in that circuit.

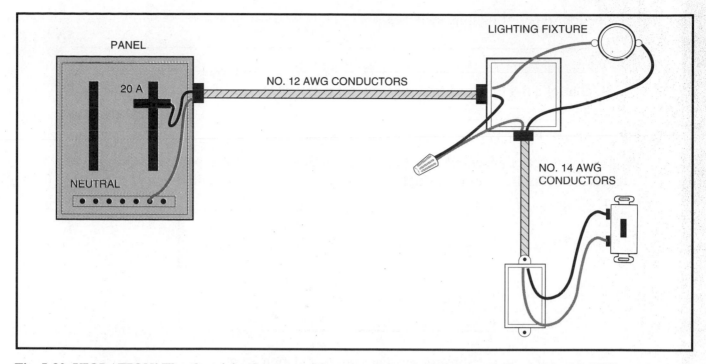

Fig. 5-30 VIOLATION! The electrician has run a 20-ampere branch-circuit consisting of No. 12 AWG conductors to the outlet box where a lighting fixture will be installed. No. 14 AWG conductors are run as the switch legs between the lighting outlet and the switch. The switch legs are part of the branch-circuit and must have an ampacity equal to that of the branch-circuit ampere rating, 20 amperes. Switch legs are not "taps." See text for details. For the same reasons, it is a Code violation to use short "pigtails" of No. 14 AWG wire to connect receptacles to the No. 12 AWG conductors fed from a 20-ampere branch-circuit.

REVIEW

Note: Refer to the Code or the plans where necessary.

1. The identified grounded circuit conductor must be _____ or _____ in color.

2. Explain how lighting switches are rated. _____

   _____

   _____

   _____

   _____

3. A T-rated switch may be used to its _____ current capacity when controlling an incandescent lighting load.

4. What switch type and rating is required to control five 300-watt tungsten filament lamps on a 120-volt circuit? Show calculations. _____

   _____

   _____

   _____

5. List four types of lighting switches.

    a. _____   c. _____

    b. _____   d. _____

6. To control a lighting load from one control point, what type of switch would be used?

_____

7. Single-pole switches are always connected to the _____ wire.

8. Complete the connections in the following arrangement so that both ceiling light outlets are controlled from the single-pole switch. Assume the installation is in cable.

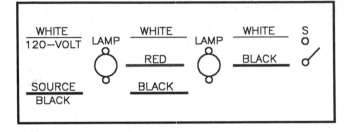

9. Complete the connections for the diagram. Installation is cable.

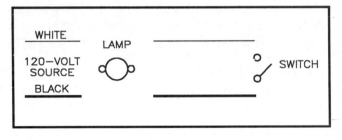

10. A three-way switch may be compared to a _____ switch.

11. What type of switch is installed to control a lighting fixture from two different control points? How many switches are needed and what type are they? _____

_____

12. Complete the connections in the following arrangement so that the lamp may be controlled from either three-way switch.

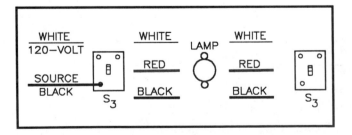

13. When connecting four-way switches, care must be taken to connect the travelers to the _____ terminals.

14. Show the connections for a ceiling outlet that is to be controlled from any one of three switch locations. The 120-volt feed is at the light, *Section 200-7*. Label the conductor colors. Assume the installation is in cable.

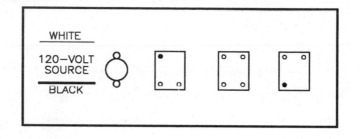

15. Match the following switch types with the correct number of terminals for each.

    Three-way switch   Two terminals
    Single-pole switch   Four terminals
    Four-way switch   Three terminals

16. When connecting single-pole, three-way, and four-way switches, they must be wired so that all switching is done in the _____ circuit conductor.

17. What section of the Code emphasizes the fact that all circuiting must be done so as to avoid the damaging effects of induction heating? _____

18. If you had to install an underground three-wire feeder to a remote building using three individual conductors, which of the following installations "meet Code"? Circle correct installation.

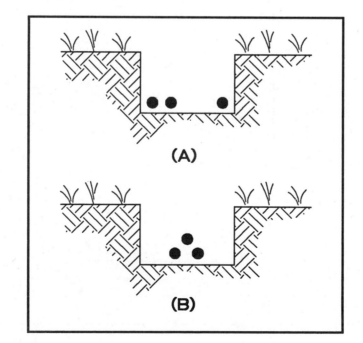

19. Is it always necessary to attach the bare equipment grounding conductor of a nonmetallic-sheathed cable to the green hexagon-shaped grounding screw on a receptacle? Explain.

_____

_____

_____

_____

20. List the methods by which an equipment grounding conductor is connected to a device box.

_____

_____

_____

_____

_____

21. When two nonmetallic-sheathed cables (Romex) enter a box, is it permitted to bring both of the bare equipment grounding conductors directly to the grounding terminal of a receptacle, using the terminal as a splice point? _____

22. This installation is in electrical metallic tubing. Because receptacle outlet "B" is only a few feet from switch "C," and switch "D" is only a few feet from receptacle outlet "A," the electrician saved a considerable amount of wire by picking up the feed for switch "C" from receptacle "B." The switch leg return is at switch "D." Please comment on this installation.

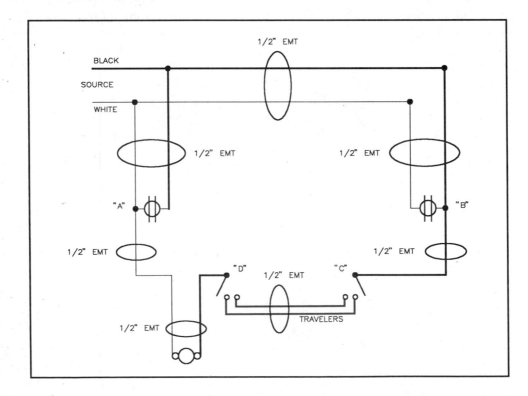

_____

_____

_____

23. Define an equipment grounding conductor. _____

_____

_____

_____

24. When metal toggle switch plates are used with nonmetallic boxes, the faceplate must be grounded according to *Section 380-12* of the *NEC*®. How is this accomplished?

_____

25. Does the Code permit the ampacity of switch legs to be less than the ampere rating of the branch-circuit? _____

26. Does the Code permit connecting receptacles to a 20-ampere branch-circuit, using short lengths of No. 14 AWG as the "pigtail" between the receptacle terminals and the No. 12 branch-circuit conductors? _____

UNIT 6

Ground-Fault Circuit Interrupters, Transient Voltage Surge Suppressors, Isolated Ground Receptacles, Immersion Detection Circuit Interrupters

OBJECTIVES

After studying this unit, the student will be able to

- understand the theory of operation of ground-fault circuit-interrupting (GFCI) devices.
- explain the operation and connection of GFCIs.
- discuss the do's and don'ts relative to the use of GFCI devices.
- explain why GFCIs are required.
- discuss locations where GFCIs must be installed in homes.
- understand the proper way to install and utilize feedthrough GFCI receptacles.
- understand testing, recording, and identification of GFCI receptacles.
- discuss the Code rules relating to the replacement of existing receptacles with GFCI receptacles.
- understand the Code requirements for GFCI protection on construction sites.
- understand the logic of the exemptions to GFCI mandatory requirements for receptacles in certain locations in kitchens, garages, and basements.
- discuss and understand the basics of transient voltage surge suppressors, and isolated ground receptacles.
- discuss and understand the principles of *immersion detection circuit interrupters*.

ELECTRICAL HAZARDS

There can be no compromise where human life is involved!

Many lives have been lost because of electrical shock from an appliance or a piece of equipment that is "hot." This means that the hot circuit conductor in the appliance is contacting the metal frame of the appliance. This condition may be due to the breakdown of the insulation because of wear and tear, defective construction, or accidental misuse of the equipment.

The shock hazard exists whenever the user can touch both the defective equipment and grounded surfaces, such as water pipes, metal sink rims, grounded metal lighting fixtures, earth, concrete in contact with the earth, water, or any other grounded surface.

Figure 6-1 shows a time-current curve that indicates *the amount of current* that a normal healthy adult can stand *for a certain time.* Just as in overcurrent protection for branch-circuits, motors, appliances, and so on, it is always a matter of *how much current is flowing and how long the current will flow.*

EFFECT OF ELECTRIC SHOCK

	Current in milliamperes @ 60 hertz	
	Men	Women
• cannot be felt	0.4	0.3
• a little tingling—mild sensation	1.1	0.7
• shock—not painful—can still let go	1.8	1.2
• shock—painful—can still let go	9.0	6.0
• shock—painful—just about to point where you can't let go—called "threshold"—you may be thrown clear	16.0	10.5
• shock—painful—severe—can't let go—muscles immobilize—breathing stops	23.0	15.0
• ventricular fibrillation (usually fatal) length of time . . . 0.03 sec. length of time . . . 3.0 sec.	1000 100	1000 100
• heart stops for the time current is flowing—heart may start again if time of current flow is short	4 amperes	4 amperes
• burning of skin—generally not fatal unless heart or other vital organs burned	5 or more amperes	5 or more amperes

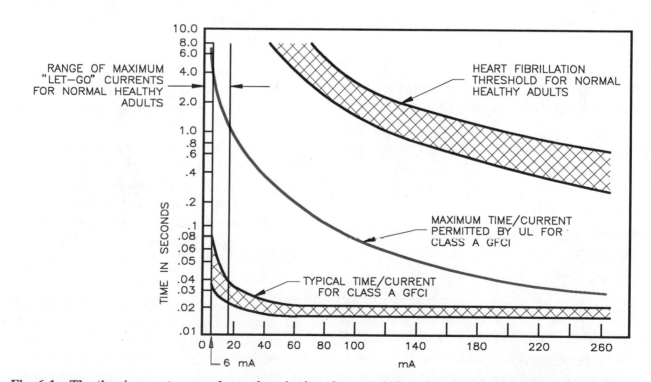

Fig. 6-1   The time/current curve shows the tripping characteristics of typical Class A GFCI. Note that if you follow the 6-mA line vertically to the crosshatched typical time/current curve, you will find that the GFCI will open in from approximately 0.035 second to just less than 0.1 second. One electrical cycle is ¹⁄₆₀ of a second (0.0167 second). An air bag in an automobile inflates in approximately ¹⁄₂₀ of a second.

Experts say that the resistance of the human body varies from a few hundred ohms to many thousand ohms, but generally agree on values of 800 to 1000 ohms as typical. Body internal resistance, body contact resistance, moisture, duration, voltage, as well as the path of current: finger to finger, hand to hand, hand holding "hot" wire, hand holding a pair of pliers, hand holding an electric drill, hand grasping a grounded metal pipe, two hands grasping a grounded metal pipe, hand in water, standing in water, hand to foot, foot to foot, etc. all have an effect on the severity of the shock.

Generally, voltages of less than 40 volts to ground are considered safe. Voltages 50 volts and greater are considered lethal. The *NEC®* in *Section 250-5(b)* requires that (with a few exceptions) all AC systems of 50 volts to 1,000 volts that supply premise wiring be grounded.

Some circuits of less than 50 volts must be grounded according to *Section 250-5(a)*:

- if the circuit is supplied by a transformer where the primary exceeds 150 volts to ground, or
- if the circuit is supplied by a transformer where the primary supply is ungrounded, or
- if the circuit is overhead wiring outside of a building.

CODE REQUIREMENTS FOR GROUND-FAULT CIRCUIT INTERRUPTERS *(SECTION 210-8)*

To protect against electric shocks, *Section 210-8(a)* of the *National Electrical Code®* requires that personnel ground-fault protection be provided for *all* 125-volt, 15- and 20-ampere receptacle outlets in dwellings as follows:

- in bathrooms. See figure 10-7 for definition of a bathroom.
- outdoors. ◄
- in kitchens, for all receptacles that serve the countertop surfaces. This includes receptacles installed on islands and peninsulas, figure 6-2. ◄

  GFCI protection is *not* required for:
  – a receptacle installed solely for a clock.
  – a receptacle installed inside of an upper kitchen cabinet for plugging in a microwave oven that is fastened to the underside of the cabinet.

  – a receptacle installed below the kitchen sink for plugging in a food-waste disposer or dishwasher.

- within 6 feet (1.83 m) from the edge of a wet-bar sink, figure 6-2. ◄

- in attached or detached garages.
  GFCI protection is *not* required for:
  – receptacles that are not readily accessible, such as a receptacle installed on the ceiling for an overhead garage door opener.
  – a single receptacle, or a duplex receptacle is installed for cord- and plug-connected appliances in a dedicated space where the appliance is not easily moved from one location to another, such as a freezer or a refrigerator. ◄

- in other buildings (sheds, workshops, or storage buildings).
  GFCI protection is *not* required for:
  – receptacles that are not readily accessible.
  – a single receptacle, or a duplex receptacle is installed for cord- and plug-connected appliances in a dedicated space where the appliance is not easily moved from one location to another. ◄ Examples are a freezer or a refrigerator.

- in crawl spaces that are at or below grade level, and in unfinished basements. GFCI protection is not required if the receptacle is *not* readily accessible.
  GFCI protection is *not* required for:
  – a single receptacle, or a duplex receptacle installed for cord- and plug-connected appliances in a dedicated space where the appliance is not easily moved from one location to another. ◄ Examples are a sump pump, a gas-fired electric controlled water heater, or a water softener.

Some individuals try to get around the requirement for GFCI personnel protection in basements by calling the basement "finished," even though it is clearly evident that it is not "finished." If the basement is intended to be "finished," then the Code requirements for the spacing of receptacles, required lighting outlets, and switch controls would have to be conformed to. This is covered in unit 3.

- be careful when installing porcelain pull-chain or keyless lampholders. If they have a receptacle, the receptacle outlet must be GFCI protected.

- residential *underground* circuits rated 120 volts or less, 20 amperes or less, and buried no less than 12 inches below grade level. Refer to *Table 300-5* of the *NEC®* An example of this would be the Type UF cable supplying the post light in front of the residence. Type UF cable would have to be buried no less than 24 inches if the circuit is not GFCI protected. If the underground circuit is installed in rigid or intermediate metal conduit, a depth of not less than 6 inches is required.

- for swimming pools, see unit 30.

The Code requirements for ground-fault circuit protection can be met in many ways. Figure 6-3 illustrates a GFCI circuit breaker installed on a

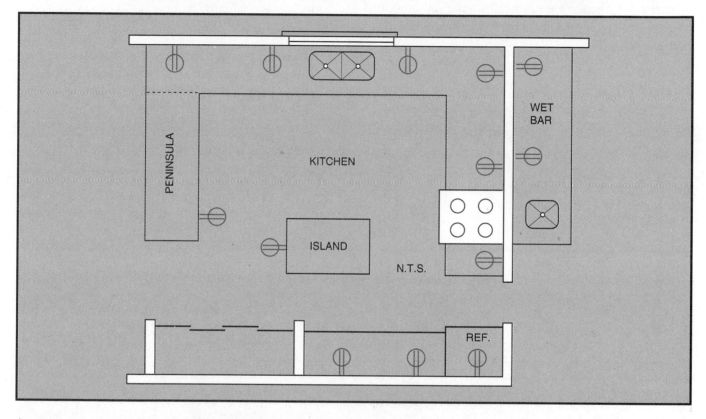

► Fig. 6-2   In kitchens, *all* 125-volt, single-phase, 15- and 20-ampere receptacles that serve countertop surfaces (walls, islands, or peninsulas) must be GFCI protected. The key words are "that serve countertop surfaces." GFCI personnel protection is *not* required for the refrigerator receptacle, a receptacle under the sink for plugging-in a food waste disposer, or a receptacle installed inside cabinets above the countertops for plugging-in a microwave oven attached to the underside of the cabinets. Where wet-bar sinks are installed outside of the kitchen area, *all* 125-volt, single-phase, 15- and 20-ampere receptacles that serve the countertops must be GFCI protected if they are within 6 feet (1.83 m) measured from the outer edge of the wet-bar sink. See *Section 210-8(a)(5)*. ◄ Drawing is not to scale.

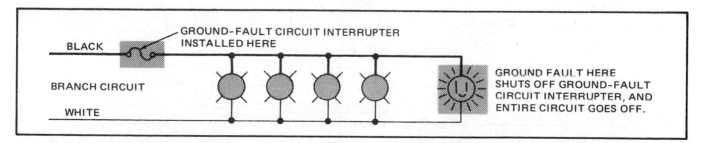

Fig. 6-3   **Ground-fault circuit interrupter as a part of the branch-circuit overcurrent device.**

specified circuit on Panel B. A ground fault in the range of 4 to 6 milliamperes or more shuts off the entire circuit. For example, a ground fault at any point on circuit B14 will shut off the garage lighting, garage receptacles, rear garage door outside bracket fixture, post light, and overhead door opener.

When a GFCI receptacle is installed, then only that receptacle is shut off when a ground fault greater than 6 milliamperes occurs, figure 6-4.

Figure 6-5 shows the effect of a GFCI installed as part of a feeder supplying 15- and 20-ampere receptacle branch-circuits. This is rarely used in residential wiring.

What a GFCI Does!

A GFCI monitors the current balance between the ungrounded "hot" conductor and the grounded "neutral" conductor. As soon as the current flowing through the "hot" conductor is in the range of 4 to 6 milliamperes or more than the current flowing in the

"return" grounded conductor, the GFCI senses this unbalance and trips (opens) the circuit off. The unbalance indicates that part of the current flowing in the circuit is being diverted to some path other than the normal return path along the grounded return conductor. If the "other" path is through a human body, as illustrated in figure 6-6, the outcome could be fatal.

- Class "A" GFCI devices are designed to trip when a ground fault in the range of 4 to 6 milliamperes ($4/1000$ to $6/1000$ of an ampere) or more occurs.

- Class "B" GFCI devices are designed to trip when a ground fault of 20 milliamperes ($20/1000$ of an ampere) or more occurs. According to the NEMA Standard No. 280, these would be used only for swimming pool underwater lighting installed before the adoption of the 1965 edition of the *National Electrical Code.*® A Class "A" GFCI would nuisance trip.

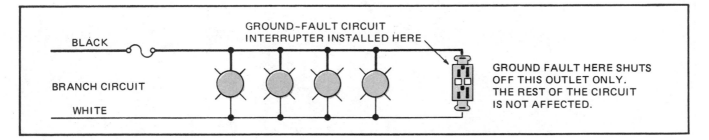

Fig. 6-4   Ground-fault circuit interrupter as an integral part of a receptacle outlet.

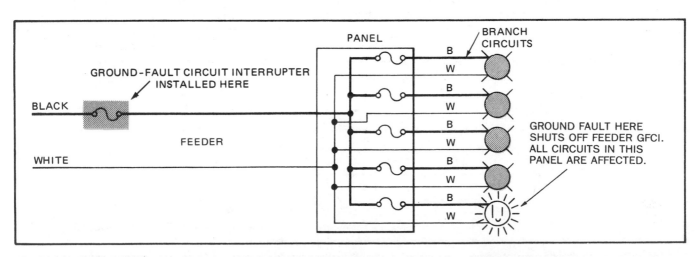

Fig. 6-5   Ground-fault circuit interrupter as part of the feeder supplying a panel that serves a number of 15- and 20-ampere branch-circuits. This is permitted by *Section 215-9* of the Code, but is rarely—if ever—used in residential wiring.

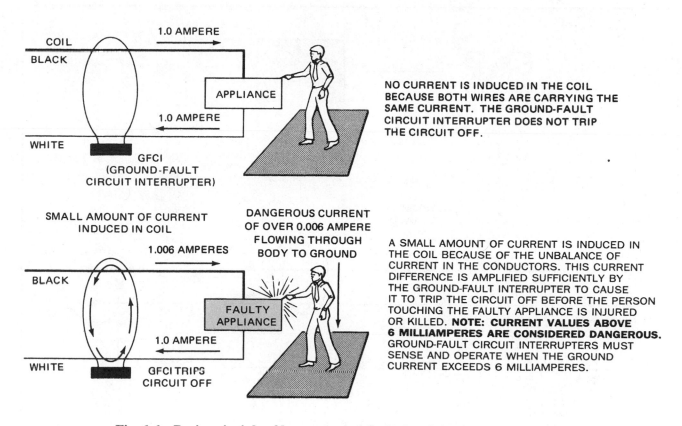

Fig. 6-6   Basic principle of how a ground-fault circuit interrupter operates.

What a GFCI Does Not Do!

- It *does not* protect against electrical shock when a person touches both circuit conductors at the same time (two "hot" wires, or one "hot" wire and one grounded neutral wire) because the current flowing in both conductors is the same. Thus, there is no unbalance of current for the GFCI to sense and trip.

- It *does not* limit the magnitude of ground-fault current. It does limit the time that a ground fault will be allowed to flow. See figure 6-1.

- It *does not* sense solid short-circuits between the "hot" conductor and the grounded "neutral" conductor. The branch-circuit fuse or circuit breaker provides this protection.

- It *does not* sense solid short-circuits between two "hot" conductors. The branch-circuit fuse or circuit breaker provides this protection.

- It *does not* sense "arcing" faults between the "hot" conductor and the grounded "neutral" conductor. The branch-circuit fuse or circuit breaker provides this protection. An example of an arcing fault is common with frayed exten-

sion cords. The heat generated by the arc can easily result in a fire, yet the amount of current flowing can be much less than the rating of the branch-circuit fuse or circuit breaker.

Figure 6-6 is a pictorial view of how a ground-fault circuit interrupter operates. Figure 6-7 shows the actual internal wiring of a GFCI. If the GFCI trips off, it is an indication of a possible shock hazard from a line-to-ground fault.

Nuisance tripping of GFCIs is known to occur. Sometimes this can be attributed to extremely long runs of cable for the protected circuit. Consult the manufacturer's instructions to determine if maximum lengths for a protected circuit are recommended. Some electricians advise the use of nonmetallic staples (see figure 4-13C) or nonmetallic straps instead of metallic as a means of preventing nuisance tripping of GFCIs.

Do not reverse the line and load connections on a feedthrough GFCI. This would result in the receptacle still being "live" even though the GFCI mechanism has tripped.

A GFCI receptacle *does not* provide overload protection for the circuit conductor. It provides *ground-fault protection only*.

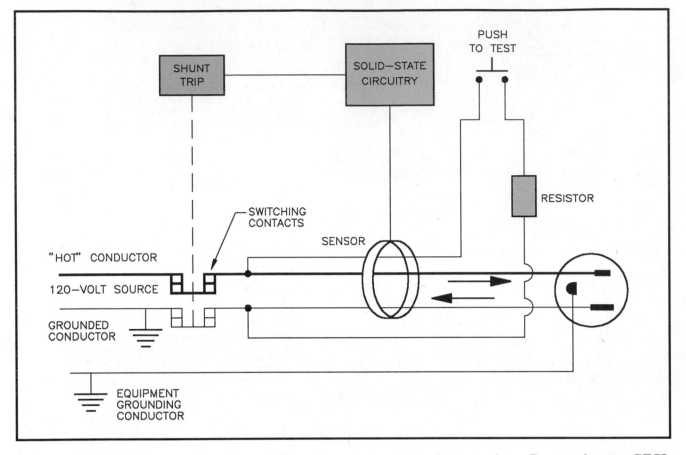

Fig. 6-7   Ground-fault circuit interrupter internal components and connections. Receptacle-type GFCIs switch both the phase (hot) and grounded conductors. Note that when the test button is pushed, the test current passes through the test button, the sensor, then back around (bypasses, outside of) the sensor, then back to the opposite circuit conductor. This is how the "unbalance" is created, then monitored by the solid-state circuitry to signal the GFCI's contacts to open. Note that because both normal "load" currents pass through the sensor, no unbalance is present.

GROUND-FAULT CIRCUIT INTERRUPTER IN RESIDENCE CIRCUITS

▶ In this residence, GFCI personnel protection must be provided for receptacle outlets installed outdoors, in the bathrooms, in the garage, in the workshop, specific receptacles in the basement area, in the kitchen, and adjacent to the wet-bar in the recreation room. ◀ The GFCI protection can be provided by GFCI circuit-breakers or by GFCI receptacles of the type shown in figure 6-8. This is a design issue that confronts the electrician. The electrician must decide how to provide the GFCI personnel protection required by *Section 210-8* of the *National Electrical Code®.* These receptacle outlets are connected to the circuits listed in Table 6-1.

Swimming pools also have special requirements for GFCI protection. These requirements are covered in unit 30.

GFCIs operate properly only on grounded electrical systems, as is the case in all residential, condo,

CIRCUITS	GFCI RECEPTACLE LOCATION
A19	Outdoor receptacle on rear of residence outside of Master Bedroom
A15	Front porch receptacle
A16	Outdoor receptacle on rear of residence outside of Front Bedroom
A18	Workbench receptacles
A20	Workshop receptacles on window wall
A22	Master Bath receptacle
A23	Bath (off hall) receptacle
B9	Wet-bar (Recreation Room) receptacles
B13	Kitchen receptacles
B14	Garage receptacles
B15	Kitchen receptacles
B16	Kitchen receptacles
B17	Outdoor receptacle outside of Living Room next to sliding door
B20	Outdoor receptacle outside of laundry
B21	Powder Room receptacle

Table 6-1   Circuit number and general location of GFCI receptacles in this residence.

apartment, commercial, and industrial wiring. The GFCI will operate on a two-wire circuit even though an equipment grounding conductor is not included with the circuit conductor. In this residence, equipment grounding conductors are in the cables. In the case of metal conduit, the equipment grounding conductor is the metal conduit.

Never ground a system neutral conductor except at the service equipment. A GFCI would be inoperative.

Long branch-circuit runs can cause nuisance tripping of the GFCI due to leakage currents in the circuit wiring. GFCI receptacles at the point of utilization tend to minimize this problem. A circuit supplied by a GFCI circuit breaker in the main panel could cause nuisance tripping if the branch circuit is long—some electricians say 50 feet or more can cause nuisance tripping of the GFCI breaker.

Never connect the neutral of one circuit to the neutral of another circuit.

When a GFCI feeds an isolation transformer (separate primary winding and separate secondary winding), as might be used for swimming pool underwater fixtures, the GFCI *will not* detect any ground faults on the secondary of the transformer.

Ground-fault circuit interrupters may be installed on other circuits, in other locations, and even when rewiring existing installations where the Code does not specifically call for GFCI protection. See figure 6-9.

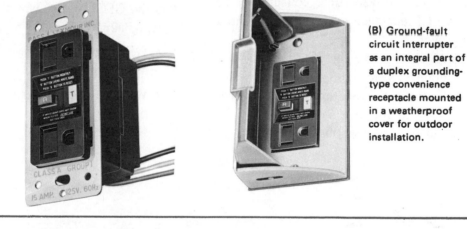

(A) Ground-fault circuit interrupter as an integral part of a duplex grounding-type convenience receptacle.

(B) Ground-fault circuit interrupter as an integral part of a duplex grounding-type convenience receptacle mounted in a weatherproof cover for outdoor installation.

Fig. 6-8   Ground-fault circuit interrupters (GFCIs).

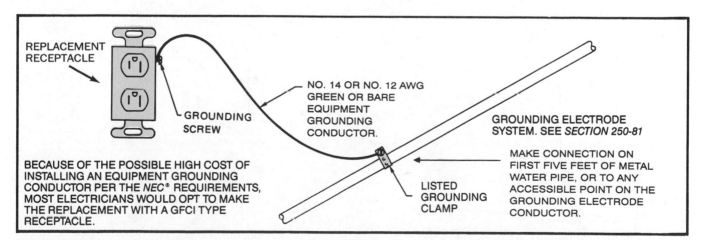

REPLACEMENT RECEPTACLE

GROUNDING SCREW

NO. 14 OR NO. 12 AWG GREEN OR BARE EQUIPMENT GROUNDING CONDUCTOR.

LISTED GROUNDING CLAMP

GROUNDING ELECTRODE SYSTEM. SEE *SECTION 250-81*

MAKE CONNECTION ON FIRST FIVE FEET OF METAL WATER PIPE, OR TO ANY ACCESSIBLE POINT ON THE GROUNDING ELECTRODE CONDUCTOR.

BECAUSE OF THE POSSIBLE HIGH COST OF INSTALLING AN EQUIPMENT GROUNDING CONDUCTOR PER THE *NEC* REQUIREMENTS, MOST ELECTRICIANS WOULD OPT TO MAKE THE REPLACEMENT WITH A GFCI TYPE RECEPTACLE.

▶ **Fig. 6-9   In existing installations that do not have an equipment grounding conductor as part of the branch-circuit wiring, *Section 210-7(d)* permits replacing an old-style nongrounding type receptacle with a grounding type receptacle. The grounding terminal on the receptacle must be properly grounded. One of the acceptable ways to do this is to run an equipment grounding conductor to any accessible point on the grounding electrode system. Another way is to run an equipment grounding conductor to any accessible point on the grounding electrode conductor. See *Section 250-50(b), Exception*. ◀**

FEEDTHROUGH GROUND-FAULT CIRCUIT INTERRUPTER

The decision to use more GFCIs rather than trying to protect many receptacles through one GFCI becomes one of economy and practicality. GFCI receptacles are more expensive than regular receptacles. Here is where knowledge of material and labor costs comes into play. The decision must be made separately for each installation, keeping in mind that GFCI protection is a safety issue recognized and clearly stated in the Code. However, the actual circuit layout is left up to the electrician.

Figure 6-10 illustrates how a feedthrough GFCI receptacle supplies many other receptacles. Should a ground fault occur anywhere on this circuit, all 11 receptacles lose power—not a good circuit layout. Attempting to locate the ground-fault problem, unless obvious, can be very time consuming. The use of more GFCI receptacles is generally the more practical approach.

Figure 6-11 shows how the feedthrough GFCI receptacle is connected into the circuit. When connected in this manner, the feedthrough GFCI receptacle and all downstream outlets on the same circuit will have ground-fault protection, figure 6-12.

The most important factors to be considered are the continuity of electrical power and the economy of the installation. The decision must be made separately for each installation.

TESTING AND RECORDING OF TEST DATA FOR GFCI RECEPTACLES

Refer to figure 6-8 and note that these GFCI receptacles have "T" (Test) and "R" (Reset) buttons. Pushing the test button places a small ground fault on the circuit. If operating properly, the GFCI receptacle should trip to the OFF position. The operation is the same for the GFCI circuit breakers. Pushing the reset button will restore power.

Detailed information and testing instructions are furnished with GFCI receptacles and GFCI circuit breakers. Monthly testing is recommended to insure that the GFCI mechanism will operate properly if a human being is subjected to an electrical shock.

- Will the homeowner remember to do the monthly testing?
- Will the homeowner recognize and understand the purpose and function of the GFCI receptacle?

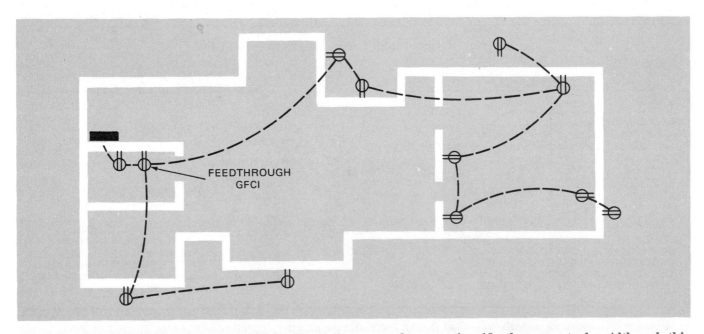

FEEDTHROUGH GFCI

Fig. 6-10   Illustration showing one GFCI feedthrough receptacle protecting 10 other receptacles. Although this might be an economical way to protect many receptacle outlets with only one GFCI feedthrough receptacle, there is great likelihood that the GFCI will nuisance trip because of leakage currents. Use common sense and manufacturer's recommendations as to how many devices should be protected from one GFCI.

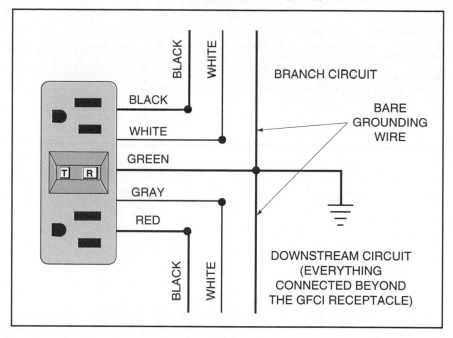

Fig. 6-11   Connecting feedthrough ground-fault circuit interrupter into circuit.

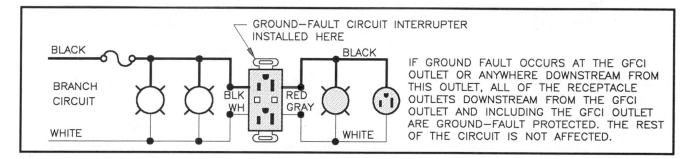

Fig. 6-12   Feedthrough ground-fault circuit interrupter as an integral part of the convenience receptacle outlet.

- Should all of the basement GFCI receptacles be identified?

- Will the homeowner recognize a receptacle if protected by a GFCI circuit breaker located in an electrical panel far from the actual receptacle?

These are important questions and a possible reason why Underwriters Laboratories requires detailed installation and testing instructions to be included in the packaging for the GFCI receptacle or circuit breaker. Instructions are not only for the electrician but also for homeowners, and must be left in a conspicuous place so homeowners can familiarize themselves with the receptacle, its operation, and its need for testing. Figure 6-13 shows a chart homeowners can use to record monthly GFCI testing.

REPLACING EXISTING RECEPTACLES

The *National Electrical Code*® is very specific on the type of receptacle permitted to be used as a replacement for an existing receptacle. These Code rules are found in *Sections 210-7(d)* and *250-50*.

Section 210-7(d)(2) is a retroactive ruling requiring that when an existing receptacle is replaced in *any* location where the present code requires GFCI protection (bathrooms, kitchens, outdoors, unfinished basements, garages, etc.), the replacement receptacle must be GFCI protected. This could be a GFCI-type receptacle, or the branch-circuit could be protected by a GFCI-type circuit breaker.

It is important to note that on existing wiring systems, such as "knob and tube," or older non-metallic-sheathed cable wiring that did not have an

OCCUPANT'S TEST RECORD

TO TEST, depress the "TEST" button, the "RESET" button should extend. Should the "RESET" button not extend, the GFCI will not protect against electrical shock. Call qualified electrician.

TO RESET, depress the "RESET" button firmly into the GFCI unit until an audible click is heard. If reset properly, the RESET button will be flush with the surface of the test button.

This label should be retained and placed in a conspicuous location to remind the occupants that for maximum protection against electrical shock, each GFCI should be tested monthly.

YEAR	JAN	FEB	MAR	APR	MAY	JUN	JUL	AUG	SEP	OCT	NOV	DEC

Fig. 6-13   Homeowner's testing chart for recording GFCI testing dates.

equipment grounding conductor, that a GFCI receptacle will still properly function. It does not need an equipment grounding conductor. Thus, protection against lethal shock is still provided. See figure 6-7.

Replacing Existing Two-Wire Receptacles Where Grounding Means Does Exist
(Refer to figure 6-14)

When replacing an existing receptacle (A) where the wall box is properly grounded (E) or where the branch-circuit wiring contains an equip-

ment grounding conductor (D), at least two easy choices are possible for the replacement receptacle:

1. The replacement receptacle must be of the grounding type (B) unless . . .

2. The replacement receptacle is of the GFCI type (C).

Important: ▶ When replacing a receptacle in *any* location where the Code requires GFCI protection in a new installation, the replacement receptacle must be of the GFCI type, or GFCI protected by a GFCI circuit breaker protecting that particular circuit, *Section 210-7(d)(2).* ◀

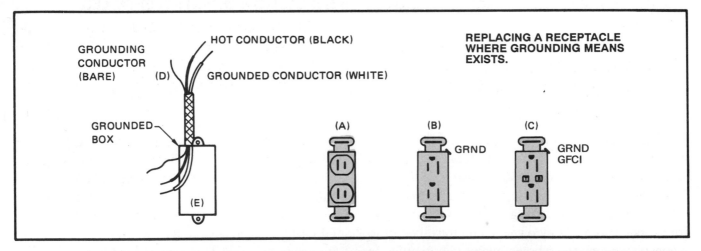

Fig. 6-14   Where box (E) is properly grounded by any of the methods specified in *Section 250-91(b)*, an existing receptacle of the type shown in (A) *must* be replaced with a grounding-type receptacle (B). It is also OK to replace (A) with a GFCI receptacle (C). See text for further discussion. Also see *Section 210-7(d)*.

▶ If the enclosure is not grounded, the *Exception to Section 250-50(b)* permits an equipment grounding conductor to be run from the enclosure to be grounded, or from the green hexagon grounding screw of a grounding-type receptacle to any accessible point on the grounding electrode system or to any accessible point on the grounding electrode conductor. Note however, that this permission is only for existing installations. See figure 6-9. ◀

Be sure that the equipment grounding conductor of the circuit is connected to the receptacle's green hexagon-shaped grounding terminal.

Do not connect the white grounded circuit conductor to the green hexagon-shaped grounding terminal of the receptacle.

Do not connect the white grounded circuit conductor to the wall box.

Replacing Existing Two-Wire Receptacles Where Grounding Means Does Not Exist
(Refer to figures 6-15 and 6-15A)

When replacing an existing nongrounding-type receptacle (A), where the box is not grounded (E), or where an equipment grounding conductor has not been run with the circuit conductors (D), four choices are possible for selecting the replacement receptacle:

1. The replacement receptacle may be a non-grounding type (A).
2. The replacement receptacle may be a GFCI type (C).

▶ • The GFCI replacement receptacle must be marked "*No Equipment Ground.*" ◀

 • The green hexagon grounding terminal of the GFCI replacement receptacle does not have to be connected to any grounding means. It can be left "unconnected." The GFCI's trip mechanism will operate properly when ground faults occur anywhere on the load side of the GFCI replacement receptacle. Ground-fault protection is still there. Refer to figure 6-7.

 • Do not connect an equipment grounding conductor from the green hexagon grounding terminal of a replacement GFCI receptacle (C) to any other downstream receptacles that are fed through the replacement GFCI receptacle.

The reason this is not permitted is that if at a later date someone saw the conductor connected to the green hexagon grounding terminal of the downstream receptacle, there would be an immediate

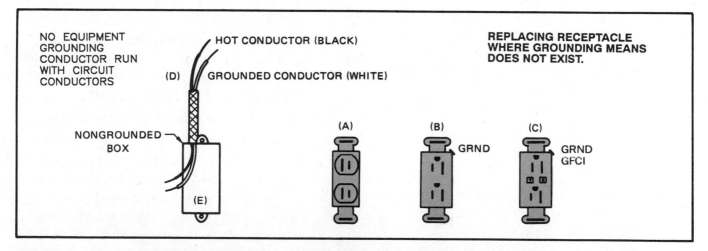

Fig. 6-15 Where box (E) is *not* grounded, or where an equipment grounding conductor has *not* been run with the circuit conductors (D), an existing nongrounding-type receptacle (A) may be replaced with a nongrounding-type receptacle (A), a GFCI-type receptacle (C), a grounding-type receptacle (B) if supplied through a GFCI receptacle (C), or a grounding-type receptacle (B) if a separate equipment grounding conductor is run from the receptacle to any accessible point on the grounding electrode system, *Section 250-50(a)* and *(b)*, *Exception.* Most electricians would replace an existing nongrounding-type receptacle with a GFCI type rather than install an equipment grounding conductor, which could be very labor intensive. See *Section 210-7(d)*.

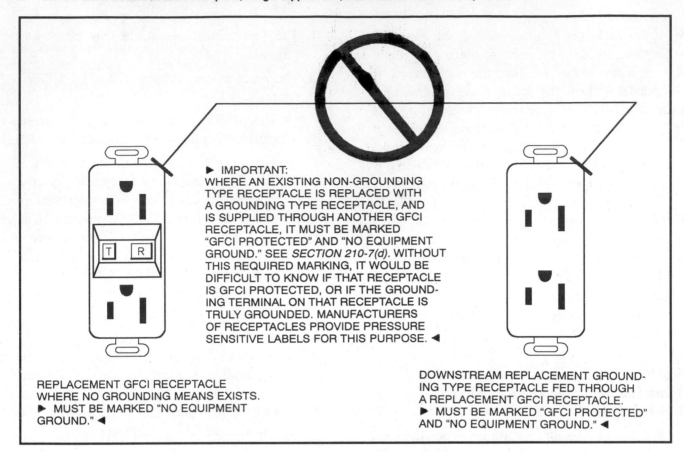

▶ IMPORTANT:
WHERE AN EXISTING NON-GROUNDING
TYPE RECEPTACLE IS REPLACED WITH
A GROUNDING TYPE RECEPTACLE, AND
IS SUPPLIED THROUGH ANOTHER GFCI
RECEPTACLE, IT MUST BE MARKED
"GFCI PROTECTED" AND "NO EQUIPMENT
GROUND." SEE *SECTION 210-7(d).* WITHOUT
THIS REQUIRED MARKING, IT WOULD BE
DIFFICULT TO KNOW IF THAT RECEPTACLE
IS GFCI PROTECTED, OR IF THE GROUND-
ING TERMINAL ON THAT RECEPTACLE IS
TRULY GROUNDED. MANUFACTURERS
OF RECEPTACLES PROVIDE PRESSURE
SENSITIVE LABELS FOR THIS PURPOSE. ◀

REPLACEMENT GFCI RECEPTACLE
WHERE NO GROUNDING MEANS EXISTS.
▶ MUST BE MARKED "NO EQUIPMENT
GROUND." ◀

DOWNSTREAM REPLACEMENT GROUND-
ING TYPE RECEPTACLE FED THROUGH
A REPLACEMENT GFCI RECEPTACLE.
▶ MUST BE MARKED "GFCI PROTECTED"
AND "NO EQUIPMENT GROUND." ◀

▶ **Fig. 6-15A**   **Where a grounding means *does not* exist, there are certain marking requirements as presented in this diagram. *Do not* connect an equipment grounding conductor between these receptacles.** ◀

assumption that the other end of that conductor had been properly connected to an acceptable grounding point of the electrical system. The fact is that the so-called equipment grounding conductor had been connected to the replacement GFCI receptacle's green hexagon grounding terminal that was not grounded in the first place. We now have a false sense of security, a real shock hazard.

3. The replacement receptacle may be a grounding type (B) if it is supplied through a GFCI-type receptacle (C).

   ▶ • The replacement receptacle must be marked "*GFCI Protected*" and "*No Equipment Ground.*" ◀

   • The green hexagon equipment grounding terminal of the replacement grounding-type receptacle (B) need not be connected to any grounding means. It may be left "unconnected." The upstream feedthrough GFCI receptacle (C) trip mechanism will work properly when ground faults occur

anywhere on its load-side. Ground-fault protection is still there. Refer to figure 6-7.

   ▶ • Do not connect an equipment grounding conductor from the green hexagon grounding terminal of a replacement GFCI receptacle (C) to any other downstream receptacles that are fed through the replacement GFCI receptacle. ◀

4. The replacement receptacle may be a grounding type (B) if:

   • an equipment grounding conductor, sized per *Table 250-95*, is run from the replacement receptacles green hexagon grounding screw and properly connected to a listed ground clamp at some point on the first 5 feet of the metal water piping as it enters the building. You are *not* permitted to make the equipment ground connection to the interior metal water piping anywhere beyond the first 5 feet. See the *Exception* and *FPN* to *Section 250-50*. Also read *Section 250-81*.

PERSONNEL GROUND-FAULT PROTECTION FOR ALL TEMPORARY WIRING

Because of the nature of construction sites, there is a continual presence of shock hazard that can lead to serious personal injury or death through electrocution. Workers are standing in water, standing on damp or wet ground, or in contact with steel framing members. Electric cords and cables are lying on the ground, subject to severe mechanical abuse. All of these conditions spell "DANGER."

Therefore, *Section 305-6(a)* of the Code requires that all 125-volt, single-phase, 15- and 20-ampere receptacle outlets that are not part of the permanent wiring of the building and that will be used by the workers on the construction site must be GFCI protected. ▶ Receptacle outlets that are part of the actual permanent wiring of a building and are used by personnel for temporary power are also required to be GFCI protected. ◀

▶ For receptacles *other* than the 125-volt, single-phase, 15- and 20-ampere receptacles, *Section 305-6(b)* permits a written *Assured Equipment Grounding Conductor Program* that must be continuously enforced on the construction site. ◀ A designated person must keep a written log, insuring that all electrical equipment is properly installed and maintained according to the applicable requirements of *Sections 210-7(c), 250-45, 250-59*, and *305-4(d)*. This option has proved to be difficult to enforce. Thus, most "authorities having jurisdiction" require GFCI protection as stated in *Section 305-6(a)*.

IMMERSION DETECTION CIRCUIT INTERRUPTERS

The danger of electrical shock is always present when using electrical appliances. *Ground Fault Circuit Interrupters* (GFCIs) are required in specific locations in new homes. But something had to be done to protect users from this shock hazard in older homes that did not require GFCIs at the time they were built. The major culprit was found to be personal grooming appliances.

To address the problem, a new type of circuit interrupter was developed. This device is called an *Immersion Detection Circuit Interrupter* (IDCI). It can be an integral part of the attachment plug cap found on personal grooming appliances that are commonly used near water and basins, such as hair dryers, heated stylers, heated air combs, heated air curlers, curling iron–hair dryer combinations, and so on.

These electrical appliances are most commonly used at a vanity having a wash basin with water in it, or even worse, next to a water-filled bathtub with a person (child or adult) in it. Accidentally dropping or pulling the appliance into the tub when reaching for a towel can be lethal.

Electricity and water do not mix! The shock hazard protection of these appliances is permitted to be in the form of a GFCI device or in the form of an immersion detection device. If any of these appliances fall into a basin or tub that is filled with water, the IDCI will sense the leakage current and will trip to the OFF position. Both GFCI and IDCI devices disconnect both conductors of the branch-circuit. These devices are also required to open the circuit regardless of whether the appliance switch is in the ON or OFF position.

The operation of a GFCI device has been discussed previously. The operation of an IDCI depends upon a "third" wire in the cord, referred to as a *probe*, which is connected to a sensor within the appliance. It is this probe that will detect leakage current flow when conductive liquid enters the appliance and makes contact with any live part inside the appliance and the internal "sensor." When this happens, the IDCI is instructed to trip to the OFF position.

It is not permitted to replace the special GFCI or IDCI attachment plug cap on these appliances because that would make the appliance absolutely unsafe. The UL requirement is that the plug cap shall be marked: "WARNING: TO REDUCE THE RISK OF ELECTRIC SHOCK, DO NOT REMOVE, MODIFY, OR IMMERSE THIS PLUG."

An IDCI must open the circuit when the sensor detects a leakage current to ground in the range of 4 to 6 milliamperes. An IDCI on a hair dryer or similar appliance must be manually reset. *Section 422-24* of the *National Electrical Code®* requires protection against electrocution for freestanding hydromassage appliances and hand-held hair dryers. This is intended to include heated stylers, heated air combs, heated air curlers, and curling iron–hair dryer combinations.

An Overview

Because most older homes do not have GFCI protection, codes and standards are slowly but surely requiring the manufacturers of appliances to provide the protection as an integral part of certain appliances. In addition to GFCIs and IDCIs, other devices are being used to protect human life. *Appliance Leakage Circuit Interrupters* (ALCI) trip at 6 milliamperes, *Equipment Leakage Circuit Interrupters* (ELCI) trip at 50 milliamperes, and *Leakage Monitoring Receptacles* (LMR) trip at 50 milliamperes. As time goes on, we will see more and more of these types of "people protectors."

TRANSIENT VOLTAGE SURGE SUPPRESSION (TVSS)

In today's homes, we find many electronic appliances (television, stereo, personal computers, facsimile equipment, word processors, video cassette recorders, digital stereo equipment, microwave ovens), all of which contain many sensitive, delicate electronic components (printed circuit boards, chips, microprocessors, transistors, etc.).

Voltage transients, called *surges* or *spikes*, can stress, degrade, and/or destroy these components; they can cause loss of memory in the equipment or "lock up" the microprocessor.

The increased complexity of electronic integrated circuits makes this equipment an easy target for "dirty" power that can and will affect the performance of the equipment.

Voltage transients cause abnormal current to flow through the sensitive electronic components. This energy is measured in joules. A *joule* is the unit of energy when one ampere passes through a one-ohm resistance for one second (like wattage, only on a much smaller scale).

Line surges can be line-to-neutral, line-to-ground, and line-to-line.

Transients

Transients are generally grouped into two categories:

- Ring wave—These transients originate within the building and can be caused by copiers; computers; printers; HVAC cycling; spark igniters on furnaces, water heaters, dryers, gas ranges, and ovens; motors; and other inductive loads.

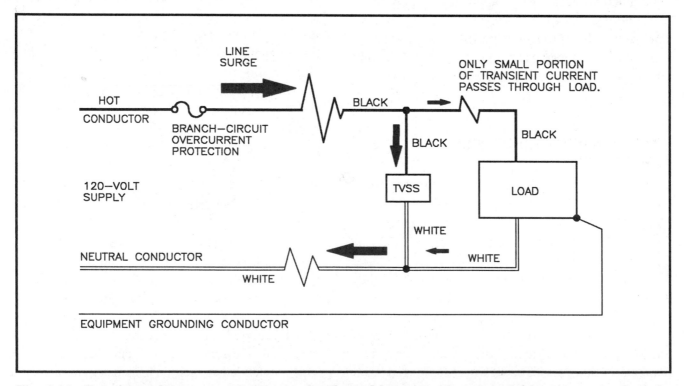

Fig. 6-16   Transient voltage surge suppressor absorbs and bypasses (shunts) transient currents around the load. The M.O.V. dissipates the surge in the form of heat.

- Impulse—These transients originate outside of the building and are caused by utility company switching, lightning, and so on.

To minimize the damaging results of these transient line surges, service equipment, panels, load centers, feeders, branch-circuits, and individual receptacle outlets can be protected with a device called a *transient voltage surge suppressor* (TVSS), figure 6-16.

A TVSS contains one or more metal-oxide varistors (M.O.V.) that clamp the transients by absorbing the major portion of the energy (joules) created by the surge, allowing only a small, safe amount of energy to enter the actual connected load.

The M.O.V. clamps thc transient in times of less than 1 nanosecond, which is one billionth of a second, and keeps the voltage spike passed through to the connected load to a maximum range of from 400 to 500 peak volts.

Typical TVSS devices for homes are available as an integral part of receptacle outlets that mount in the same wall boxes as do regular receptacles. They look the same as a normal receptacle, and may have an audible alarm that sounds when a M.O.V. has failed. The TVSS device may also have a visual indication, such as an LED (light-emitting diode) that glows continuously until an M.O.V. fails. Then the LED starts to flash on and off. See figure 6-17.

TVSS devices are also available in plug-in strips, figure 6-18.

One TVSS on a branch-circuit will provide surge suppression for other receptacles nearby that are on the same circuit.

Noise

Low-level nondamaging transients can also be present. These can be caused by fluorescent lamps and ballasts, other electronic equipment in the area, x-ray equipment, motors running or being switched on and off, improper grounding, and so forth. Although not physically damaging to the electronic equipment, computers can lose memory or perform wrong calculations. Their intended programming can malfunction.

"Noise" comes from *electromagnetic interference* (EMI) and *radio frequency interference* (RFI). EMI is usually caused by ground currents of very low values from motors, utility switching loads, lightning, and so forth, and are transmitted through metal conduits. RFI is noise, like the buzzing heard on a car radio when driving under a high-voltage transmission line. This interference "radiates" through the air from the source and is picked up by the grounding system of the building.

Fig. 6-17   Surge protector. *Courtesy* of Leviton.

Fig. 6-18   Surge suppressor. *Courtesy* of Leviton.

This undesirable noise can be reduced by reducing the number of ground reference points on a system. This can be accomplished by installing an *isolated ground receptacle*.

ISOLATED GROUND RECEPTACLE

In a standard conventional receptacle outlet, figure 6-19, the green grounding hexagon screw, the grounding contacts, the yoke (strap), and the metal wall box are all "tied" together to the building's equipment grounding system. Picture then the many receptacles in a building, all having and creating the multiple-ground situation.

In an isolated ground receptacle, the green grounding hexagon screw and the grounding contacts of the receptacle are isolated from the metal yoke (strap) of the receptacle and also from the building's equipment grounding system. A separate green insulated grounding conductor is then installed from the green hexagon screw on the receptacle all the way back to the main service disconnect. This separate green grounding conductor does *not* connect to any panels, load centers, or other ground reference points in between. The system grounding is now "clean," and the result is less transient noise (disturbances) transmitted to the connected load.

Isolated ground receptacles are identified by an orange triangle imprinted on the face of the receptacle; or they may be made entirely of orange material, in which case the triangle is of some other color.

In this residence, a TVSS and/or isolated grounding-type receptacle could be installed in the study/bedroom at the request of the owner. This might be desirable should the owner have a personal computer (PC), a word processor, or other electronic equipment subject to the problems that arise when transients appear on the wiring.

As more and more electronic equipment is brought onto the marketplace, the subject of transient voltage surge suppression and isolated grounding receptacles will be given even more attention than it is today.

Underwriters Laboratories Standards 1449 and 498 cover transient voltage surge suppressors and isolated ground receptacles.

The *National Electrical Code®* in *Section 210-7(c), Fine Print Note* refers us to *Section*

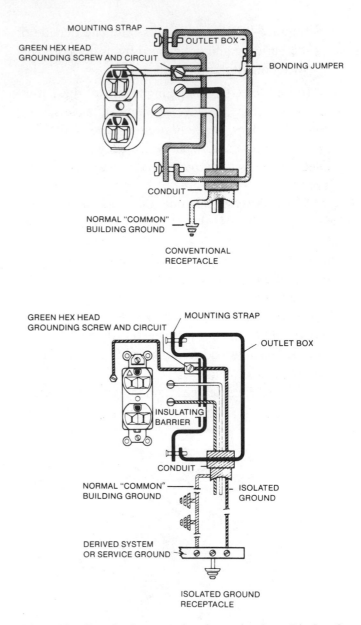

Fig. 6-19   Standard conventional receptacle and isolated ground receptacle. *Courtesy* of Hubbell Corporation.

250-74, Exception 4. Here we find the permission to install isolated ground receptacles to reduce noise. ▶ Permission is granted in *Section 384-20, Exception* to pass the separate insulated equipment grounding conductor through one or more panels without connecting it to that panel's equipment grounding terminal bar. ◀ The FPN of this Code reference reminds us, however, that the metal raceway and metal outlet both must still be grounded.

This separate grounding conductor is carried all the way back to the main service, where it is connected to the point where the neutral of the system is bonded to the grounding electrode conductor and the equipment grounding bus. In a residence, this would be the terminal bus provided for the connection of equipment grounding conductors as shown in figure 28-10.

When a transformer is installed between the equipment and the main service, the isolated grounding conductor is then run to the panelboard connected to the secondary of the transformer where the conductor is connected to the terminal. At the terminal, the grounding electrode conductor, equipment grounding conductor, and neutral conductors are bonded together.

REVIEW

1. Explain the operation of a ground-fault circuit interrupter. Why are GFCI devices used? Where are GFCI receptacles required? _____

   _____

   _____

   _____

   _____

   _____

   _____

   _____

   _____

   _____

   _____

   _____

   _____

2. Residential GFCI devices are set to trip a ground-fault current above _____ milliamperes.

3. Where must GFCI receptacles be installed in residential garages? _____

   _____

   _____

   _____

   _____

   _____

4. The Code requires GFCI protection for certain receptacles in the kitchen. Explain where these are required. _____

   _____

   _____

5. Is it a Code requirement to install GFCI receptacles in a fully carpeted, finished recreation room in the basement? _____

6. A homeowner calls in an electrical contractor to install a separate circuit in the basement (unfinished) for a freezer. Is a GFCI receptacle required? _____

7. GFCI protection is available as (a) branch-circuit breaker GFCI, (b) feeder circuit breaker GFCI, (c) individual GFCI receptacles, (d) feedthrough GFCI receptacles. In your opinion, for residential use, what type would you install? _____

8. Extremely long circuit runs connected to a GFCI branch-circuit breaker might result in
   a. nuisance tripping of the GFCI
   b. loss of protection
   c. the need to reduce the load on the circuit
   Circle the correct answer.

9. If a person comes in contact with the hot and grounded conductors of a two-wire branch-circuit that is protected by a GFCI, will the GFCI trip "off"? Why?

   _____

   _____

   _____

   _____

10. What might happen if the line and load connections of a feedthrough GFCI receptacle were reversed? _____

    _____

    _____

    _____

    _____

    _____

    _____

    _____

11. May a GFCI receptacle be installed as a replacement in an old installation where the 2-wire circuit has no equipment grounding conductor? _____

12. What two types of receptacles may be used to replace a defective receptacle in an older home that is wired with knob-and-tube wiring where no equipment grounding means exists in the box? _____

    _____

13. You are asked to replace a receptacle. Upon checking the wiring, you find that the wiring method is conduit and that the wall box is properly grounded. The receptacle is of the older style two-wire type that does not have a grounding terminal. You remove the old receptacle and replace it with (Circle the letter of the correct answer.)
    a. the same type of receptacle as the type being removed.
    b. a receptacle that is of the grounding type.
    c. a GFCI receptacle.

14. What color are the terminals of a standard grounding-type receptacle? _____

    _____

    _____

15. What special shape are the grounding terminals of receptacle outlets and other devices? _____

    _____

16. Construction sites can be dangerous because of the manner in which extension cords, portable electrical tools, and other electrical equipment are used and abused. *Article 305* of the Code offers two options, either of which may be followed to reduce the hazards of electrical shock. In your own words, what are these options?

_____

_____

_____

_____

_____

_____

17. In your own words, explain why the Code does not require certain receptacle outlets in kitchens, garages, and basements to be GFCI protected. _____

_____

_____

_____

_____

_____

_____

18. Circuit A1 supplies GFCI receptacles ① and ②. Circuit A2 supplies GFCI receptacles ③ and ④. Receptacles ① and ③ are feedthrough type. Using colored pencils or marking pens, complete all connections.

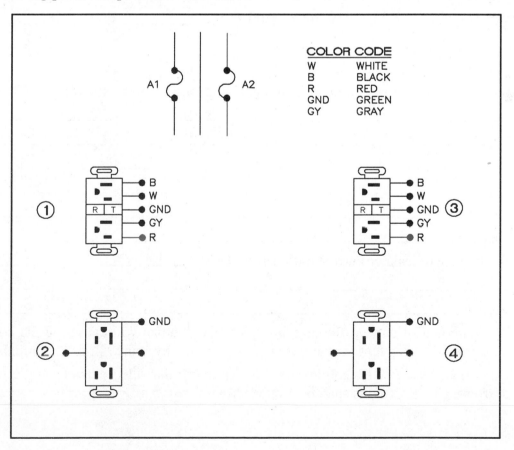

19. The term TVSS is becoming quite common. What do the letters stand for? _____
_____
_____

20. Transients (surges) on a line can cause spikes or surges of energy that can damage delicate electronic components. A TVSS device contains one or more _____
_____ _____ that bypass and absorb the energy of the transient.

21. Undesirable "noise" on a circuit can cause computers to lock up, lose their memory, and/or cause erratic performance of the computer. This "noise" does not damage the equipment. The two types of this "noise" are EMI and RFI. What do these letters mean?
_____
_____
_____
_____

22. Can TVSS receptacles be installed in standard device boxes? _____

23. Some line transients are not damaging to electronic equipment, but can cause the equipment to operate improperly. The effects of these transients can be minimized by installing _____ _____ _____ .

24. When an isolated ground receptacle is installed, the Code permits us to carry the separate equipment grounding conductor back through the raceway, through one or more panels, back to the point in the panel where the grounding electrode conductor, the equipment grounding conductors, and the neutral circuit conductors are connected together.

    What sections of the Code reference this topic? _____
    _____
    _____

25. Briefly explain the operation of an immersion detection circuit interrupter (IDCI).
_____
_____
_____
_____
_____
_____

26. What range of leakage current must trip an IDCI? _____
_____
_____

27. What amount of current flowing through a male human will cause muscle contractions that will keep that person from letting go of the live wire? _____ milliamperes

28. Other than for a few exceptions, the *NEC*® requires that all systems having a voltage of over _____ volts shall be grounded.

29. In an old house, an existing non-grounding non-GFCI-type receptacle in a bathroom needs to be replaced. According to *Section 210-7(d)*, this replacement receptacle shall be:

a. a non-grounding-type receptacle.

b. a GFCI-type receptacle or be GFCI protected in some other way, such as changing out the breaker that supplies the receptacle to a GFCI type.

30. Do all receptacle outlets in kitchens have to be GFCI protected? Explain. _____

_____

_____

_____

UNIT 7

Lighting Fixtures and Ballasts

OBJECTIVES

After studying this unit, the student will be able to

- understand fixture terminology, such as Type IC, Type NON IC, suspended ceiling fixtures, recessed fixtures, and surface-mounted fixtures.
- connect recessed fixtures, both prewired and nonprewired types, according to Code requirements.
- specify insulation clearance requirements.
- discuss the importance of temperature effects while planning recessed fixture installations.
- describe thermal protection for recessed fixtures.
- understand the cautions to be observed when installing outdoor lighting.
- understand class P and other types of fluorescent ballasts.

TYPES OF LIGHTING FIXTURES

This unit begins an in-depth coverage of the various types of recessed fixtures commonly installed in homes. It also considers other types of available lighting fixtures. ▶ As more and more lighting fixtures are imported into the United States, you will begin to see cartons marked "luminaire," the international term for lighting fixture, *Section 410-1, NEC.®* ◀

The Code in *Article 410* sets forth the requirements for the installation of lighting fixtures. These requirements are discussed throughout this text as necessary. Although not involved in the actual manufacture of lighting fixtures, the electrician must "meet Code" when installing fixtures, including mounting, supporting, grounding, live-parts exposure, insulation clearances, supply conductor types, maximum lamp wattages, and so forth.

The homeowner, interior designer, architect, or electrician has literally thousands of different types of fixtures from which to choose to satisfy certain needs, space requirements, and price considerations, among other factors.

It is absolutely essential for the electrician to know early in the roughing-in stage of wiring a house what types of fixtures are to be installed. This is particularly true for recessed-type fixtures. The electrician must work closely with the general building contractor, carpenter, plumber, heating contractor, and the other building trades people. This is to insure that such factors as location, proper and adequate framing, clearances from piping and ducts, clearances from combustible material, and insulation restrictions as concern the fixture installations are being complied with during construction.

Underwriters Laboratories provides the safety standards to which a fixture manufacturer must conform. Common fixture types are:

FLUORESCENT	INCANDESCENT
-surface	-surface
-recessed	-recessed
-suspended ceiling	-suspended ceiling

IMPORTANT: Always carefully read the label on the fixture. The information found on the label, together with conformance to *Article 410* of the Code, should result in an essentially safe installation. The label will provide such information as:

- For wall mount only
- Ceiling mount only
- Maximum lamp wattage
- Type of lamp
- Access above ceiling required
- Suitable for air handling use
- For chain or hook suspension only
- Suitable for operation in ambient temperatures not exceeding _____ °F (°C)
- Suitable for installation in poured concrete
- For installation in poured concrete only
- For line volt-amperes, multiply lamp wattage by 1.25
- Suitable for use in suspended ceilings
- Suitable for use in noninsulated ceilings
- Suitable for use in insulated ceilings
- Suitable for damp locations (such as bathrooms and under eaves)
- Suitable for wet locations
- Suitable for use as a raceway
- Suitable for mounting on low-density cellulose fiberboard
- For supply connections, use wire rated for at least _____ °C
- Not for use in dwellings
- Thermally protected fixture
- Type IC
- Type NON IC
- Inherently protected

The Underwriters Laboratories *Electrical Construction Materials Directory* (Green Book), and *General Information* (White Book), and the fixture manufacturers' catalogs and literature are excellent sources of information.

Underwriters Laboratories list, test, and label fixtures for conformance to their standards and to the *National Electrical Code.*®

Recessed fixtures, figure 7-1, in particular, have an inherent heat problem. Therefore, they must be suitable for the application and must be properly installed.

To protect against overheating, Underwriters Laboratories requires that recessed fixtures be equipped with an integral thermal protector, figure 7-2. These devices will cycle ON and OFF repeatedly until the heat problem is removed.

Figure 7-3 shows a fluorescent fixture mounted on low-density cellulose fiberboard. Because of the

Fig. 7-1 Typical recessed fixture.

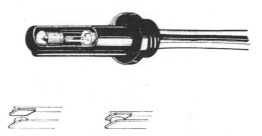

Fig. 7-2 Recessed fixture thermal cutout.

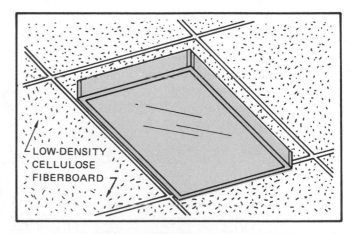

Fig. 7-3 When mounted on *low-density ceiling fiberboard*, surface-mounted fluorescent fixtures must be marked "Suitable for Surface Mounting on Low-Density Cellulose Fiberboard."

potential fire hazard when mounting lighting fixtures on combustible materials, *Section 410-76(b)*
states that these fixtures must be listed for such
mounting, or must be spaced at least 1½ inches
(38 mm) from the fiberboard surface. It is important
to refer to the *Fine Print Note (FPN)* below *Section
410-76(b)*. This FPN explains combustible low-density cellulose fiberboard as sheets, panels, and tiles
that have a density of 20 pounds per cubic foot or
less that are formed of bonded plant fiber material.
Solid or laminated wood does not come under the
definition of "combustible low-density cellulose
fiberboard." It does not include fiberboard that has a
density of over 20 pounds per cubic foot or material
that has been integrally treated with fire-retarding
chemicals to meet specific standards.

CODE REQUIREMENTS FOR INSTALLING RECESSED FIXTURES

The Code requirements for the installation and
construction of recessed fixtures are given in
Sections 410-64 through *410-72*. Of particular
importance are the Code restrictions on conductor
temperature ratings, fixture clearances from combustible materials, and maximum lamp wattages.
Recessed fixtures generate a great deal of heat within the enclosure. **Thus, these fixtures are a fire
hazard if they are not wired and installed properly**, figure 7-4. Figure 7-5 shows the roughing-in box
of a recessed fixture with mounting brackets and
junction box.

Recessed Fixtures

Most residential recessed lighting fixtures come
equipped with a junction box. With some types, regular
90°C building wire such as that found in nonmetallic-
sheathed cable may be run directly to this junction
box. For others, higher temperature conductors might
be required. A label on the fixture will provide this
required conductor temperature information.

Where the recessed lighting fixture does not
have a junction box on it, a junction box must be
installed near the fixture—and this box must be
accessible. The branch-circuit conductors are run to

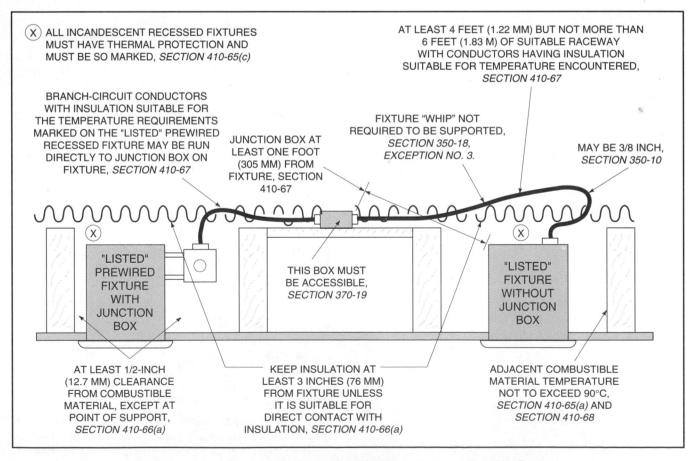

Fig. 7-4   Clearance requirements for installing recessed lighting fixtures.

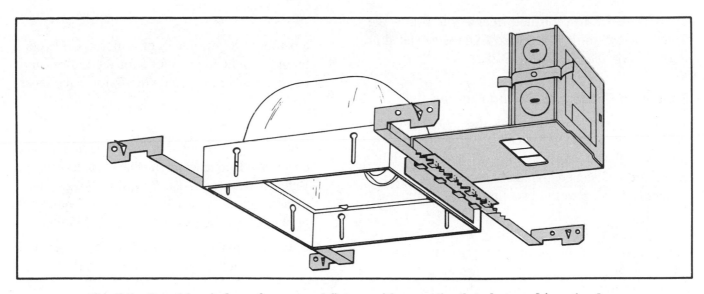

Fig. 7-5   Roughing-in box of a recessed fixture with mounting brackets and junction box.

this junction box. In this junction box, the branch-circuit conductors are spliced to conductors that are run to the recessed fixture. The conductors between the junction box and recessed lighting fixture must be suitable for the temperature rating as marked on the recessed fixture. The junction box is placed at least one foot (305 mm) from the fixture. Thus, heat radiated from the fixture cannot overheat the wires in the junction box. Conductors rated for higher temperatures must run through at least 4 feet (1.22 m) of suitable raceway, but not to exceed 6 feet (1.83 m) between the fixture and the junction box. As a result, any heat conducted from the fixture along the metal raceway will be reduced before reaching the junction box. Many recessed fixtures are factory equipped with a flexible metal raceway containing high-temperature (150°C) wires that meet the requirements of *Section 410-67.*

Recessed Fixture Trims

▶ Recessed incandescent fixtures are listed by Underwriters Laboratories in combination with specific trims. Fixture/trim combinations are marked on the label inside the fixture. The type of lamp and maximum wattage for a specific combination of fixture/trim is also indicated on the label. To install trims *not listed* for use with a specific recessed incandescent fixture can result in overheating and possible tripping (ON-OFF cycling) of the fixture's thermal protective device. The next step that an untrained person might do is to defeat (by-pass) the

thermal protective device, setting the stage for a fire. ◀ Read the label. Make sure that the fixture/trim combination is a listed combination!

Mismatching fixtures and trims is a violation of *Section 110-3(b)* of the *National Electrical Code®,* which states that "Listed or labeled equipment shall be used or installed in accordance with any instructions included in the listing or labeling."

Suspended Ceiling Lay-in Fixtures
(Figure 7-6)

The recreation room of the residence plans has "dropped" suspended acoustical paneled ceiling. The fixtures installed in this ceiling bear a label stating "SUSPENDED CEILING FIXTURE." Note that these fixtures are *not* classified as recessed fixtures because, in most cases, there is a great deal of open space above and around these fixtures.

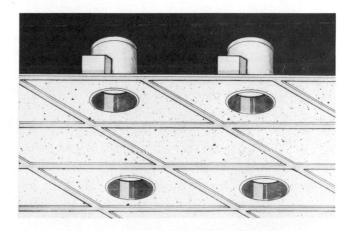

Fig. 7-6   Suspended ceiling fixtures.

Because of this, it is not necessary to follow the Code rules relating to recessed fixtures as stated in *Part M* of *Article 410* of the Code.

Support of Suspended Ceiling Fixtures

Section 410-16(c) requires that when framing members of a suspended ceiling grid are used to support lighting fixtures, then all members must be securely fastened together and to the building structure. It is up to the installer of the ceiling grid to do this.

All lay-in suspended ceiling fixtures must be securely fastened to the ceiling grid members by bolts, screws, or rivets. Most fixture manufacturers supply special clips that are suitable for fastening the fixture to the grid.

The logic behind all of the "securely fastening" requirements is that in the event of a major problem (i.e., earthquake or fire), the lighting fixtures will not fall down and injure someone.

Connecting Suspended Ceiling Fixtures

The most common way to connect suspended ceiling lay-in lighting fixtures is to complete all of the wiring above the suspended ceiling using normal wiring methods, such as electrical metallic tubing, nonmetallic-sheathed cable, armored cable, or whatever type of wiring method is acceptable by the governing electrical code. Then from outlet boxes strategically placed above the ceiling near the intended location of the lay-in lighting fixtures, a flexible connection is made between the outlet box and the fixture.

The flexible connection is usually made by installing

- a 6-foot (1.83 m) length of ⅜-inch flexible metal conduit using conductors suitable for the temperature requirement as stated on the fixture label, usually at least 90°C. See *Section 350-10(a)(2), Exception No. 2*, of the Code. Also refer to *Section 250-91(b), Exception No. 1*, which states that flexible metal conduit in lengths not over 6 feet (1.83 m) and protected by overcurrent devices not over 20 amperes, and using fittings listed for grounding is acceptable as a means of grounding the fixture. See figure 7-7.

or

- a 6-foot (1.83 m) length of armored cable containing conductors suitable for the temperature requirements as stated on the fixture label, usually 90°C.

or

- a 6-foot (1.83 m) length of ⅜ flexible nonmetallic conduit with conductors suitable for at least 90°C. Refer to *Section 351-24, Exception No. 2*. A nonmetallic raceway would require an equipment grounding conductor sized per *Table 250-95*.

In the electrical trade, these flexible connections between an outlet box and a fixture are called *fixture whips*.

A factor to be considered by the electrician when installing recessed fixtures is the necessity of working with the installer of the insulation to be sure that the clearances, as required by the Code, are maintained.

As previously stated, the installation of recessed fixtures in both new work and rework (remodel) must conform to the *National Electrical Code®* and to the instructions furnished with the fixtures. Figure 7-8(A) shows one type of recessed fixture that must maintain the 3-inch clearance from thermal insulation. If there is less than 3 inches of clearance between the fixture and the insulation, the thermal protector built into the fixture will cycle on and off until the problem is corrected. Maintaining the necessary 3-inch clearance around the fixture results in a considerable amount of heat loss.

Figure 7-8(B) is a recessed fixture that has been tested and listed to be completely buried in insulation. This type of recessed fixture is recommended where insulation is to be installed to minimize heat loss. All fixtures are marked with their maximum lamp wattage. If a higher than the maximum wattage lamp is installed, the built-in thermal protector will cycle on and off until the problem is corrected.

Figure 7-8(C) shows the clearances for a recessed lighting fixture installed near thermal insulation, *Section 410-66*.

Recessed fixtures of the type installed in ceilings usually have a box mounted on the side of the fixture. See figure 7-5. The branch-circuit conductors can be run directly into this box where they are connected to the conductors entering the fixture.

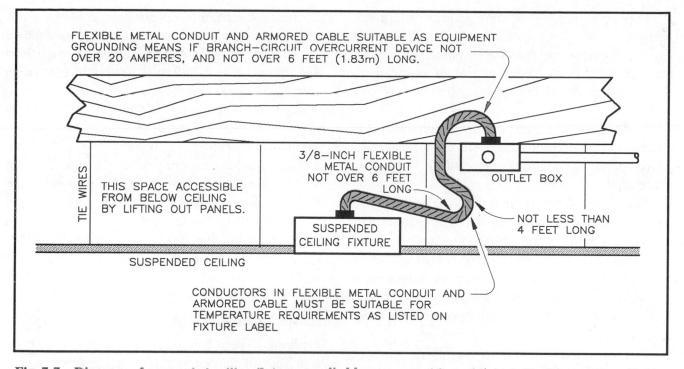

Fig. 7-7   Diagram of suspended ceiling fixture supplied by not over 6 feet of ⅜-inch flexible metal conduit or armored cable. Although not required to be secured within 12 inches of the outlet box per *Section 350-18, Exception 3*, and *Section 333-7, Exception 3*, many electricians do secure (strap) the flex within 12 inches of the outlet box. See figures 4-30A and 4-33 where the "combined" length of flexible metal conduit exceeds 6 feet (1.83 m). The fixture whip between the outlet box and the fixture is permitted to be either metallic or non-metallic, provided they are suitable for the purpose as required in *Section 410-67(b)*.

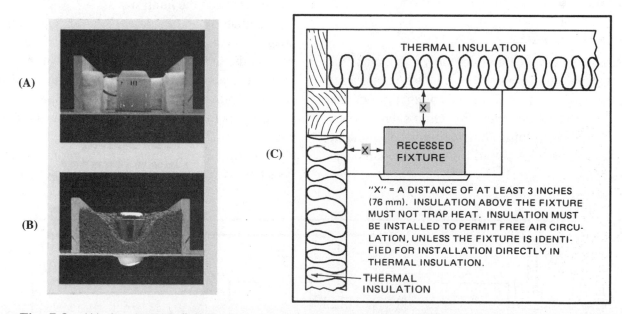

Fig. 7-8   (A) A recessed fixture that requires 3-inch clearance from thermal insulation. (B) A recessed fixture that may be completely covered with thermal insulation. (C) Clearances for recessed lighting fixture installed near thermal insulation, *Section 410-66*.

Prewired fixtures do not require additional wiring, figure 7-9. *Section 410-11* states that branch-circuit wiring shall not be passed through an outlet box that is an integral part of an incandescent fixture unless the fixture is identified for through wiring.

For a recessed fixture that is not prewired, the electrician must check the fixture for a label indicating what insulation temperature rating is required.

The cables to be installed in this residence have 90°C (194°F) insulation. If the temperature will exceed this value, conductors with other types of insulation must be installed, *Table 310-13*.

Refer to unit 17 for additional information regarding recessed fixture connections. See figure 7-12 for a summary of the requirements pertaining to recessed fixtures.

FLUORESCENT BALLASTS AND LAMPS

The following text is intended to introduce the student to the basics of fluorescent ballasts and lamps. This topic is extremely complex. The student is urged to send for the technical brochures, bulletins, and booklets mentioned in the Instructor's Guide. These brochures contain a wealth of information on ballasts, lamps, wiring diagrams, troubleshooting hints, application suggestions, and so on.

Fluorescent Ballasts

An incandescent lamp contains a filament that has a specific "hot" resistance value when operating at rated voltage. The current flowing through the lamp filament is governed by Ohm's law, where Amperes = Volts/Ohms. When energized, the lamp will provide the light output for which the lamp was designed.

Fluorescent lamps do not have a filament running from end-to-end. Instead, they have a filament at each end. The filaments are connected to a ballast. See figure 7-9A. A ballast is needed to control the energy, control voltage to heat the filaments, control voltage across the lamp to start the arc within the tube (end-to-end) needed to ionize the gas and vaporize the droplets of mercury inside the lamp, and to "limit" the current flowing through the lamp. Without a ballast, the fluorescent lamp would probably not start. But, if it did, then the internal resistance (impedance) would get lower and lower, and the current flowing through the lamp would get higher and higher, and in a very short time, the lamp would be destroyed.

Ballast Types (Circuitry)

Preheat.   These ballasts are connected in a simple series circuit configuration. Easily identified because the fixture will have a "starter." One type of starter is automatic, and looks like a small roll of life-savers with two "buttons" on one end. When the fixture is turned on, the starter is in series with the lamp's filaments and the ballast. The lamp filaments glow for a few seconds. The starter's bimetal element heats up and "warps," causing the starter contacts to automatically open the series circuit. At the instant the starter contacts open, an arc is established inside the lamp, end-to-end, and the lamp starts.

Another type of starter is a manual ON/OFF switch that has a momentary "make" position just beyond the ON position. When you push the switch on, and hold it there for a few seconds, the lamp filaments glow. When the switch is released, the start contacts open, an arc is initiated within the lamp—and the lamp lights up.

The fluorescent lamps for use with preheat ballasts have two pins (bi-pin) on each end of the lamp.

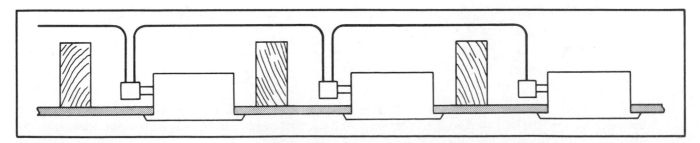

Fig. 7-9   Installation permissible only with prewired recessed fixtures with approved junction box, *Section 410-11*.

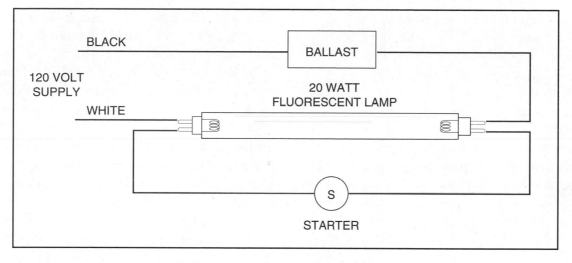

Fig. 7-9A   Simple series circuit.

Preheat lamps and ballasts are not used for dimming applications.

Rapid Start.   This is probably the most common type used today. The lamps start in less than 1 second. The filaments remain energized with a low-voltage circuit from the ballast. The lamps start up much faster than the preheat type. To assure proper starting, ballast manufacturers recommend that there be a grounded reflector close (within ½ inch) to the lamp and running the full length of the lamp, that the ballast be grounded, and that the supply circuit originate from a grounded system. Of course, virtually all branch-circuits today are derived from a grounded system, such as the 240/120-volt systems in homes. Commercial installations might be 208/120 volt or 480/277 volts.

Fluorescent lamps for use with rapid start ballast have two pins (bi-pin) on each end of the lamp.

Another version of a rapid start ballast is the *trigger start* ballast. Trigger start ballasts are used with preheat fluorescent lamps up to 32 watts. The trigger start ballast was the forerunner of the rapid start ballast.

Rapid start lamps can be dimmed using special dimming ballast. See unit 13.

Instant Start.   These ballasts provide a high voltage "kick" to start the lamp instantly. They require special fluorescent lamps that do not require preheating of the lamp filaments. Because instant start fluorescent lamps are started by brute force, they have a shorter life (as much as 40% less) than rapid start lamps. Most instant start fluorescent lamps have a single pin on each end of the lamp, although some

have two pins (bi-pin), in which case a short-circuiting arrangement is designed into the lampholder.

Instant start ballasts/lamps cannot be dimmed.

Ballast Types (Applications)

The three types of ballast discussed previously are available in:

High Efficiency.   These are highly recommended because they save energy. Energy-saving ballast can be of the *electronic type* or *magnetic type* (core & coil). If they are of the magnetic type, then they will probably have steel laminations and copper wire instead of iron laminations and aluminum wire. High efficiency, energy-saving ballasts have a high power factor rating. Power factor is the ratio of watts to volt-amperes.

For example, a 40-watt fluorescent lamp is installed in a fixture that has a ballast marked 42 VA. The power factor of the complete fixture is:

$$\text{Power Factor} = \frac{\text{Watts}}{\text{Volt-amperes}} = \frac{40}{42} = 0.95 \ (95\%)$$

Energy-saving ballasts cost more to buy initially than low power factor ballasts, but over time, the savings in energy consumption far exceeds the higher initial costs. A standard fluorescent ballast might consume 14 to 16 watts. An energy-efficient ballast will use 8 to 10 watts. Combined with an energy-efficient fluorescent lamp that uses 32 or 34 watts instead of 40 watts, there is a considerable energy savings. You are buying lighting—not heat. Always install fixtures that have a high power factor rating.

This data is found on the nameplate of the ballast. Today, electronic ballasts are the most energy efficient when compared to magnetic ballasts containing coils, capacitors, and transformers.

The federal government's National Appliance Energy Conservation Amendment of 1988, Public Law 100-357 prohibited manufacturers from producing certain types of commonly used low power factor (less than 90%) ballasts as of January 1990. Among these were ballasts for single and two-lamp ballasts for 4-foot T12 rapid start lamps. Fixture manufacturers had to start using high power factor ballasts (90% or greater) as of April 1991. Ballasts that "meet or exceed" the requirements of the federal law are marked with an "E" in a circle. Dimming ballasts and ballasts designed specifically for residential use were exempted.

Following the ballast legislation, the National Energy Policy Act of 1992 enacted restrictions on lamps. Lamp manufacturers could no longer produce certain 8-foot fluorescent lamps as of April 1994. In October 1995, the common 4-foot T12 linear and U-tube fluorescent lamps, and the R30, R40, and PAR38 incandescent lamps were eliminated. These were replaced by energy-efficient lamps that are direct replacements for the discontinued lamps. For example, the typical PAR38 reflector flood lamp could be replaced with a 60PAR/HIR incandescent lamp. An F40T12WW fluorescent lamp could be replaced with an F40SP30RS/WM fluorescent lamp that consumes 34 watts instead of 40 watts for just about the same amount of light (lumen output).

It has been estimated that the total electric bill savings across the country will exceed $250 billion over the next 15 years. Figure 7-9B clearly illustrates how much energy saving is possible when using high-efficiency (high power factor) ballast. It compares a standard F40T12 fluorescent lamp when used with different types of ballasts.

It is possible to overload a branch-circuit if too many low power factor ballasts are connected to the circuit. In this residence the use of low cost, low power factor ballasts for the recessed fluorescent lighting fixtures in the recreation room could result in the need for two branch-circuits instead of one—as well as higher energy (bigger light bills) consumption. High-efficiency, high power factor ballasts should always be used.

Fluorescent loads are a good example of why volt-amperes, rather than watts, must be considered in order to find the current draw on the circuit.

Let's make a comparison of possibilities for the recreation room lighting where there are six recessed fluorescent fixtures, each containing two 2-lamp ballasts. The nameplate on each ballast indicates a line current rating of 0.70 ampere at 120 volts, which is 84 volt-amperes. The lamps are marked 40-watts.

The total amperes is:

$$12 \times 0.70 = 8.4 \text{ amperes}$$

The total volt-amperes is:

$$8.4 \times 120 = 1008 \text{ volt-amperes}$$

The total lamp wattage is:

$$40 \times 24 = 960 \text{ watts}$$

If we had calculated the load for these fixtures based upon wattage, the result would have been:

$$\frac{W}{E} = \frac{960}{120} = 8 \text{ amperes}$$

Not much different than the 8.4 amperes using data from the ballast nameplate.

Ballast	Line Current	Line Voltage	Line Volt/Amperes	Lamp Wattage	Line Power Factor
No. 1	0.35	120	42	40	0.95 (95%)
No. 2	0.45	120	54	40	0.74 (74%)
No. 3	0.55	120	66	40	0.61 (61%)
No. 4	0.85	120	102	40	0.39 (39%)

Fig. 7-9B   Table showing various line currents, volt-amperes, wattages, and overall power factor using different ballasts.

Had we used low cost, low-efficiency ballasts like No. 2 shown in figure 7-9B, the results would be entirely different.

The total amperes is:

$$12 \times 1.70 = 20.4 \text{ amperes}$$

The total volt-amperes is:

$$20.4 \times 120 = 2448 \text{ volt-amperes}$$

The total lamp wattage is:

$$40 \times 24 = 960 \text{ watts}$$

$$\frac{W}{E} = \frac{960}{120} = 8 \text{ amperes}$$

But we really have a current draw of 20.4 amperes, which would overload the 15-ampere branch-circuit B12. In fact, a load of 20.4 amperes is too much for a 20-ampere branch-circuit. Two 15-ampere branch-circuits would have been needed to hook-up the recreation room recessed fluorescent fixtures. Remember, *do not* load any circuit to more than 80% of the branch-circuit's rating. That is 16 amperes of a 20-ampere branch-circuit and 12 amperes for a 15-ampere branch-circuit. And that is the maximum.

It is apparent that the *installed* cost using cheap fixtures is greater and more complicated than if high-quality fixtures using energy-efficient ballasts and lamps had been used. After the initial installation, the energy savings adds up significantly.

Dimming Ballast. These special ballasts are discussed in unit 13. They are used where the dimming of rapid start fluorescent lamps is desired. Instant start lamps and ballasts cannot be used for dimming applications. Dimming ballasts are more costly and are less energy efficient than standard ballasts.

Noise Levels. All ballasts will hum, but some do so more than others. Ballasts containing transformers and choke coils hum when the metal laminations vibrate because of the alternating current reversals. This hum can be magnified by the fixture and/or the surface the fixture is mounted on. Electronic ballasts have little if any hum.

Do not insert spacers, washers, or shims between a ballast and the fixture in attempts to make the ballast more quiet. This will cause the ballast to run much hotter, and could result in shortened ballast life, and possible fire hazard. Instead, replace the noisy ballast with a quiet, sound-rated ballast. Sometimes checking and tightening the many nuts, bolts, and screws of the fixture will solve the problem.

Ballasts are sound rated. Again, cost is a consideration. Ballasts are marked with letters *A* through *F*. *A* is the quietest, and *F* is the noisiest.

A—homes, churches, libraries, hospital rooms.

B—classrooms, private offices.

C—open office areas.

D—retail stores.

E & F—outdoors and in factories.

Lamp Diameters. There is a simple way to determine the diameter of a lamp (light bulb or tube) at its widest measurement. The industry uses an "eighths of an inch rule." For example, a T5 fluorescent lamp is ⅝ inch in diameter. A T8 fluorescent lamp is ⅞ inches, or 1 inch in diameter. A T12 lamp is ¹²⁄₈ inches, or 1½ inches in diameter. Incandescent lamps follow the same system. A PAR38 lamp is ³⁸⁄₈ inches, or 4¾ inches in diameter. An R30 lamp is ³⁰⁄₈ inches, or 3¾ inches in diameter.

Other Considerations

- Because heat trapped by insulation around and on top of a fixture can shorten the life of a ballast, always follow the manufacturer's installation requirements and the requirements found in *Article 410* of the *National Electrical Code,*® which were discussed earlier in this unit. A 10°C temperature rise above the ballast's rated temperature (90°C) can reduce the ballast's life to one-half of its expected life. This "half-life rule" is true for conductors, motors, transformers, and other electrical equipment.

- Fluorescent lamps that are intensely blackened on both ends should be replaced. Operating a two lamp ballast with only one lamp working will cause the ballast to run hot, and will shorten the life of the ballast. Severe blackening of one end of the lamp can also ruin the ballast. A flickering lamp should be replaced.

- Poor starting of a fluorescent lamp can be caused by poor contact in the lamp holder, poor grounding, excessive moisture on the outside of the tube, cold temperature (approximately 50°F), as well as a dirt, dust, and grime on the lamp.

- Lamp life generally is rated in "X" number of hours of operation based on 3 hours per start. Frequent switching results in shortened expected lamp life. Inversely, leaving the lamps on for long periods of time extends the expected lamp life. Vibration, rough handling, cleaning, etc., shortens lamp life.

Incandescent Lamp Life at Different Voltages

Operating an incandescent lamp at other than rated voltage will result in longer—or shorter—lamp life. The following formula can predict the approximate expected lamp life at different voltage. For example, assume that a gas-filled 120-volt incandescent lamp has a published lamp life of 1,000 hours. The calculations show the expected lamp life of this lamp when operated at 130 volts, and the expected lamp life when operated at 110 volts.

- At 130 volts:

$$\text{Expected Life} = \left(\frac{\text{Rated Volts}}{\text{Actual Volts}}\right)^{13.1} \times \text{Rated Life}$$

$$= \left(\frac{120}{130}\right)^{13.1} \times 1,000$$

$$= 0.350 \times 1,000$$

$$= 350 \text{ hours}$$

- At 110 volts:

$$\text{Expected Life} = \left(\frac{\text{Rated Volts}}{\text{Actual Volts}}\right)^{13.1} \times \text{Rated Life}$$

$$= \left(\frac{120}{110}\right)^{13.1} \times 1,000$$

$$= 3.126 \times 1,000$$

$$= 3,126 \text{ hours}$$

Incandescent Lamp Lumen Output at Different Voltages

Notice that operating an incandescent lamp at a voltage lower than the lamp's voltage rating will result in longer lamp life. A strong case might be made to install 130-volt lamps, particularly where they are hard to reach, such as flood lights located high up. However, the lumen output is reduced. For example, calculate the approximate lumen output of a 100-watt, 130-volt incandescent lamp that has an initial lumen output of 1,750 lumens at rated voltage. The lamp is to be operated at 120 volts.

$$\begin{aligned}\frac{\text{Expected}}{\text{Lumens}} &= \left(\frac{\text{Actual Volts}}{\text{Rated Volts}}\right)^{3.38} \times \text{Rated Lumens} \\ &= \left(\frac{120}{130}\right)^{3.38} \times 1,750 \\ &= 0.763 \times 1,750 \\ &= 1,335 \text{ Lumens}\end{aligned}$$

Formulas such as these are found in lamp manufacturer's catalogs. These formulas are useful in determining the affect of applied voltage to lamp wattage, line current, lumen output, lumens per watt, and lamp life. A calculator that has a y^x power function is needed to solve these equations.

Similar formulas found in lamp manufacturer's catalogs are used to determine the changes in lamp wattage, lumen output, and line current.

Lamp Efficiency/Dimming/Colors/Applications/Life/Shapes

Lighting efficiency is measured in *Lumens Per Watt*. A lumen is a unit of light measurement. One lumen falling on one square foot of surface produces one *foot-candle*. The following is a very brief comparison of different types of lamps used in residences.

The more lumens per watt, the greater the efficiency of the lamp. See figure 7-9C.

Section 410-73(e) states that all fluorescent ballasts installed indoors (except simple reactance-type ballasts), both for new and replacement installations, must have thermal protection built into the ballast by the manufacturer of the ballast, figure 7-10. Ballasts provided with built-in thermal protection are listed by the Underwriters Laboratories as Class P ballasts. Under normal conditions, the Class P ballast has a case temperature not exceeding 90°C. The thermal protector must open within two hours when the case temperature reaches 110°C.

Some Class P ballasts also have a nonresetting fuse integral with the capacitor to protect against capacitor leakage and violent rupture. The Class P ballast's internal thermal protector will disconnect the ballast from the circuit in the event of over-temperature. Excessive temperatures can be caused by abnormal voltage and improper installation, such as being covered with insulation.

	Lumens/ Watt	Dimming	Color and Application	Life (Approx. Hours)	Typical Shapes
Incandescent	18	Yes	Warm and Natural. Great for general lighting.	500, 1000, 1500 Depends on type of lamp.	Standard, spots, floods, decorative, tubes, globes, PAR (similar to standard spots and floods but much stronger). Use rough service where there is vibration, like garage door openers and ceiling fans.
Halogen	22	Yes	Brilliant white. Excellent for accent and task lighting. Are filled with halogen gas and have an inner lamp allowing the filament to run hotter (whiter).	2000 to 4000	PAR spots and floods, mini-reflector spots and floods, double and single end quartz.
Fluorescent	105	Yes, but only 40-watt rapid start lamps using special dimming ballast. See unit 13.	Warm and deluxe warm white, cool and deluxe cool white plus many other shades of white. Great for general lighting, like the Recreation Room in this residence.	6000 to 24000. Compact fluorescents can last 9 to 13X longer than incandescent lamps. Compact fluorescents are complete units having both ballast and lamp. When lamp burns out, ballast is also replaced. Compacts with screw-in bases can replace most incandescents.	Straight tubular, U-shaped tubular, circular tubular, compacts. Compact screw-in can have twin, double twin, and triple twin tubes. This refers to the number of "loops" in the lamp. Compacts can save $20 to $30 over its lifetime when compared to an equivalent incandescent lamp. Compacts last longer than incandescent lamps.

Fig. 7-9C

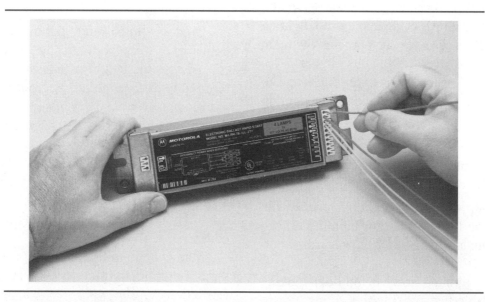

Fig. 7-10   An electronic Class P ballast, thermally protected as required in *Section 410-73(e)*. Electronic ballasts provide improved performance in fluorescent lighting installations. Electronic ballast lighting systems are 25–40% more energy efficient than conventional magnetic (core & coil) ballast fluorescent systems. Motorola and Ⓜ are registered trademarks of Motorola Inc. All rights reserved.

The reason for thermal protection is to reduce the hazards of possible fire due to an overheated ballast when the ballast becomes shorted, grounded, covered with insulation, lacking in air circulation, and so on. Ballast failure has been a common cause of electrical fires.

Additional backup protection can be provided by an in-line fuseholder connected in series with the black line lead and the black ballast lead, figure 7-11. The in-line fuseholder acts as a disconnect for the ballast. Should a ballast fail (short out), the individual fuse opens, thus isolating the faulty ballast but not affecting the rest of the circuit. When the ballast is to be replaced, the fuse is removed to disconnect the ballast, but the entire circuit need not be turned off. In general, the ballast manufacturer will recommend the correct type and ampere rating of the fuse to be used for the particular ballast.

See figure 7-12 for some important requirements for fluorescent and incandescent lighting fixtures.

LIGHTING FIXTURE VOLTAGE LIMITATIONS

The maximum voltage allowed for residential lighting fixtures is 120 volts between conductors per *Section 210-6* of the Code.

Section 410-80(b) makes a further restriction stating that for dwelling occupancies, no lighting equipment shall be used if it operates with an open-circuit voltage over 1000 volts, figure 7-13. This section more or less focuses on the desire to use neon lighting for more decorative lighting purposes.

Cautions to Be Observed When Installing Outdoor Lighting

Before installing outdoor lighting, check with the local electrical inspector and/or building official to find out if there are any restrictions regarding outdoor lighting.

A virtually unlimited array of outdoor lighting fixtures is available that provide uplighting, downlighting, diffused lighting, moonlighting, shadow and texture lighting, accent lighting, silhouette lighting, and bounce lighting. After the lighting fixtures are selected, the type and color of the lamp is then selected. Some lighting fixtures have specific light

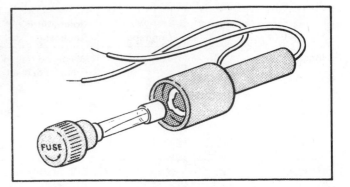

Fig. 7-11 In-line fuseholder for ballast provides added protection.

"cut-off" data that is useful in determining whether or not the emitting light will spill over onto the neighbor's property. The method of control must also be considered. Switch control, timer control, dusk-to-dawn control, and motion sensors are ways of turning outdoor lighting on and off.

In recent years, more and more complaints are coming from neighbors claiming that they are being bothered by glare, brightness, and light spillover from their neighbor's outdoor lighting fixtures. Security lighting, yard flood lights, driveway lighting, "moonlighting" in trees, and shrub lighting are examples of sources of light that might cross over the property line and be a "nuisance" to the next door neighbor.

Outdoor "lightscape" lighting considered by a homeowner to be aesthetically wonderful might be offensive to the neighbor. Nuisance lighting is also referred to as *light pollution*, *trespassing*, *intrusion*, *glare*, *spill-over*, and *brightness*. Some quiet residential neighborhoods are beginning to look like commercial areas, used car lots, and airport runways because of no restrictions relative to outdoor lighting methods. This is not a safety issue, and is not addressed in the *National Electrical Code®.* However, the issue might be found in local building codes.

Many communities are being forced to legislate strict outdoor lighting laws, specifying various restrictions on the type, size, wattage, and/or footcandles for outdoor lighting fixtures. Checking building codes in your area might reveal requirements such as:

Light Source: The source of light (the lamp) must not be seen directly.

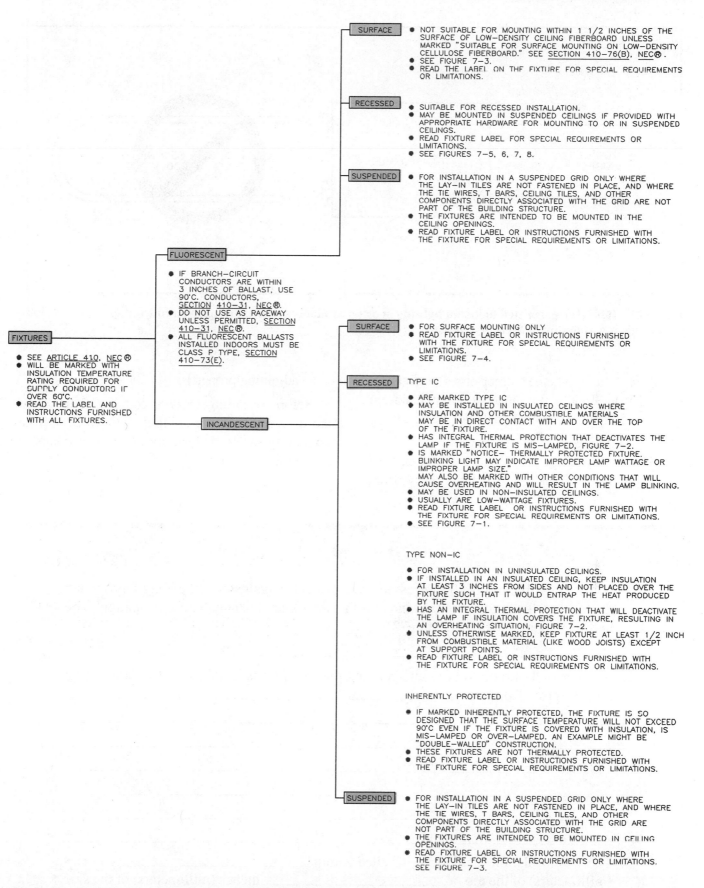

FIXTURES
- SEE ARTICLE 410, NEC®
- WILL BE MARKED WITH INSULATION TEMPERATURE RATING REQUIRED FOR SUPPLY CONDUCTORS IF OVER 60°C.
- READ THE LABEL AND INSTRUCTIONS FURNISHED WITH ALL FIXTURES.

FLUORESCENT
- IF BRANCH-CIRCUIT CONDUCTORS ARE WITHIN 3 INCHES OF BALLAST, USE 90°C. CONDUCTORS, SECTION 410-31, NEC®.
- DO NOT USE AS RACEWAY UNLESS PERMITTED, SECTION 410-31, NEC®.
- ALL FLUORESCENT BALLASTS INSTALLED INDOORS MUST BE CLASS P TYPE, SECTION 410-73(E).

SURFACE
- NOT SUITABLE FOR MOUNTING WITHIN 1 1/2 INCHES OF THE SURFACE OF LOW-DENSITY CEILING FIBERBOARD UNLESS MARKED "SUITABLE FOR SURFACE MOUNTING ON LOW-DENSITY CELLULOSE FIBERBOARD." SEE SECTION 410-76(B), NEC®.
- SEE FIGURE 7-3.
- READ THE LABEL ON THE FIXTURE FOR SPECIAL REQUIREMENTS OR LIMITATIONS.

RECESSED
- SUITABLE FOR RECESSED INSTALLATION.
- MAY BE MOUNTED IN SUSPENDED CEILINGS IF PROVIDED WITH APPROPRIATE HARDWARE FOR MOUNTING TO OR IN SUSPENDED CEILINGS.
- READ FIXTURE LABEL FOR SPECIAL REQUIREMENTS OR LIMITATIONS.
- SEE FIGURES 7-5, 6, 7, 8.

SUSPENDED
- FOR INSTALLATION IN A SUSPENDED GRID ONLY WHERE THE LAY-IN TILES ARE NOT FASTENED IN PLACE, AND WHERE THE TIE WIRES, T BARS, CEILING TILES, AND OTHER COMPONENTS DIRECTLY ASSOCIATED WITH THE GRID ARE NOT PART OF THE BUILDING STRUCTURE.
- THE FIXTURES ARE INTENDED TO BE MOUNTED IN THE CEILING OPENINGS.
- READ FIXTURE LABEL OR INSTRUCTIONS FURNISHED WITH THE FIXTURE FOR SPECIAL REQUIREMENTS OR LIMITATIONS.

INCANDESCENT

SURFACE
- FOR SURFACE MOUNTING ONLY
- READ FIXTURE LABEL OR INSTRUCTIONS FURNISHED WITH THE FIXTURE FOR SPECIAL REQUIREMENTS OR LIMITATIONS.
- SEE FIGURE 7-4.

RECESSED

TYPE IC
- ARE MARKED TYPE IC
- MAY BE INSTALLED IN INSULATED CEILINGS WHERE INSULATION AND OTHER COMBUSTIBLE MATERIALS MAY BE IN DIRECT CONTACT WITH AND OVER THE TOP OF THE FIXTURE.
- HAS INTEGRAL THERMAL PROTECTION THAT DEACTIVATES THE LAMP IF THE FIXTURE IS MIS-LAMPED, FIGURE 7-2.
- IS MARKED "NOTICE- THERMALLY PROTECTED FIXTURE. BLINKING LIGHT MAY INDICATE IMPROPER LAMP WATTAGE OR IMPROPER LAMP SIZE." MAY ALSO BE MARKED WITH OTHER CONDITIONS THAT WILL CAUSE OVERHEATING AND WILL RESULT IN THE LAMP BLINKING.
- MAY BE USED IN NON-INSULATED CEILINGS.
- USUALLY ARE LOW-WATTAGE FIXTURES.
- READ FIXTURE LABEL OR INSTRUCTIONS FURNISHED WITH THE FIXTURE FOR SPECIAL REQUIREMENTS OR LIMITATIONS.
- SEE FIGURE 7-1.

TYPE NON-IC
- FOR INSTALLATION IN UNINSULATED CEILINGS.
- IF INSTALLED IN AN INSULATED CEILING, KEEP INSULATION AT LEAST 3 INCHES FROM SIDES AND NOT PLACED OVER THE FIXTURE SUCH THAT IT WOULD ENTRAP THE HEAT PRODUCED BY THE FIXTURE.
- HAS AN INTEGRAL THERMAL PROTECTION THAT WILL DEACTIVATE THE LAMP IF INSULATION COVERS THE FIXTURE, RESULTING IN AN OVERHEATING SITUATION, FIGURE 7-2.
- UNLESS OTHERWISE MARKED, KEEP FIXTURE AT LEAST 1/2 INCH FROM COMBUSTIBLE MATERIAL (LIKE WOOD JOISTS) EXCEPT AT SUPPORT POINTS.
- READ FIXTURE LABEL OR INSTRUCTIONS FURNISHED WITH THE FIXTURE FOR SPECIAL REQUIREMENTS OR LIMITATIONS.

INHERENTLY PROTECTED
- IF MARKED INHERENTLY PROTECTED, THE FIXTURE IS SO DESIGNED THAT THE SURFACE TEMPERATURE WILL NOT EXCEED 90°C EVEN IF THE FIXTURE IS COVERED WITH INSULATION, IS MIS-LAMPED OR OVER-LAMPED. AN EXAMPLE MIGHT BE "DOUBLE-WALLED" CONSTRUCTION.
- THESE FIXTURES ARE NOT THERMALLY PROTECTED.
- READ FIXTURE LABEL OR INSTRUCTIONS FURNISHED WITH THE FIXTURE FOR SPECIAL REQUIREMENTS OR LIMITATIONS.

SUSPENDED
- FOR INSTALLATION IN A SUSPENDED GRID ONLY WHERE THE LAY-IN TILES ARE NOT FASTENED IN PLACE, AND WHERE THE TIE WIRES, T BARS, CEILING TILES, AND OTHER COMPONENTS DIRECTLY ASSOCIATED WITH THE GRID ARE NOT PART OF THE BUILDING STRUCTURE.
- THE FIXTURES ARE INTENDED TO BE MOUNTED IN CEILING OPENINGS.
- READ FIXTURE LABEL OR INSTRUCTIONS FURNISHED WITH THE FIXTURE FOR SPECIAL REQUIREMENTS OR LIMITATIONS. SEE FIGURE 7-3.

Fig. 7-12   This figure shows some of the most important Underwriters Laboratories and *National Electrical Code®* requirements for fluorescent and incandescent fixtures. Always refer to the UL Standards, the *NEC,®* and the label and/or instructions furnished with the fixture.

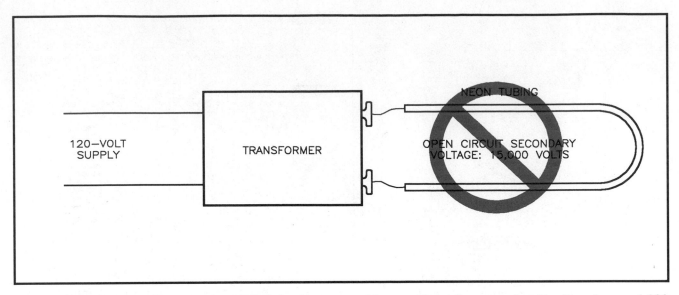

Fig. 7-13   It is NOT permitted to install lighting in or on residences where the open-circuit voltage is over 1,000 volts. This is a violation of *Section 410-80(b)*.

Glare: Glare, whether direct or reflected, such as from floodlights, and as differentiated from general illuminations, shall not be visible at any property line.

Exterior Lighting: Any lights used for exterior illuminations shall direct light away from adjoining properties.

Before installing exterior lighting fixtures, be sure to check with the building authorities in your community to see if there are any requirements or restrictions for outdoor lighting.

REVIEW

1. Is it permissible to install a recessed fixture directly against wood ceiling joists when the label on the fixture does not indicate that the fixture is "suitable for insulation to be in direct contact with the fixture"?

   _____

2. If a recessed fixture without an approved junction box is installed, what extra wiring must be provided? _____

   _____

   _____

   _____

   _____

   _____

   _____

3. Thermal insulation shall not be installed within _____ inches (millimeters) of the top or _____ inches (millimeters) of the side of a recessed fixture unless the fixture is identified for use in direct contact with thermal insulation.

4. Recessed fixtures are available for installation in direct contact with thermal insulation. These fixtures bear the UL mark "Type _____ ."

5. Unless specifically designed, all recessed incandescent fixtures must be provided with factory-installed _____ .

6. Plans require the installation of a surface-mounted fluorescent fixture on the ceiling of a recreation room that is finished with low-density ceiling fiberboard. What sort of mark would you look for on the label of the fixture? _____

_____

_____

7. If a recessed fixture bears no marking that it is listed for branch-circuit feedthrough wiring, is it permitted to run the circuit conductors from fixture to fixture? What section of the Code covers this? _____

8. Fluorescent ballasts for all indoor applications must be _____ type. These ballasts contain internal _____ protection to protect against overheating.

9. Additional backup protection for ballasts can be provided by connecting a(an) _____ with the proper size fuse as recommended by the ballast manufacturer.

10. You are called upon to install a number of lighting fixtures in a suspended ceiling. The ceiling will be dropped approximately 8 inches from the ceiling joists. Briefly explain how you might go about wiring these fixtures. _____

_____

_____

_____

_____

_____

_____

_____

_____

_____

_____

_____

_____

11. The Code places a maximum open-circuit voltage on lighting equipment in or on homes. This maximum voltage is (600) (750) (1000). (Circle the correct answer.) Where in the *National Electrical Code*® is this voltage maximum referenced?

_____

_____

12. The letter **E** in a circle on a ballast nameplate indicates that the ballast _____

_____

_____

13. A 2-lamp fluorescent ballast for two 40-watt lamps is marked 85 volt-amperes. What is the power factor of the ballast? _____

_____

_____

14. Can an incandescent lamp dimmer be used to control a fluorescent lamp load? _____

15. A good "rule-of-thumb" to estimate the expected life of a motor, a ballast, or other electrical equipment is that for every _____°C above rated temperature, the expected life will be cut in _____ .

16. A post light has a 12-volt, 60-watt lamp installed. The lamp has an expected lamp life of 1,000 hours. The homeowner installed a dimmer ahead of the post light, and leaves the dimmer set so the output voltage is 100 volts. The lamp burns slightly dimmer when operated at 100 volts, but this is not a problem. What is the expected lamp life when operated at 100 volts? _____

_____

_____

_____

17. Does your community have exterior outdoor lighting restrictions? _____
   If yes, what are they? _____

_____

_____

UNIT 8

Lighting Branch-Circuit for Front Bedroom

OBJECTIVES

After studying this unit, the student will be able to

- explain the factors that influence the grouping of outlets into circuits.
- understand the meaning of general, accent, task, and security lighting.
- estimate loads for the outlets of a circuit.
- draw a cable layout and a wiring diagram based on information given in the residence plans, the specifications, and Code requirements.
- select the proper wall box for a particular installation.
- explain how wall boxes can be grounded.
- list the requirements for the installation of fixtures in clothes closets.

This unit is longer than the other units because it is the first exposure to real-life installations. Most issues confronting the electrician for a typical residential installation are covered in this unit. Repetition in later units is kept to a minimum.

GROUPING OUTLETS

The grouping of outlets into circuits must conform to *National Electrical Code®* standards and good wiring practices. There are many possible combinations or groupings of outlets. In most residential installations, circuit planning is usually done by the electrician, who must insure that the circuits conform to Code requirements. In larger, more costly residences, the circuit layout may be completed by the architect and included on the plans.

Because many circuit arrangements are possible, there are few guidelines for selecting outlets for a particular circuit. An electrician plans circuits that are economical without sacrificing the quality of the installation.

For example, some electricians prefer to have more than one circuit feed a room. Should one circuit have a problem, the second circuit would still continue to supply power to the other outlets in that room. See figure 8-1.

Another excellent way to provide good wiring and at the same time economize on wiring materials is to connect receptacle outlets back-to-back, as illustrated in figure 8-2.

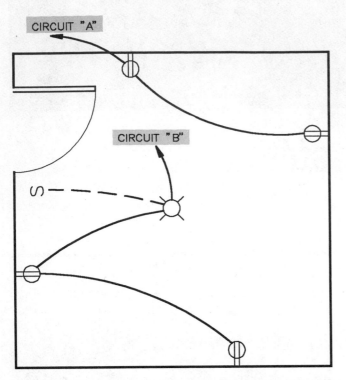

Fig. 8-1   Wiring layout showing one room that is fed by two different circuits. If one of the circuits goes out, the other circuit will still provide electricity to the room.

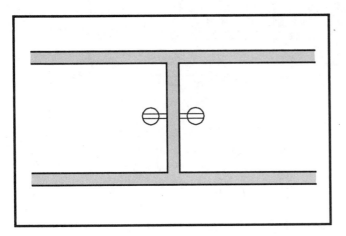

Fig. 8-2   Receptacle outlets connected back-to-back. This can reduce the cost of the installation because of the short distance between the outlets.

Some electricians consider it poor practice to include outlets on different floors on the same circuit. Here, too, the decision can be a matter of personal choice. Some local building codes limit this type of installation to lights at the head and foot of a stairway.

Residential Lighting

Residential lighting is a personal thing. The homeowner, builder, and electrical contractor must meet to decide on what types of lighting fixtures are to be installed in the residence. Many variables (cost, personal preference, construction obstacles, etc.) must be taken into consideration.

Residential lighting can be segmented into four groups.

General Lighting. General lighting provides over-all illumination for a given area, such as the front hall, the bedroom hall, the workshop, or the garage lighting in this residence. General lighting can be very basic, which means just getting the job done by providing adequate lighting for the area involved. Or, it can take the form of decorative lighting, such as a chandelier over a dining room table or other decorative fixtures that can be attached to ceiling paddle fans.

Accent Lighting. Accent lighting provides focus and attention to an object or area in the home. Examples are the recessed "spots" over the fireplace. Another example is the track lighting in the living room of this residence, which can accent a picture, photo, painting, or sculpture that might be hung on the wall. To be effective, accent lighting should be at least five times that of the surrounding general lighting.

Task Lighting. This is sometimes referred to as "activity" lighting. Task lighting provides proper lighting where "tasks" or "activities" are performed. The fluorescent fixtures above the workbench, the kitchen range hood, and the recessed fixture over the kitchen sink are all examples of "task" lighting. To avoid too much contrast, task lighting should not exceed three times that of the surrounding general lighting.

Security Lighting. Security lighting generally includes outdoor lighting, such as post lamps, wall fixtures, walkway lighting, and all other lighting that serves the purpose of providing lighting for security and safety reasons. Security lighting in many instances is provided by the normal types of lighting fixtures found in the typical residence. In the residence discussed in this text, the outdoor bracket fixtures in front of the garage and next to the entry doors, as well as the post light, might be considered to be security lighting even though they add to the beauty of the residence. Security lighting can be controlled manually with regular wall switches, timers, light sensors, or motion detectors.

An organization called the American Lighting Association offers excellent reading materials with residential lighting suggestions and other materials relating to residential lighting. This organization is made up of the many manufacturers of lighting fixtures, as well as electrical distributors that maintain extensive lighting showrooms across the country. A visit to one of these showrooms offers the prospective buyer of lighting fixtures an opportunity to select from hundreds (in some cases thousands) of lighting fixtures. Most of the finer lighting showrooms are staffed with individuals highly qualified to make recommendations to homeowners so that they will get the most value for their money.

The lighting provided throughout this residence conforms to the residential lighting standards as recommended by the American Lighting Association. Certainly many other variations are possible. For the purposes of studying an entire wiring installation for this residence, all of the load calculations, wiring diagrams, and so on, have been accomplished using the fixture selection as indicated on the plans and in the specifications.

CABLE RUNS

A student studying the great number of wiring diagrams throughout this text will find many different ways to run the cables and make up the circuit connections. Sizing of boxes that will conform to the Code for the number of wires and devices is covered in great detail in this text. When a circuit is run "through" a recessed fixture like the type shown in figure 7-1, that recessed fixture must be identified for "through wiring." This subject is discussed in detail in unit 7. When in doubt as to whether or not the fixture is suitable for running wires to and beyond it to another fixture or part of

the circuit, then it is recommended that the circuitry be designed so as to end up at the recessed fixture with only the two wires that will connect to that fixture.

To summarize, the grouping of outlets into circuits must satisfy the requirements of the *National Electrical Code,*® local building codes, good wiring practices, and common sense. A good wiring practice is to divide all loads as evenly as possible among the circuits, as required by *Section 220-4(d).*

ESTIMATING LOADS FOR OUTLETS

Although this text is not intended to be a basic electrical theory text, it should be mentioned that, when calculating loads,

$$VOLTS \times AMPERES = VOLT\text{-}AMPERES$$

Yet, many times we say that

$$VOLTS \times AMPERES = WATTS$$

What we really mean to say is that

$$VOLTS \times AMPERES \times POWER\ FACTOR = WATTS$$

In a pure resistive load such as a simple light bulb, a toaster, a flat iron, or a resistance electric heating element, the power factor is 100%. Then

$$VOLTS \times AMPERES \times 1 = WATTS$$

When we are involved with transformers, motors, ballasts, and other "inductive" loads, wattage is not necessarily the same as volt-amperes.

EXAMPLE: Calculate the wattage and volt-amperes of a 120-volt, 10-ampere resistive load.

SOLUTIONS:
   a. $120 \times 10 \times 1 = 1200$ watts
   b. $120 \times 10 = 1200$ volt-amperes

EXAMPLE: Calculate the wattage and volt-amperes of a 120-volt, 10-ampere motor load at 50% power factor.

SOLUTIONS:
   a. $120 \times 10 \times 0.5 = 600$ watts
   b. $120 \times 10 = 1200$ volt-amperes

Therefore, to be sure that adequate ampacity is provided for in branch-circuit wiring, feeder sizing, and service-entrance calculations, the Code requires that we use the term **VOLT-AMPERES**. This allows us to ignore power factor, and address the *true* current draw that will enable us to determine the correct ratings of electrical equipment.

However, in some instances the terms **WATTS** and **VOLT-AMPERES** can be used interchangeably without creating any problems. For instance, the Code, in *Section 220-19*, recognizes that for electric ranges and other cooking equipment, the kVA and kW ratings shall be considered to be equivalent for the purpose of branch-circuit and feeder calculations.

Examples No. 2(a) and (b) in *Chapter 9* of the Code indicate that for wall-mounted ovens, counter-mounted cooking units, water heaters, dishwashers, and combination clothes dryers, their kW ratings are equivalent to kVA values. Therefore, throughout this text the terms **WATTAGE** and **VOLT-AMPERES** are used when calculating and/or estimating loads.

Building plans typically do not specify the ratings in watts for the outlets shown. When planning circuits, the electrician must consider the types of fixtures that may be used at the various outlets. To do this, the electrician must know the general uses of the receptacle outlets in the typical dwelling.

Unit 3 shows that the general lighting load of a residence is determined by allowing a load of 3 volt-amperes for each square foot (0.093 m²) of floor area, *Section 220-3(b)*. For the residence in the plans, it is shown that six lighting circuits meet the minimum standards set by the Code. However, to provide sufficient capacity, 13 lighting circuits are to be installed in this residence.

How Many Outlets Are Permitted on One Branch-Circuit?

The *National Electrical Code®* does not specify the maximum number of receptacle outlets or lighting outlets that may be connected to one 120-volt lighting or small appliance branch-circuit in a residence. It may seem to be ridiculous and illogical that 10, 20, or more receptacle outlets and lighting outlets can be connected to one branch-circuit and not be in violation of the Code. Consider the fact, however, that having many "convenience" receptacles is "safer" because many receptacle outlets will virtually eliminate the use of extension cords, one of the highest reported causes of electrical fires. Rarely, if ever, would all receptacle outlets and lighting outlets be fully loaded at the same time. There is much diversity of load in residential occupancies.

Section 220-3(c)(7) indicates that "other outlets" are to be figured at 180 VA per outlet. However, this requirement does not pertain to house wiring, *Section 220-13*. The asterisk (*) note following *Section 220-3(c)(7)* tells us that the 180 VA requirement does not apply to the receptacles connected to the small appliance branch-circuits in a residence.

The load for small appliance branch-circuits in homes is figured at 1,500 watts for each 2-wire small appliance branch-circuit, *Section 220-16*.

The general-use receptacle outlets in homes are considered to be general lighting, and their load is included in the general lighting load calculation, and that no additional load calculations are necessary, *Table 220-3(b)*, asterisk (*) footnote.

Circuit Loading Guidelines

A good rule to follow in residential wiring (in fact *any* wiring) is *never load the circuit to more than 80% of its rating*. A 15-ampere, 120-volt branch-circuit would be calculated as

$$15 \times 0.80 = 12 \text{ amperes}$$
$$\text{OR}$$
$$12 \text{ amperes} \times 120 \text{ volts} = 1440 \text{ volt-amperes}$$

Although 20-ampere lighting circuits are generally not installed in residences, the maximum allowable load in volt-amperes for such a circuit is

$$20 \times 0.80 = 16 \text{ amperes}$$
$$\text{OR}$$
$$16 \text{ amperes} \times 120 \text{ volts} = 1920 \text{ volt-amperes}$$

This is one method that can be used to *estimate* residential lighting loads.

Certain fixtures, such as recessed lights and fluorescent lights, are marked with their maximum lamp wattage and ballast current (for fluorescent fixtures only). Other fixtures, however, are not marked. Furthermore, the electrician does not know the exact load that will be connected to the receptacle outlets. Also unknown is the size of the lamps that will be installed in the lighting fixtures (other than the recessed and fluorescent types). In other words, it is difficult to anticipate what the homeowner may do after the installation is complete. The electrician should remember that the room in which the outlets are located does give some indications as to their possible uses. The circuits should be planned accordingly.

Load Estimation

The lamp loads in the lighting fixtures can be estimated by assuming the lamp wattages that will probably be needed in each fixture to provide adequate lighting for the area involved. Recommendations are found in American Lighting Association publications and in various manufacturers' publications. It is recommended that the student write to these organizations and request copies of the latest lighting publications. The Instructor's Guide for Residential Wiring lists the names, addresses, and titles of available publications.

Estimating Number of Outlets by Assigning an Amperage Value to Each

One method of determining the number of lighting and receptacle outlets to be included on one circuit is to assign a value of 1 to 1½ amperes to each outlet to a total of 15 amperes. Thus, a total of 10 to 15 outlets can be included in a 15-ampere circuit.

As stated previously, the *National Electrical Code®* does not limit the number of outlets on one circuit. However, many local building codes do specify the maximum number of outlets per circuit. Before planning any circuits, the electrician must check the local building code requirements.

All outlets will not be required to deliver 1½ amperes. For example, closet lights, night lights, and clocks will use only a small portion of the allowable current. A 60-watt closet light will draw less than 1 ampere.

$$I = \frac{W}{E} = \frac{60}{120} = 0.5 \text{ ampere}$$

If low-wattage fixtures are connected to a cir-

cuit, it is quite possible that 15 or more lighting outlets (many times referred to as "openings") could be connected to the circuit without a problem. On the other hand, if the load consists of high-wattage lamps, the number of outlets would be less. In this residence, estimated loads were figured at 120 volt-amperes (1 ampere) for those receptacle outlets obviously intended for general use. Receptacle outlets in the garage and workshop were figured at 180 volt-amperes (1½ amperes) because of the use these outlets will be supplying (i.e., tools, drills, table saws, etc.).

Receptacle outlets in the kitchen and laundry have been included in the small appliance circuit calculations.

If an electrician were to *estimate* the number of 15-ampere lighting branch-circuits desired for a new residence where the total count of lighting and receptacle outlets is, for example, 80, then

$$\frac{80}{10} = 8 \quad \text{(eight 15-ampere lighting circuits)}$$

If the circuit consists of low-wattage loads, then

$$\frac{80}{15} = 5.3 \quad \text{(six 15-ampere lighting circuits)}$$

In residential occupancies there is great diversity in the loading of lighting branch-circuits.

CAUTION: The foregoing procedure for estimating the number of lighting and receptacle outlets is for branch-circuits only. Obviously, where the small appliance circuits in the kitchen, laundry area, and other areas supply heavy concentrations of plug-in appliances, the 1½ amperes per receptacle outlet would not be applicable. These small appliance circuits are discussed later on in this text.

Review of How to Determine the Minimum Number of Lighting Circuits

1. Use the volt-amperes per square foot method as required by the Code and discussed in unit 3.

2. Estimate the probable loading for each lighting outlet and receptacle outlet (do not include the small appliance receptacle outlets). Try to have a total VA not over 1440 for a 15-ampere lighting circuit, and not over 1920 for a 20-ampere lighting circuit.

3. Connect approximately 10 to 15 outlets per circuit—more than 10 if the probable connected load is small, less than 10 if the probable connected load is large. For instance, the wattage of a 120-volt smoke detector is approximately ½ watt.

Whereas item 1 is the method required by the Code, items 2 and 3 are guidelines and are not to be considered "cast-in-concrete" rules, because general lighting loads in homes are so diversified. As the electrician lays out the circuits of a residence, these guidelines will generally provide adequate circuitry. This would include both lighting outlets and receptacle outlets that are intended for general lighting, *not* appliance circuit receptacle outlets. Be practical. Consider the wide diversity factor that is present.

Divide Loads Evenly

Section 220-4(d) makes it mandatory to divide loads evenly between the various circuits. The obvious reason is not to experience overload conditions on some circuits, whereas other circuits might be lightly loaded. This is common sense. At the Main Service, don't connect the branch-circuits and feeders to result in, for instance, 120 amperes on phase A and 40 amperes on phase B. Look at the probable and/or calculated loads to attain as close a balance as possible, like 80 amperes on each A and B phase.

SYMBOLS

The symbols used on the cable layouts in this text are the same as those found on the actual electrical plans for the residence. Refer to figure 2-4.

Pictorial illustrations are used on all wiring diagrams in this text to make it easy for the reader to complete the wiring diagrams, figure 8-3.

DRAWING THE WIRING DIAGRAM OF A LIGHTING CIRCUIT

The electrician must take information from the building plans and convert it to the forms that will be most useful in planning an installation that is economical, conforms to all Code regulations, and follows good wiring practice. To do this, the electrician first makes a cable layout. Then a wiring diagram is prepared to show clearly each wire and all connections in the circuit.

A skilled electrician, after many years of experience, will not prepare wiring diagrams for most residential circuitry because the skilled electrician has the ability to "think" the connections through

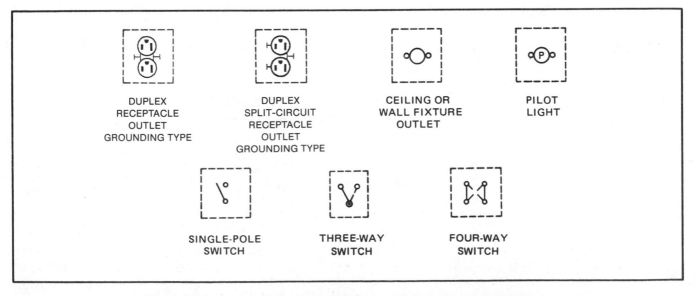

Fig. 8-3   Pictorial illustrations used on wiring diagrams in this text.

mentally. An unskilled individual should make detailed wiring diagrams.

The following steps will guide the student in the preparation of wiring diagrams. The student is required to draw wiring diagrams in later units of this text. As the student works on the wiring diagrams of the many circuits provided in this text, note that the receptacles are positioned exactly as though you were standing in the center of the room. Then, as you turn to face each wall, you will find the receptacles positioned in that wall with the grounded conductor slot on the top, as shown in figure 8-4.

1. Refer to plans and make a cable layout of all lighting and receptacle outlets, as in figure 8-5.

2. Draw a wiring diagram showing the traveler conductors for all three-way switches, if any, as in figure 8-6.

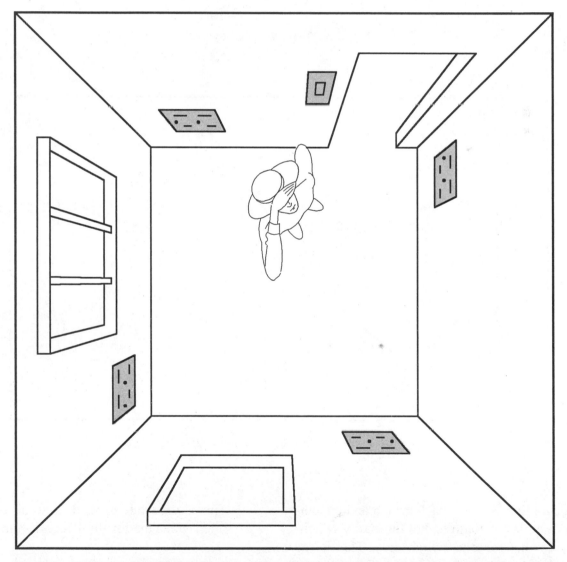

Fig. 8-4   This top view of a room shows the relative positioning of receptacle outlets so that the wide grounded terminal slot is at the top. This provides for added safety if a metal face plate loosens or if a metal clip or other object falls on top of the blades of an attachment plug cap that is not plugged in all the way. All of the wiring diagrams throughout this text will illustrate the receptacles in this manner.

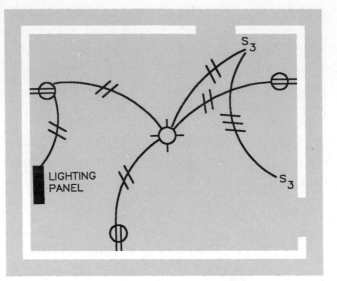

Fig. 8-5   Typical cable layout.

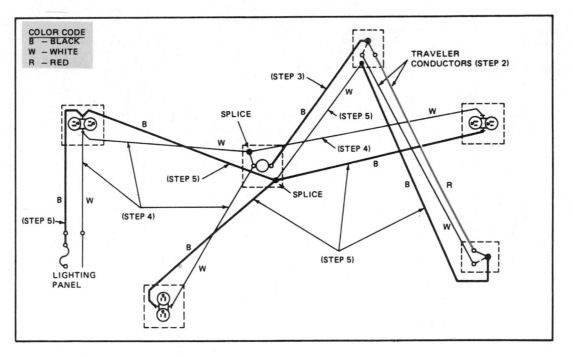

Fig. 8-6   Wiring diagram of circuit shown in figure 8-5.

3. Draw a line between each switch and the outlet or outlets it controls. For three-way switches, do this for one switch only. This is the "switch leg."

4. Draw a line from the grounded terminal on the lighting panel to each current-consuming outlet. This line may pass through switch boxes, but must not be connected to any switches.

This is the white grounded circuit conductor, often called the "neutral" conductor.

NOTE: An exception to step 4 may be made for double-pole switches. For these switches, all conductors of the circuit are opened simultaneously. They are rarely used in residential wiring.

5. Draw a line from the ungrounded "hot" terminal on the panel to connect to each switch and unswitched outlet. Connect to one three-way switch only.

6. Show splices as small dots where the various wires are to be connected together. In the wiring diagram, the terminal of a switch or outlet may be used for the junction point of wires. In actual wiring practice, however, the Code does not permit more than one wire to be connected to a terminal unless the terminal is a type identified for use with more than one conductor. The standard screw-type terminal is *not* acceptable for more than one wire, *Section 110-14*.

7. The final step in preparing the wiring diagram is to mark the color of the conductors, as in figure 8-6. Note that the colors selected—black (B), white (W), and red (R)—are the colors of two- and three-conductor cables (refer to unit 5). It is suggested that the student use colored pencils or markers for different conductors when drawing the wiring diagram. Yellow can be used to indicate white conductors to provide better contrast with white paper. A note to this effect should be placed on the diagram, *Section 200-7*.

LIGHTING BRANCH-CIRCUIT A16 FOR FRONT BEDROOM

This is a 15-ampere branch-circuit. As we look at this circuit, we find five split-circuit receptacle outlets. These will provide the general lighting through the use of table or swag lamps that will be plugged into the switched receptacles. See figure 8-7.

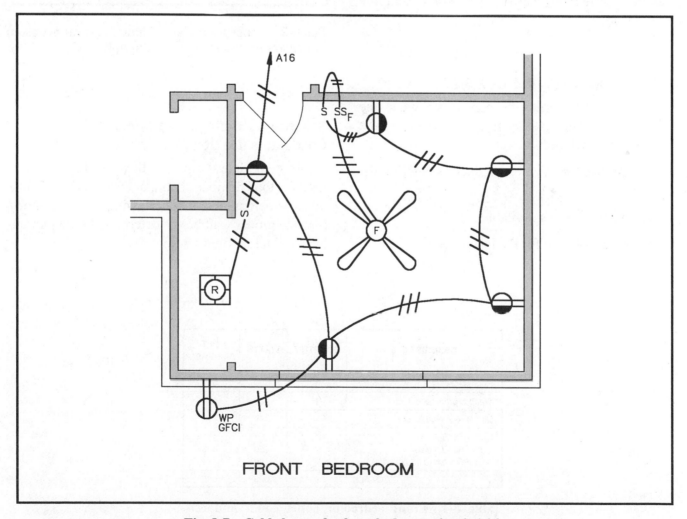

FRONT BEDROOM

Fig. 8-7   Cable layout for front bedroom circuit A16.

Radio, television, clocks, and other electrical and electronic items not intended to be controlled by the wall switch will be plugged into the "hot continuously" receptacle.

Next to the wall switch we find the control for the ceiling fan, which also has a lighting fixture as an integral part of the fan, figure 8-8. A single-pole switch controls the ON/OFF and speed of the fan motor.

The recessed closet light is controlled by a single-pole switch outside and to the right of the closet door.

Note that one outside weatherproof receptacle is also connected to circuit A16. Outdoor receptacles must have GFCI protection. See unit 6 for a complete discussion regarding GFCI protection.

Checking the actual electrical plans, we find one television outlet and one telephone outlet are to be installed in the front bedroom. Television and telephones are discussed in unit 25. Table 8-1 summarizes the outlets in the front bedroom and the estimated load.

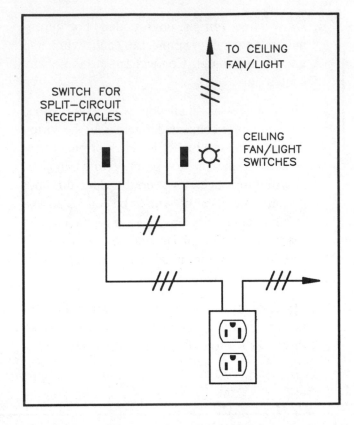

Fig. 8-8   Conceptual view of how the front bedroom switch arrangement is to be accomplished.

DETERMINING THE WALL BOX SIZE

Several factors must be considered when the electrician determines the size of the wall box. These factors include:

- the number and size of conductors entering the box.
- the types of boxes available.
- the space allowed for the installation of the box.

Box Size According to the Number of Conductors in a Box *(Section 370-16)*

To determine the proper box size for any location, the total number of conductors entering the box must be determined. The following example shows how the proper wall box is determined for a particular installation, figure 8-9.

DESCRIPTION	QUANTITY	WATTS	VOLT-AMPERES
Receptacles @ 120 watts each	5	600	600
Weatherproof receptacle	1	120	120
Closet recessed fixture	1	75	75
Ceiling fan/light 3 - 50-W lamps Fan motor (0.75 A @ 120 V)	1	 150 80	 150 90
TOTALS	8	1,025	1,035

Table 8-1   Front bedroom: outlet count and estimated load. Circuit A16.

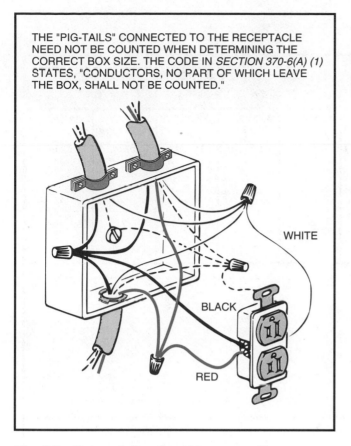

THE "PIG-TAILS" CONNECTED TO THE RECEPTACLE NEED NOT BE COUNTED WHEN DETERMINING THE CORRECT BOX SIZE. THE CODE IN *SECTION 370-6(A) (1)* STATES, "CONDUCTORS, NO PART OF WHICH LEAVE THE BOX, SHALL NOT BE COUNTED."

WHITE

BLACK

RED

Fig. 8-9 Determining size of box according to number of conductors.

1. Add the No. 14 AWG circuit conductors:

   2 + 3 + 3 =   8

2. Add equipment grounding wires (count one only)   1

3. Add two conductors for the receptacle   2

   Total   11 Conductors

Note that the pigtails connected to the receptacle, figure 8-9, need *not* be counted when determining the correct box size. ▶ The Code in *Section 370-16(b)(1)* states, "A conductor, no part of which leaves the box, *shall not* be counted." ◀

▶ Once the total number of conductors, plus the volume count required for wiring devices, fixture studs, hickeys, and clamps is known, refer to *Table 310-16(a)* of the *National Electrical Code*® or Table 8-2 in this unit to find a box that has sufficient space. ◀ For example, a 4" × 2⅛" square box with a suitable plaster ring can be used.

The volume of the box plus the space provided by plaster rings, extension rings, and raised covers may be used to determine the total available volume. In addition, it is desirable to install boxes with external cable clamps. Remember that if the box contains one or more devices, such as cable clamps, fixture studs, or hickeys, the number of conductors permitted in the box shall be one less than shown in *Table 370-16(a)* for *each type* of device contained in the box. (See the example under "Number of Conductors" in unit 2.)

GROUNDING OF WALL BOXES

The specifications for the residence state that *all* metal boxes are to be grounded. The means of grounding is armored cable or nonmetallic-sheathed cable containing an extra equipment grounding conductor. This conductor is used only to ground the metal box. This bare equipment grounding conductor must *not* be used as a current-carrying conductor because severe shocks can result.

According to *Section 210-7(a)*, grounding-type receptacles must be installed on 15-ampere and 20-ampere branch-circuits. The methods of attaching the equipment grounding conductor to the proper terminal on the convenience receptacle are covered in unit 5.

POSITIONING OF SPLIT-CIRCUIT RECEPTACLES

The receptacle outlets shown in the front bedroom are called two-circuit or split-circuit receptacles. The top portion of such a receptacle is hot at all times, and the bottom portion is controlled by the wall switch, figure 8-10. It is recommended that the electrician wire the bottom section of the receptacle as the switched section. As a result, when the

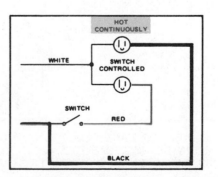

Fig. 8-10 Split-circuit wiring for receptacles in bedroom.

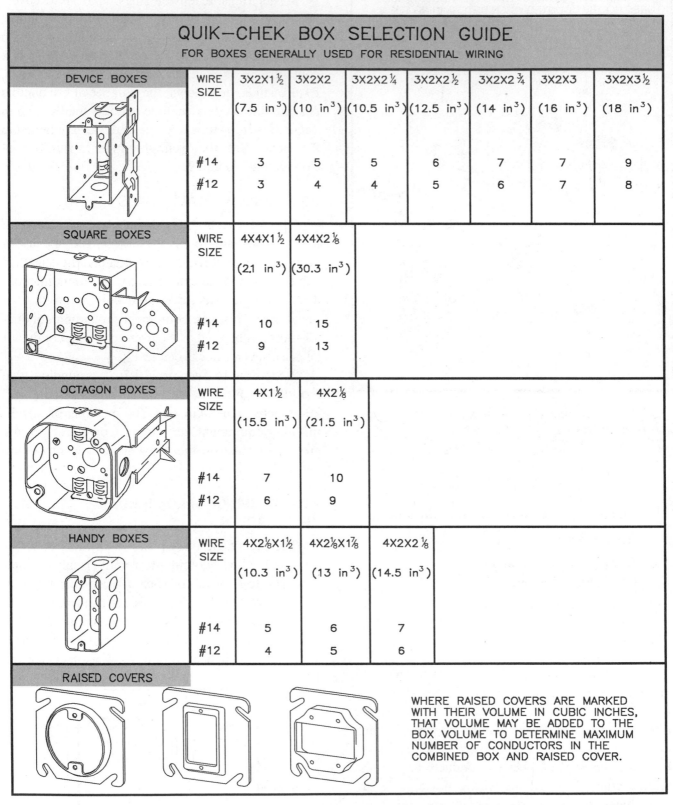

QUIK–CHEK BOX SELECTION GUIDE
FOR BOXES GENERALLY USED FOR RESIDENTIAL WIRING

DEVICE BOXES

WIRE SIZE	3X2X1½ (7.5 in³)	3X2X2 (10 in³)	3X2X2¼ (10.5 in³)	3X2X2½ (12.5 in³)	3X2X2¾ (14 in³)	3X2X3 (16 in³)	3X2X3½ (18 in³)
#14	3	5	5	6	7	7	9
#12	3	4	4	5	6	7	8

SQUARE BOXES

WIRE SIZE	4X4X1½ (21 in³)	4X4X2⅛ (30.3 in³)
#14	10	15
#12	9	13

OCTAGON BOXES

WIRE SIZE	4X1½ (15.5 in³)	4X2⅛ (21.5 in³)
#14	7	10
#12	6	9

HANDY BOXES

WIRE SIZE	4X2⅛X1½ (10.3 in³)	4X2⅛X1⅞ (13 in³)	4X2X2⅛ (14.5 in³)
#14	5	6	7
#12	4	5	6

RAISED COVERS

WHERE RAISED COVERS ARE MARKED WITH THEIR VOLUME IN CUBIC INCHES, THAT VOLUME MAY BE ADDED TO THE BOX VOLUME TO DETERMINE MAXIMUM NUMBER OF CONDUCTORS IN THE COMBINED BOX AND RAISED COVER.

NOTE: BE SURE TO MAKE DEDUCTIONS FROM THE ABOVE MAXIMUM NUMBER OF CONDUCTORS PERMITTED FOR WIRING DEVICES, CABLE CLAMPS, FIXTURE STUDS, AND GROUNDING CONDUCTORS. THE CUBIC INCH (IN³) VOLUME IS TAKEN DIRECTLY FROM *TABLE 370–6A* OF THE CODE.

Table 8-2   Quik-Chek Box Selection Guide.

attachment plug cap of a lamp is inserted into the bottom switched portion of the receptacle, the cord does not hang in front of the unswitched section. This unswitched section can be used as a receptacle for clock, vacuum cleaner, radio, stereo, compact disc player, personal computer, television, or other appliances where a switch control is not necessary or desirable.

When split-circuit receptacles are horizontally mounted, which is common when 4" square boxes are used because the plaster ring is easily fastened to the box in either a vertical or horizontal position, locate the switched portion to the right.

POSITIONING OF RECEPTACLES NEAR ELECTRIC BASEBOARD HEATING

This unit is the first unit in this text where we begin to discuss actual circuit layout of receptacle outlets, lighting outlets, and switches.

Later we discuss electric heat. Electric heating for a home can be accomplished with an electric furnace, electric baseboard heating units, heat pump, or resistance heating cables embedded in plastered ceilings or sandwiched between two layers of dry-wall sheets on the ceiling.

The important thing to remember at this point is

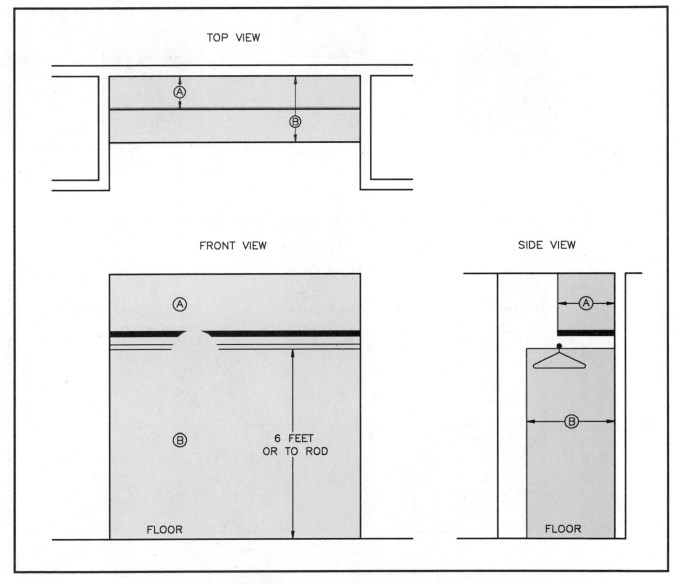

Fig. 8-11   Typical closet with one shelf and one rod. The shaded area defines "storage space." Dimension A is width of shelf or 12 inches from wall, whichever is greater. Dimension B is below rod, 24 inches from wall. See *Section 410-8(a).*

that there are stringent *National Electrical Code*® rules governing the positioning of wall electrical baseboards in relationship to the location of wall receptacle outlets. See *Section 210-52(a), Exception* and *FPN*.

See unit 23 for a detailed discussion of installing electric baseboard heating units and their relative positions below receptacle outlets.

The front bedroom does have a ceiling fan/light on the ceiling. Ceiling fans are discussed in unit 9.

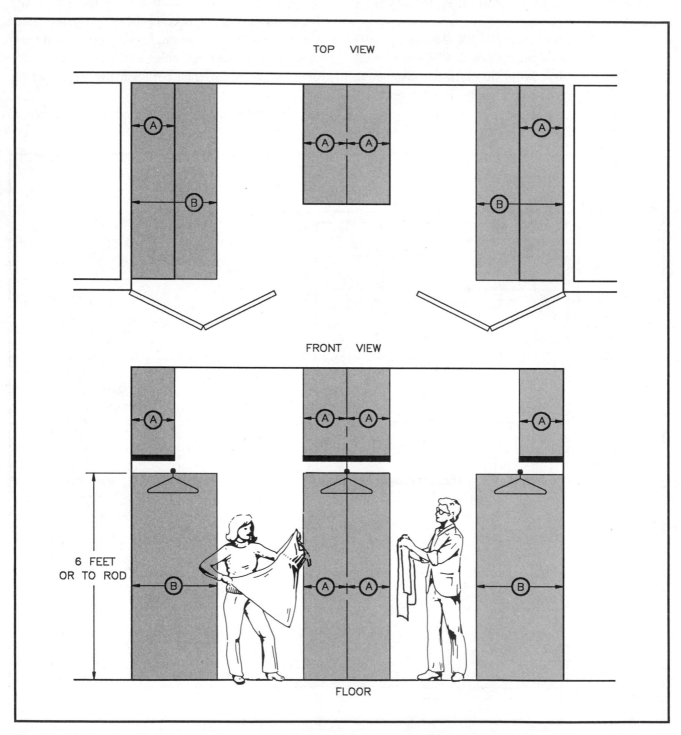

Fig. 8-12   Large walk-in closet where there is access to the center rod from both sides. The shaded area defines "storage space." Dimension A is width of shelf or 12 inches, whichever is greater. Dimension B is 24 inches from wall. See *Section 410-8(a)*.

FIXTURES IN CLOTHES CLOSETS

The *National Electrical Code*® does not require lighting fixtures in clothes closets, *Section 210-70*. However, some local codes and some building codes do. When lighting fixtures are installed in clothes closets, they must be installed properly.

Clothing, boxes, and other material normally stored in clothes closets are a potential fire hazard. These items may ignite on contact with the hot surface of exposed lamps. Incandescent lamps have a hotter surface temperature than fluorescent lamps. *Section 410-8* of the Code gives very specific rules relative to the location and types of lighting fixtures permitted to be installed in clothes closets. Figures 8-11 through 8-15 illustrate the requirements of *Section 410-8*.

Figure 8-16 shows typical bedroom-type lighting fixtures.

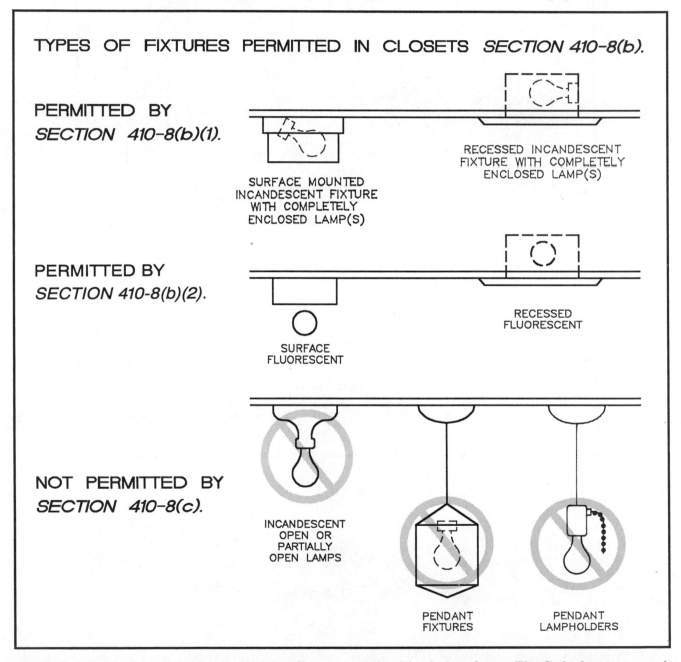

Fig. 8-13 Illustrations of the types of lighting fixtures permitted in clothes closets. The Code does not permit bare incandescent lamps, pendant fixtures, or pendant lampholders to be installed in clothes closets.

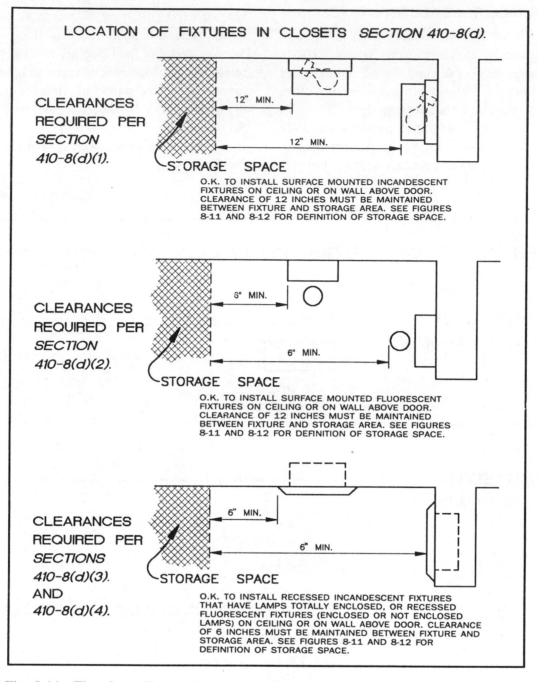

LOCATION OF FIXTURES IN CLOSETS *SECTION 410-8(d).*

CLEARANCES REQUIRED PER *SECTION 410-8(d)(1).*

12" MIN.

12" MIN.

STORAGE SPACE

O.K. TO INSTALL SURFACE MOUNTED INCANDESCENT FIXTURES ON CEILING OR ON WALL ABOVE DOOR. CLEARANCE OF 12 INCHES MUST BE MAINTAINED BETWEEN FIXTURE AND STORAGE AREA. SEE FIGURES 8-11 AND 8-12 FOR DEFINITION OF STORAGE SPACE.

CLEARANCES REQUIRED PER *SECTION 410-8(d)(2).*

6" MIN.

6" MIN.

STORAGE SPACE

O.K. TO INSTALL SURFACE MOUNTED FLUORESCENT FIXTURES ON CEILING OR ON WALL ABOVE DOOR. CLEARANCE OF 12 INCHES MUST BE MAINTAINED BETWEEN FIXTURE AND STORAGE AREA. SEE FIGURES 8-11 AND 8-12 FOR DEFINITION OF STORAGE SPACE.

CLEARANCES REQUIRED PER *SECTIONS 410-8(d)(3). AND 410-8(d)(4).*

6" MIN.

6" MIN.

STORAGE SPACE

O.K. TO INSTALL RECESSED INCANDESCENT FIXTURES THAT HAVE LAMPS TOTALLY ENCLOSED, OR RECESSED FLUORESCENT FIXTURES (ENCLOSED OR NOT ENCLOSED LAMPS) ON CEILING OR ON WALL ABOVE DOOR. CLEARANCE OF 6 INCHES MUST BE MAINTAINED BETWEEN FIXTURE AND STORAGE AREA. SEE FIGURES 8-11 AND 8-12 FOR DEFINITION OF STORAGE SPACE.

Fig. 8-14 The above illustrations show the minimum required clearances between lighting fixtures and the storage space.

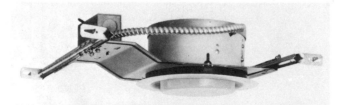

Fig. 8-15 Recessed incandescent closet lighting fixture with pull-chain switch that may be used where a separate wall switch is not installed.

Fig. 8-16   Typical bedroom-type lighting fixtures.

REVIEW

Note: Refer to the Code or the plans where necessary.

1. Can the outlets in a circuit be arranged in different groupings to obtain the same result? Why? _____

_____

_____

2. Is it good practice to have outlets on different floors on the same circuit? Why?

_____

_____

3. What usually determines the grouping of outlets into a circuit? _____

_____

_____

_____

_____

_____

4. A good rule to follow is to never load a circuit to more than _____ percent of the branch-circuit rating.

5. The *NEC*® does not limit the number of lighting and receptacle outlets permitted on one branch-circuit for residential installations. A guideline that is often used by electricians for house wiring is to allow _____ to _____ amperes per outlet. For a 15-ampere circuit, this results in _____ to _____ outlets on the branch-circuit.

6. For this residence, what are the estimated wattages used in determining the loading of branch-circuit A16?

   Receptacles _____ watts (volt-amperes)
   Closet recessed fixture _____ watts (volt-amperes)

7. What is the ampere rating of circuit A16? _____

8. What size wire is used for the lighting circuit in the front bedroom?_____

9. How many receptacles are connected to this circuit?_____

_____

_____

_____

10. What main factor influences the choice of wall boxes?_____

_____

_____

11. How is a wall box grounded? _____

_____

_____

12. What is a split-circuit receptacle? _____

_____

_____

_____

_____

13. Is the switched portion of an outlet mounted toward the top or the bottom? Why?

_____

_____

_____

14. The following questions pertain to lighting fixtures in clothes closets.
    a. Does the Code allow bare incandescent lamp fixtures such as porcelain keyless or porcelain pull-chain lampholders to be installed? _____

    b. Does the Code allow bare fluorescent lamp fixtures to be installed?_____

    c. Does the Code permit pendant fixtures or pendant lampholders to be installed?

    _____

    d. What is the minimum clearance from the storage area to surface-mounted incandescent fixtures? _____

    e. What is the minimum clearance from the storage area to surface-mounted fluorescent fixtures? _____

    f. What is the minimum distance between recessed incandescent or recessed fluorescent fixtures and the storage area? _____

    g. Define the "storage area."

    _____

    _____

    _____

    _____

    _____

    _____

    h. If a clothes hanging rod is installed where there is access from both sides, such as might be found in a large walk-in closet, define the storage area under that rod.

    _____

    _____

    _____

    _____

15. How many switches are in the bedroom circuit and of what type are they? _____

_____

_____

16. The following is a layout of the lighting circuit for the front bedroom. Using the cable layout shown in figure 8-8, make a complete wiring diagram of this circuit. Indicate the color of each conductor.

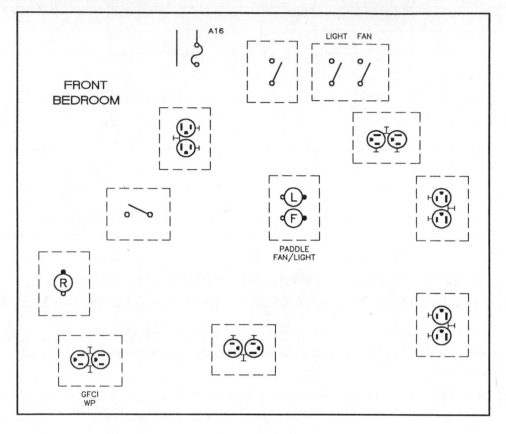

17. When planning circuits, what common practice is followed regarding the division of loads? _____

18. The Code uses the terms *watts, volt-amperes, kW,* and *kVA.* Explain their significance in calculating loads. _____

19. How many No. 14 AWG conductors are permitted in a device box that measures 3" × 2" × 2¾"? _____

20. A 4" × 1½" octagon box has one cable clamp and one fixture stud. How many No. 14 AWG conductors are permitted? _____

UNIT 9

Lighting Branch-Circuit for Master Bedroom

OBJECTIVES

After studying this unit, the student will be able to

- draw the wiring diagram of the cable layout for the master bedroom.
- study Code requirements for the installation of ceiling fans.
- estimate the probable connected load for a room based on the number of fixtures and outlets included in the circuit supplying the room.
- gain more practice in determining box sizing based upon the number of conductors, devices, and clamps in the box.
- make the connections for three-way switches.

The discussion in unit 3 of grouping of outlets, estimating loads, selecting wall box sizes, and drawing wiring diagrams can also be applied to the circuit for the master bedroom.

The residence panel schedules show that the master bedroom is supplied by Circuit A19. Because Panel A is located in the basement below the wall switches next to the sliding doors, the home run for Circuit A19 is brought into the outdoor weatherproof receptacle. This *results* in six conductors in the outdoor receptacle box. The home run could have been brought into the corner receptacle outlet in the bedroom. Again, it is a matter of studying the circuit to determine the best choice for conservation of cable or conduit in these runs and to economically select the correct size of wall boxes.

LIGHTING BRANCH-CIRCUIT A19 FOR MASTER BEDROOM

Figure 9-1 and Table 9-1 and the electrical plans show that Circuit A19 has four split-circuit receptacle outlets in this bedroom, one outdoor weatherproof GFCI receptacle outlet, two recessed closet fixtures, each on a separate switch, plus a ceiling fan/light fixture, one telephone outlet, and one television outlet.

In addition, an outdoor bracket fixture is located adjacent to the sliding door and is controlled by a single-pole switch just inside the sliding door.

The split-circuit receptacle outlets are controlled by two three-way switches. One is located next to the sliding door. As in the front bedroom, living room, and study/bedroom, the use of split-circuit receptacles offers the advantage of having switch

DESCRIPTION	QUANTITY	WATTS	VOLT-AMPERES
Receptacles @ 120 watts each	4	480	480
Weatherproof receptacle	1	120	120
Outdoor bracket fixture	1	150	150
Closet recessed fixtures One 75-W lamp each	2	150	150

Ceiling fan/light
   Three 50-W lamps
   Fan motor (0.75 @ 120 V)
	1	150
		80
			150
			90

TOTALS	9	1130	1140

Table 9-1 Master bedroom outlet count and estimated load. Circuit A19.

control of one of the receptacles at a given outlet, while the other receptacle remains "live" at all times. See figures 3-5 and 12-13 for definition of a receptacle.

Next to the bedroom door we find one three-way switch, plus the ceiling fan/light controls, which are installed in a separate two-gang box. See figure 9-2.

SLIDING GLASS DOORS

See unit 3 for a discussion on how to space receptacle outlets on walls where sliding glass doors are installed.

SELECTION OF BOXES

As discussed in unit 2, the selection of outlet boxes and switch boxes is made by the electrician. These decisions are based on Code requirements, space allowances, good wiring practices, and common sense.

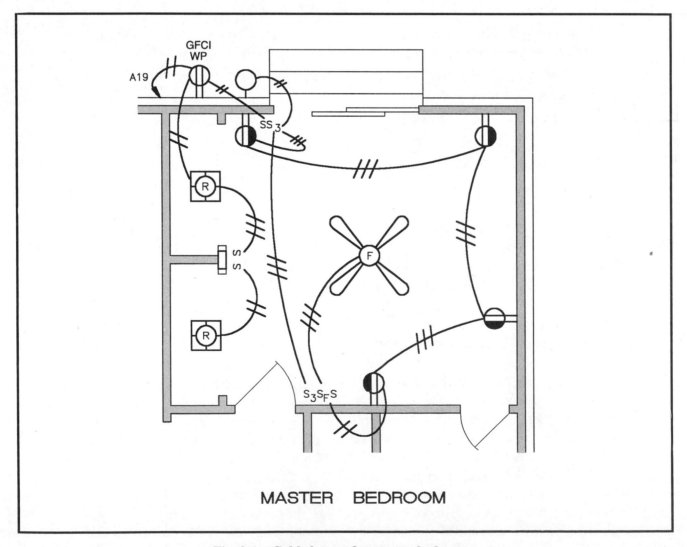

Fig. 9-1 Cable layout for master bedroom.

For example, in the master bedroom, the electrician may decide to install a 4-inch square box with a two-gang raised plaster cover at the box location next to the sliding door. Or two sectional switch boxes ganged together may be installed at that location. See figures 9-3 and 9-4.

The type of box to be installed depends upon the number of conductors entering the box. The suggested cable layout, figure 9-1, shows that four cables enter this box for a total of 14 conductors (10 No. 14 AWG circuit conductors and 4 equipment grounding conductors). ▶ The details for calculating box fill is found in *Section 370-16* of the Code. Also refer to figure 2-26 for a summary of box fill requirements. In this example, we have ten circuit conductors, four equipment grounding conductors (counted as one conductor), four cable clamps (counted as one conductor), one single-pole switch (counted as two conductors), and one three-way switch (counted as two conductors), for a total of 16 conductors. ◀

1. Count the circuit conductors
   $2 + 2 + 3 + 3 =$ 10

2. Add one for one or more
   equipment grounding conductors   1

3. Add two for each switch (2 + 2)   4

4. Add one for one or more cable clamps   1
                   Total   16

Now look at the Quik-Chek Box Selection Guide, Table 8-2, and select a box or a combination of gangable device boxes that are permitted to hold 16 conductors.

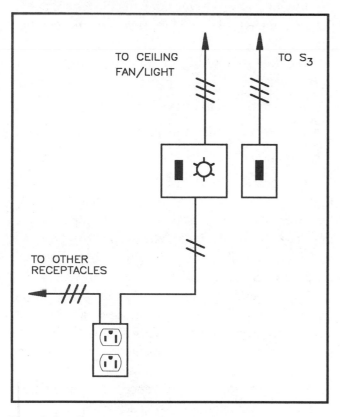

Fig. 9-2 Conceptual view of how the switching arrangement is to be accomplished in the master bedroom Circuit A19.

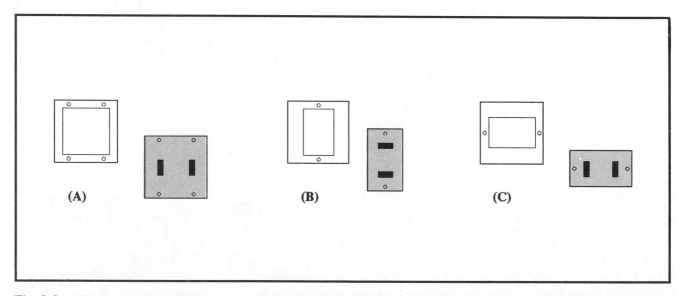

Fig. 9-3 (A) shows two switches and wall plate that attach to a two-gang, 4" square raised plaster cover or to two device boxes that have been ganged together. (B) and (C) show two interchangeable-type switches and wall plate that attach to a one-gang, 4" square raised plaster cover or to a single device (switch) box. These types of wiring devices may be mounted vertically or horizontally.

No. of Gangs	Height	Width
1	4 1/2" (114.3 mm)	2 3/4" (69.85 mm)
2	4 1/2" (114.3 mm)	4 9/16" (115.9 mm)
3	4 1/2" (114.3 mm)	6 3/8" (161.9 mm)
4	4 1/2" (114.3 mm)	8 3/16" (207.9 mm)
5	4 1/2" (114.3 mm)	10" (254 mm)
6	4 1/2" (114.3 mm)	11 13/16" (300 mm)

Fig. 9-4   Height and width of standard wall plates.

Possibilities are:

1. Gang two 3" × 2" × 3½" device boxes together (9 × 2 = 18 No. 14 AWG conductors allowed).

2. Install a 4" × 4" × 2⅛" square box (15 No. 14 AWG conductors allowed). Note that the raised plaster cover, if marked with its cubic-inch volume, can increase the maximum number of conductors permitted for the combined box and raised cover. See figure 2-27.

ESTIMATING CABLE LENGTHS

The length of cable needed to complete an installation can be estimated roughly. It may be possible to run the cable in a straight line directly from one wall outlet to another. In some cases, obstacles such as steel columns, sheet-metal ductwork, and plumbing may require the cable routing to follow a longer path.

To insure that the estimate is not short, all measurements are to be made "square." For example, measure from the ceiling outlet straight to the wall, and finally measure straight over to the wall outlet. Add one foot (305 mm) of cable at each cable termination. This is an allowance to permit the outer jacket of the cable to be stripped at all junction and outlet boxes. Check the plans carefully as an aid in determining where the cables can be routed.

CEILING FANS

For appearance and added comfort, ceiling fans have become extremely popular in recent years, figure 9-5. These fans rotate slowly (60 r/min to 250 r/min for home-type fans). The air currents they create can save energy during the heating season because they destratify the warm air at the ceiling, bringing it down to living levels. In the summer, even with air conditioning, a slight circulation of air

creates a wind-chill factor and causes a person to feel cooler.

Residential ceiling fans usually extend approximately 12 inches (305 mm) below the ceiling; so, for 8-foot (2.44 m) ceilings, the fan blades are about 9 to 10 inches (229–254 mm) below the ceiling, figure 9-6.

The safe supporting of a ceiling fan involves three issues.

* the actual weight of the fan.

* the twisting and turning motion when started.

* vibration.

To address these issues, the 1996 National Electrical Code®, *Section 370-27(c)* states that outlet boxes shall not be used as the sole support of the fan. The fan could be supported by the building structure, in which case, the outlet box does not actually support the fan.

How does the electrician or inspector know where the homeowner might install a fan after the wiring has been completed? This is a major problem, since most ceiling fans are installed after the wiring has been completed and the homeowner has moved in. There have been many reports of ceiling fans falling down on people, causing serious injury.

Ceiling outlet boxes listed as "suitable for fan support" should be installed in all habitable rooms, such as kitchens, family rooms, dining rooms, parlors, libraries, dens, sun rooms, bedrooms, recreation rooms, stairways, or similar rooms or areas.

Ceiling boxes in closets, or boxes installed close to a wall for accent lighting, or boxes installed specifically for such things as security systems, fire

Fig. 9-5   Typical home-type ceiling fan.

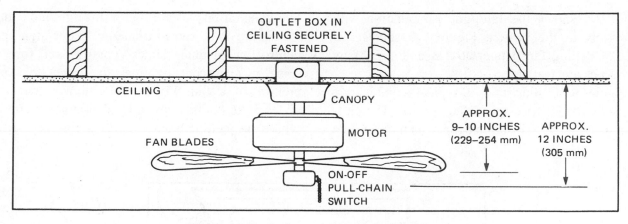

Fig. 9-6   Typical mounting of a ceiling fan.

alarms, and smoke detectors are locations where ceiling fans would not be installed, and therefore would not require the "suitable for fan support" making.

Requirements relating to the support of ceiling fans are also found in the UL White Book, and in the UL Standards 507 (electric fans), 514A (metallic boxes), and 514C (nonmetallic boxes).rooms, parlors, libraries, dens, sun rooms, bedrooms, recreation rooms, stairways, or similar rooms or areas in dwellings shall be considered as likely to support a fan, even though a ceiling fan is not installed now.

Requirements relating to the mounting of ceiling fans are also found in UL Standards 507, 514A, and 514C.

▶ *Section 422-18* states that listed ceiling fans that weigh not over 35 pounds (15.8 kg) may be supported by an outlet box that is identified as suitable for fan support. Ceiling fans weighing more than 35 pounds must be supported independently of the outlet box. The combined weight of the fan and any accessories (such as a light kit) must be considered when determining compliance with the 35-pound requirement. ◀

Always check the instructions furnished with the ceiling fan to be sure that the mounting and supporting methods will be safe. Boxes that are permitted to support a fan are marked ACCEPTABLE FOR FAN SUPPORT.

Section 410-16 of the Code requires that a lighting fixture must be supported independently of the outlet box if the fixture weighs more than 50 pounds (22.7 kg).

Figure 9-7 shows the wiring for a fan/light combination with the supply at the fan/light unit.

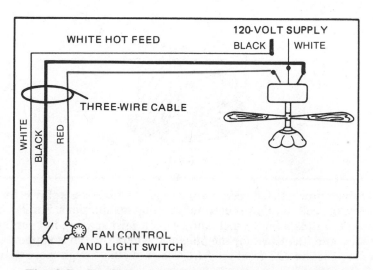

Fig. 9-7   Fan/light combination with the supply at the fan/light unit.

Figure 9-8 shows the fan/light combination with the supply at the switch. Figure 9-9 shows two types of ceiling fan hanger/box assemblies. Many other types are available.

Figure 9-10 illustrates a typical combination light switch and three-speed fan control. The electrician is required to install a deep 4-inch box with a two-gang raised plaster ring for this light/fan control.

Figure 9-11 shows a single "slider" fan speed control and a dual "slider" control, one for dimming incandescent lighting, the other for speed control of a fan. Figure 9-12 shows two other styles of fan/light controls. Dual controls are needed where the ceiling fan also has a lighting fixture.

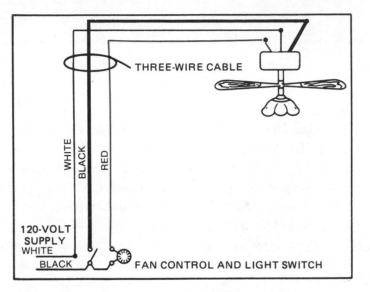

Fig. 9-8 Fan/light combination with the supply at the switch.

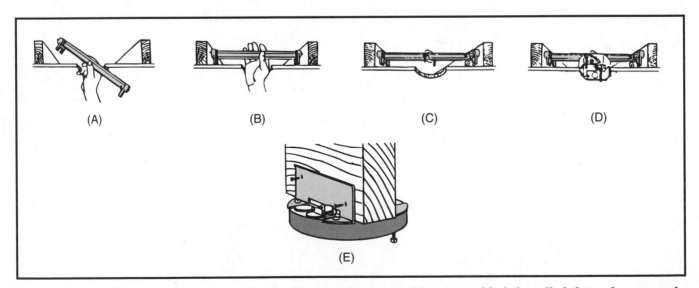

▶ **Fig. 9-9** (A)(B)(C)(D) show how a listed ceiling fan hanger and box assembly is installed through a properly sized and carefully cut hole in the ceiling. This is done for existing installations. Similar hanger/box assemblies are used for new work. The hanger adjusts for 16-inch and 24-inch joist spacing but can be cut shorter if necessary. (E) shows a type of box (listed and identified for the purpose) where the fan is supported from the joist, independent of the box, as required by *Section 422-18* of the *NEC®* for fans that weigh more than 35 pounds. ◀ *Courtesy Reiker Enterprises Inc.*

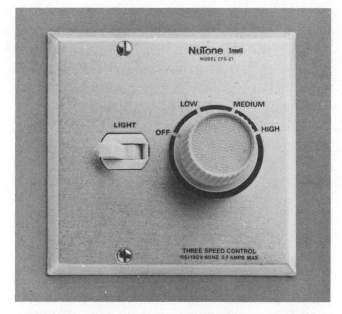

Fig. 9-10   Combination light switch and three-speed fan switch. *Courtesy* of NuTone.

Fig. 9-11   The "single slider" switch is for speed control of a fan. The "dual slider" switch is for speed control of a fan and dimming of an incandescent lamp. *Courtesy* of Pass & Seymour/Legrand.

Fig. 9-12   These photos show two types of fan/light controls. One has a toggle switch for turning a ceiling fan OFF and ON, plus a "slider" that provides 3-speed control of the fan. The other has a toggle switch for turning a light OFF and ON, plus two "sliders," one for ON/OFF and 3-speed control of a ceiling fan, the other for full range dimming of lighting. *Courtesy* of Lutron.

These controls fit into a deep single-gang device box, or preferably into a 4-inch square box with single-gang plaster cover. Be careful of "box fill." Make sure the wall box is large enough. Dimmer and speed control devices are quite large when compared to regular wall switches. Do not jam the control into the wall box.

A typical home-type ceiling fan motor draws 50 to 100 watts (50 to 100 volt-amperes). This is approximately 0.4 to slightly over 0.8 ampere.

Ceiling fan/light combinations increase the load requirements somewhat. Some ceiling fan/light units have one lamp socket, whereas others have four or five lamp sockets.

For the ceiling fan/light unit in the master bedroom, 240 volt-amperes were included in the load calculations. The current draw is

$$I = \frac{VA}{V} = \frac{240}{120} = 2 \text{ amperes}$$

For most fans, switching on and off may be accomplished by the pull-switch, an integral part of the fan (OFF-HIGH-MED-LOW), or with a solid-state switch having an infinite number of speeds. Some of the more expensive models offer a remote control similar to that used with television, VCRs, stereos, and cable channel selectors.

Most ceiling fan motors have an integral reversing switch so as to blow air downward or upward.

Do not use a standard incandescent lamp dimmer to control fan motors, fluorescent ballasts, transformers, low-voltage lighting systems (they use transformers to obtain the desired low voltage), motor-operated appliances, or other "inductive" loads. Serious overheating and damage to the motor can result. Fan speed controllers have internal circuitry that is engineered specifically for use on inductive load circuits. Fan speed controllers virtually eliminate any fan motor "hum." Read the label and the instructions furnished with a dimmer to be sure it is suitable for the application.

REVIEW

Note: Refer to the Code or the plans where necessary.

1. What circuit supplies the master bedroom? _____

2. What lighting besides the bedroom is supplied by this circuit? _____
_____

3. What type of receptacles are provided in this bedroom? How many receptacles are there? _____

4. How many ceiling outlets are included in this circuit? _____
_____

5. What wattage for the recessed closet fixtures was used for calculating their contribution to the circuit? _____

6. What is the current draw for the ceiling fan/light fixture? _____

7. What is the estimated load in volt-amperes for the circuit supplying the master bedroom? _____

8. Which is installed first, the switch and outlet boxes or the cable runs? _____
_____

9. How many conductors enter the ceiling fan/light wall box? _____

10. What type and size of box may be used for the ceiling fan/light wall box? _____

_____

_____

_____

11. What type of covers are used with 4-inch square outlet boxes? _____

_____

_____

12. a. Does the circuit for the master bedroom have a grounded circuit conductor?

   b. Does it have an equipment grounding conductor? _____

   c. Explain the difference between a "grounded" conductor and a "grounding" conductor. _____

_____

_____

_____

_____

_____

13. Approximately how many feet (meters) of two-wire cable and three-wire cable are needed to complete the circuit supplying the master bedroom? Two-wire cable _____ feet (_____ meters). Three-wire cable _____ feet (_____ meters).

14. If the cable is laid in notches in the corner studs, what protection for the cable must be provided? _____

_____

_____

_____

15. How high are the receptacles mounted above the finish floor in this bedroom?

_____

16. Approximately how far from the bedroom door is the first receptacle mounted? (See the plans.) _____

17. What is the distance from the finish floor to the center of the wall switches in this bedroom? _____

18. The master bedroom features a sliding glass door. For the purpose of providing the proper receptacle outlets, answer the following statements true or false.

   • Sliding glass panels are considered to be wall space.   _____

   • Fixed panels of glass doors are considered to be wall space.   _____

19. What type of receptacle will be installed outdoors, just outside of the master bedroom?

_____

_____

_____

20. Show your calculations of how to select a proper wall box for the closet fixture switch. Keep in mind that the available space between the wood casings is small. See figure 9-1. _____

   _____

   _____

   _____

   _____

   _____

   _____

   _____

   _____

21. When an outlet box is to be installed to support a ceiling fan, how must it be marked?

   _____

   _____

22. The following is a layout of the lighting circuit for the master bedroom. Using the cable layout shown in figure 9-1, make a complete wiring diagram of this circuit. Indicate the color of each conductor.

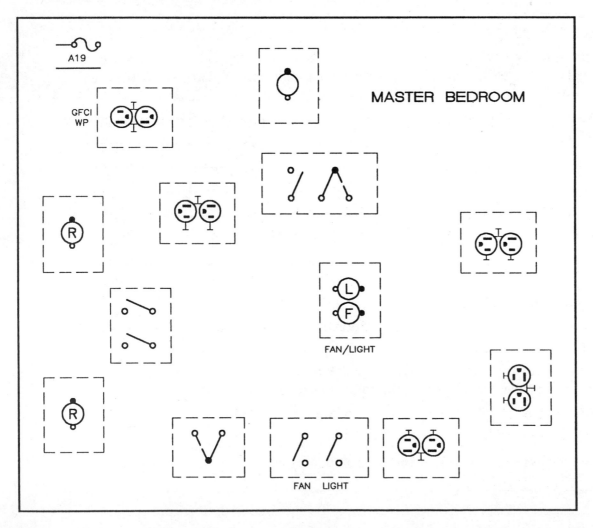

23. Connect the 3-wire cable, fan, light, fan speed control and light switch. Refer to *Section 200-7* of the *NEC®* to review the permitted use of the white conductor in the cable.

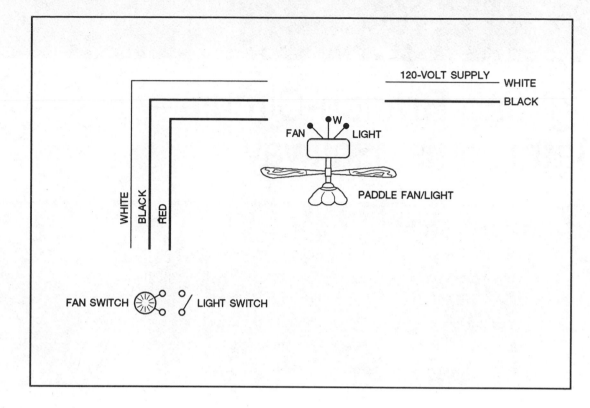

24. What is the width of a two-gang wall plate? _____

25. May a standard electronic dimmer be used to control the speed of a fan motor?

    _____

26. Outlet boxes that pass the tests at Underwriters Laboratories for the support of ceiling fans are listed and marked ACCEPTABLE FOR FAN SUPPORT or SUITABLE FOR FAN SUPPORT. When roughing-in the electrical wiring for a new home, this type of outlet box must be installed wherever there is a likelihood of installing a ceiling fan. Name those locations in homes that the Code considers likely to support a ceiling fan.

    _____

    _____

    _____

UNIT 10

Lighting Branch-Circuit— Bathrooms, Hallway

OBJECTIVES

After studying this unit, the student will be able to

- list equipment grounding requirements for bathroom installations.
- draw a wiring diagram for the bathroom and hallway.
▶ • understand Code requirements for receptacles installed in bathrooms. ◀
- understand Code requirements relating to receptacle outlets in hallways.
- discuss fundamentals of proper lighting for bathrooms.

Circuit A14 supplies the lighting outlets and receptacle outlet in the bedroom hall and both bathrooms. ▶ The receptacle outlets in the bathrooms are supplied by separate circuits A22 and A23. This is a Code requirement, and is discussed later on in this unit and in unit 15. ◀

Table 10-1 summarizes outlets and estimated load for the bathrooms and bedroom hall.

Note that each bathroom shows a ceiling heater/light/fan that is connected to separate circuit ▲J and ▲K. These are discussed in detail in unit 22.

A hydromassage tub is located in the bathroom serving the master bedroom. It is connected to a separate circuit ▲A and is also discussed in unit 22.

The attic exhaust fan in the hall is supplied by a separate circuit ▲L, also covered in unit 22.

DESCRIPTION	QUANTITY	WATTS	VOLT-AMPERES
Receptacles @ 120 W each	1	120	120
Vanity fixtures @ 200 W each	2	400	400
Hall fixture	1	100	100
TOTALS	4	620	620

Table 10-1   Bathroom and bedroom hall: outlet count and estimated load. Circuit A14. ▶ The receptacles in the master bathroom and hall bathroom are not included in this table as they are connected to separate 20-ampere branch-circuits A22 and A23. ◀

LIGHTING BRANCH-CIRCUIT A14 FOR HALLWAY AND BATHROOMS

Figure 10-1 and the electrical plans for this area of the home show that each bathroom has a fixture above the vanity mirror. Some typical fixtures are shown in figure 10-2. Of course, the homeowner might decide to purchase a medicine cabinet complete with a self-contained lighting fixture. See figure 10-3(A) and (B), which illustrate how to rough-in the wiring for each type. These fixtures are controlled by single-pole switches at the doors.

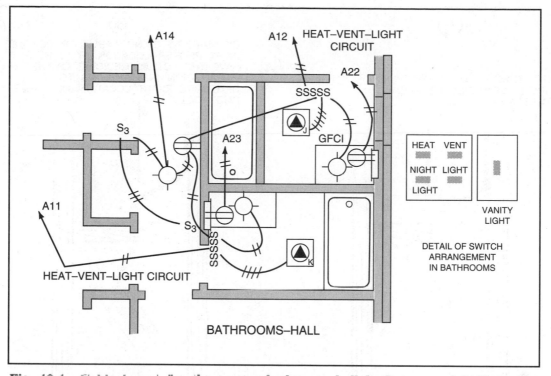

Fig. 10-1   Cable layout for the master bathroom, hall bathroom and hall. Layout includes lighting Circuit A14, two circuits (A11 and A12) for the heat/vent/lights, ▶ and two circuits (A22 and A23) for the receptacles in the bathrooms. ◀ The special purpose outlets for the hydromassage tub, attic exhaust fan, and smoke detector are covered elsewhere in this text.

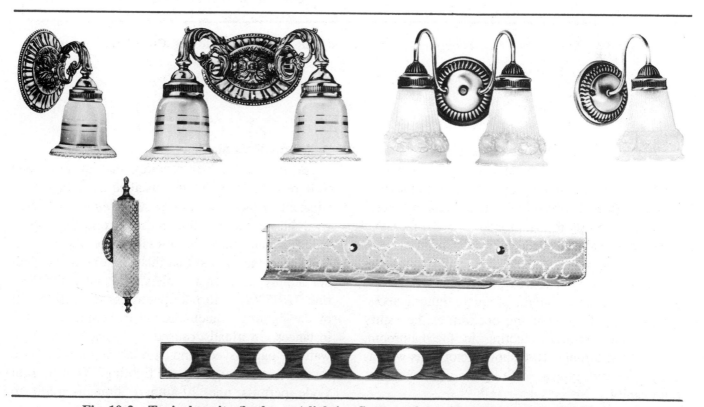

Fig. 10-2   Typical vanity (bathroom) lighting fixtures of the side bracket and strip types.

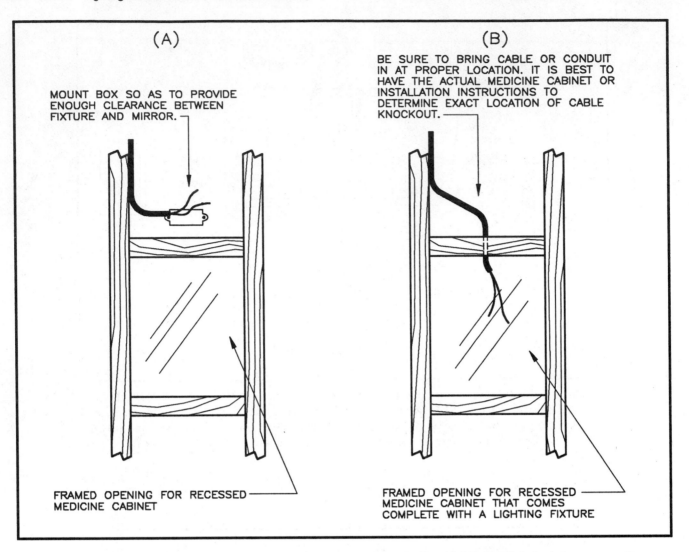

Fig. 10-3   Two methods most commonly used for roughing-in the wiring for lighting above a vanity.

You must remember that a bathroom (powder room) should have proper lighting for shaving, combing hair, grooming, and so on. Mirror lighting can accomplish this, because a mirror will reflect what it "sees." If the face is poorly lit, with shadows on the face, that is precisely what will be reflected in the mirror.

A lighting fixture directly overhead will light the top of one's head, but will cause shadows on the face. Mirror lighting and/or adequate lighting above and forward of the standing position at the vanity can provide excellent lighting in the bathroom. Figures 10-4 through 10-6 show pictorial as well as section views of typical soffit lighting above a bathroom vanity.

Bathroom Receptacles

Section 210-52(d) requires that at least one wall receptacle outlet be installed in a bathroom adjacent to the basin (sink). ▶ *Section 210-52(d)* goes on to require that at least one 20-ampere branch-circuit must be provided for these bathroom receptacles, and that this (these) circuits shall not supply any other outlets. The intent of *Section 210-52(d)* is to take bathroom receptacles off of the lighting branch-circuits in homes. In many instances, overloads caused by plugging in high wattage hair dryers and similar appliances have resulted in total loss of lighting. The present Code permits one 20-ampere branch-circuit to

Fig. 10-4   Positioning of bathroom lighting fixtures. Note the wrong way and the right way to achieve proper lighting.

feed the receptacles in all three bathrooms (one is the powder room) in the residence presented in this text. In this text, we have chosen to run a separate 20-ampere branch-circuit A22 to the receptacle in the master bedroom bathroom, a separate 20-ampere branch-circuit A23 to the receptacle in the front bedroom, and another separate 20-ampere branch-circuit B21 to the receptacle in the powder room located near the laundry. These separate circuits are included in the general lighting load calculations, so no additional load need be added. ◄

Bathroom receptacles are required to be GFCI protected per *Section 210-8(a)(1)* as discussed in unit 6. ► The Code in *Article 100* defines a *bathroom* as an area including a basin, with one or more of the following: a toilet, a tub, or a shower. See figure 10-7. ◄

Receptacles in Bathtub and Shower Spaces

► Because of the obvious hazards associated with water and electricity, receptacles are *not* permitted to be installed in bathtub and shower spaces, *Section 410-57(c)*. ◄

GENERAL COMMENTS ON LAMPS AND COLOR

Incandescent lamps (light bulbs) provide pleasant color tones, bringing out the warm red flesh tones similar to those of natural light. This is particularly true for the "soft" white lamps. Fluorescent lamps available today provide a wide range of "coolness" to "warmth." Rated in degrees Kelvin (i.e., 2500°K, 3000°K, 3500°K, 4000°K, 5000°K, etc.), the lower the Kelvin degrees, the "warmer" the color tone. Conversely, the higher the Kelvin degrees, the "cooler" the color tone. Warm fluorescent lamps bring out the red tones, whereas cool fluorescent lamps tend to give a person's skin a pale appearance. Thus, you will find many types of fluorescent lamps, each providing different color tones and different efficiencies. These might be marked daylight **D** *(very cool)*, cool white

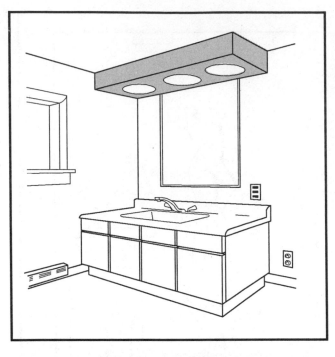

TYPICAL INCANDESCENT RECESSED SOFFIT LIGHTING
OVER BATHROOM VANITY

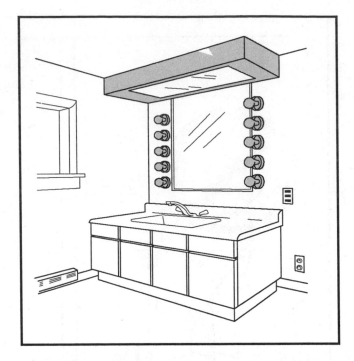

TYPICAL FLUORESCENT RECESSED SOFFIT LIGHTING
OVER BATHROOM VANITY. NOTE ADDITIONAL
INCANDESCENT "SIDE—OF—MIRROR" LIGHTING.

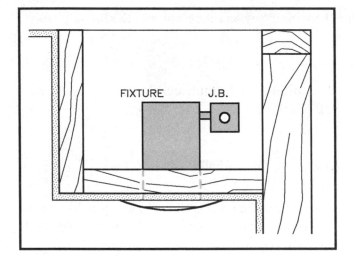

END CUTAWAY OF SOFFIT SHOWING RECESSED
INCANDESCENT FIXTURES IN TYPICAL SOFFIT ABOVE
BATHROOM VANITY. TWO OR THREE FIXTURES GENERALLY
INSTALLED TO PROVIDE PROPER LIGHTING.

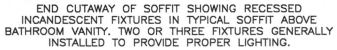

Fig. 10-5   Incandescent soffit lighting.

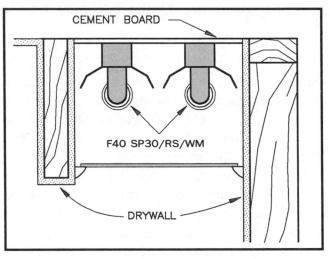

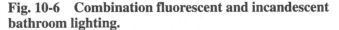

END CUTAWAY VIEW OF SOFFIT ABOVE BATHROOM
VANITY SHOWING RECESSED FLUORESCENT FIXTURES
CONCEALED ABOVE TRANSLUCENT ACRYLIC LENS

**Fig. 10-6   Combination fluorescent and incandescent
bathroom lighting.**

CW *(cool)*, white **W** *(moderate)*, and warm white **WW** *(warm)*. These categories break down further into a *deluxe* **X** series, *specification* **SP** series, and *specification deluxe* **SPX** series. Unit 7 of this text discusses some basics of lighting. The Instructor's Guide lists some lamp manufacturers' publications that cover the subject of lighting in greater detail.

HANGING FIXTURES IN BATHROOMS

▶ *Section 410-4(d)* of the Code states that no parts of cord-connected fixtures, hanging fixtures, lighting track, pendants, or ceiling fans shall be located within a zone measuring 3 feet (0.91 m) horizontally and 8 feet (2.44 m) vertically from the top of the

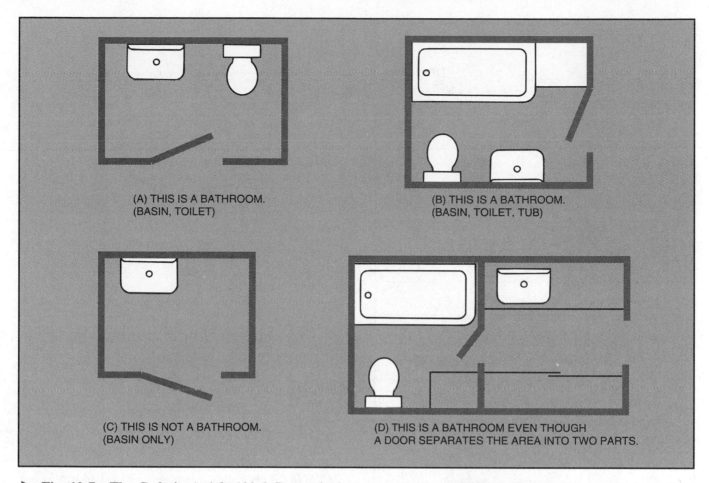

(A) THIS IS A BATHROOM.
(BASIN, TOILET)

(B) THIS IS A BATHROOM.
(BASIN, TOILET, TUB)

(C) THIS IS NOT A BATHROOM.
(BASIN ONLY)

(D) THIS IS A BATHROOM EVEN THOUGH
A DOOR SEPARATES THE AREA INTO TWO PARTS.

▶ Fig. 10-7   The Code in *Article 100* defines a *bathroom* as an area including a basin with one or two of the following: a toilet, a tub, or a shower. ◀

bathtub rim. ◀ Figure 10-8 projects the top view of the 3-foot restriction. Figure 10-9 shows both the allowable installation and Code violations of the given dimensions.

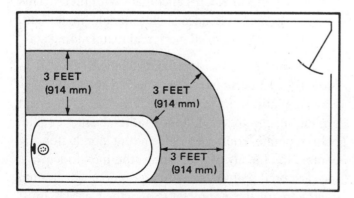

3 FEET
(914 mm)

3 FEET
(914 mm)

3 FEET
(914 mm)

▶ Fig. 10-8   No parts of cord-connected fixtures, hanging fixtures, lighting track, pendants, or ceiling fans shall be located within 3 feet (0.91 m) measured horizontally from the top of the bathtub rim (top view), *Section 410-4(d)*. ◀

HALLWAY LIGHTING

The hallway lighting is provided by one ceiling fixture that is controlled with two three-way switches located at either end of the hall. The home run to Main Panel A has been brought into this ceiling outlet box.

RECEPTACLE OUTLETS IN HALLWAYS

One receptacle outlet has been provided in the hallway as required in *Section 210-52(h)*, which states that "for hallways of 10 feet (3.05 m) or more in length, at least one receptacle outlet shall be required."

For the purpose of determining the length of a hallway, the measurement is taken down the center-line of the hall, turning corners if necessary, but not passing through a doorway, figure 10-10.

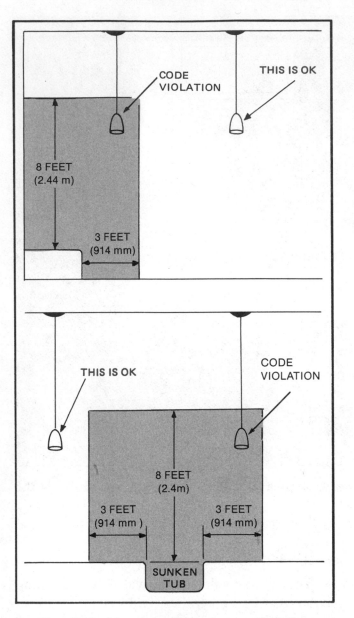

► Fig. 10-9 No parts of cord-connected fixtures, hanging fixtures, lighting track, pendants, or ceiling fans shall be allowed in shaded areas, *Section 410-4(d)*. ◄

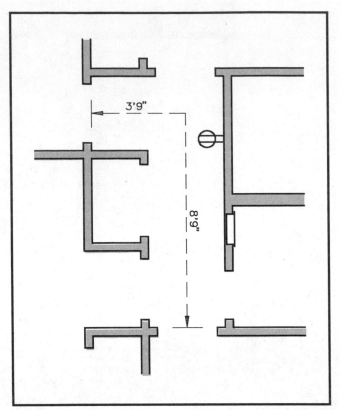

Fig. 10-10 The centerline measurement of the bedroom hallway in this residence is 12'6", which requires at least one wall receptacle outlet, *Section 210-52(h)*. This receptacle may be installed anywhere in the hall. The Code does not specify a location.

EQUIPMENT GROUNDING REQUIREMENTS FOR A BATHROOM CIRCUIT

All exposed metal equipment, including fixtures, electric heaters, faceplates, and similar items must be grounded. Equipment grounding requirements are found in *Section 250-42*.

In general, all exposed noncurrent-carrying metal parts of electrical equipment must be grounded:

• if they are within 8 feet (2.44 m) vertically or 5 feet (1.52 m) horizontally of the ground or other grounded metal objects.

• if the ground and the electrical equipment can be touched at the same time.

• if they are located in wet or damp locations, such as in bathrooms, showers, and outdoors.

• if they are in electrical contact with metal. This requirement includes such things as metal lath, aluminum foil insulation, and metal sidings.

Equipment is considered grounded when it is properly and permanently connected to metal raceway, the armor of armored cable, the equipment grounding conductor in nonmetallic-sheathed cable, or a separate equipment grounding conductor. Of course, the means of grounding (the metal raceway, the armored cable, or the equipment grounding conductor in the nonmetallic-sheathed cable) must itself be properly grounded.

Section 250-45(c) lists the types of cord- and plug-connected appliances in residences that must be grounded. The exception to *Section 250-45(c)* accepts "double insulation" or equivalent in lieu of grounding. Many hand-held appliances, such as electric drills, electric razors, and electric toothbrushes make use of the "double insulation" technique. Such appliances have a 2-wire cord and 2-wire plug cap instead of 3-wire cord that has an equipment grounding conductor and a 3-wire plug cap that has a ground prong.

Double-insulated appliances are clearly marked to indicate that they are double insulated.

Immersion Detection Circuit Interrupters

Another way to protect people from electrical shock is to use grooming appliances that are equipped with a special attachment plug-cap that is a liquid immersion detection circuit interrupter, referred to as an IDCI. IDCIs are discussed in unit 6.

▶ Do *not* install switches in wet locations, such as in bathtubs or showers, unless they are part of a listed tub or shower assembly, in which case the manufacturer would have taken all of the proper precautions, and would have submitted the assembly to a recognized testing laboratory to undergo exhaustive testing to establish the safety of the equipment. ◀

To field install switches in these locations would be a violation of *Section 380-4*.

REVIEW

1. List the number and types of switches and receptacles used in Circuit A14. _____

   _____

2. There is a three-way switch in the bedroom hallway leading into the living room. Show your calculation of how to determine the box size for this switch. Refer to the Quik-Chek Box Selection Guide or *Section 370-16* and *Table 370-16(a) and (b)*. The box will contain cable clamps.

3. What wattage was used for each vanity fixture to calculate the estimated load on Circuit A14? _____

4. What is the current draw for the answer given in question 3? _____

   _____

5. Exposed noncurrent-carrying metallic parts of electrical equipment must be grounded if installed within _____ feet (_____ meters) vertically or _____ feet (_____ meters) horizontally of bathtubs, plumbing fixtures, pipes, or other grounded metal work or grounded surfaces.

6. What color are the faceplates in the bathrooms? Refer to the specifications.

   _____

7. Most appliances of the type commonly used in bathrooms, such as hair dryers, electric shavers, and curling irons, have two-wire cords. These appliances are _____ insulated or _____ _____ _____ _____ protected.

8. a. The *NEC®* in *Section* _____ requires that all receptacles in bathrooms be _____ protected.

   b. The *NEC®* in *Section* _____ requires that all receptacles in bathrooms be connected to one or more separate 20-ampere branch-circuits that serve no other outlets.

   c. Would it be in conformance to the *NEC®* to connect the receptacles in bathrooms to a lighting circuit that is different than the lighting circuit in the bathroom? (Yes) (No). Circle the correct answer.

   d. The *NEC®* in *Section* _____ prohibits mounting receptacles in bathrooms face-up in the countertops near basins.

9. Hanging lighting fixtures must be kept at least _____ feet (_____ meters) from the edge of the tub as measured horizontally. In bathrooms with high ceilings, where the hanging fixture is installed directly over the tub, it must be kept at least _____ feet (_____ meters) above the edge of the tub.

10. The following is a layout of a lighting circuit for the bathroom and hallway. Using the cable layout shown in figure 10-1, make a complete wiring diagram of this circuit. Use colored pencils to indicate the conductors.

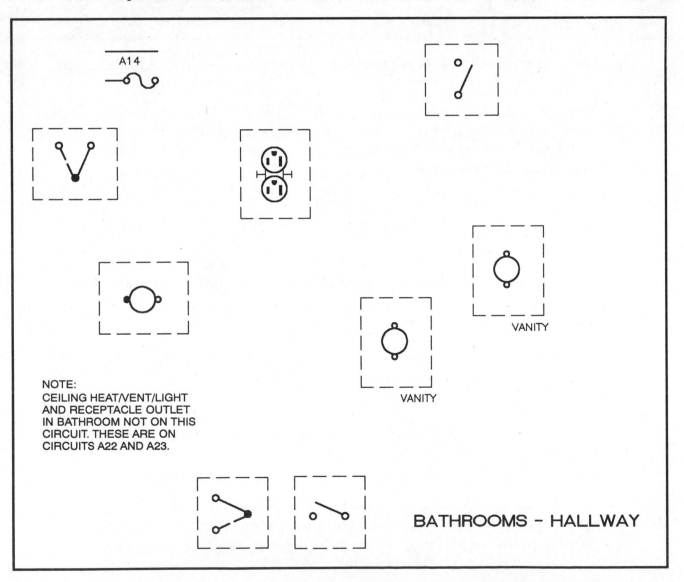

A14

NOTE:
CEILING HEAT/VENT/LIGHT
AND RECEPTACLE OUTLET
IN BATHROOM NOT ON THIS
CIRCUIT. THESE ARE ON
CIRCUITS A22 AND A23.

VANITY

VANITY

BATHROOMS – HALLWAY

11. Circle the correct answer as to whether or not a receptacle outlet is required in the following hallways.

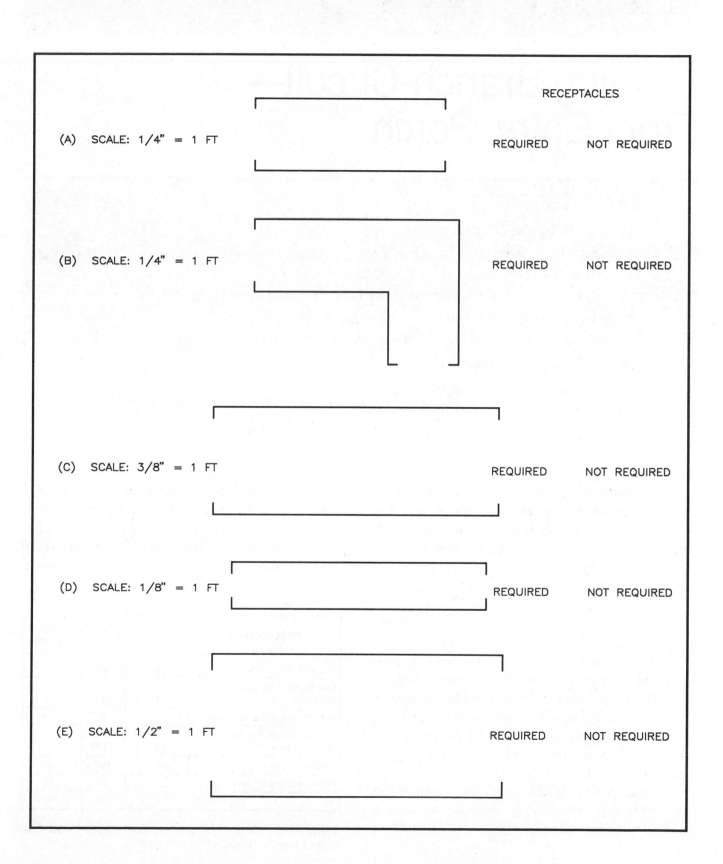

RECEPTACLES

(A)   SCALE: 1/4" = 1 FT   REQUIRED   NOT REQUIRED

(B)   SCALE: 1/4" = 1 FT   REQUIRED   NOT REQUIRED

(C)   SCALE: 3/8" = 1 FT   REQUIRED   NOT REQUIRED

(D)   SCALE: 1/8" = 1 FT   REQUIRED   NOT REQUIRED

(E)   SCALE: 1/2" = 1 FT   REQUIRED   NOT REQUIRED

UNIT 11

Lighting Branch-Circuit—
Front Entry, Porch

OBJECTIVES

After studying this unit, the student will be able to

- understand how to install a switch in a doorjamb for automatic ON/OFF when the door is opened or closed.
- discuss type of fixtures recommended for porches and entries.
- complete the wiring diagram for the entry-porch circuit.
- discuss the advantages of switching outdoor receptacles from indoors.
- define wet and damp locations.
- understand box fill for sectional ganged device boxes.

The front entry-porch is connected to Circuit A15. The home run enters the ceiling box in the front entry. From this box, the circuit spreads out, feeding the recessed closet light, the porch bracket fixture, the two bracket fixtures on the front of the garage, one receptacle outlet in the entry, and one outdoor weatherproof GFCI receptacle on the porch. See figure 11-1. Table 11-1 summarizes the outlets and estimated load for the entry and porch.

Note that the receptacle on the porch is controlled by a single-pole switch just inside the front door. This allows the homeowner to plug in outdoor lighting, such as strings of ornamental Christmas lights, or decorative lighting, and have convenient switch control of that receptacle from inside the house, a nice feature.

Typical ceiling fixtures commonly installed in front entryways, where it is desirable to make a good first impression on guests, are shown in figure 11-2.

Typical outdoor ceiling-mounted and wall-bracket-style porch and entrance lighting fixtures are shown in figure 11-3.

According to the Code, any location exposed to the weather is considered to be a wet location.

DESCRIPTION	QUANTITY	WATTS	VOLT-AMPERES
Receptacles @ 120 W	1	120	120
Weatherproof receptacles @ 120 W	1	120	120
Outdoor porch bracket fixture	1	100	100
Outdoor garage bracket fixtures @ 100 W each	2	200	200
Ceiling fixture	1	150	150
Closet recessed fixture	1	75	75
TOTALS	7	765	765

Table 11-1 Entry and porch: outlet count and estimated load. Circuit A15.

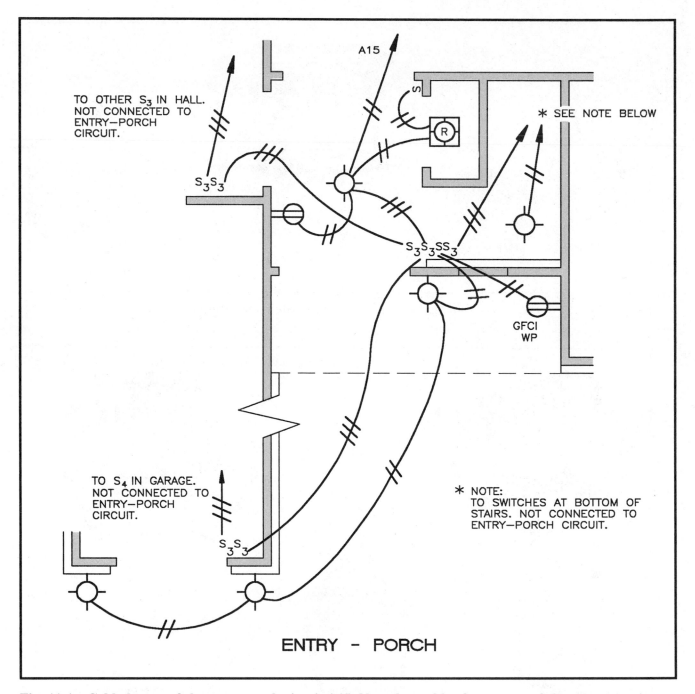

TO OTHER S₃ IN HALL.
NOT CONNECTED TO
ENTRY—PORCH
CIRCUIT.

A15

* SEE NOTE BELOW

GFCI
WP

TO S₄ IN GARAGE.
NOT CONNECTED TO
ENTRY—PORCH
CIRCUIT.

* NOTE:
TO SWITCHES AT BOTTOM OF
STAIRS. NOT CONNECTED TO
ENTRY—PORCH CIRCUIT.

ENTRY - PORCH

Fig. 11-1 Cable layout of the entry-porch circuit A15. Note that cables from some of the other circuits are shown so that you can get a better idea of exactly how many wires will be found at the various locations.

Section 410-4(a) states that outdoor fixtures must be constructed so that water cannot enter or accumulate in lampholders, wiring compartments, or other electrical parts. These fixtures must be marked "Suitable for Wet Locations."

Partially protected areas under roofs, open porches, or areas under canopies are defined by the Code as damp locations. Fixtures to be used in these locations must be marked "Suitable for Damp Locations." Many types of fixtures are available. Therefore, it is recommended that the electrician check the UL label on the fixture to determine the suitability of the fixture for a wet or damp location.

CIRCUIT A15

This circuit is a rather simple circuit that has two sets of three-way switches: one set controlling the

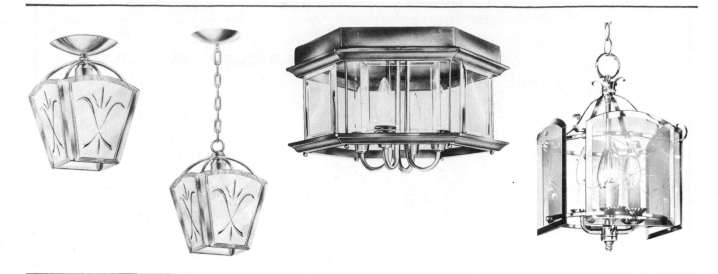

Fig. 11-2   Hall lighting fixtures: ceiling mount and chain mount.

Fig. 11-3   Outdoor porch and entrance lighting fixtures: ceiling mount and wall bracket styles.

front entry ceiling fixture, the other set controlling the porch bracket fixture and the two bracket fixtures on the front of the garage. Also one single-pole switch controls the weatherproof porch receptacle outlet, and another single-pole switch controls the recessed closet light. In the Review questions you will be asked to follow the suggested cable layout for making up all of the circuit connections.

Probably the most difficult part of this circuit is planning and making up the connections at the switch location just inside the front door. Here we find four toggle switches: (1) a three-way switch for ceiling fixture; (2) a three-way switch for porch and garage fixtures; (3) a single-pole switch for weatherproof receptacle on porch; and (4) a three-way switch for the fixture over the stairway leading to the basement (connected to the Recreation Room lighting circuit B12, not the entry-porch circuit A15).

This location is another challenge for determining the proper size wall box. Let's give it a try.

1. Add the circuit conductors
   $2 + 2 + 3 + 3 + 3 + 3 =$ 16

2. Add for equipment grounding wire
   (count 1 only)   1

3. Add eight for the four switches   8

4. Add for cable clamps (count 1 only)   1

                Total   26

Checking figure 2-18, the Quik-Chek Box Selector Guide, and figures 2-24 and 2-25, we find many possibilities. For example, four $3 \times 2 \times 3\frac{1}{2}$ device boxes could be ganged together ($4 \times 9 = 36$ conductors).

An interesting possibility presents itself for the front entry closet. Although the plans show that the recessed closet fixture is turned on and off by a standard single-pole switch to the left as you face the closet, a doorjamb switch could have been installed. A doorjamb switch is usually mounted about 6 feet (1.83 m) above the floor, on the inside of the 2×4 framing for the closet door. The electrician will run the cable to this point, and let it hang out until the carpenter can cut the proper size opening into the doorjamb for the box. After the finished woodwork is completed, the electrician then finishes installing the switch. Figure 11-4 illustrates a door switch. Note that the plunger on the switch is pushed inward when the edge of the door pushes on it as the door is closed, shutting off the light. The plunger can be adjusted in or out to make the switch work properly. These doorjamb switches come complete with their own special wall box.

The integral box furnished with door switches is generally suitable for one cable only. Wiring space is very limited. Do not plan on using this box for splices other than those necessary to make up the connections to the switch.

Fig. 11-4   Door switch.

REVIEW

1. a. How many circuit wires enter the entry ceiling box? _____
   b. How many equipment grounding conductors enter the entry ceiling box? _____

2. Assuming that an outlet box for the entry ceiling has a fixture stud but has external cable clamps, what size box could be installed?

   _____

   _____

   _____

   _____

   _____

   _____

3. How many receptacle outlets and lighting outlets are supplied by Circuit A15?

   _____

4. Outdoor fixtures directly exposed to the weather must be marked as:

   a. suitable for damp locations

   b. suitable for dry locations only

   c. suitable for wet locations

   Circle correct answer.

5. Make a material list of all types of switches and receptacles connected to Circuit A15.

   _____

   _____

   _____

   _____

6. From left to right, facing the switches, what do the switches next to the front door control?

   _____

   _____

   _____

   _____

7. Who is to select the entry ceiling fixture? _____

8. The following layout is for Lighting Circuit A15, the entry, porch, and front garage lights. Using the cable layout shown in figure 11-1, make a complete wiring diagram of this circuit. Use colored pencils to indicate the color of the conductors' insulation.

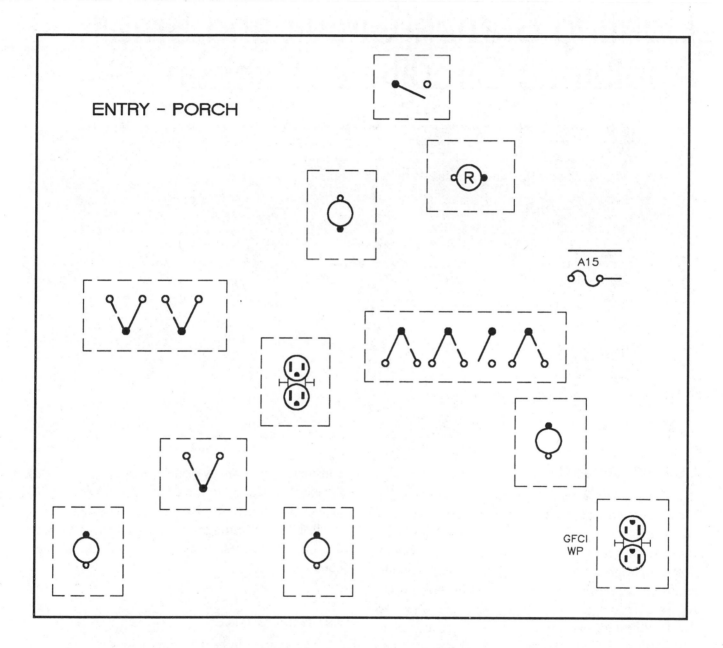

UNIT 12

Lighting Branch-Circuit and Small Appliance Circuits for Kitchen

OBJECTIVES

After studying this unit, the student will be able to

- discuss the features of an exhaust fan for the removal of kitchen cooking odors and humidity.
- explain the Code requirements for small appliance circuits in kitchens.
- understand the Code requirements for ground-fault protection for receptacles that serve countertops.
- discuss split-circuit receptacles.
- discuss multiwire circuits.
- discuss exhaust fan noise ratings.
- understand the overall general concept of grounding electrical equipment.
- know the difference between a receptacle outlet and a lighting outlet.
- discuss typical kitchen, dining room, undercabinet, and swag fixtures.

LIGHTING CIRCUIT B7

The lighting circuit supplying the kitchen and nook originates at Panel B. The circuit number is B7. The home run is very short, leading from Panel B in the corner of the recreation room to the switch box to the right of the kitchen sink. Here, the circuit continues upward through the junction box on the recessed fixture above the sink, which is approved for feedthrough wiring (see unit 7 for complete explanation of Code requirements for recessed fixtures) through the clock outlet box, and then across the kitchen ceiling connecting the ceiling light fixtures, the track light, the outdoor bracket fixture, and range hood exhaust fan/light unit, figure 12-1.

Another way to run this circuit would be to run the branch-circuit home run to the clock outlet box, then over to the recessed fixture. This is a matter of

considering how many conductors enter the boxes, the ease in making up the connections, and whether or not the junction box on the recessed fixture is "listed" for "through" wiring. Table 12-1 shows the outlets and estimated loads for the kitchen/ nook area.

KITCHEN LIGHTING

The general lighting for the kitchen is provided by two four-lamp fluorescent surface-mounted fixtures of the type shown in figure 12-2. These fixtures are controlled by a single-pole switch located next to the doorway leading to the living room.

An adjustable chain swag fixture, figure 12-3A, is mounted on a track to provide lighting above the breakfast nook table. This fixture is controlled by

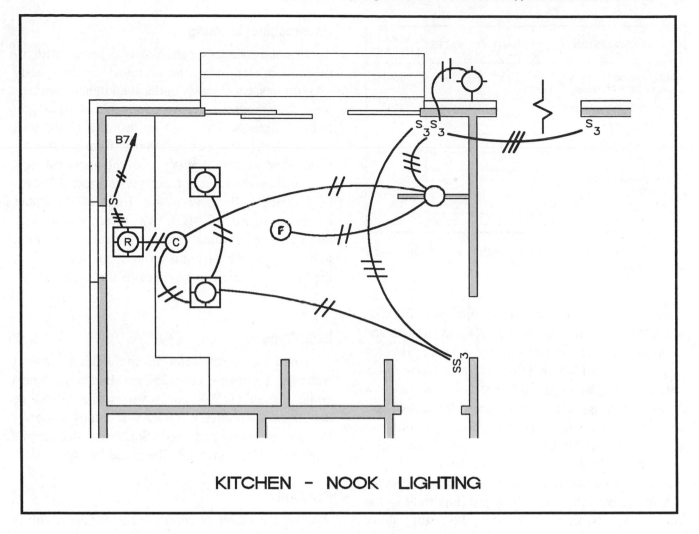

KITCHEN - NOOK LIGHTING

Fig. 12-1   Cable layout for the kitchen lighting. This is Circuit B7. The appliance circuit receptacles are connected to 20-ampere circuits B13, B15, and B16 and are not shown on this lighting circuit layout.

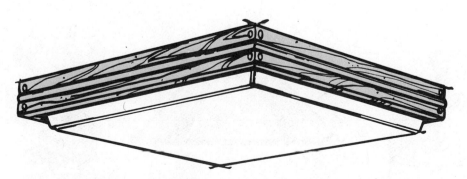

Fig. 12-2   Typical ceiling fixture of the type installed in the kitchen of this residence. This type of fixture might have four 20-watt fluorescent lamps, or it might have two 40-watt U-shaped lamps.

DESCRIPTION	QUANTITY	WATTS	VOLT-AMPERES
Ceiling fixtures (eight 20-W FL lamps)	2	160	180
Recessed fixture over sink	1	100	100
Track light (nook)	1	100	100
Outdoor rear bracket fixture	1	100	100
Clock outlet	1	15	15
Range hood fan/light Two 60-W lamps Fan motor (1 1/3 A @ 120 V)	1	120 160	120 180
TOTALS	7	755	795

Table 12-1   Kitchen/Nook: outlet count and estimated load. Circuit B7.

three-way switches; one located adjacent to the sliding door, the other at the doorway leading to the living room. Many fixtures of this type are provided with a three-position switch that allows one to select low, medium, or high light levels. For instance, a three-way bulb might provide these levels of light, such as 50-100-150 watts. This switch is built into the fixture itself. Other types of ceiling-mount lighting fixtures are shown in figure 12-3B.

One goal of good lighting is to reduce shadows in work areas. This is particularly important in the kitchen. The two ceiling fixtures, the fixture above the sink, plus the fixture within the range hood will provide excellent lighting in the kitchen area. Of course, the track lighting in the nook area offers good lighting for that area.

Undercabinet Lighting

In some instances, particularly in homes with little outdoor exposure, the architect might specify strip fluorescent fixtures under the kitchen cabinets. Several methods may be used to install undercabinet lighting fixtures. They can be fastened to the wall just under the upper cabinets, figure 12-4(A), installed in a recess that is part of the upper cabinets so that they are hidden from view, figure 12-4(B), or fastened under and to the front of the upper cabinets, figure 12-4(C). All three possibilities require close coordination with the cabinet installer to be sure that the wiring is brought out of the wall at the proper location to connect to the undercabinet fixtures.

Lamp Type

Good color rendition in the kitchen area is achieved by using incandescent lamps or warm white deluxe (WWX) or warm white (WW) or deluxe cool white (CWX) 3000-K lamps wherever fluorescent lamps are used. See unit 7 for details regarding energy saving ballasts and lamps.

Fan Outlet

The fan outlet is connected to branch lighting circuit B7. Fans can be installed in a kitchen either to exhaust the air to the outside of the building or to filter and recirculate the air. Both types of fans can be used in a residence. The ductless hood fan does

Fig. 12-3A   Fixtures that can be hung either single or on a track over dinette tables, such as in the nook area of this kitchen.

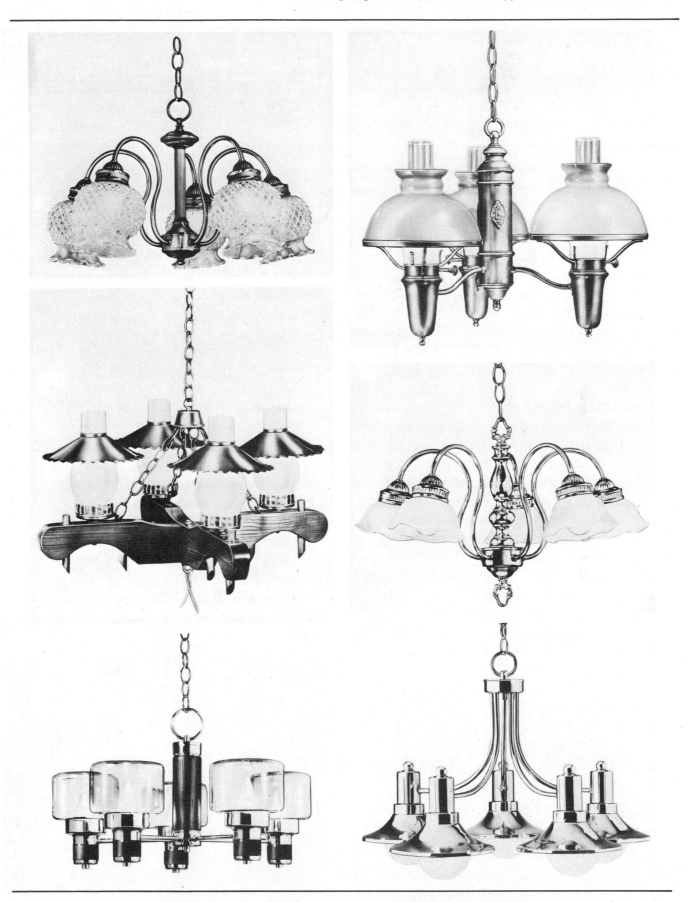

Fig. 12-3B   Types of fixtures used over dining room tables.

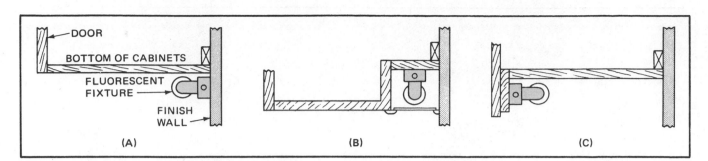

Fig. 12-4   Methods of installing undercabinet lighting fixtures.

not exhaust air to the outside and does not remove humidity from the air.

Homes heated electrically with resistance-type units usually have excess humidity. The heating units neither add nor remove moisture from the air unless humidity control is provided. The proper use of weather stripping, storm doors, and storm windows, and a vapor barrier of polyethylene plastic film tends to retain humidity.

Vapor barriers are installed to keep moisture out of the insulation. This protection normally is installed on the warm side of ceilings, walls, and floors, under concrete slabs poured directly on the ground, and as a ground cover in crawl spaces. Insulation must be kept dry to maintain its effectiveness. For example, a 1% increase of moisture in insulating material can reduce its efficiency 5%. Humidity control using a humidistat on the attic exhaust fan is covered in unit 22.

An exhaust fan simultaneously removes odors and moisture from the air and exhausts heated air, figure 12-5. The choice of the type of fan to be installed is usually left to the owner. The electrician, builder, and/or architect may make suggestions to guide the owner in making the choice.

For the residence in the plans, a separate wall switch is not required because the speed switch, light, and light switch are integral parts of the fan.

Fan Noise

The fan motor and the air movement through an exhaust fan generate a certain amount of noise. The manufacturers of exhaust fans assign a sound-level rating so that it is possible to get some idea as to the noise level one might expect from the fan after it is installed.

The unit used to define fan noise is the *sone* (rhymes with tone). For simplicity, one sone is the noise level of an average refrigerator. The lower the sone rating, the quieter the fan. For the technical person, the sone is a unit of loudness equal to the loudness of a sound of one kilohertz at 40 decibels above the threshold of hearing of a given listener.

All manufacturers of exhaust fans provide this information in their descriptive literature.

Clock Outlets

The clock outlet is connected to lighting circuit B7. The clock outlet in the face of the soffit may be installed in any sectional switch box or 4-inch square box with a single-gang raised plaster cover. A deep sectional box is recommended because a recessed clock outlet takes up considerable room in the box.

Some decorative clocks have the entire motor recessed so that only the hands and numbers are

Fig. 12-5   Typical range hood exhaust fan. Speed control and light are integral part of the unit.

exposed on the surface of the soffit or wall. Some of these clocks require special outlet boxes usually furnished with the clock. Other clocks require a standard 4-inch square outlet box. Accurate measurements must be taken to center the *clock* between the ceiling and the bottom edge of the soffit. If the clock is available while the electrician is roughing in the wiring, the dimensions of the clock can be checked to help in locating the clock outlet.

If an electrical clock outlet has not been provided, battery-operated clocks are available.

SMALL APPLIANCE BRANCH-CIRCUITS FOR RECEPTACLES IN KITCHEN

The Code requirements for small appliance circuits and the spacing of receptacles in the kitchens of dwellings are covered in *Sections 210-52(b), 220-4,* and *220-16.* This was discussed in detail in unit 3. Highlights are:

- At least two 20-ampere small appliance circuits shall be installed, figure 12-6.

- Either (or both) of the two circuits required in the kitchen is permitted to supply receptacle outlets in other rooms, such as the dining room, breakfast room, or pantry. Figure 12-6 illustrates this portion of the Code.

- These small appliance circuits shall be assigned a load of 1500 volt-amperes each when calculating feeders and service-entrance requirements.

▶• The receptacle for the refrigerator in the kitchen is permitted to be supplied by a separate 15- or 20-ampere branch-circuit rather than connecting it to one of the two required 20-ampere small appliance circuits. This permission is found in *Section 210-52(b)(1).* Because this tends to diversify the load, *Exception No. 1* to *Section 220-16(a)* permits this separate branch-circuit to be omitted from the calculations for services and feeders. Stated another way, an additional 1500 volt-amperes *does not* have to be included in the load calculations. However, there is nothing in the Code that prohibits adding another 1500 volt-amperes to the load calculations for this additional branch-circuit, as this would result in more capacity in the electrical system. ◀

- The clock outlet may be connected to an appliance circuit when this outlet is to supply and support the clock only (see figure 3-3). In the residence in the plans, the clock outlet is connected to the lighting circuit.

- Countertop receptacles must be supplied by at least two 20-ampere small appliance circuits, figure 12-6.

▶• All 125-volt, single-phase, 15- and 20-ampere receptacles installed in kitchens to serve the countertops shall be GFCI protected, *Section 210-8(a)(5).* See figure 12-7. Prior to the adoption of the 1996 *National Electrical Code®,*

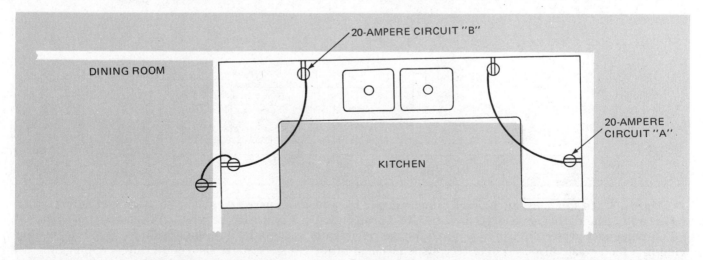

Fig. 12-6   A receptacle outlet in a dining room may be connected to one of the two required small appliance circuits in the kitchen, *Section 210-52(b)(1).* Countertop receptacles must be supplied by at least two 20-ampere small appliance circuits, *Section 210-52(b)(3).*

GFCI protected receptacles were required within 6 feet of the kitchen sink. A person can receive an electrical shock by touching a "live" wire and a grounded surface. The kitchen sink (plumbing, water) was always considered to be the hazard. The Code now recognizes that there are other grounded "things" a person might touch, such as a range, oven, cooktop, refrigerator, microwave oven, or countertop food processor. ◄

• A receptacle may be installed for plugging in a gas appliance to serve the ignition and/or clock devices on the appliance. See figure 3-3.

• The small appliance circuits are provided to serve plug-in portable appliances. They are not permitted by the Code to serve lighting fixtures or appliances such as dishwashers, garbage disposals, or exhaust fans.

►• The receptacle for a refrigerator may be supplied by one of the required 20-ampere small-appliance branch-circuits, or it may be supplied by an individual 15- or 20-ampere branch-circuit. Some local electrical codes require that the receptacle outlet for the refrigerator be supplied by a separate branch-circuit. ◄

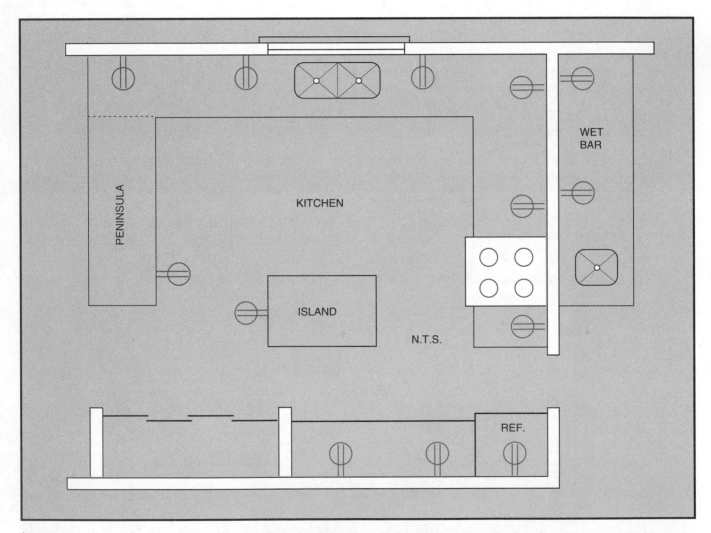

► Fig. 12-7   In kitchens, *all* 125-volt, single-phase, 15- and 20-ampere receptacles that serve countertop surfaces (walls, islands, or peninsulas) must be GFCI protected. The key words are "that serve countertop surfaces." GFCI personnel protection is *not* required for the refrigerator receptacle, a receptacle under the sink for plugging-in a food waste disposer, or a receptacle installed inside cabinets above the countertops for plugging-in a microwave oven attached to the underside of the cabinets. Where wet-bar sinks are installed outside of the kitchen area, *all* 125-volt, single-phase, 15- and 20-ampere receptacles that serve the countertops must be GFCI protected if they are within 6 feet (1.83 m) measured from the outer edge of the wet-bar sink. See *Section 210-8(a)(5)*. ◄ Drawing is not to scale.

The small appliance circuits prevent circuit overloading in the kitchen as a result of the heavy concentration and use of electrical appliances. For example, one type of cord-connected microwave oven is rated 1500 watts at 120 volts. The current required by this appliance alone is:

$$I = \frac{W}{E} = \frac{1500}{120} = 12.5 \text{ amperes}$$

Section 90-1(b) states that the Code requirements are considered necessary for safety. According to the Code, compliance with these rules will not necessarily result in an efficient, convenient, or adequate installation for good service or future expansion of electrical use.

Figure 12-8 shows three 20-ampere small appliance circuits feeding the kitchen/nook receptacles, of which there are nine. These circuits are B13, B15, B16.

Obviously, there are many ways that receptacles can be connected to the small appliance circuits. The intent is to arrange the circuitry so as to provide the best availability of electrical power for appliances that will be used in the heavily concentrated work areas in the kitchen.

Figure 12-8 shows one way to arrange the kitchen small appliance circuits. Note that there is a receptacle outlet on the side of the kitchen cooktop island. This is required by the Code. The receptacle is connected to Circuit B16. It is *not* connected to the cooktop circuit.

When a receptacle is installed below the sink for easy plug-in connection of a food waste disposer, it is *not* to be included in the Code requirements for small appliance circuits, and it is *not* required to be GFCI protected, *Section 210-8(a)(5)*. Refer to unit 3 for spacing requirements for receptacles located in kitchens.

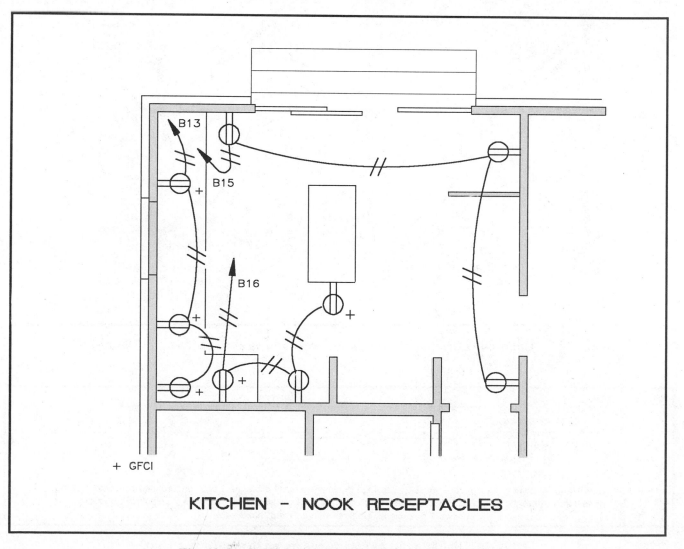

KITCHEN - NOOK RECEPTACLES

Fig. 12-8   Cable layout for kitchen/nook receptacles.

Sliding Glass Doors

See unit 3 for a discussion on how to space receptacle outlets on walls where sliding glass doors are installed.

Small Appliance Branch-Circuit Load Calculations

The receptacle load for the three small appliance branch-circuits in the kitchen is calculated using a load of 1500 volt-amperes per circuit. Thus, the computed load is

1500 volt-amperes × 3 = 4500 volt-amperes

Section 220-16(a) requires that *feeder* loads be calculated on the basis of 1500 volt-amperes for each two-wire small appliance circuit. Using the same basis for estimating the branch-circuit loading will result in ample circuit capacity. See unit 29 for complete coverage of load calculations.

SPLIT-CIRCUIT RECEPTACLES AND MULTIWIRE CIRCUITS

Split-circuit receptacles may be installed where a heavy concentration of plug-in load is anticipated, in which case each receptacle is connected to a separate circuit as in figures 12-9 and 12-10.

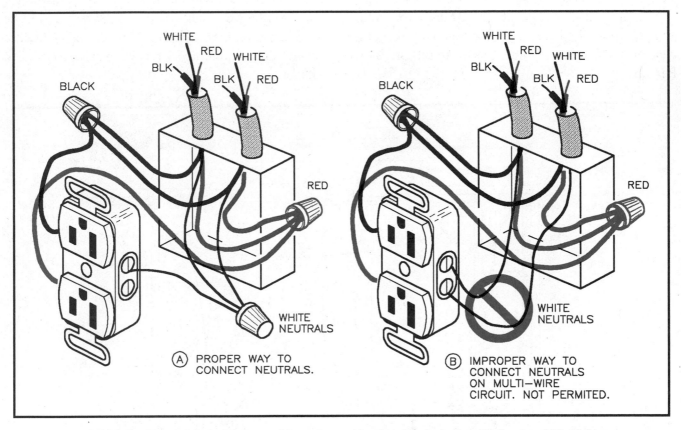

Fig. 12-9   Connecting the neutral in a three-wire (multiwire) circuit, *Section 300-13(b)*.

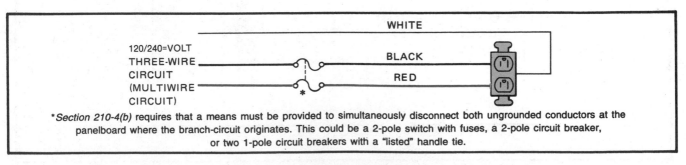

*Section 210-4(b) requires that a means must be provided to simultaneously disconnect both ungrounded conductors at the panelboard where the branch-circuit originates. This could be a 2-pole switch with fuses, a 2-pole circuit breaker, or two 1-pole circuit breakers with a "listed" handle tie.

Fig. 12-10   Split-circuit receptacle connected to multiwire circuit.

A popular use of split-circuit receptacles is to control one receptacle with a switch and leave one receptacle "hot" at all times, as is done in the bedrooms and living room of this residence.

Split-circuit GFCI receptacles are not available in the marketplace. Refer to unit 6 where GFCIs are discussed in great detail.

A feedthrough GFCI receptacle could be installed to the right of the sink, with regular grounding-type receptacles installed for the other two receptacles on Circuit B13.

Another feedthrough GFCI receptacle could be installed for the first receptacle fed by Circuit B16. Another way to connect Circuit B16 would be to use regular grounding-type receptacles for the first two receptacles fed by Circuit B16, then install a GFCI receptacle on the side of the range cooktop island where the plans indicate a receptacle outlet.

Circuit B15 does not require any GFCI receptacles because these receptacles do not serve countertop areas.

Therefore, the use of split-circuit receptacles is rather difficult to incorporate where GFCI receptacles are required. A multiwire branch-circuit could be run to a box where proper connections are made, with one circuit feeding one feedthrough GFCI receptacle and its downstream receptacles.

This would be used when it becomes more economical to run one three-wire circuit rather than installing two two-wire circuits. Again, it is a matter of knowing what to do, and when it makes sense to do it. See figure 12-11.

When multiwire circuits are connected, the neutral conductor must not be broken at the receptacles. *NEC® Section 300-13(b)* states that where more than one conductor is to be spliced, all connections for

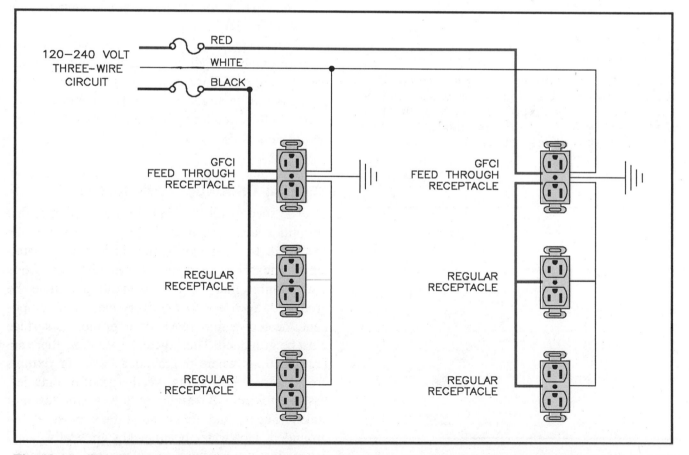

Fig. 12-11 **This diagram shows how a multiwire circuit can be used to carry two circuits to a box, then split the three-wire circuit into two two-wire circuits. This circuitry does not require simultaneous disconnect of the ungrounded conductors as shown in figure 12-10.**

the grounded conductors in multiwire circuits must be made up independently of the receptacles, lampholders, and so on. This requirement is illustrated in figure 12-9. Note that the screw terminals of the receptacle *must* not be used to splice the neutral conductors. The hazards of an open neutral are discussed in unit 17. Figure 12-12 shows two types of grounding receptacles commonly used for installations of this type.

When multiwire branch-circuits supply more

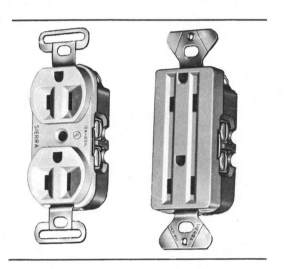

Fig. 12-12 Grounding-type duplex receptacles, 15A–125V rating. Most duplex receptacles can be changed into a "split-circuit" receptacle by breaking off the small metal tab on the terminals. The above photo clearly shows this feature.

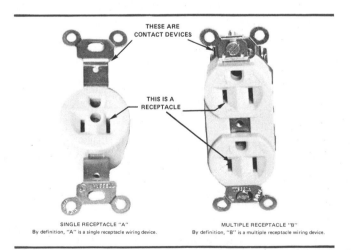

SINGLE RECEPTACLE "A"
By definition, "A" is a single receptacle wiring device.

MULTIPLE RECEPTACLE "B"
By definition, "B" is a multiple receptacle wiring device.

THESE ARE CONTACT DEVICES

THIS IS A RECEPTACLE

Fig. 12-13 Receptacles: See *NEC*® definitions, *Article 100*.

than one receptacle on the same yoke, a means must be provided to disconnect simultaneously all of the "hot" conductors at the panelboard where the branch-circuit originates. Figure 12-10 shows that 240 volts are present on the wiring device, only 120 volts are connected to each receptacle on the wiring device. See *Sections 210-6(a)(2)* and *210-6(b)(3)*.

The "simultaneous disconnect" rule applies to any receptacles, lampholders, or switches mounted on one yoke. See *Section 210-4(b)*.

RECEPTACLES AND OUTLETS

Article 100 of the *NEC*® gives the definition of a receptacle and an outlet, as illustrated in figures 3-5, 12-13, and 12-14.

Receptacles are selected for circuits according to the following guidelines. A single receptacle connected to a circuit must have a rating not less than the rating of the circuit, *Section 210-21(b)*. In a residence, typical examples for this requirement for single receptacles are the clothes dryer outlet (30 amperes), or the range outlet (50 amperes), or the freezer outlet (15 amperes).

Circuits rated 15 amperes supplying two or more receptacles shall not contain receptacles rated over 15 amperes. For circuits rated 20 amperes supplying two or more receptacles, the receptacles connected to the circuit may be rated 15 or 20 amperes. See *Table 210-21(b)(3)*.

GENERAL GROUNDING CONSIDERATIONS

The specifications for the residence require that all outlet boxes, switch boxes, and fixtures be grounded. In other words, the entire wiring installation must be a grounded system. The Code also requires that the outside lighting fixtures be grounded, *Section 410-17*. Every electrical component located within reach of a grounded surface must be grounded. Thus, it can be seen that there are few locations where ungrounded boxes or fixtures are permitted. Post lights, weatherproof receptacles, basement wiring, boxes and fixtures within reach of sinks, ranges, and range hood fans must all be grounded. Hot water heating, hot air heating, and steam heating registers are all grounded surfaces. *Section 250-42* requires that any boxes installed near

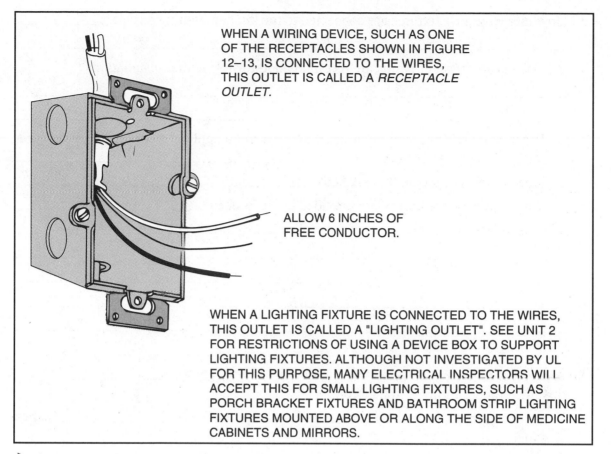

WHEN A WIRING DEVICE, SUCH AS ONE OF THE RECEPTACLES SHOWN IN FIGURE 12–13, IS CONNECTED TO THE WIRES, THIS OUTLET IS CALLED A *RECEPTACLE OUTLET.*

ALLOW 6 INCHES OF FREE CONDUCTOR.

WHEN A LIGHTING FIXTURE IS CONNECTED TO THE WIRES, THIS OUTLET IS CALLED A "LIGHTING OUTLET". SEE UNIT 2 FOR RESTRICTIONS OF USING A DEVICE BOX TO SUPPORT LIGHTING FIXTURES. ALTHOUGH NOT INVESTIGATED BY UL FOR THIS PURPOSE, MANY ELECTRICAL INSPECTORS WILL ACCEPT THIS FOR SMALL LIGHTING FIXTURES, SUCH AS PORCH BRACKET FIXTURES AND BATHROOM STRIP LIGHTING FIXTURES MOUNTED ABOVE OR ALONG THE SIDE OF MEDICINE CABINETS AND MIRRORS.

▶ **Fig. 12-14   For ease in working with wiring devices and splicing conductors, the Code in *Section 300-14* requires that the branch-circuit conductors have at least 6 inches of free conductor length left at each outlet, junction, and switch point for splices or the connection of fixtures or devices. The 6-inch length is generally measured from the outer edge of the box. Leaving too much conductor length will result in having to jam the wires into the box. ◀**

metal or metal lath, tinfoil, or aluminum insulation must also be grounded.

As previously discussed, grounding is accomplished through the proper use of nonmetallic-sheathed cable, armored cable, and metal conduit.

In the case of nonmetallic-sheathed cable, do not confuse the bare *grounding* conductor with the *grounded* circuit conductor. The *grounding* conductor is used to ground equipment, whereas the white *grounded* conductor is one of the branch-circuit conductors. They are *not* to be connected together except at the main service-entrance equipment.

REVIEW

Note: Refer to the Code or plans where necessary.

1. If everything on Circuit B7 were turned on, what would be the total current draw?

   _____

   _____

2. From what panel does the kitchen lighting circuit originate? What size conductors are used? _____

   _____

3. How many lighting fixtures are connected to the kitchen lighting circuit? _____

_____

4. What color fluorescent lamps are recommended for residential installations?

_____

_____

5. a. What is the minimum number of 20-ampere small appliance circuits required for a kitchen according to the Code? _____

   b. How many are there in this kitchen? _____

6. How many receptacle outlets are provided in the kitchen? _____

7. What is meant by the term two-circuit (split-circuit) receptacles? _____

_____

_____

_____

_____

_____

8. A single receptacle connected to a circuit must have a rating _____ than the ampere rating of the circuit.

9. In kitchens, a receptacle must be installed at each counter space _____ inches or wider.

10. A fundamental rule regarding the grounding of metal boxes, fixtures, and so on, is that they must be grounded when "in reach of _____."

11. How many circuit conductors enter the box

    a. where the clock will be installed? _____

    b. in the ceiling box over which the track will be installed? _____

    c. at the switch location to the right of the sliding door? _____

12. How much space is there between the countertop and upper cabinets? _____

13. Where is the speed control for the range hood fan located? _____

14. Who is to furnish the range hood? _____

15. List the appliances in the kitchen that must be connected. _____

_____

_____

_____

_____

_____

16. Complete the wiring diagram, connecting feedthrough GFCI 2 to also protect receptacle 1, both to be supplied by Circuit A1. Connect feedthrough GFCI 3 to also protect receptacle 4, both to be supplied by Circuit A2. Use colored pencils or markers to show proper color. Assume that the wiring method is electrical metallic tubing, where more freedom in the choice of insulation colors is possible.

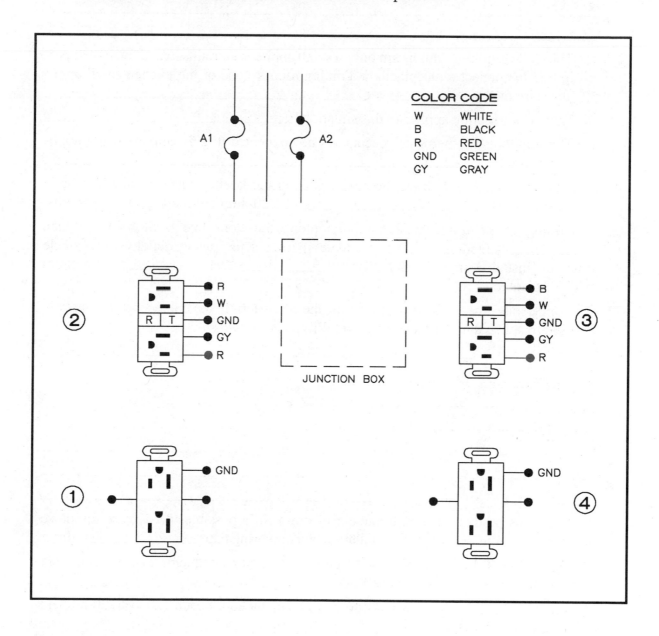

17. Each 20-ampere small appliance branch-circuit load demand shall be determined at _____ . Choose one.

   a. 2400 volt-amperes   b. 1500 volt-amperes   c. 1920 volt-amperes

18. Is it permitted to connect an outlet supplying a clock receptacle to a 20-ampere small appliance circuit? _____

_____

_____

_____

19. a. The Code requires a minimum of two small appliance circuits in a kitchen. Is it permitted to connect a receptacle in a dining room to one of the kitchen small appliance circuits? _____

What Code section applies to this situation? _____

b. May outdoor weatherproof receptacles be connected to a 20-ampere small appliance circuit? _____

c. Receptacles located above the countertops in the kitchen must be supplied by at least _____ 20-ampere small appliance circuits.

20. According to *Section 210-52*, no point along the floor line shall be more than _____ feet (_____ meters) from a receptacle outlet. A receptacle must be installed in any wall space _____ feet (_____ meters) wide or greater.

21. The Code states that in multiwire circuits, the screw terminals of a receptacle must *not* be used to splice the neutral conductors. Why?

_____

_____

_____

_____

_____

_____

_____

22. Electric fans produce a certain amount of noise. It is possible to compare the noise levels of different fans prior to installation by comparing their _____ ratings.

23. Is it permitted to connect the white grounded circuit conductor to the equipment grounding terminal of a receptacle? _____

24. A load of _____ volt-amperes must be included for each 120-volt, 20-ampere small appliance circuit.

25. a. All 125-volt, single-phase 15- and 20-ampere receptacles installed in a kitchen that serve countertop surfaces must be GFCI protected. (True) (False) Circle the correct answer.

b. All 125-volt, single-phase 15- and 20-ampere receptacles installed within 6 feet of a wet-bar sink outside of the kitchen that serve countertop surfaces must be GFCI protected. (True) (False) Circle the correct answer.

26. The kitchen/nook features a sliding glass door. For the purpose of providing the proper receptacle outlets, answer the following statements *true* or *false*.

   a. Sliding glass panels are considered wall space. _____

   b. Fixed panels of glass doors are considered wall space. _____

27. The following is a layout for the lighting circuit for the Kitchen/Nook. Complete the wiring diagram using colored pens or pencils to show the conductors' insulation color.

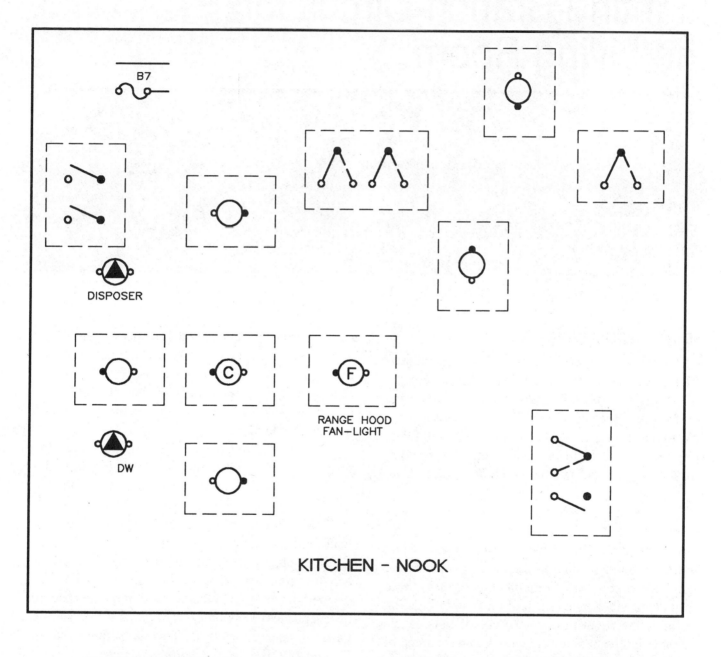

KITCHEN - NOOK

28. When roughing in the wiring, *Section* _____ requires that at least (4) (6) (8) inches of free length of conductor be left at an outlet box. This measurement is generally taken from (the outer edge of the box) (where the cable enters the box). Circle the correct answers.

UNIT 13

Lighting Branch-Circuit for the Living Room

OBJECTIVES

After studying this unit, the student will be able to

- understand various types of dimmer controls.
- connect dimmers to incandescent lamp loads and fluorescent lamp loads.
- understand the phenomenon of incandescent lamp inrush current.
- discuss Class P ballasts and ballast overcurrent protection.
- discuss the basics of track lighting.

LIGHTING CIRCUIT OVERVIEW

The feed for the living room lighting branch-circuit is connected to Circuit B17, a 15-ampere circuit. The home run is brought from Panel B to the weatherproof receptacle outside of the living room next to the sliding door. The circuit is then run to the split-circuit receptacle just inside the sliding door, almost back-to-back with the outdoor receptacle, figure 13-1.

A three-wire cable is then carried around the living room feeding in and out of the eight receptacles. This three-wire cable carries the black and white circuit conductors plus the red wire, which is the switched conductor. Remember, because these receptacles will provide most of the general lighting for the living room through the use of floor lamps and table lamps, it is advantageous to have some of these lamps controlled by the three switches—two three-way switches and one four-way switch.

Note that the receptacle below the four-way switch is required according to the Code ruling that states that a receptacle outlet shall be installed in any wall space 2 feet (611 mm) or more in width, *Section 210-52(a)*. It is highly unlikely that a split-circuit receptacle connected for switch control, as are the other receptacles in the living room, will ever be needed, plus the fact that the distance to the Study/Bedroom receptacle is so short it makes economic sense to make up the connections as shown on the cable layout.

Accent lighting above the fireplace is in the form of two recessed fixtures controlled by a single-pole dimmer switch. See unit 7 for installation data and Code requirements for recessed fixtures.

Table 13-1 summarizes the outlets and estimated load for the living room circuit.

The spacing requirements for receptacle outlets and location requirements for lighting outlets are covered in unit 3, as is the discussion on the spacing of receptacle outlets on exterior walls where sliding glass doors are involved. This is pointed out in *Section 210-52(a)* of the *National Electrical Code.®*

TRACK LIGHTING *(Article 410, Part R)*

The plans show that track lighting is mounted on the ceiling of the living room on the wall opposite the fireplace.

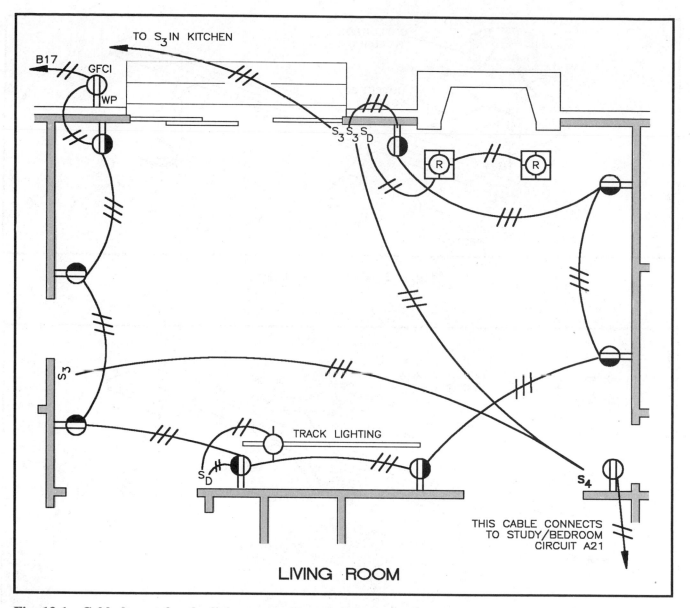

TO S₃ IN KITCHEN

B17

GFCI

WP

S₃ S₃ S_D

R   R

S₃

S_D

TRACK LIGHTING

S₄

THIS CABLE CONNECTS
TO STUDY/BEDROOM
CIRCUIT A21

LIVING ROOM

Fig. 13-1   Cable layout for the living room. Note that the receptacle on the short wall where the four-way switch is located is fed from the study/bedroom Circuit A21.

DESCRIPTION	QUANTITY	WATTS	VOLT-AMPERES
Receptacles @ 120 W Note: Receptacle under S₄ is connected to Study/Bedroom circuit.	8	960	960
Weatherproof receptacle	1	120	120
Track light five lamps @ 40 W each	1	200	200
Fireplace recessed fixtures @ 75 W each	2	150	150
TOTALS	12	1430	1430

Table 13-1   Living room outlet count and estimated load. Circuit B17.

Track lighting provides accent lighting for a fireplace, a painting on the wall, some item (sculpture or collection) that the homeowner wishes to focus attention on, or it may be used to light work areas such as counters, game tables, or tables in such areas as the kitchen nook.

Differing from recessed fixtures and individual ceiling fixtures that occupy a very definite space, track lighting offers flexibility because the actual lampholders can be moved and relocated on the track as desired.

Lampholders from track lighting selected from hundreds of styles are inserted into the extruded

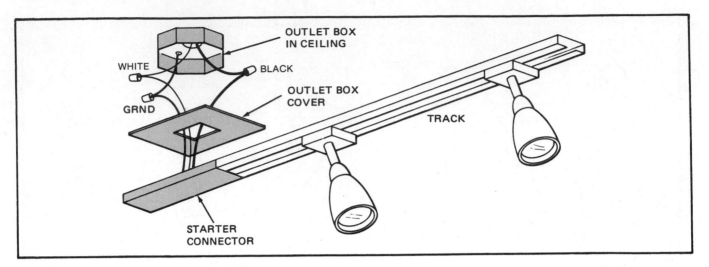

Fig. 13-2   End-feed track light.

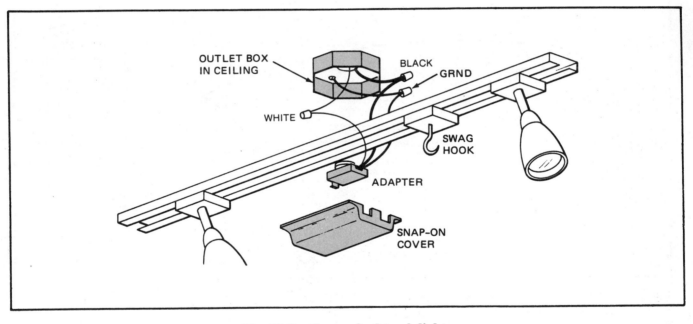

Fig. 13-3   Center-feed track light.

aluminum track at any point (the circuit conductors are in the track), and the plug-in connector on the lampholder completes the connection. In addition to being fastened to the outlet box, the track is generally fastened to the ceiling with toggle bolts or screws. Various track light installations are shown in figures 13-2, 13-3, and 13-4.

Note on the plans for the residence that the living room ceiling has wood beams. This presents a challenge when installing track lighting that is longer than the space between the wood beams.

There are a number of things the electrician can do:

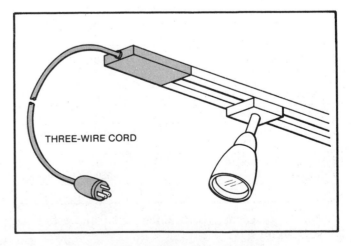

Fig. 13-4   Plug-in track light.

1. Install pendant kit assemblies that will allow the track to hang below the beam, figure 13-5.

2. Install conduit conductor fittings on the ends of sections of the track, then drill a hole in the beam through which ½-inch electrical metallic tubing can be installed and connected to the conduit connector fittings, figure 13-6.

Running track lighting through walls or partitions is prohibited by the *National Electrical Code,*® *Section 410-101(c)(7)*.

3. Run the branch-circuit wiring concealed above the ceiling to outlet boxes located at the points where the track is to be fed by the branch-circuit wiring, figure 13-7.

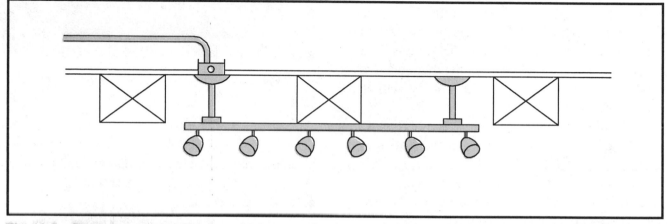

Fig. 13-5   Track lighting mounted on the bottom of a wood beam. The supply wires to the track assembly are fed through the pendant.

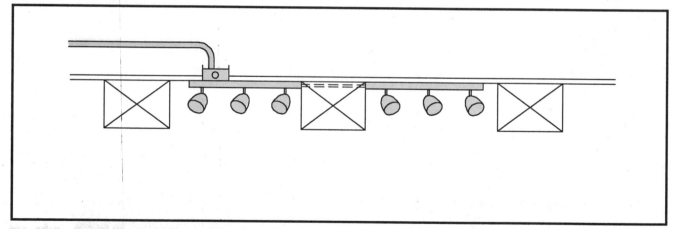

Fig. 13-6   Track lighting mounted on the ceiling between the wood beams. The two track assembly sections are connected together by means of a conduit that has been installed through a hole drilled in the beam.

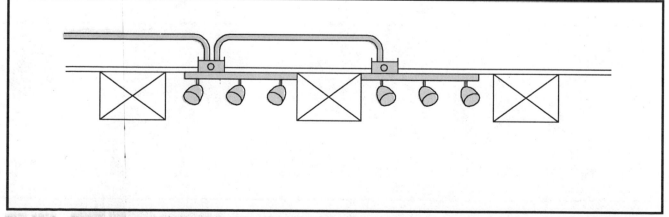

Fig. 13-7   Track lighting mounted on the ceiling between the wood beams. Each of the two track assembly sections are connected to outlet boxes in the ceiling in the same manner that a regular lighting fixture is hung.

Where to Mount Track Lighting

The ceiling height, the aiming angle, the type of lampholder, the type of lamp, and the type of lighting to be achieved are all important factors that must be considered in order to install the lighting track at the proper distance from the wall.

For example, if we are designing an installation that will illuminate an oil painting that is to be centered 5'5" (average eye level) from the floor, where the ceiling height is 9'0", and the lampholder and lamp information tell us that the aiming angle should be 60°, we would find in the manufacturer's installation data that the track should be mounted 2 feet from the wall.

If, in this example, the ceiling height was 8'0", then the distance to the track from the wall would be 18".

Because all of the factors are variables that affect the positioning of the track, it is absolutely necessary to consult the track lighting manufacturer's catalog for recommendations that will achieve the desired results.

Track lighting shall not be installed less than 5 feet (1.52 m) above the floor unless it is of the low-ge type that has an open-circuit voltage of less 30 volts RMS, or if the track is protected from physical damage. See *Section 410-101(c)(8)*.

The track itself must be grounded properly, *Section 410-105(b)*. The plug-in connector also has a ground contact to insure that the track lighting fixture is safely grounded. Figure 13-8 shows tracks in cross-sectional views.

Typical fixtures that attach to track lighting systems are shown in figure 13-9.

The track lighting in the living room is turned on and off by a single-pole dimmer switch.

Fig. 13-8   Tracks in cross-sectional views.

Some sections of track systems are permitted to be cut to length in the field by the installer. Be sure to check the manufacturer's instructions on where and how to make the cut, and how to close off the cut with a proper end cap.

Portable Lighting Tracks

Portable lighting tracks per Underwriters Laboratories' standards must:

- not be longer than 4 feet.
- have a cord no longer than 9 feet.
- have integral switching in the lampholders.
- have integral overcurrent protection in the lampholders.
- not have its length altered in the field. To do so would void the UL listing of the track.

Remember, *Section 110-3(b)* of the *NEC®* states that "listed or labeled equipment shall be used or installed in accordance with any instructions included in the listing or labeling."

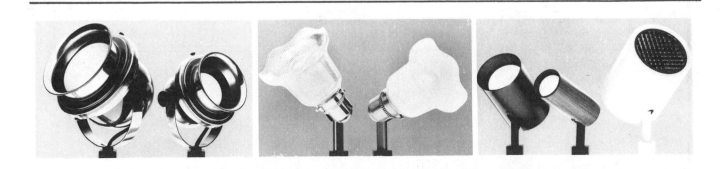

Fig. 13-9   Typical fixtures that attach to track lighting systems.

Supporting Track Lighting

Lighting tracks that are 4 feet or less in length must be fastened at two places minimum. Then, if more track is needed, install at least one additional support for each individual extension that is 4 feet long or less. See *Section 410-104, NEC.*®

Load Calculations for Lighting Track

For track lighting installed in homes, it is not necessary to add additional loads for the purpose of branch-circuit, feeder, or service-entrance calculations. This is confirmed in *Section 220-3(c)(6)*, which refers us to *Section 410-102*. The exception found in *Section 410-102* exempts track lighting installed in homes from being considered as an additional load to the basic VA per square foot of area for the general lighting load calculations.

▶ However, if the lighting track is to be installed in a commercial building, a loading factor of 150 VA for each 2 feet of lighting track must be added to the calculations for branch-circuit, feeder, and service-entrance computations. ◀

For further information regarding track lighting, refer to *Part R* of *Article 410* of the *National Electrical Code.*®

DIMMER CONTROLS FOR HOMES

Dimmers are used in homes to lower the level of light. There are two basic types.

Electronic Dimmers

Figure 13-10 shows different types (slide, touch, rotate) of electronic dimmers. These are very popular. They "fit" in a standard switch (device) box. They can replace standard single-pole or three-way toggle switches. See figure 13-11 for the typical hook-up of a single-pole and a three-way dimmer. Because dimmer and fan speed controls are physically larger than most wiring devices (plugs and switches), make sure there is enough room in the wall box to accommodate the dimmer, taking into consideration the number of conductors that are in the switch box. To "jam" the dimmer in may cause trouble, such as short-circuits and ground faults. The most common type of electronic dimmer for residential use is rated 125 volts, 60-hertz alternating current, 600 watts. Also available are 1000-watt, 1500-watt, and 2000-watt dimmer switches. The 1500-watt and 2000-watt dimmers fit into a single-gang wall box, but require a two-gang plate because of the need to cover the heat sinking "fins." To

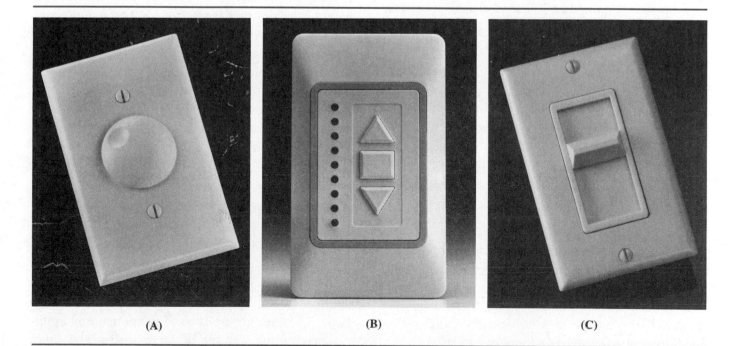

(A)   (B)   (C)

Fig. 13-10   Some typical electronic dimmers. All perform the same function of varying the voltage to the connected incandescent lamp load. (A) rotating knob. (B) push buttons. (C) slider knob. *Courtesy* Pass & Seymour/Legrand.

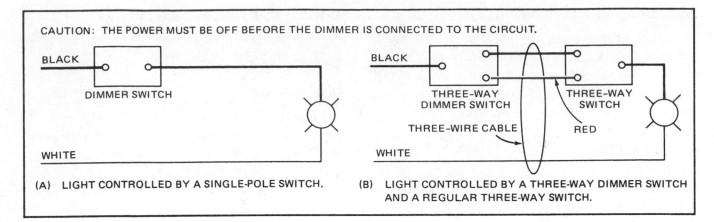

Fig. 13-11   Use of single-pole and three-way electronic solid-state dimmer controls in circuits.

prevent an overload, check the total wattage to be controlled with the rated capacity of the dimmer. The maximum wattage that can safely be connected is shown on the nameplate of the dimmer.

Most instructions furnished with an electronic dimmer will state: "Do not use to control receptacle outlets." The reason for this is twofold:

1. Serious overloading could result, because it may not be possible to determine or limit what other loads might be plugged in the receptacle.

2. Dimmers reduce voltage. Reduced voltage supplying a television, radio, stereo components, personal computer, vacuum cleaner, etc., can result in costly damage to the appliance.

Electronic dimmers for use *only* to control incandescent lamps are marked with a *wattage* rating. Serious overheating and damage to the transformer and to the electronic dimmer will result if an "incandescent only" dimmer is used to control inductive loads.

For example, a dimmer marked 600 watts @ 120 volts is capable of safely carrying 5 amperes.

$$\text{Amperes} = \frac{\text{Watts}}{\text{Volts}} = \frac{600}{120} = 5 \text{ amperes}$$

Dimmers that are suitable for control of "inductive" loads, such as lighting systems that incorporate a transformer as in low voltage outdoor lighting and low voltage track lighting, will be marked in *volt-amperes*. The instructions will clearly indicate that the dimmer can be used on inductive loads.

See unit 7 for discussion on watts versus volt-amperes.

Do *not* hook up a dimmer with the circuit "hot." Not only is it dangerous because of the possibility of receiving an electrical shock, but the electronic solid-state dimmer can be destroyed when the splices are being made as a result of a number of "make and breaks" before the actual final connection is completed. The momentary inrush of current each time there is a "make and break" can cause the dimmer's internal electronic circuitry to heat up beyond its thermal capability. There is the possibility that as one lead is being connected, the other lead could come in contact with "ground." Unless the dimmer has some sort of "built-in" short-circuit protection, the dimmer is destroyed. The manufacturer's warranty will probably be null and void if the dimmer is worked "hot." It is well worth the time to take a few minutes to de-energize the circuit.

Electronic dimmers use triacs to "chop" the ac sine wave, so some incandescent lamps might "hum" because the lamp filament vibrates. This hum can usually be eliminated or greatly reduced by installing "rough service" lamps, or changing the dimmer.

Autotransformer Dimmer Control

Figure 13-12 shows an autotransformer (one winding)-type dimmer. These are physically larger than the electronic dimmer, and require a special wall box. They are not ordinarily used for residential applications unless the load to be controlled is very large. They are generally used in commercial applications for the control of large loads. The lamp intensity is controlled by varying the voltage to the lamp load. When the knob is rotated, a brush contact moves over a bared portion of the transformer winding. As with electronic dimmers, they will not work on direct current, but that is not a problem because all homes are supplied with alternating current. The maximum wattage to be controlled is marked on the nameplate of the dimmer. Overcurrent protection is built into the dimmer control to prevent burnout due to an overload.

Figure 13-13 illustrates the connections for an autotransformer dimmer that controls incandescent lamps.

Dimming Fluorescent Lamps

To dim incandescent lamps, the voltage to the lamp filament is varied (raised or lowered) by the dimmer control. The lower the voltage, the less intense is the light. The dimmer is simply connected "in series" with the incandescent lamp load.

Fluorescent lamps are a different story. The connection is not a simple "series" circuit. To dim fluorescent lamps, special dimming ballasts and special dimmer switches are required. The dimmer control

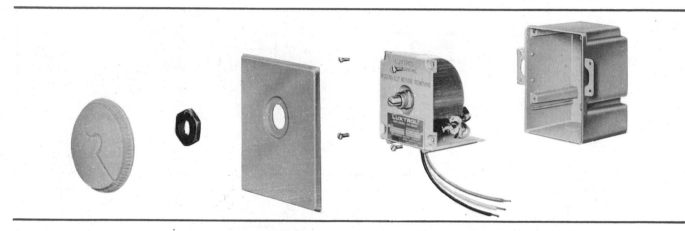

Fig. 13-12   Dimmer control, autotransformer type.

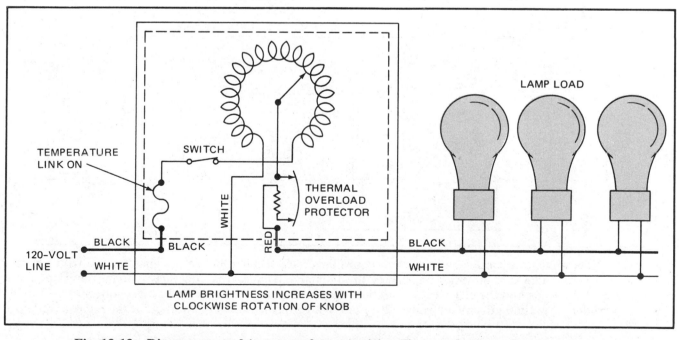

Fig. 13-13   Dimmer control (autotransformer) wiring diagram for incandescent lamps.

allows the dimming ballast to maintain a voltage to the cathodes that will maintain the cathodes' proper operating temperature, and also allows the dimming ballast to vary the current flowing in the arc. This in turn varies the intensity of light coming from the fluorescent lamp.

Figure 13-14 illustrates the connections for an autotransformer-type dimmer for the control of fluorescent lamps using a special dimming ballast. The fluorescent lamps are F40T12 rapid start lamps. With the newer electronic dimming ballasts, 32-watt, rapid start T8 lamps are generally used.

Figure 13-15 shows an electronic dimmer for the control of fluorescent lamps using a special dimming ballast.

One manufacturer of electronic ballasts offers a "50% Power Level Dimming Ballast." This special dimming ballast is connected with a special three-position toggle switch.

Position 1: off

Position 2: full light output

Position 3: 40% light output (50% power)

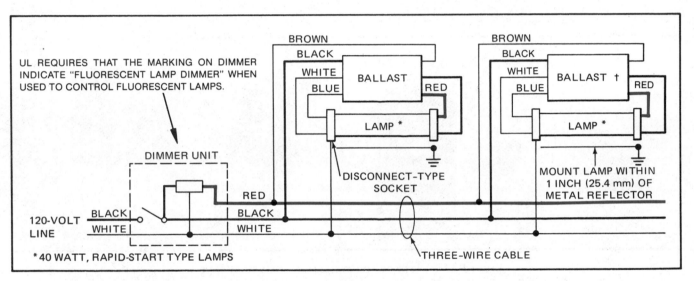

Fig. 13-14   Dimmer control wiring diagram for fluorescent lamps.

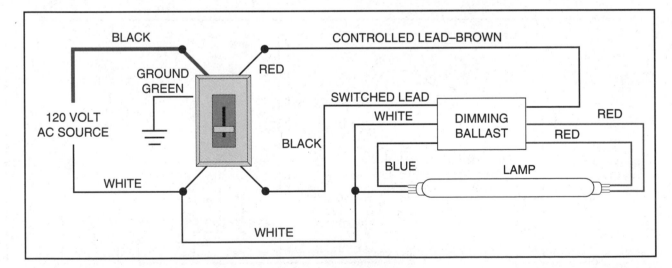

Fig. 13-15   This wiring diagram shows the connections for an electronic dimmer designed for use with a special dimming ballast for rapid start lamps. Always check the dimmer and ballast manufacturers' wiring diagrams, as the color coding of the leads and the electrical connections may be different than shown in this diagram. The dimmer shown is a "slider" type, and offers wide-range (20% to 100%) adjustment of light output for the fluorescent lamp.

Another manufacturer has a dimming ballast that can be controlled with a wide adjustment range of 20% to 100% light output. The end result of installing these types of dimming ballasts and associated controls is energy savings, and lighting levels similar to that of turning off some of the fixtures, or turning off one ballast of a two ballast, four-lamp fluorescent fixture. However, the big difference is uniform light, rather than having some fixtures off and some fixtures on. Although primarily used in commercial installations, the recreation room recessed fluorescent fixtures in the residence discussed in this text could have been connected up using this method.

To assure proper dimming performance, the lamp reflector and the ballast case must be solidly grounded.

Because incandescent lamps and fluorescent lamps have different characteristics, they cannot be controlled simultaneously with a single dimmer control. Nor should electronic ballasts and magnetic ballasts be controlled simultaneously with a single dimmer control.

Always read and follow the manufacturer's instructions furnished with dimmers and ballasts.

INCANDESCENT LAMP LOAD INRUSH CURRENTS

An unusual action occurs when a circuit is energized to supply a tungsten filament lamp load. A tungsten filament lamp has a very low resistance when it is cold. When the lamp is connected to the proper voltage, the resistance of the filament increases very quickly. This increase occurs within $1/240$ second or one-quarter cycle after the circuit is energized. During this period, there is an inrush current that is 10 to 20 times greater than the normal operating current of the lamp.

T-rated switches have this momentary high inrush current taken into consideration. T-rated switches were discussed in unit 5. Branch-circuit conductors are not affected by this momentary high surge of current. Circuit breakers have been known to nuisance trip when the incandescent lamp load is controlled through a dimmer, or when the branch-circuit is heavily loaded.

Let's look at what might happen when a lamp load is controlled through a dimmer. Assume that the dimmer is turned to a preset low (dim) setting and the circuit is turned on with a separate switch. The resulting inrush current lasts slightly longer due to the slower heating of the lamp filament.

This prolonged surge of current may cause certain types of circuit breakers to trip. These breakers have a highly sensitive magnetic tripping mechanism. When this problem occurs, the sensitive breaker must be replaced with one that has high magnetic tripping characteristics. Most manufacturers whose breakers react to the inrush current surge of tungsten filament lamp loads can supply the high magnetic tripping circuit breaker. Branch-circuit fuses are not affected by this momentary inrush current because they have sufficient time delay to override the inrush.

The nuisance tripping problem is generally experienced on heavy dimming load applications, such as in restaurants, libraries, and similar occupancies. It is not a common occurrence in residential installations.

REVIEW

Note: Refer to the Code or the plans where necessary.

1. To what circuit is the living room connected? _____

2. How many convenience receptacles are connected to the living room circuit? _____

3. a. How many wires enter the switch box at the four-way switch location? _____

   b. What type and size of box may be installed at this location? _____

      _____

      _____

      _____

4. How many wires must be run between an incandescent lamp and its dimmer control?

   _____

5. Complete the wiring diagram for the dimmer and lamp.

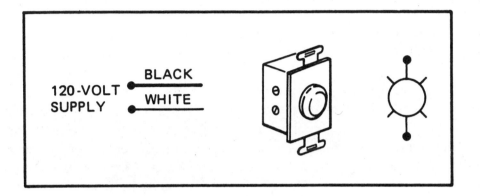

6. What is meant by incandescent lamp load inrush current? What type of switch is required? _____

   _____

   _____

   _____

   _____

   _____

7. Is it possible to dim standard fluorescent ballasts? _____

8. a. How many wires must be run between a dimming-type fluorescent ballast and the dimmer control? _____

   b. Is a switch needed in addition to the dimmer control? _____

      _____

      _____

      _____

      _____

9. Explain why fluorescent lamps having the same wattage draw different current values.

   _____

   _____

   _____

   _____

   _____

10. What is the total current consumption of the track lighting and recessed fixtures above the fireplace? Show your calculations.

11. How many television outlets are provided in the living room? _____

12. Where is the telephone outlet located in the living room? _____

   _____

13. Solid-state electronic dimmers of the type sold for residential use (shall) (shall not) be used to control fluorescent fixtures. These dimmers (are) (are not) intended for speed control of small motors. Circle the correct answers.

14. When calculating a dwelling lighting load on the "volt-ampere per square foot" basis, the receptacle load used for floor lamps, table lamps, and so forth,
    a. must be added to the general lighting load at 1½ amperes per receptacle.
    b. must be added to the general lighting load at 1,500 volt-amperes for every 10 receptacles.
    c. is already included in the calculations as part of the volt-amperes per square foot. Circle the correct letter.

15. A layout of the outlets, switches, dimmers, track lighting, and recessed fixtures is shown in the following diagram. Using the cable layout of figure 13-1, make a complete wiring diagram of this circuit. Use colored pencils or marking pens to indicate conductors.

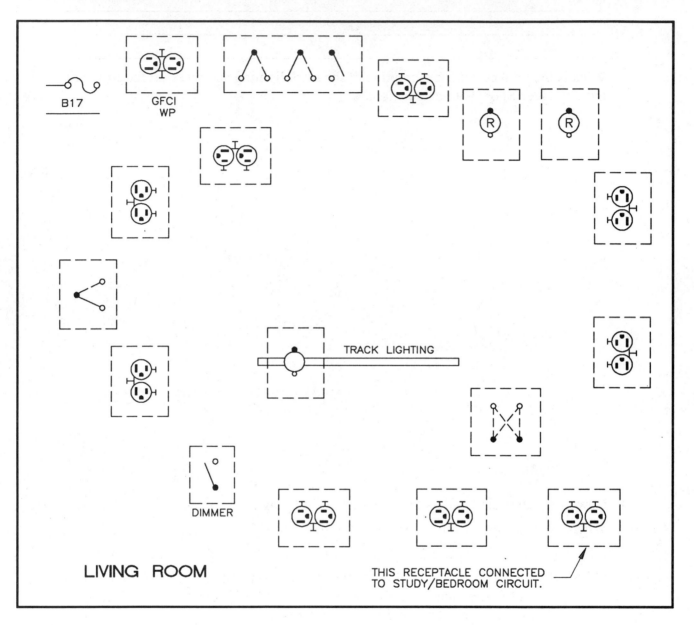

16. Prepare a list of television outlets, telephone outlets, and wiring devices shown on the plans and cable layout for the living room area. Include the number of each type present.

_____

_____

_____

_____

_____

17. The living room features a sliding glass door. For the purpose of providing the proper receptacle outlets, answer the following statements *true* or *false*.

   a. Sliding glass pancls arc considered wall space. _____

   b. Fixed panels of glass doors are considered wall space. _____

18. a. May lengths of track lighting be added when the track is permanently connected?

     _____

   b. May lengths of track lighting be added when the track is cord-connected?

     _____

19. Must track lighting always be fed (connected) at one end of the track? _____

     _____

20. May a standard electronic dimmer be used to control low-voltage lighting that incorporates a transformer, or for controlling the speed of a ceiling fan motor?

   Yes _____   No _____ .

21. Does the Code permit cutting lengths of portable track lighting to change its length?

   Yes _____   No _____ .

22. Why should the power be turned off when hooking up a dimmer? _____

     _____

UNIT 14

Lighting Branch-Circuit for the Study/Bedroom

OBJECTIVES

After studying this unit, the student will be able to

- discuss valance lighting.
- make all connections in the study/bedroom for the receptacles, switches, fan, and lighting.

CIRCUIT OVERVIEW

The study/bedroom circuit is so named because it can be used as a study for the present, providing excellent space for home office, personal computer, or den. Should it become necessary to have a third bedroom, the changeover is easily done.

Circuit A21 originates at Main Panel A. This circuit feeds the study/bedroom at the receptacle in the living room just outside of the study. From this point, the circuit feeds the split-circuit receptacle just inside the door of the study. A three-wire cable runs around the room to feed the other four split-circuit receptacles. Figure 14-1 shows the cable layout for Circuit A21.

The black and white conductors carry the "live" circuit, and the red conductor is the "switch return" for the control of one portion of the split-circuit receptacle.

There are four switched split-circuit receptacles in the study/bedroom controlled by two three-way switches. A fifth receptacle on the short wall leading to the bedroom hallway is not switched. The ready access to this receptacle makes it ideal for plugging in vacuum cleaners or similar appliances.

A ceiling fan/light fixture is mounted on the ceiling. This installation is covered in unit 9. Table 14-1 summarizes the outlets and estimated load for the study/bedroom circuit.

VALANCE LIGHTING

Above the windows, indirect fluorescent lighting is provided. See figure 14-2, which illustrates one method of installing fluorescent valance lighting.

DESCRIPTION	QUANTITY	WATTS	VOLT-AMPERES
Receptacles @ 120 W each	6	720	720
Closet recessed fixture	1	75	75
Paddle fan/light Three 50-W lamps Fan motor (0.75 A @ 120 V)	1	150 80	150 90
Valance lighting Two 40-W fluorescent lamps	1	80	90
TOTALS	9	1105	1125

Table 14-1 Study/bedroom outlet count and estimated load. Circuit A21.

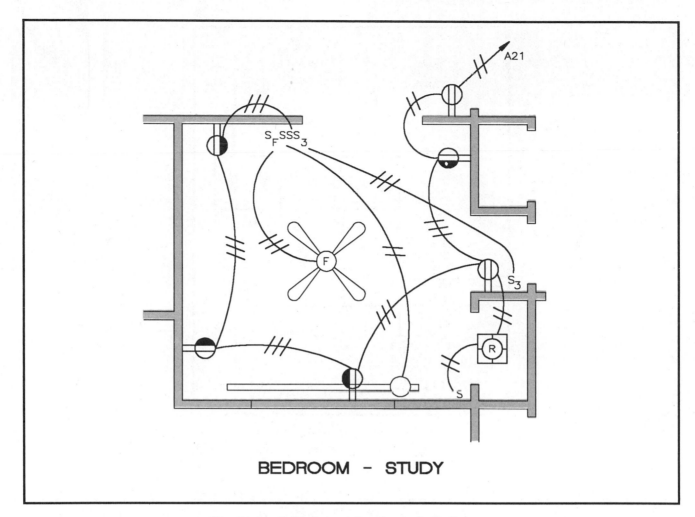

Fig. 14-1   Cable layout for the study/bedroom.

If the fluorescent valance lighting is to have a dimmer control, then special dimming ballasts would be required. See discussion on fluorescent lamp dimming in unit 13.

The study/bedroom wiring is rather simple, most of which has been covered in previous units in this text and need not be repeated here.

Figure 14-3 suggests one way to arrange the wiring for the ceiling fan/light control, the valance lighting control, and the receptacle outlet control.

SURGE SUPPRESSORS

Because this room is intended initially to be utilized as a study (home office), it is very likely that personal computers will be located here. Transient voltage surge suppressors are covered in unit 6.

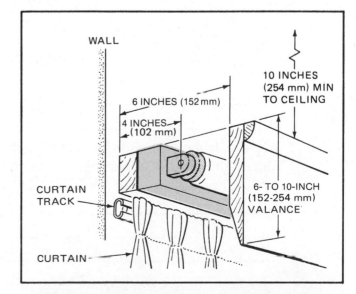

Fig. 14-2   Fluorescent valance with draperies.

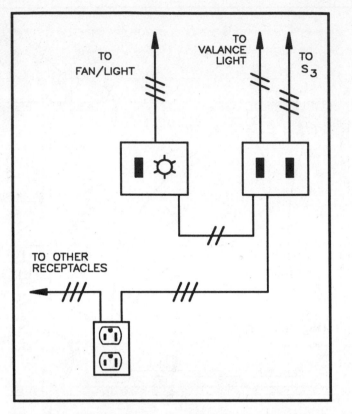

Fig. 14-3   Conceptual view of how the study/bedroom switch arrangement is to be accomplished, Circuit A21.

REVIEW

Note: Refer to the Code or plans where necessary.

1. Based upon the total estimated load calculations, what is the current draw on the study/bedroom circuit? _____

2. The study/bedroom is connected to circuit _____

3. The conductor size for this circuit is _____

4. Why is it necessary to install a receptacle in the wall space leading to the bedroom hallway? _____

   _____

   _____

5. Show calculations needed to select a properly sized box for the receptacle outlet mentioned in question 4. Nonmetallic-sheathed cable is the wiring method. Refer to the Quik-Chek Box Selector Guide, unit 2.

   _____

   _____

   _____

   _____

   _____

6. Show calculations necessary to select a properly sized box for the receptacle outlet mentioned in question 4. The wiring method is electrical metallic tubing (EMT), usually referred to as "thin-wall conduit."

_____

_____

_____

_____

_____

7. Prepare a list of *all* wiring devices used in Circuit A21.

_____

_____

_____

_____

_____

_____

8. Using the cable layout shown in figure 14-1, make a complete wiring diagram of Circuit A21.

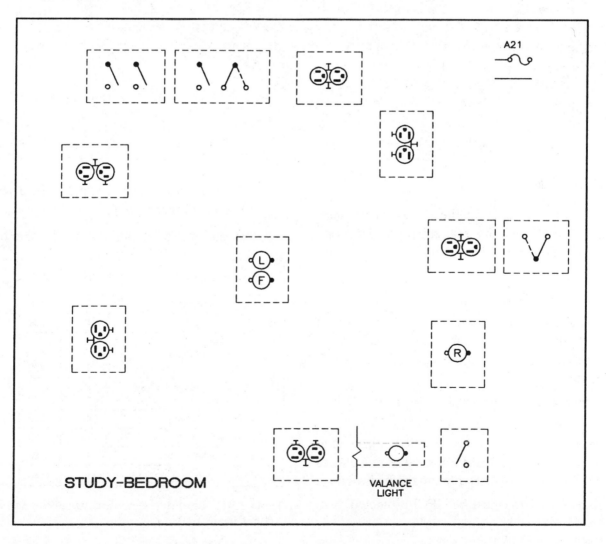

STUDY-BEDROOM

VALANCE
LIGHT

A21

UNIT 15

Dryer Outlet and Lighting Circuit for the Laundry, Powder Room, Rear Entry Hall, and Attic

OBJECTIVES

After studying this unit, the student will be able to
- ► understand the Code requirements for receptacles installed in bathrooms. ◄
- make the proper wiring and grounding connections for large appliances, based on the type of wiring method used.
- make load calculations for electric dryers.
- ► understand how to connect clothes dryers with cord- and plug-connections. ◄
- understand the Code requirements governing the receptacle outlet(s) for laundry areas.
- calculate proper conduit sizing for conductors of the same size and for conductors of different sizes in one conduit.
- discuss the subject of "reduced size neutrals."
- discuss the Code rules pertaining to wiring methods in attics.
- demonstrate the proper way to connect pilot lights and pilot light switches.

LIGHTING CIRCUIT B10

The estimated loads for the various receptacles and fixtures in the laundry, rear entry, powder room, and attic are shown in Table 15-1.

RECEPTACLE CIRCUIT B21

► Because of today's high wattage grooming appliances such as hair dryers and portable electric space heaters, *Section 210-52(d)* of the *NEC®* requires that:

- at least one receptacle shall be installed in a bathroom.

- one receptacle must be installed adjacent to the basin. In large bathrooms, other receptacles could be installed elsewhere in the bathroom.

- the receptacle(s) must be GFCI protected, *Section 210-8(a)(1)*.

- the receptacle(s) must be supplied by a separate 20-ampere branch-circuit.

- this separate circuit must not supply other outlets.

DESCRIPTION	QUANTITY	WATTS	VOLT-AMPERES
Rear entry hall receptacle	1	120	120
Powder room vanity fixture	1	200	200
Rear entrance hall recessed ceiling fixtures @ 100 W each	2	200	200
Laundry room fluorescent fixture Four 40-W lamps	1	160	180
Attic lights @ 75 W each	4	300	300
Laundry exhaust fan	1	80	90
Powder room exhaust fan	1	80	90
TOTALS	11	1140	1180

Table 15-1   Laundry-Rear Entrance-Powder Room-Attic. Circuit B10.

- this separate circuit is permitted to supply more than one bathroom.

Since Panel B is relatively close to the powder room, we have run a separate 20-ampere Circuit B21 to supply the receptacle in the powder room.

Although this receptacle(s) is required to be on a separate circuit, no additional load is added for purposes of service-entrance and feeder load calculations. Because diversity comes into play, the load is considered to be included in the general lighting load value of 3 volt-amperes per square foot as referred to in *Table 220-3(b)* for dwellings. ◄

A *bathroom* is defined in the *National Electrical Code*® as an area including a basin with one or more of the following: a toilet, a tub, or a shower.

DRYER CIRCUIT ▲D

A separate circuit is provided in the laundry room for the electric clothes dryer. This appliance demands a large amount of power. The special circuit provided for the dryer is indicated on the plans by the symbol ▲D. The dryer is connected to Circuit B(1-3).

The Clothes Dryer

Clothes dryer manufacturers make many dryer models with different wiring arrangements and connection provisions. All electric dryers have electric heating units and a motor-operated drum that tumbles the clothes as the heat evaporates the moisture. Dryers also have thermostats to regulate the temperature of the air inside the dryer. Timers are used to regulate the lengths of the various drying cycles. Drying time is determined by the type of fabric and can be set to a maximum of about 85 minutes. Dampness controls are often provided to stop the drying process at a preselected stage of dryness.

Clothes dryers create large amounts of humid air. Therefore, be sure that proper ventilation is provided. This residence has a through-the-wall exhaust fan located in the laundry room.

Electric clothes dryers operate from a 120/240-volt, single-phase, three-wire circuit. Be sure that the circuit has adequate capacity to handle the load.

Dryer Connection Methods

Electric clothes dryers can be connected in several ways. The method used depends largely on local codes. The local electrical inspector can provide information about the proper wiring procedure. Figure 15-1 shows the internal components and wiring of a typical dryer circuit.

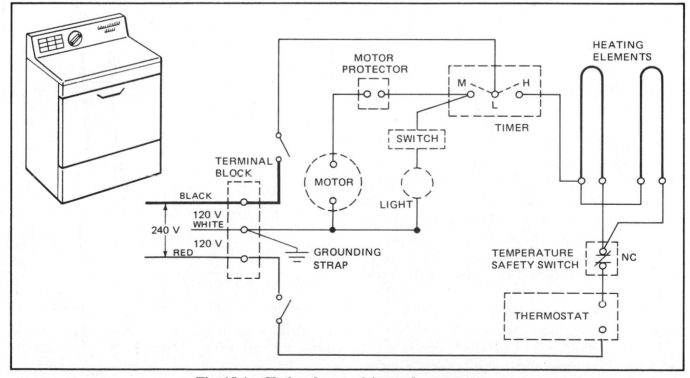

Fig. 15-1   Clothes dryer: wiring and components.

One method of connecting the dryer is to run a three-wire armored cable to a junction box on the dryer provided by the manufacturer, figure 15-2. The conductors are connected to their corresponding dryer terminals in this junction box.

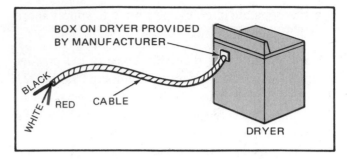

Fig. 15-2   Dryer connected by armored cable.

Another method of connecting a clothes dryer is to run conduit (EMT) to a point just behind the dryer. Flexible conduit from 24 to 36 inches (610 mm to 914 mm) in length is installed between the dryer junction box and the EMT, figure 15-3. The transition from the EMT to the flexible conduit is made using a fitting such as the combination coupling shown in figure 15-3. This flexible connection allows the dryer to be moved for cleaning and servicing without disconnecting the wiring.

For the "hard wiring" connection methods illustrated in figures 15-2 and 15-3, a disconnecting means must be provided for the dryer. This disconnecting means must be within sight of the dryer, or must be capable of being locked in the OFF position. This requirement is found in *Section 422-21(b)* of the *NEC®*. Obviously, having a means of disconnecting the dryer from the power source when working on the appliance is important from a safety standpoint.

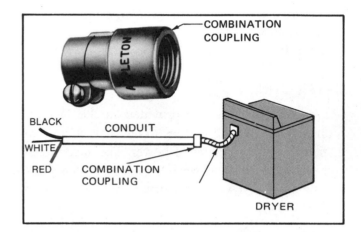

Fig. 15-3   Dryer connected by EMT and flexible metal conduit.

A circuit breaker in the panel can serve as the required disconnecting means if it is within sight of the dryer, or if it can be locked in the OFF position. Breaker manufacturers supply these "lock-off" devices that fit over the circuit breaker.

Still another method of connecting a clothes dryer is to install a receptacle on the wall behind the dryer. A cord set is connected to the dryer, figure 15-4, and is plugged into the receptacle. This method satisfies the requirements of *Section 422-22(a)* for a disconnecting means.

▶ Figure 15-4 shows the 30-ampere, 4-wire receptacle configuration for dryer installation made in conformance to the 1996 *National Electrical Code®*. Figure 15-5 illustrates the 30-ampere, 3-wire receptacle configuration used for dryer installations made prior to adoption of the 1996 *National Electrical Code®*

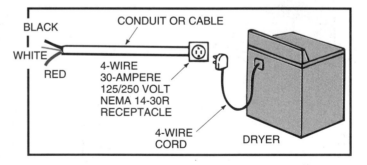

Fig. 15-4   Dryer connection using a cord set.

For residential electric clothes dryers, dryer receptacles and cord sets rated 30 amperes, 125/250 volts are in most cases large enough. The L-shaped slot is the neutral. The horseshoe-shaped slot is the equipment ground. A 30-ampere cord set contains:

- **3-pole, 3-wire:** three No. 10 AWG conductors
- **3-pole, 4-wire:** four No. 10 AWG conductors

If for some reason a 30-ampere device is too small, the next size is 50 amperes, the type generally used for electric ranges. A 50-ampere cord set contains:

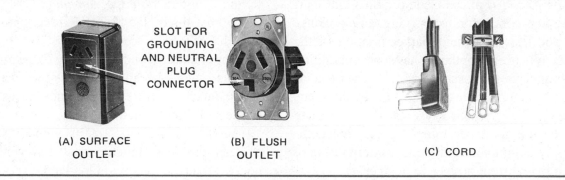

(A) SURFACE OUTLET   (B) FLUSH OUTLET   (C) CORD

SLOT FOR GROUNDING AND NEUTRAL PLUG CONNECTOR

▶ **Fig. 15-5** Prior to adoption of the 1996 *National Electrical Code,*® 3-wire receptacles and cords were used to connect electric clothes dryers. The grounded circuit conductor served as both the neutral and the equipment grounding conductor. This practice is no longer permitted. The 1996 *NEC*® made it mandatory that a separate equipment grounding conductor be installed to ground the frames of electric clothes dryers. It became necessary to install 4-wire receptacles and cords with plug- and cord-connecting electric clothes dryers. Figure 15-4 shows a 4-wire connection. Refer to *Sections 250-60* and *250-61* of the *NEC.* Also see figure 20-1. ◀

- **3-pole, 3-wire:** two No. 6 AWG, one No. 8 AWG conductors

- **3-pole, 4-wire:** two No. 6 AWG, two No. 8 AWG conductors

Surface mounted dryer outlets can be wired directly with nonmetallic-sheathed cable, armored cable, or conduit. Flush mounted dryer outlets are quite large. The receptacle plus the four No. 10 AWG conductors (three circuit conductors and one equipment grounding conductor) will probably require a 4-inch square, 1½-inch or 2⅛-inch deep outlet box with suitable single-gang or two-gang plaster ring. You're looking for trouble if you try to jam the wires and receptacle into a smaller outlet box or device box. Correct box sizing is determined by applying the rules of *Section 370-16* and the tables located in this section.

The use of 3-wire and 4-wire receptacles and cord sets relative to grounding issues is discussed later on in this unit under *Grounding Frames of Ranges and Dryers.* Also refer to unit 20 for additional material relating to the use of 3-wire and 4-wire hook-ups. ◀

Code Requirements for Connecting a Dryer
(Article 422)

The information necessary to install a dryer or any appliance is contained in *Article 422* of the *National Electrical Code.*®

In this residence, the dryer is installed in the laundry room. The schedule of special-purpose out-

lets in the specifications shows that the dryer is rated at 5700 watts and 120/240 volts. The schematic wiring diagram, figure 15-1, shows that the heating elements are connected to the 240-volt terminals of the terminal block. The dryer motor and light are connected between the "hot" wire and the neutral terminal, 120 volts.

The motor of the dryer has integral thermal protection. This thermal protector prevents the motor from reaching dangerous temperatures as the result of an overload or failure to start. *Section 422-28* lists overcurrent protection requirements. This section also refers to parts of *Article 430.*

To determine the feeder or branch-circuit rating, *Section 220-18* requires that the minimum load included for a dryer be 5000 watts (volt-amperes) or the nameplate rating of the dryer, whichever is larger.

The electric clothes dryer in this residence has a nameplate rating of 5700 watts. Thus

$$I = \frac{W}{E} = \frac{5700}{240} = 23.75 \text{ amperes}$$

Conductor Size

The dryer circuit B(1-3) in this residence is a 3-wire, 30-ampere, 240-volt circuit. Nonmetallic-sheathed cable is run concealed in the walls from Panel B to a flush mounted 4-inch square box behind the dryer location in the laundry room. The cable consists of three No. 10 AWG conductors plus one No. 10 AWG equipment grounding conductor.

Section 422-4(a) of the Code requires that individual circuit conductors supplying an appliance must not be less than the "marked" rating of the appliance. When the appliance has load in addition to the motor, as in the case of a clothes dryer (motor and heater), then *Section 422-4(a)* makes reference to *Section 422-32*, which says that the nameplate marking on the appliance must "specify the minimum supply circuit size and the maximum rating of the circuit overcurrent protective device."

Because the manufacturer of the appliance provides the circuit requirements on the nameplate and in the installation manual, it is a simple matter to determine the proper wire size based upon this rating.

Section 210-22(a) further states that for circuits supplying motor-operated equipment plus "other loads," the computation is based on: 125% of the motor's full-load ampere rating plus the other loads.

On a clothes dryer it is very difficult for the installer to check the motor rating and the heating element rating. It is much easier to follow the rating requirements found on the nameplate of the appliance. If the branch-circuit rating is *not* marked on the appliance, *Section 422-4(a), Exception No. 1*, refers the reader to *Part B* of *Article 430* (Motors). Here again we find the fundamental 125% rule. That is, take the current rating times 125% to find the required circuit ampacity. Therefore,

$$23.75 \text{ amperes} \times 1.25 = 29.7 \text{ (30) amperes}$$

According to *Table 310-16*, the allowable ampacity for a No. 10 THHN conductor is 35 amperes. The overcurrent device is sized at not to exceed 30 amperes for this conductor. One of the most common exceptions for determining the maximum size overcurrent protection is a motor branch-circuit. A motor is required to have overload, branch-circuit, and ground-fault protection. This motor overload protection serves a dual purpose of also protecting the conductors supplying the motor from overloads (overheating). When a motor has proper overload protection, the motor branch-circuit overcurrent protection may exceed the ampacity of the conductor supplying the motor. See footnote to *Table 310-16*.

The dryer Circuit B(1-3) is a three-wire, 30-ampere, 240-volt circuit. The conductors in the nonmetallic-sheathed cable are No. 10 AWG.

► Where electrical metallic tubing is used as the wiring method, *Table C2, Appendix C* of the *NEC®* shows that three No. 10 THHN conductors require ½-inch EMT. ◄

The neutral conductor may be smaller than the phase wires because it carries the current of the motor, lamp, and timer only. In most cases, these are connected "line-to-neutral."

Section 210-19(a), first sentence, states that "branch-circuit conductors shall have an ampacity not less than the maximum load to be served."

Section 220-22 permits the neutral load for "feeders" to household electric ranges, wall-mounted ovens, counter-mounted cooking units, and electric dryers to be based upon 70% of the load on the ungrounded conductors.

By Code definition, a branch-circuit refers to the circuit conductors between the final overcurrent device and the outlet. Therefore, the conductors between the overcurrent device in Panel B and the dryer outlet constitute the dryer's branch-circuit. Knowing that the neutral conductor of a branch-circuit supplying an electric clothes dryer carries only the motor, lamp, and timer loads, most electricians apply the 70% rule as stated in *Section 220-22* for sizing the neutral conductor.

In this example,

$$23.75 \times 0.70 = 16.63 \text{ amperes}$$

Therefore, the neutral conductor could be a No. 12 THHN wire. See *Table 310-16* and the *Footnotes* to the table.

Accordingly, the dryer in this residence could be connected as follows:

Phase (hot) wires—
    two No. 10 THHN

Neutral wire—
    one No. 12 THHN

} ½-inch electrical metallic tubing

Overcurrent Protection

Section 422-5 states that the branch-circuit overcurrent device rating is not to exceed the overcurrent device rating marked on the appliance. Here, again, the manufacturer of the appliance will so mark the circuit overcurrent device rating. If the appliance is not marked, then the overcurrent protection shall be in accordance with *Section 240-3* for the conductors—30 amperes for our example, as discussed previously.

Grounding Frames of Ranges and Clothes Dryers

▶ New Installations

For protection against electrical shock hazard, clothes dryers must be grounded according to the rules found in *Sections 250-57* and *250-59*. See figure 15-6. ◀ An accepted grounding means is to connect the appliance with a metal raceway such as electrical metallic tubing (EMT), flexible metal conduit (Greenfield), armored cable (BX), or other means as listed in *Section 250-91(b)*. There are limitations on the use of flexible metal conduit as an equipment grounding conductor. See *Sections 250-91(b)* and *350-5*. This topic is covered in unit 4.

The separate equipment grounding conductor found in nonmetallic-sheathed cable is also an acceptable grounding means. The minimum size of an equipment grounding conductor is found in *Table 250-95* of the *NEC*® The equipment grounding conductor in nonmetallic-sheathed cable is sized according to *Table 250-95*.

▶ Type SE service-entrance cable may also be used provided all of the circuit conductors are insulated. The equipment grounding conductor may be insulated (green) or bare. See *Section 338-3*. Type SE cable installed as interior wiring must be installed according to the installation requirements found in *Article 336* for nonmetallic-sheathed cable, *Section 338-4*. ◀

▶ When electric clothes dryers are cord- and plug-connected, the receptacle and the cord set must be four-wire, the fourth wire being the separate equipment grounding conductor. ◀

The small copper bonding strap (link) that is furnished with an electric clothes dryer *must not* be connected between the neutral terminal and the metal frame of the appliance.

It is interesting to note that the 4-wire cord/ 4-wire receptacle requirement for electric ranges, wall-mounted ovens, surface mounted cooking units, and clothes dryers in mobile homes has been in effect since 1965, when *Article 550* relating to mobile homes first appeared in the *NEC*®

▶ Existing Installations

Prior to the 1996 edition of the *National Electrical Code*, *Section 250-60* permitted the frames of electric ranges, wall-mounted ovens, surface cooking units, and clothes dryers to be grounded to the neu-

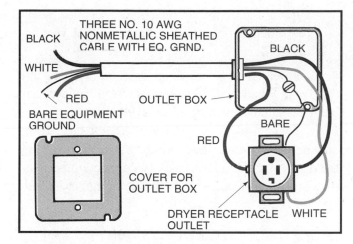

▶ **Fig. 15-6   New installations of branch-circuit wiring for electric ranges, wall-mounted ovens, surface cooking units, and clothes dryers require that all equipment grounding be done according to *Sections 250-57* and *250-59*. Receptacles and cord sets must be 4-wire (three circuit conductors *plus* equipment grounding conductor).** ◀

tral conductor. By way of *Tentative Interim Amendment No. 53*, this special permission was put into effect in July of 1942, and was supposedly an effort to conserve raw materials during World War II. In effect, this special permission allowed the neutral conductor to serve a dual purpose: (1) the neutral conductor and (2) the equipment grounding conductor. This special permission remained in effect until the 1996 *National Electrical Code*®

Section 250-60 now applies *only* to existing branch-circuit installations. For those installations wired according to the rules of *Section 250-60* prior to the 1996 *NEC*, nothing need be changed. The required appliance grounding was accomplished by the small copper bonding strap furnished by the appliance manufacturer that connected between the neutral terminal and the metal frame of the appliance. ◀

For existing installations, the requirements were:

- the supply circuit had to be 120/240 volt, single-phase, 3-wire or 208Y/120 volt derived from a 3-phase, 4-wire wye-connected system.

- the neutral had to be no smaller than No. 10 AWG copper or No. 8 AWG aluminum.

- the neutral conductor had to be insulated. The exception to this was for Type SE cable where the circuit originated at the main service, in which case the neutral conductor did not have to be insulated but only covered by the outer jacket of the cable.

- a metal outlet box used to hold a receptacle was permitted to be grounded to the neutral. See figure 15-7.

▶ If an electric clothes dryer is to be installed in a residence where the dryer branch-circuit wiring had been installed according to pre-1996 *National Electrical Code®* rules, the wiring *does not* have to be changed. Merely use the grounding strap furnished by the dryer manufacturer to make a connection between the neutral conductor and the frame of the dryer. ◀

RECEPTACLE OUTLETS—LAUNDRY

The Code in *Section 220-4(c)* requires that a separate 20-ampere branch-circuit be provided for the laundry equipment and this circuit "shall have no other outlets."

Section 210-50(c) requires that a receptacle outlet must be installed within 6 feet of the intended location of the washer.

Section 210-52(f) states that *at least* one receptacle outlet must be installed for the laundry equipment.

Section 220-16(b) requires that 1500 volt-amperes must be included for each two-wire branch-circuit serving laundry equipment. This is for the purpose of computing the size of service-entrance and/or feeders.

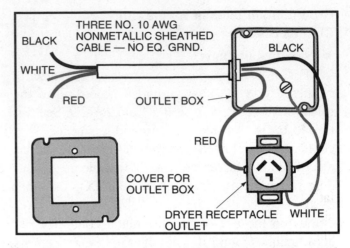

▶ **Fig. 15-7   Prior to the 1996** *National Electrical Code®,* **it was permitted to ground the junction box to the neutral only when the box was part of the circuit for electric ranges, wall-mounted ovens, surface cooking units, and clothes dryers,** *Section 250-60.* **The receptacle was permitted to be a 3-wire type because the appliance grounding was accomplished by making a connection between the neutral conductor and the metal frame of the appliance. This is no longer permitted.** ◀

See the complete calculations in unit 28.

In the laundry room in this residence, the 20-ampere Circuit B18 supplies the receptacle

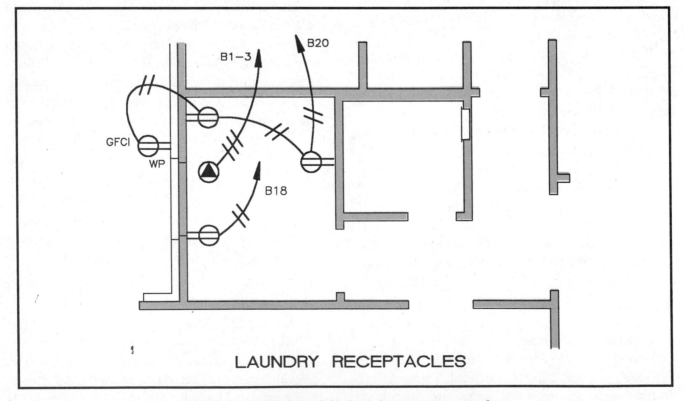

LAUNDRY RECEPTACLES

Fig. 15-8   Cable layout for laundry room receptacles.

for the washer. Circuit B18 serves no other outlets. The receptacle is not required to be GFCI protected. Figure 15-8 shows the cable layout for the laundry room.

Circuit B20 connects to the two receptacles in the laundry, which will serve an iron, sewing machine, or other small portable appliances. Circuit B20 also serves the weatherproof receptacle on the outside wall of the laundry area.

Circuits B18 and B20 are included in the calculations for service-entrance and feeder conductor sizing at a computed load of 1500 volt-amperes per circuit *(Section 220-16(b))*. This load is included with the general lighting load for the purpose of computing the service-entrance and feeder conductor sizing and is subject to the demand factors that are applicable to these calculations. See the complete calculations in unit 28.

COMBINATION WASHER/DRYERS

Combination washer/dryers are often found in multifamily dwellings where space is an issue.

Combination washer/dryers take up half the floor space of individual conventional washers and dryers. If you are involved in an electrical installation, verify and resolve the electrical requirements before you start the project. These combination appliances might require a 30-ampere branch-circuit. Installation manuals will provide this information. The electrical inspector might *not* require the traditional 20-ampere branch-circuit and receptacle that is required by *Sections 210-52(f)* and *220-4(c)*.

LIGHTING CIRCUIT

The general lighting circuit for the laundry, powder room, and hall area is supplied by Circuit B10. Note on the cable layout, figure 15-9, that in order to help balance loads evenly, the attic lights are also connected to Circuit B10.

The types of vanity lights, ceiling fixtures, and wall receptacle outlets, as well as the circuitry and switching arrangements, are similar to types already discussed in other units of this text and will not be repeated.

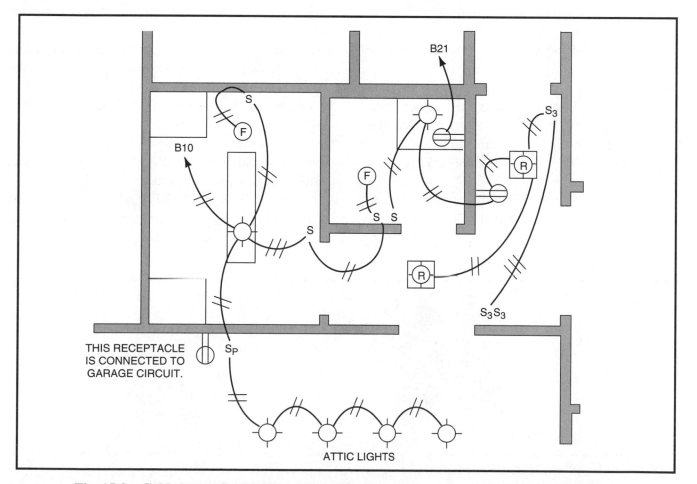

Fig. 15-9   Cable layout for laundry, rear entrance, powder room, and attic, Circuit B10.

▶ The receptacle in the powder room is supplied by Circuit B21, a separate 20-ampere branch-circuit as required by *Section 210-52(d)*. This has been discussed in unit 10. ◀

Exhaust Fans

Ceiling exhaust fans are installed in the laundry and in the powder room. These exhaust fans are connected to Circuit B10.

The exhaust fan in the laundry will remove excess moisture resulting from the use of the clothes washer.

Exhaust fans may be installed in walls or ceilings, figure 15-10. The wall-mounted fan can be adjusted to fit the thickness of the wall. If a ceiling-mounted fan is used, sheet-metal duct must be installed between the fan unit and the outside of the house. The fan unit terminates in a metal hood or grille on the exterior of the house. The fan has a shutter that opens as the fan starts up and closes as the fan stops. The fan may have an integral pull-chain switch for starting and stopping, or it may be used with a separate wall switch. In either case, single-speed or multispeed control is available. The fan in use has a very small power demand, 90 volt-amperes.

To provide better humidity control, both ceiling-mounted and wall-mounted fans may be controlled with a humidistat. This device starts the fan when the humidity reaches a certain value. When the humidity drops to a preset level, the humidistat turns the fan OFF.

Some of the more expensive upscale exhaust fans have a built-in sensor that will automatically turn on the exhaust fan at a predetermined humidity level. The fan will automatically turn off when the humidity has been lowered to a preset level. This eliminates the need for a separate wall-mounted humidity control (humidistat).

ATTIC LIGHTING AND PILOT LIGHT SWITCHES

Section 210-70(a) requires that at least one lighting outlet must be installed in an attic, and that this lighting outlet must be controlled by a switch located near the entry to the attic. This section of the Code addresses attics that are used for storage or attics that contain equipment that might need servicing. An example of such equipment would be air-conditioning equipment. *Section 210-70(a)* requires that the lighting outlet(s) be installed near this equipment.

Section 210-63 requires that a 125-volt, single-phase, 15- or 20-ampere receptacle outlet be installed in an accessible location within 25 feet (7.62 m) of air-conditioning or heating equipment in attics, in crawl space, or on the roof. Connecting this receptacle outlet to the load side of the equipment's disconnecting means is not permitted because this would mean that if you turned off the equipment, the power to the receptacle would also be off and thus would be useless for servicing the equipment. The exception to *Section 210-63* exempts requiring a receptacle outlet on the roof of one- and two-family dwellings, although installing a receptacle outlet would certainly be a good idea.

Reading *Section 210-70(a)* tells us that a switch controlled lighting outlet is required *only* if the attic, underfloor space, utility room, and basement are

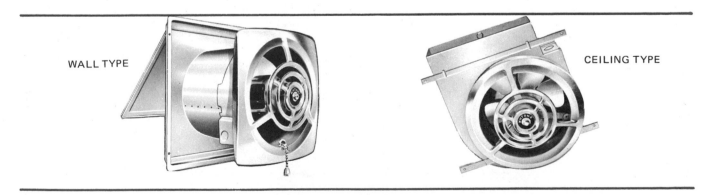

WALL TYPE   CEILING TYPE

Fig. 15-10   Exhaust fans.

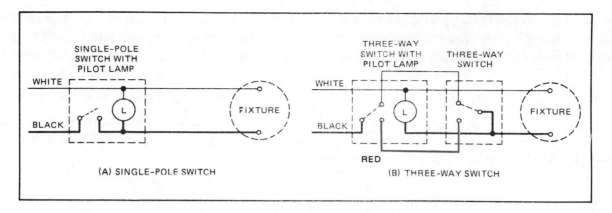

Fig. 15-11   Pilot lamp connections.

used for storage or contain equipment that might require servicing.

Reading *Section 210-63* tells us that a 125-volt, single-phase, 15- or 20-ampere receptacle outlet is required on rooftops, in attics, and in crawl spaces *only* if these areas contain heating, air-conditioning, and/or refrigeration equipment.

The residence discussed in this text does not have air-conditioning, heating, or refrigeration equipment in the attic or on the roof.

The porcelain lampholders installed in the attic of the residence are available with a receptacle outlet for convenience in plugging in an extension cord. However, most electrical inspectors (authority having jurisdiction) would not accept the porcelain lampholder's receptacle in lieu of the required receptacle outlet as stated in *Section 210-63*.

The four porcelain lampholders in the attic are turned ON and OFF by a single-pole switch on the garage wall close to the attic storable ladder.

Associated with this single-pole switch is a pilot light. The pilot light may be located in the handle of the switch or it may be separately mounted. Figure 15-11 shows how pilot lamps are connected in circuits containing either single-pole or three-way switches.

Because the attic in this residence is served by a folding storable ladder, switch control in the attic is not required. Where a permanent stairway of six or more steps is installed, *Section 210-70(a)* requires switch control at both levels.

If a neon pilot lamp in the handle (toggle) of a switch does not have a separate grounded conductor connection, then it will glow only when the switch is in the OFF position, as the neon lamp will then be in series with the lamp load, figure 15-12.

The voltage across the load lamp is virtually zero, so it does not burn, and the voltage across the neon lamp is 120 volts, allowing it to glow. When the switch is turned on, the neon lamp is bypassed (shunted),

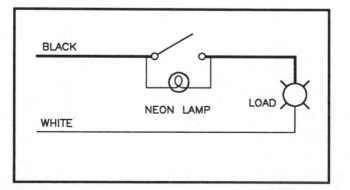

Fig. 15-12   Example of neon pilot lamp in handle of toggle switch.

causing it to turn off and the lamp load to turn on.

Use this type of switch when it is desirable to have a switch glow in the dark to make it easy to locate.

Installation of Cable in Attics

The wiring in the attic is to be done in cable and must meet the requirements of *Section 336-13* for nonmetallic-sheathed cable. This section refers the reader directly to *Section 333-12*, which describes how the cable is to be protected, figure 15-13.

In accessible attics, figure 15-13(A), cables must be protected by guard strips, figure 15-14, when:

- they are run across the top of floor joists ①.

- they are run across the face of studs ② or rafters ③ within 7 feet (2.13 m) of the floor or floor joists.

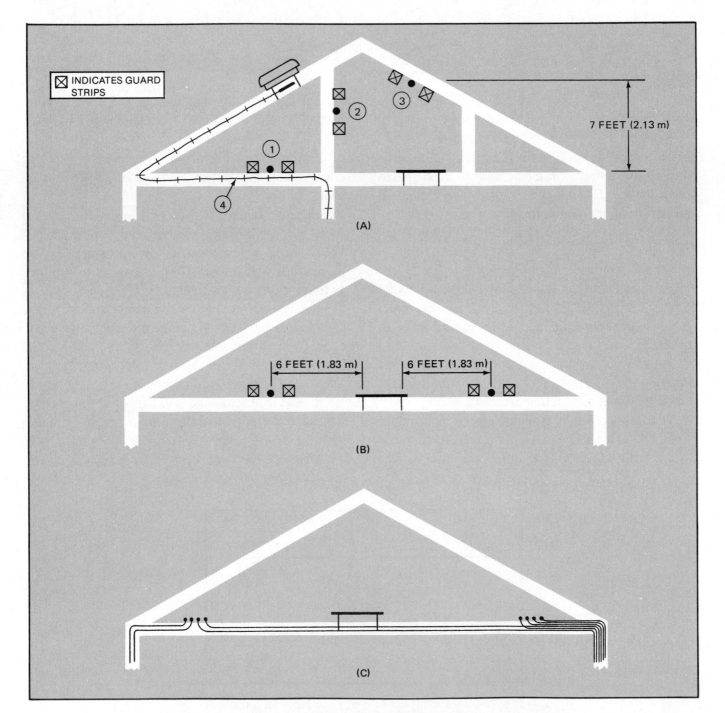

Fig. 15-13   Protection of cable in attic.

Guard strips are *not* required if the cable is run along the sides of rafters, studs, or floor joists ④.

In attics not accessible by permanent stairs or ladders, figure 15-13(B), guard strips are required only within 6 feet (1.83 m) of the nearest edge of the scuttle hole or entrance.

Figure 15-13(C) illustrates a cable installation that most electrical inspectors consider to be safe. Because the cables are installed close to the point where the ceiling joists and the roof rafters meet, they are protected from physical damage. It would be very difficult for a person to crawl into this space, or store cartons in an area with such a low clearance. Although the plans for this residence show a 2-foot-wide catwalk in the attic, the owner may decide to install additional flooring in the attic to obtain more storage space. Because of the large number of cables required to complete the circuits, it would interfere with flooring to install guard strips wherever the cables run across the tops of the joists, figure 15-14. However, the cables can be run through holes bored in the joists and along the sides of the joists and rafters. In this way, the cables do not interfere with the flooring.

When running cables parallel to framing members, be careful to maintain at least 1¼ inches (31.8 mm) between the cable and the edge of the framing member. This is a Code requirement referenced in *Section 300-4(d)* to minimize the possibility of driving nails into the cable. The entire *Section 300-4* of the Code is devoted to the subject of "protection against physical damage."

See unit 4 for more detailed discussion and illustrations relating to the mechanical protection of cables.

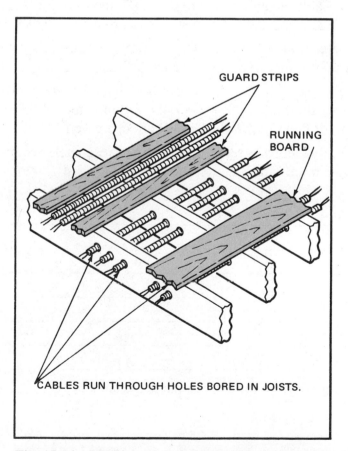

Fig. 15-14   Methods of protecting cable installations in attics. See *Sections 333-12* and *336-13*.

REVIEW

Note: Refer to the Code or plans where necessary.

1. List the switches, receptacles, and other wiring devices that are connected to Circuit B10. _____

   _____

   _____

   _____

   _____

   _____

   _____

2. What section of the Code states that the receptacle in the powder room is to be GFCI protected? _____

   _____

3. What special type of switch is controlling the attic lights? _____

   _____

4. When installing cables in an attic along the top of the floor joists _____ _____ must be installed to protect the cables.

5. The total estimated volt-amperes for Circuit B10 has been calculated to be 1300 volt-amperes. How many amperes is this at 120 volts? _____

6. If an attic is accessible through a scuttle hole, guard strips are installed to protect cables run across the top of the joists only within 6 feet (1.83 m), 7 feet (2.13 m), 12 feet (3.66 m) of the scuttle hole. Circle correct answer.

7. What section(s) of the Code refers (refer) to the receptacle required for the laundry equipment? State briefly the requirements.

   _____

   _____

   _____

   _____

   _____

   _____

   _____

8. What is the current draw of the exhaust fan in the laundry? _____

9. The following is a layout of the lighting circuit for the laundry, powder room, rear entry hall, and attic. Complete the wiring diagram using colored pens or pencils to indicate conductor insulation color. The GFCI receptacle in the powder room is not shown in the diagram because it is connected to a separate Circuit B21.

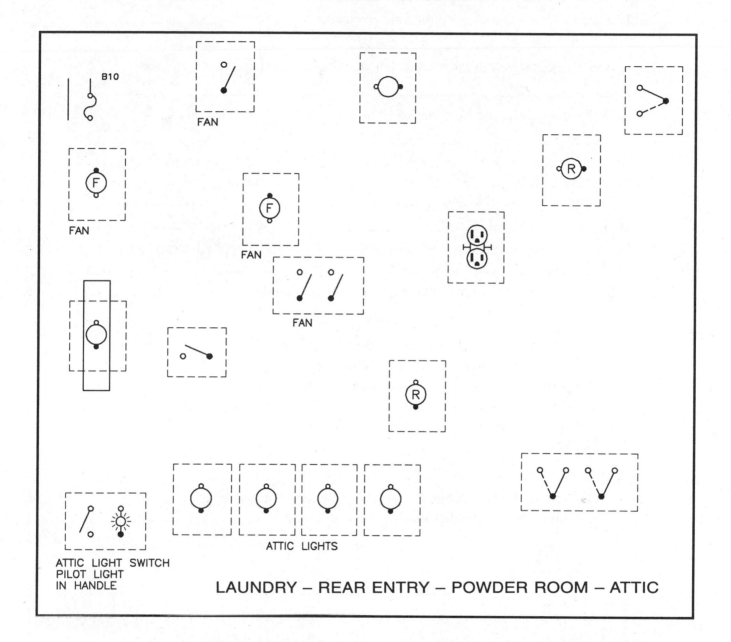

10. The receptacle outlets are connected to small appliance circuits that have been included in the residential load calculations at a value of _____ volt-amperes per circuit. Circle one: 1000, 1500, 2000 volt-amperes.

11. a. What regulates the temperature in the dryer? _____

_____

   b. What regulates drying time? _____

_____

12. List the various methods of connecting an electric clothes dryer. _____

_____

13. a. On a 125/250-volt dryer receptacle, what does the L-shaped slot represent?

_____

   b. What does the horseshoe-shaped slot represent?

_____

14. a. What is the minimum power demand allowed by the Code for an electric dryer if
       no actual rating is available for the purpose of calculating feeder and service-
       entrance conductor sizing? _____

   b. How much current is used? _____

15. What is the maximum permitted current rating of a portable appliance on a 30-ampere
    branch-circuit? _____

16. What provides motor running overcurrent protection for the dryer? _____

_____

17. a. Must the metal frame of an electric clothes dryer be grounded? _____

   b. May the metal frame of an electric clothes dryer be grounded to the neutral
       conductor? _____

18. An electric dryer is rated at 7.5 kW and 120/240 volts, three-wire, single-phase. The
    terminals on the dryer and panelboard are marked 75°C.

   a. What is the wattage rating? _____

   b. What is the current rating? _____

   c. What size type THHN copper conductors are required? For ease in calculating, size
       the neutral one size smaller than the hot conductors. _____

_____

_____

_____

_____

   d. If electrical metallic tubing is used, what minimum size is required? _____

_____

_____

_____

_____

_____

_____

19. When a metal junction box is installed as part of the cable wiring to a clothes dryer or electric range, may this box be grounded to the circuit neutral? _____

20. A residential air-conditioning unit is installed in the attic.

    a. Is a lighting outlet required? _____

    b. What Code section? _____

    c. If a lighting outlet is required, how shall it be controlled? _____

    _____

    d. What Code section? _____

    e. Is a receptacle outlet required? _____

    f. What Code section? _____

21. *Section 422-20* of the Code states that a means must be provided to disconnect an appliance. In the case of an electric clothes dryer, the disconnecting means can be: Circle the correct statement(s).

    a. a separate disconnect switch located within sight of the appliance.

    b. a separate disconnect switch located out of sight from the appliance. The switch is capable of being locked in the OFF position.

    c. a cord- and plug-connected arrangement.

    d. a circuit breaker in a panel that is within sight of the dryer or capable of being locked in the OFF position.

UNIT 16

Lighting Branch-Circuit for the Garage

OBJECTIVES

After studying this unit, the student will be able to

- understand the fundamentals of providing proper lighting in residential garages.
- understand the application of GFCI protection for receptacles in residential garages.
- understand the Code requirements for underground wiring, both conduit and underground cable.
- complete the garage circuit wiring diagram.
- discuss typical outdoor lighting and Code requirements for same.
- describe how conduits and cables are brought through cement foundations to serve loads outside of the building structure.
- understand the application of GFCI protection for loads fed by underground wiring.
- make a proper installation for a residential overhead garage door opener.
- select the proper overload protective devices based on the ampere rating of the connected motor load.

LIGHTING BRANCH-CIRCUIT

Circuit B14 originates at Panel B in the recreation room, and is brought into the wall receptacle box on the front wall of the garage. Figure 16-1 shows the cable layout for Circuit B14. From this point, the circuit is then carried to the switch box adjacent to the side door of the garage. From there the circuit conductors plus the switch leg are carried upward to the ceiling fixtures, and then to the wall receptacle outlets on the right-hand wall of the garage.

The garage ceiling's porcelain lampholders are controlled from switches located at all three entrances to the garage.

The post light in the front of the house is fed from the receptacle located just inside the overhead door. The post light turns on and off by a photocell, an integral part of the post light.

Note that the fixtures on the outside next to the overhead garage door are controlled by two three-way switches, one just inside the overhead door and the other in the front entry. These fixtures are not connected to the garage lighting circuit.

All of the wiring in the garage will be concealed. Building codes for one- and two-family dwellings require that the walls separating habitable areas and a garage have a fire resistance rating. In this residence, the walls and ceilings have a fire resistance rating of 1-hour. Fire resistance rating requirements can vary in different localities. Figures 16-1A, 16-1B, 16-1C and 16-1D show the basic requirements for attaining a fire resistance rating where electrical outlet boxes are mounted on both sides of the wall. In this residence, this occurs in the wall partition where the 4-way switch in the garage and the two 3-way switches located in the rear-entry hall are mounted on opposite sides of the wall. Refer to unit 2 for additional information relative to fire resistance ratings.

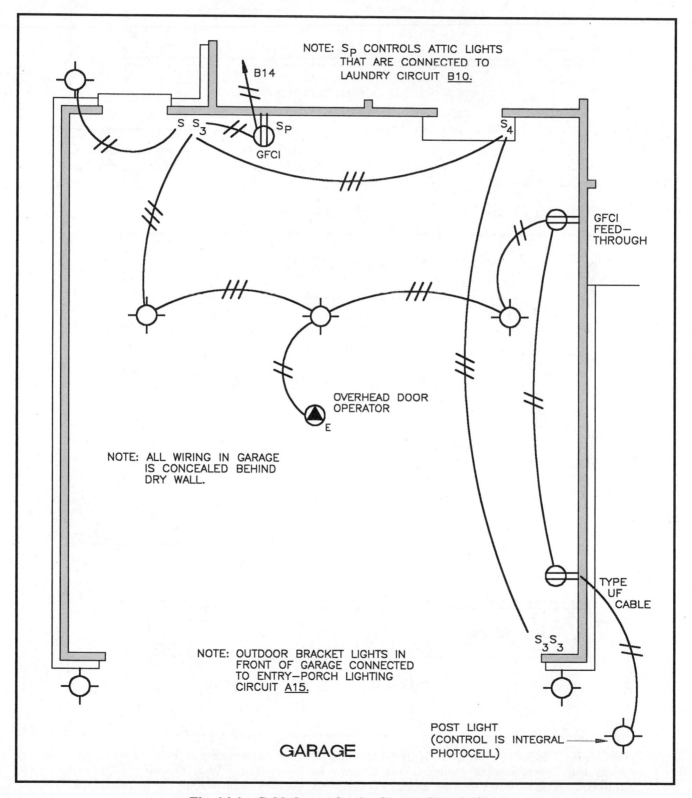

NOTE: S_P CONTROLS ATTIC LIGHTS THAT ARE CONNECTED TO LAUNDRY CIRCUIT B10.

B14

S S_3 S_P
GFCI

S_4

GFCI FEED-THROUGH

OVERHEAD DOOR OPERATOR
E

NOTE: ALL WIRING IN GARAGE IS CONCEALED BEHIND DRY WALL.

TYPE UF CABLE

$S_3 S_3$

NOTE: OUTDOOR BRACKET LIGHTS IN FRONT OF GARAGE CONNECTED TO ENTRY—PORCH LIGHTING CIRCUIT A15.

POST LIGHT (CONTROL IS INTEGRAL PHOTOCELL)

GARAGE

Fig. 16-1   Cable layout for the Garage Circuit B14.

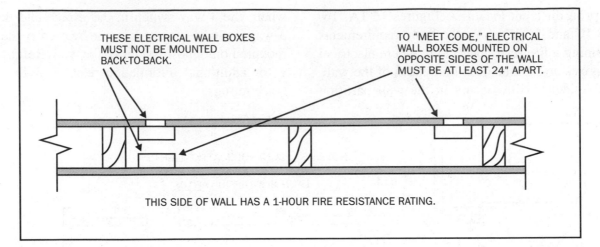

Fig. 16-1A Building codes do not permit electrical wall boxes to be mounted back-to-back. Electrical wall boxes must be kept at least 2 inches apart. This is to maintain the integrity of the fire resistance rating of the wall.

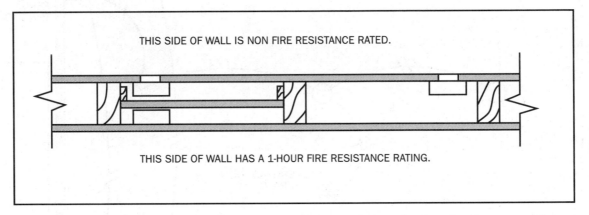

Fig. 16-1B Building codes permit boxes to be mounted back-to-back or in the same wall cavity when the fire resistance rating is maintained. In this detail, gypsum board has been installed in the wall cavity between the boxes, creating a new fire resistance barrier. Verify if this is acceptable to the building authority in your area.

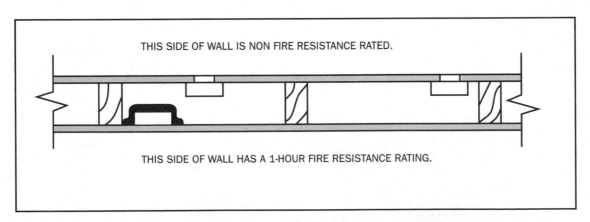

Fig. 16-1C Building codes permit electrical outlet boxes to be mounted in the same stud space when the fire resistance rating is maintained. In this detail, "Wall Opening Protective Material" is packed around the electrical outlet box, thereby maintaining the fire resistance rating of the wall. Electrical outlet boxes are not permitted to be installed back-to-back.

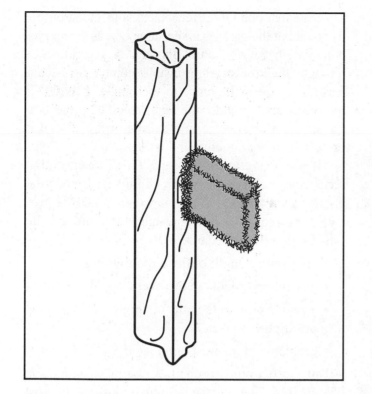

Fig. 16-1D   This illustration shows an electrical outlet box that has been covered with fire resistance "putty" to retain the wall's fire resistance rating per building codes.

Table 16-1 summarizes the outlets and the estimated load for the garage circuit.

LIGHTING A TYPICAL RESIDENTIAL GARAGE

The recommended approach to provide adequate lighting in a residential garage is to install one 100-watt lamp on the ceiling above each side of an automobile, figure 16-2. Figure 16-3 shows typical lampholders commonly mounted in garages.

DESCRIPTION	QUANTITY	WATTS	VOLT-AMPERES
Receptacles @ 180 W each	3	540	540
Ceiling fixtures @ 100 W each	3	300	300
Outdoor garage bracket fixture (rear)	1	100	100
Post light	1	100	100
Overhead door opener Two 40-W lamps Motor (1/2 A @ 120 V)	1	80 180	80 200
TOTALS	9	1300	1320

Table 16-1   Garage: outlet count and estimated load. Circuit B14.

- For a one-car garage, a minimum of two fixtures is recommended.
- For a two-car garage, a minimum of three fixtures is recommended.
- For a three-car garage, a minimum of four fixtures is recommended.

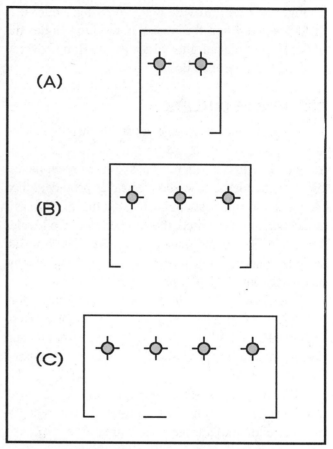

Fig. 16-2   Positioning of lights in (A) a one-car garage, (B) a two-car garage, and (C) a three-car garage.

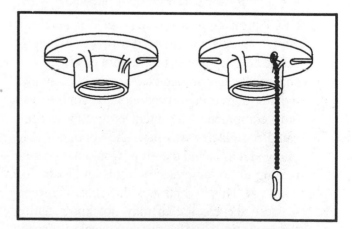

Fig. 16-3   These are typical porcelain lampholders of the type that would be installed in garages, attics, basements, crawl spaces, and similar locations. Note that one has a pull-chain and one does not.

Lighting fixtures arranged in this manner eliminate shadows between automobiles. Shadows are a hazard because they hide objects that can cause a person to trip or fall.

It is highly recommended that these ceiling fixtures be mounted toward the front end of the automobile as it is normally parked in the garage. This arrangement will provide better lighting where it is most needed when the owner is working under the hood. Do not place lights where they will be covered by the open overhead door.

RECEPTACLE OUTLETS

At least one receptacle must be installed in an attached residential garage per *Section 210-52(g)*. The garage in this residence has three receptacles. If a detached garage has no electricity provided, then obviously no rules are applicable, but just as soon as a detached garage is wired, then all pertinent Code rules become effective and must be followed. These would include rules relating to grounding, lighting outlets, receptacle outlets, GFCI protection, and so on.

The Code in *Section 210-8(a)(2)* requires that "all 125-volt, single-phase, 15- or 20-ampere receptacles installed in garages shall have ground-fault circuit-interrupter protection." GFCIs are discussed in detail in unit 6.

There are a few exceptions to *Section 210-8(a)(2)* requiring GFCI protection of receptacles in residential garages. They are:

1. GFCI protection is *not* required if the receptacle is not readily accessible, such as the receptacle installed for the overhead door opener.

▶ 2. GFCI protection is *not* required for single or duplex receptacles intended for cord- and plug-connected appliances that are not easily moved, and are located in dedicated spaces. Examples include a freezer, refrigerator, or central vacuum equipment. The intent is that these dedicated receptacles will have the appliance cord plugged into it, and therefore would not be used to plug in portable electric tools and extension cords. ◀ This exception is difficult to enforce, because the electrician may not know during the rough-in that a certain receptacle is for the sole purpose of plugging in a specific appliance, say a large freezer. Such appliances are not at the construction site during rough-in.

Note that one GFCI receptacle is installed on the front wall of the garage, in the same box as the single-pole switch for the attic lighting. A second feed-through GFCI receptacle is installed on the right-hand garage wall near the entry to the house. This GFCI receptacle also provides the required GFCI protection for the other receptacle on the same wall and for the protection of the underground wiring to the post light.

If we were to install one GFCI feedthrough receptacle at the point where Circuit B14 feeds the garage, then the entire circuit would be GFCI protected as required. But nuisance tripping of the GFCI might occur because of:

- the overall length of the circuit wiring.
- possible problems in the overhead door opener.
- possible problems in the underground run to the post light.
- problems in the post lamp itself.

All of these would result in a complete outage on Circuit B14. The electrician must consider the cost of GFCI receptacles and compare it with the cost of additional cable or conduit and wire needed. He must also consider the possibility of a power outage when trying to get by with fewer GFCI receptacles.

All receptacles and switches in the garage are to be mounted 48 inches to center according to a notation on the First Floor Electrical Plan. The Uniform Building Code requires that receptacles be mounted not less than 18 inches above the garage floor.

LANDSCAPE LIGHTING

▶ One of the nice things about residential wiring is the availability of a virtually unlimited variety of sizes, types, and designs of outdoor lighting fixtures for gardens, decks, patios, and for accent lighting. Fixture showrooms at electrical distributors and the electrical departments of home centers offer a wide range of outdoor lighting fixtures, both 120-volt and low-voltage types. Figures 16-4 and 16-5 illustrate landscape lighting mounted on the ground. Figure 16-9 is frequently referred to as "treescape" lighting.

Low-voltage lighting systems are covered in *Article 411* of the *National Electrical Code®* The basic requirements for these systems are that they shall:

- operate at 30 volts rms (42.4 volts peak) or less.
- have one or more secondary circuits, each limited to 25 amperes maximum.

- be supplied through an isolated power supply. An example of this is the step-down transformer shown in figure 16-4.

- be "listed."

- not have the wiring concealed or extended through a building unless the wiring meets the requirements of *Chapter 3, NEC®* An example of this is running Type NMC nonmetallic-

sheathed cable for the low-voltage portion of the wiring that is concealed or extended through the structure of the building.

- not be installed within 10 feet (3.05 m) of pools, spas, or water fountains unless specifically permitted in *Article 680*.

- not have the secondary of the isolating transformer grounded.

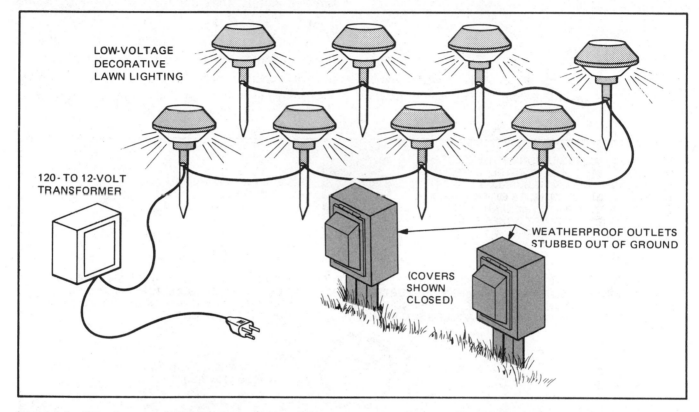

Fig. 16-4   **Weatherproof receptacle outlets stubbed out of the ground; either low-voltage decorative lighting or 120-volt PAR lighting fixtures may be plugged into these outlets.**

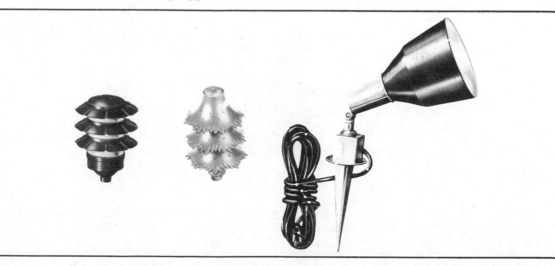

Fig. 16-5   **Fixtures used for decorative purposes outdoors under shrubs and trees, in gardens, and for lighting paths and driveways.**

• have the secondary circuit of the transformer insulated from the branch-circuit. The manufacturer of the step-down transformer used with these low-voltage systems meets this requirement in order to pass the Underwriters Laboratories standards tests. A standard step-down transformer may not meet this "shall be insulated" requirement.

• not be supplied by a branch-circuit that exceeds 20 amperes. ◄

OUTDOOR WIRING

If requested by the homeowner that 120-volt receptacles be installed away from the building structure, the electrician can provide weatherproof receptacle outlets as illustrated in figure 16-4.

Outdoor receptacles must have GFCI protection. See unit 6 for complete discussion regarding GFCI protection.

These types of boxes have ½-inch female threaded openings in which conduit fittings secure the conduit "stub-ups" to the box. Any unused openings are closed with ½-inch plugs that are screwed tightly into the threads. Refer to figure 16-6.

Section 410-57(b) requires that receptacles installed outdoors maintain their weatherproof integrity when the receptacle is in use. This requires a self-closing cover that is deep enough to shelter

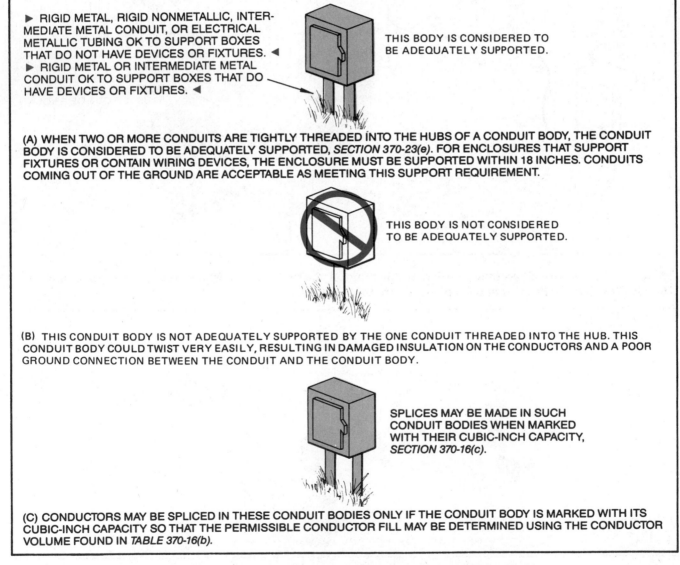

► RIGID METAL, RIGID NONMETALLIC, INTERMEDIATE METAL CONDUIT, OR ELECTRICAL METALLIC TUBING OK TO SUPPORT BOXES THAT DO NOT HAVE DEVICES OR FIXTURES. ◄
► RIGID METAL OR INTERMEDIATE METAL CONDUIT OK TO SUPPORT BOXES THAT DO HAVE DEVICES OR FIXTURES. ◄

THIS BODY IS CONSIDERED TO BE ADEQUATELY SUPPORTED.

(A) WHEN TWO OR MORE CONDUITS ARE TIGHTLY THREADED INTO THE HUBS OF A CONDUIT BODY, THE CONDUIT BODY IS CONSIDERED TO BE ADEQUATELY SUPPORTED, *SECTION 370-23(e)*. FOR ENCLOSURES THAT SUPPORT FIXTURES OR CONTAIN WIRING DEVICES, THE ENCLOSURE MUST BE SUPPORTED WITHIN 18 INCHES. CONDUITS COMING OUT OF THE GROUND ARE ACCEPTABLE AS MEETING THIS SUPPORT REQUIREMENT.

THIS BODY IS NOT CONSIDERED TO BE ADEQUATELY SUPPORTED.

(B) THIS CONDUIT BODY IS NOT ADEQUATELY SUPPORTED BY THE ONE CONDUIT THREADED INTO THE HUB. THIS CONDUIT BODY COULD TWIST VERY EASILY, RESULTING IN DAMAGED INSULATION ON THE CONDUCTORS AND A POOR GROUND CONNECTION BETWEEN THE CONDUIT AND THE CONDUIT BODY.

SPLICES MAY BE MADE IN SUCH CONDUIT BODIES WHEN MARKED WITH THEIR CUBIC-INCH CAPACITY, *SECTION 370-16(c)*.

(C) CONDUCTORS MAY BE SPLICED IN THESE CONDUIT BODIES ONLY IF THE CONDUIT BODY IS MARKED WITH ITS CUBIC-INCH CAPACITY SO THAT THE PERMISSIBLE CONDUCTOR FILL MAY BE DETERMINED USING THE CONDUCTOR VOLUME FOUND IN *TABLE 370-16(b)*.

Fig. 16-6   Supporting threaded conduit bodies, *Section 370-23.*

the attachment plug cap of the cord. See figure 16-4. This rule would apply to receptacles installed for decorative lighting, gutter heating cords, etc., where the cord will be plugged in for long periods of time.

Where flush-mounted weatherproof receptacles are desired for appearance purposes, a recessed weatherproof receptacle, as illustrated in figure 16-6A, may be used. This type of installation takes up more depth than a standard receptacle, and requires special mounting as indicated in the "exploded" view. Careful thought must be given during the "roughing-in" stages to assure that the proper depth "roughing-in" box is installed.

An exception to the previous requirement is for weatherproof receptacles where the equipment plugged into the receptacle is "attended." This would apply to receptacles installed outside of a home that are used periodically for extension cords on tools, hedge trimmers, grills, and similar appli-ances. As soon as the homeowner is finished, the appliance is unplugged, and the self-closing cover returns to a closed position. Figures 5-4 and 16-6 show typical shallow covers.

Wiring with Type UF Cable *(ARTICLE 339)*

The plans show that Type UF underground cable, figure 16-7, is used to connect the post light. The current-carrying capacity (ampacity) of Type UF cable is found in *Table 310-16*, using the 60°C column. According to *Article 339*, Type UF cable:

- is marked underground feeder cable.

- is available in sizes from No. 14 AWG through No. 4/0 (for copper conductors) and from No. 12 AWG through No. 4/0 (for aluminum con-ductors).

- may be used in direct exposure to the sun if the cable is marked "sunlight resistant."

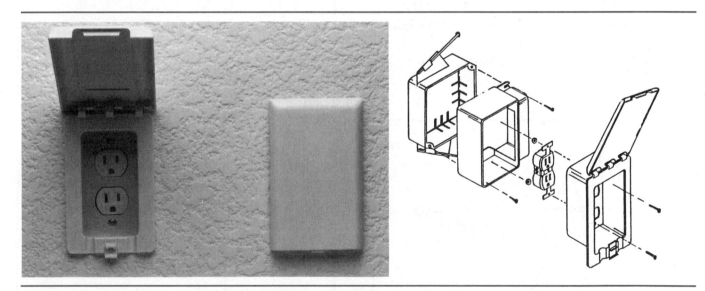

Fig. 16-6A   Photo of a recessed type of weatherproof receptacle that meets the requirements of *Section 410-57(b)* **requiring that the enclosure be weatherproof while in use if unattended, where the cord will be plugged in for long periods of time. The "exploded" view illustrates one method of "roughing-in" this type of weatherproof recepta-cle.** *Courtesy* **of TayMac Corporation.**

Fig. 16-7   Type UF underground cable.

- may be used with nonmetallic-sheathed cable fittings.
- is flame-retardant.
- is moisture-, fungus-, and corrosion-resistant.
- may be buried directly in the earth.
- may be used for branch-circuit and feeder wiring.
- may be used in interior wiring for wet, dry, or corrosive installations.
- is installed by the same methods as non-metallic-sheathed cable *(Article 336)*.
- must not be used as service-entrance cable.
- must not be embedded in concrete, cement, or aggregate.
- must be buried in the same trench where single-conductor cables are installed.
- shall be installed according to *NEC® Section 300-5*.

The grounding of equipment fed by Type UF cable is accomplished by properly connecting the bare equipment grounding conductor found in the UF cable to the equipment to be grounded, figure 16-8.

Regarding the subject of underground wiring, it is significant to note that whereas the Code does permit running up the side of a tree with conduit or cable (when protected from physical damage), figure 16-9, the Code does *not* permit supporting conductors between trees, figure 16-10. Figure 16-11 shows one type of lighting fixture that is permitted for the installation shown in figure 16-9. ▶ Splices are permitted to be made directly in the ground, but only if the connectors are "listed" by a recognized testing laboratory for such use, *Section 110-14(b)*. ◀

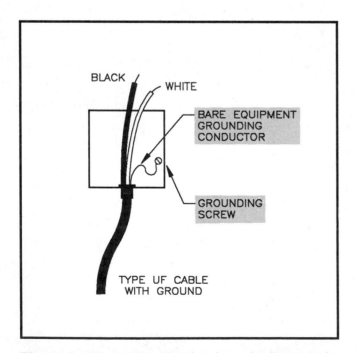

Fig. 16-8   Illustration showing how the bare equipment grounding conductor of a Type UF cable is used to ground the metal box.

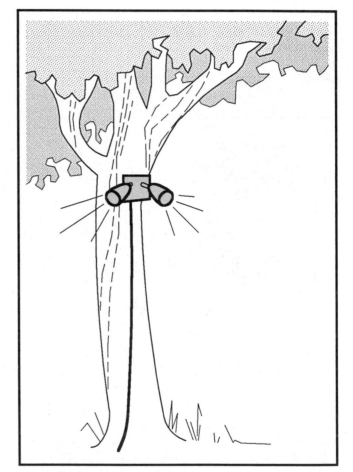

Fig. 16-9   This conforms to the Code. *Section 225-26* does not permit spans of conductors to be run between live or dead trees, but it does not prohibit installing decorative lighting as shown in this figure. *Section 410-16(h)* permits outdoor lighting fixtures and associated equipment to be supported by trees. Approved wiring methods to carry the conductors up the tree must be used.

Fig. 16-10   VIOLATION. *Section 225-26* does not permit live or dead vegetation such as trees to support overhead conductor spans. Lumber and treated poles are not considered to be vegetation for the purpose of applying the requirement of *Section 225-26*.

Fig. 16-11   Outdoor fixture with lamps.

UNDERGROUND WIRING

Underground wiring is common in residential applications. Examples include decorative landscape lighting and wiring to post lamps and detached buildings such as garages or tool sheds.

Table 300-5 of the Code shows the minimum depths required for the various types of wiring methods. Refer to figures 16-12A, 16-12B and 16-12C.

Across the top of the table are listed wiring methods such as direct burial cables or conductors, rigid metal conduit, intermediate metal conduit, rigid nonmetallic conduit that is approved for direct burial, special considerations for residential branch-circuits rated 120 volts or less having GFCI protection and overcurrent protection not over 20 amperes, and low-voltage landscape lighting supplied with Type UF cable.

Down the left-hand side of the table we find the location of the wiring methods that are listed across the top of the table.

The notes to *Table 300-5* are very important. Note 4 allows us to select the shallower of two depths when the wiring methods in Columns 1, 2, or 3 are combined with Columns 4 or 5. For example, the basic rule for direct buried cable is that it be covered by a minimum of 24 inches. However, this minimum depth is reduced to 12 inches for residential installations when the branch-circuit is not over 120-volts, is GFCI protected, and is not over 20 amperes. This is for residential installations only.

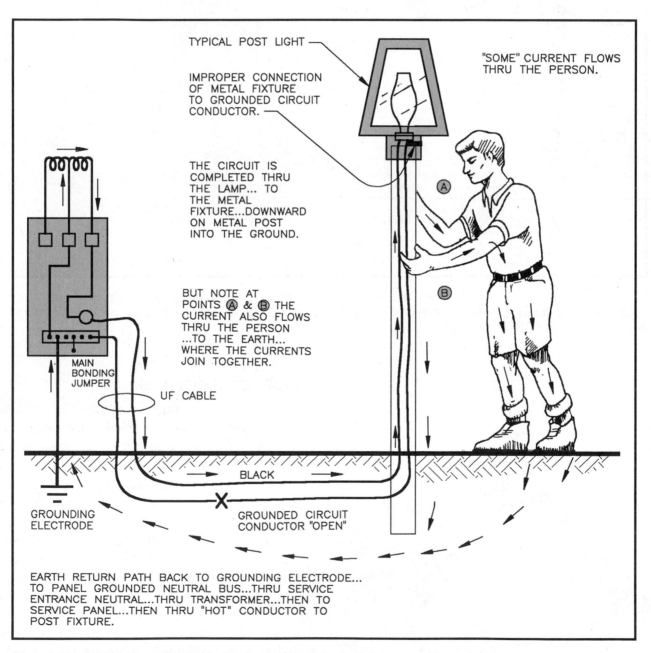

TYPICAL POST LIGHT

IMPROPER CONNECTION OF METAL FIXTURE TO GROUNDED CIRCUIT CONDUCTOR.

"SOME" CURRENT FLOWS THRU THE PERSON.

THE CIRCUIT IS COMPLETED THRU THE LAMP... TO THE METAL FIXTURE...DOWNWARD ON METAL POST INTO THE GROUND.

BUT NOTE AT POINTS Ⓐ & Ⓑ THE CURRENT ALSO FLOWS THRU THE PERSON ...TO THE EARTH... WHERE THE CURRENTS JOIN TOGETHER.

MAIN BONDING JUMPER

UF CABLE

BLACK

GROUNDING ELECTRODE

GROUNDED CIRCUIT CONDUCTOR "OPEN"

EARTH RETURN PATH BACK TO GROUNDING ELECTRODE... TO PANEL GROUNDED NEUTRAL BUS...THRU SERVICE ENTRANCE NEUTRAL...THRU TRANSFORMER...THEN TO SERVICE PANEL...THEN THRU "HOT" CONDUCTOR TO POST FIXTURE.

Fig. 16-12A   This diagram clearly illustrates the electric shock hazard associated with the code violation practice of grounding metal objects to the grounded conduit conductor. This is a violation of *Section 250-61(b).*

The measurement for the depth requirement is from the top of the raceway or cable to the top of the finished grade, concrete, or other similar cover.

Do not back-fill a trench with rocks, debris, or similar course material, *Section 300-5(f)*.

Installation of Conduit Underground

Some local electrical codes require the installation of conduit for all underground wiring, using conductors identified for wet locations, such as Type TW, THW, or THWN.

All conduit installed underground must be protected against corrosion. The manufacturer of the conduit and referring to Underwriters' Laboratories Standards will furnish the information as to whether or not the conduit is suitable for direct burial.

Additional supplemental protection of metal conduit can be accomplished by "painting" the conduit with a nonmetallic coating. See Table 16-2.

When metal conduit is used as the wiring method, the metal boxes, post lights, and so forth that are properly connected to the metal raceways are considered to be grounded because the conduit serves as the equipment ground, *Section 250-91(b)*, figure 16-13. See *Article 410, Part E* for details on grounding requirements for fixtures.

When nonmetallic raceways are used, then a separate equipment grounding conductor, either green or bare, must be installed to accomplish the adequate grounding of the equipment served. This equipment grounding conductor must be sized according to *Table 250-95* of the Code. See figure 16-14.

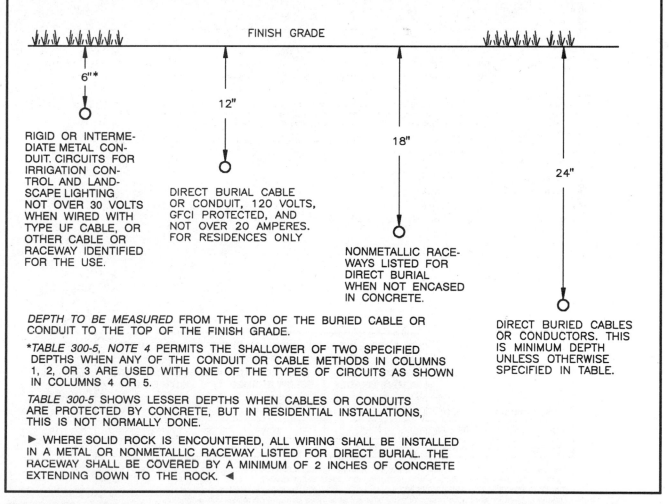

FINISH GRADE

6"*

RIGID OR INTERME-
DIATE METAL CON-
DUIT. CIRCUITS FOR
IRRIGATION CON-
TROL AND LAND-
SCAPE LIGHTING
NOT OVER 30 VOLTS
WHEN WIRED WITH
TYPE UF CABLE, OR
OTHER CABLE OR
RACEWAY IDENTIFIED
FOR THE USE.

12"

DIRECT BURIAL CABLE
OR CONDUIT, 120 VOLTS,
GFCI PROTECTED, AND
NOT OVER 20 AMPERES.
FOR RESIDENCES ONLY

18"

NONMETALLIC RACE-
WAYS LISTED FOR
DIRECT BURIAL
WHEN NOT ENCASED
IN CONCRETE.

24"

DIRECT BURIED CABLES
OR CONDUCTORS. THIS
IS MINIMUM DEPTH
UNLESS OTHERWISE
SPECIFIED IN TABLE.

DEPTH TO BE MEASURED FROM THE TOP OF THE BURIED CABLE OR
CONDUIT TO THE TOP OF THE FINISH GRADE.

*TABLE 300-5, NOTE 4 PERMITS THE SHALLOWER OF TWO SPECIFIED
DEPTHS WHEN ANY OF THE CONDUIT OR CABLE METHODS IN COLUMNS
1, 2, OR 3 ARE USED WITH ONE OF THE TYPES OF CIRCUITS AS SHOWN
IN COLUMNS 4 OR 5.

TABLE 300-5 SHOWS LESSER DEPTHS WHEN CABLES OR CONDUITS
ARE PROTECTED BY CONCRETE, BUT IN RESIDENTIAL INSTALLATIONS,
THIS IS NOT NORMALLY DONE.

► WHERE SOLID ROCK IS ENCOUNTERED, ALL WIRING SHALL BE INSTALLED
IN A METAL OR NONMETALLIC RACEWAY LISTED FOR DIRECT BURIAL. THE
RACEWAY SHALL BE COVERED BY A MINIMUM OF 2 INCHES OF CONCRETE
EXTENDING DOWN TO THE ROCK. ◄

Fig. 16-12B   Minimum depths for cables and conduits installed underground. See *Table 300-5* for depths for other conditions.

Table 300-5. Minimum Cover Requirements, 0 to 600 Volts, Nominal, Burial in Inches
(Cover is defined as the shortest distance measured between a point on the top surface of any direct
buried conductor, cable, conduit, or other raceway and the top surface of finished grade, concrete, or similar cover.)

	Type of Wiring Method or Circuit				
Location of Wiring Method or Circuit	1 Direct Burial Cables or Conductors	2 Rigid Metal Conduit or Intermediate Metal Conduit	3 Nonmetallic Raceways Listed for Direct Burial without Concrete Encasement or Other Approved Raceways	4 Residential Branch Circuits Rated 120 Volts or Less with GFCI Protection and Maximum Overcurrent Protection of 20 Amperes	5 Circuits for Control of Irrigation and Landscape Lighting Limited to Not More Than 30 Volts and Installed with Type UF or in Other Identified Cable or Raceway
All Locations Not Specified Below	24	6	18	12	6
In Trench Below 2-Inch Thick Concrete or Equivalent	18	6	12	6	6
Under a Building	0 (In Raceway Only)	0	0	0 (In Raceway Only)	0 (In Raceway Only)
Under Minimum of 4-Inch Thick Concrete Exterior Slab with No Vehicular Traffic and the Slab Extending Not Less than 6 Inches beyond the Underground Installation	18	4	4	6 (Direct Burial) 4 (In Raceway)	6 (Direct Burial) 4 (In Raceway)
Under Streets, Highways, Roads, Alleys, Driveways, and Parking Lots	24	24	24	24	24
One- and Two-Family Dwelling Driveways and Outdoor Parking Areas, and Used Only for Dwelling-Related Purposes	18	18	18	12	18
In or Under Airport Runways, Including Adjacent Areas Where Trespassing Prohibited	18	18	18	18	18

Note 1. For SI units: 1 in. = 25.4 mm.
Note 2. Raceways approved for burial only where concrete encased shall require concrete envelope not less than 2 in. thick.
Note 3. Lesser depths shall be permitted where cables and conductors rise for terminations or splices or where access is otherwise required.
Note 4. Where one of the wiring method types listed in columns 1-3 is used for one of the circuit types in columns 4 and 5, the shallower depth of burial shall be permitted.
Note 5. Where solid rock is encountered, all wiring shall be installed in metal or nonmetallic raceway permitted for direct burial. The raceways shall be covered by a minimum of 2 in. of concrete extending down to rock.

	IS SUPPLEMENTAL CORROSION PROTECTION REQUIRED?		
	IN CONCRETE ABOVE GRADE?	IN CONCRETE BELOW GRADE?	IN DIRECT CONTACT WITH SOIL?
Rigid Conduit[1]	No	No	No[2]
Intermediate Conduit[1]	No	No	No[2]
Electrical Metallic Tubing[1]	No	Yes	Yes[3]

[1] Severe corrosion can be expected where ferrous metal conduits come out of concrete and enter the soil. Here again, some electrical inspectors and consulting engineers might specify the application of some sort of supplemental nonmetallic corrosion protection.
[2] Unless subject to severe corrosive effects, different soils have different corrosive characteristics.
[3] In most instances, electrical metallic tubing is not permitted to be installed underground in direct contact with the soil because of corrosion problems.

Table 16-2

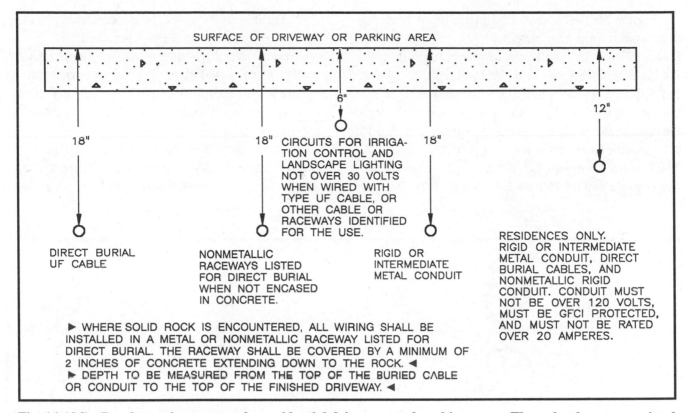

SURFACE OF DRIVEWAY OR PARKING AREA

6"

12"

18"

18"   CIRCUITS FOR IRRIGA-
TION CONTROL AND
LANDSCAPE LIGHTING
NOT OVER 30 VOLTS
WHEN WIRED WITH
TYPE UF CABLE, OR
OTHER CABLE OR
RACEWAYS IDENTIFIED
FOR THE USE.

18"

DIRECT BURIAL
UF CABLE

NONMETALLIC
RACEWAYS LISTED
FOR DIRECT BURIAL
WHEN NOT ENCASED
IN CONCRETE.

RIGID OR
INTERMEDIATE
METAL CONDUIT

RESIDENCES ONLY.
RIGID OR INTERMEDIATE
METAL CONDUIT, DIRECT
BURIAL CABLES, AND
NONMETALLIC RIGID
CONDUIT. CONDUIT MUST
NOT BE OVER 120 VOLTS,
MUST BE GFCI PROTECTED,
AND MUST NOT BE RATED
OVER 20 AMPERES.

► WHERE SOLID ROCK IS ENCOUNTERED, ALL WIRING SHALL BE
INSTALLED IN A METAL OR NONMETALLIC RACEWAY LISTED FOR
DIRECT BURIAL. THE RACEWAY SHALL BE COVERED BY A MINIMUM OF
2 INCHES OF CONCRETE EXTENDING DOWN TO THE ROCK. ◄
► DEPTH TO BE MEASURED FROM THE TOP OF THE BURIED CABLE
OR CONDUIT TO THE TOP OF THE FINISHED DRIVEWAY. ◄

Fig. 16-12C   Depth requirements under residential driveways and parking areas. These depths are permitted for one- and two-family residences.

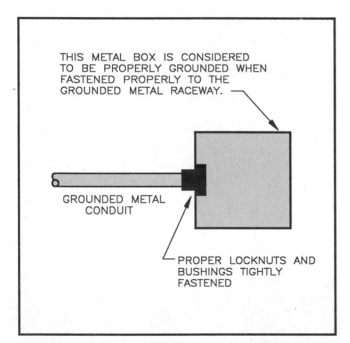

THIS METAL BOX IS CONSIDERED
TO BE PROPERLY GROUNDED WHEN
FASTENED PROPERLY TO THE
GROUNDED METAL RACEWAY.

GROUNDED METAL
CONDUIT

PROPER LOCKNUTS AND
BUSHINGS TIGHTLY
FASTENED

Fig. 16-13   This illustration shows that a box is grounded when properly fastened to the grounded metal raceway. This is acceptable in *Section 250-91(b)*. More and more Code-enforcing authorities and consulting engineers are requiring that a separate equipment grounding conductor be installed in all raceways. This insures effective grounding of the installation. Refer to unit 18.

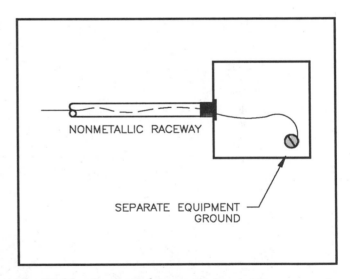

NONMETALLIC RACEWAY

SEPARATE EQUIPMENT
GROUND

Fig. 16-14   Grounding a metal box with a separate equipment grounding conductor.

Depending upon the type of soil and moisture content of the soil, the earth may be a good ground or may not be. The Code recognizes that the earth by itself is not considered to be an adequate ground. *Section 250-51*, second paragraph states that "The earth shall not serve as the sole equipment grounding conductor." See figure 16-14A.

Figure 16-14B shows two methods of bringing the conduit into the basement: (1) it can be run below ground level and then can be brought through the basement wall, or (2) it can be run up the side of the building and through the basement wall at the ceiling joist level. When the conduit is run through the basement wall, the opening must be sealed to prevent moisture from seeping into the basement. The electrician must decide which of the two methods is more suitable for each installation.

According to the Code, any location exposed to the weather is considered a wet location.

Section 410-4(a) states that outdoor fixtures must be constructed so that water cannot enter or accumulate in lampholders, wiring compartments, or other electrical parts. These fixtures must be marked "Suitable for Wet Locations."

Partially protected areas such as roofed open porches or areas under canopies are defined by the Code as damp locations. Fixtures to be used in these

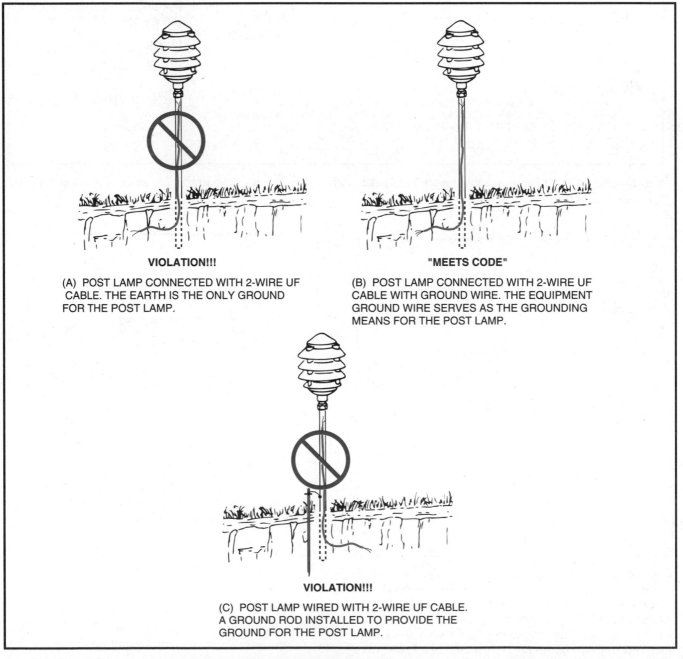

VIOLATION!!!

(A) POST LAMP CONNECTED WITH 2-WIRE UF CABLE. THE EARTH IS THE ONLY GROUND FOR THE POST LAMP.

"MEETS CODE"

(B) POST LAMP CONNECTED WITH 2-WIRE UF CABLE WITH GROUND WIRE. THE EQUIPMENT GROUND WIRE SERVES AS THE GROUNDING MEANS FOR THE POST LAMP.

VIOLATION!!!

(C) POST LAMP WIRED WITH 2-WIRE UF CABLE. A GROUND ROD INSTALLED TO PROVIDE THE GROUND FOR THE POST LAMP.

Fig. 16-14A These drawings illustrate the intent of *Section 250-51*, second paragraph, which states that "The earth shall not serve as the sole equipment grounding conductor."

locations must be marked "Suitable for Damp Locations." Many types of fixtures are available. Therefore, it is recommended that the electrician check the UL label on the fixture to determine the suitability of the fixture for a wet or damp location.

The post in figure 16-14B may or may not be embedded in concrete, depending on the consistency of the soil, the height of the post, and the size of the lighting fixture. Most electricians prefer to embed the base of the post in concrete to prevent rotting of wood posts and rusting of metal posts.

Figure 16-15 illustrates some typical post lights.

OVERHEAD GARAGE DOOR OPERATOR ⒶE

The overhead garage door operator in this residence is plugged into the receptacle provided for this purpose on the garage ceiling. This particular overhead door operator is rated 5.8 amperes at 120

volts. The actual installation of the overhead door will usually be done by a specially trained overhead door installer.

Principles of Operation

The overhead door operator contains a motor, gear reduction unit, motor reversing switch, and electric clutch. This unit is preassembled and wired by the manufacturer.

Almost all residential overhead door operators use split-phase, capacitor-start motors in sizes from $\frac{1}{4}$ to $\frac{1}{2}$ horsepower. The motor size selected depends upon the size (weight) of the door to be raised and lowered. By using springs in addition to the gear reduction unit to counterbalance the weight of the door, a small motor can lift a fairly heavy door.

There are two ways to change the direction of a split-phase motor: (1) reverse the starting winding with respect to the running winding; (2) reverse the

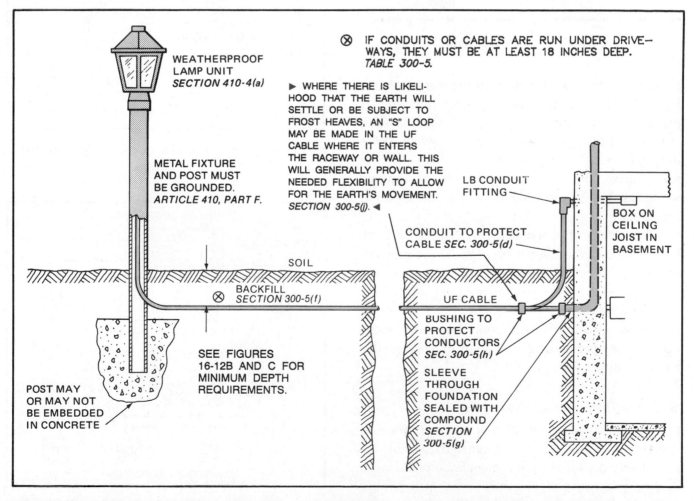

Fig. 16-14B   Methods of bringing cable and/or conduit through a concrete wall and/or upward from a concrete wall into the hollow space within a framed wall.

Fig. 16-15   Post-type lights.

running winding with respect to the starting winding. The motor can be run in either direction by properly connecting the starting and running winding leads of the motor to the reversing switch. When the motor shaft is connected to a chain drive or screw drive, the door can be raised or lowered according to the direction in which the motor is rotating.

The electrically operated reversing switch can be controlled in a number of ways. These methods include push-button stations marked "open-close-stop," "up-down-stop," or "open-close," indoor and outdoor weatherproof push buttons, and key-operated stations. Radio-operated controllers are popular because they permit overhead doors to be controlled from an automobile.

Wiring of Garage Door Operators

The wiring of overhead door operators for residential use is quite simple, because these units are completely prewired by the manufacturer. All residential-type overhead door operators come equipped with a cord that is plugged into a 120-volt receptacle outlet installed in the ceiling of the garage, near the

location of the overhead door operator. An alternate method would be to install a box-cover unit, discussed later. A receptacle for the overhead door unit does not have to be GFCI protected, *Section 210-8(a)(2), Exception 1*. The overhead door operator could be "hard-wired," as in figure 16-16, although this is not often done in residential installations.

The buttons that are pushed to operate the overhead door operator can be placed in any convenient, desirable location. These push buttons are wired with low-voltage wiring, usually 24 volts, figure 16-17. This is the same type of wire used for wiring bells and chimes. From each button, the electrician runs a two-wire, low-voltage cable back to the control unit. Because the residential openers operate

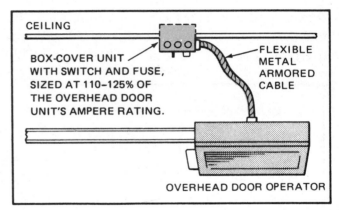

Fig. 16-16   Connections for overhead door operator.

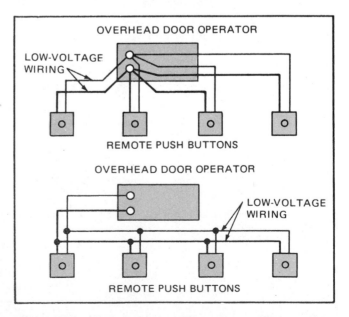

Fig. 16-17   Push-button wiring for overhead door operator.

on the momentary contact principle for push buttons, a two-wire cable is all that is necessary between the operator and the push buttons. Commercial overhead door operators can require three or more conductors.

It is desirable to install push buttons at each door leading into the garage.

The actual connection of the low-voltage wiring between the controller and push buttons is a parallel circuit connection, figure 16-17.

Overhead door operators can be connected directly by using flexible metal conduit or armored cable. The illustration in figure 16-16 shows how this can be accomplished. Note that the flexible cable is run between the door operator and an electrical box that has been mounted on the ceiling.

Box-Cover Unit

Instead of merely installing a receptacle outlet on the ceiling near the garage door operator unit, it has been found to be advantageous to provide a box-cover unit, either a switch-fuse type, figure 16-16, or a fuse-receptacle type, figures 16-18 and 16-19. This box-cover unit permits the use of dual-element Type S fuses sized at 110 to 125% of the motor rating.

This box-cover unit provides convenient disconnection of power to the overhead door unit for servicing or when the owner is away from home.

Box-cover units are readily available for mounting onto 4-inch octagon boxes, 4-inch square boxes, and handy boxes. Should the need arise to provide for two individual disconnects and individual fusing, then a box-cover unit of the style illustrated in figure 16-19 can be installed.

Overcurrent Protection

Overhead door operators have integral overload protection. The addition of the box-cover fuse device provides added protection. This added overload protection in the form of Type S, dual-element, time-delay fuses installed in the box-cover units is determined by sizing the Type S fuse at not over 125% of the full-load current rating of the motor, per *Section 430-32(c)*:

$$5.8 \times 1.25 = 7.25 \text{ amperes}$$

Using 110% sizing, we have

$$5.8 \times 1.10 = 6.38 \text{ amperes}$$

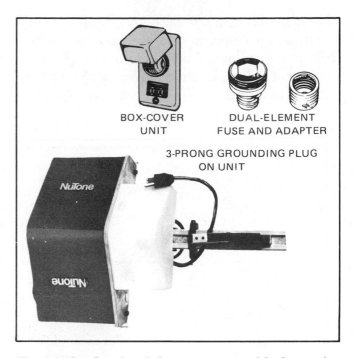

Fig. 16-18   Overhead door operator with three-wire cord connector.

Fig. 16-19   Box-cover unit that provides disconnecting means plus individual overcurrent protection.

This would give us an ampere rating range to choose from. In the previous instance, a Type S 6¼-ampere dual-element fuse would be an excellent choice for backup overload protection of the overhead door operator.

A dual-element Type S fuse installed as shown in figures 16-16 and 16-18 would sense any problem that might arise in the electrical wiring inside the overhead door unit, and would clear the problem before the branch-circuit overcurrent protective device opened, because the ampere rating of the fuse at the overhead door unit is much smaller than the rating of the branch-circuit overcurrent device, figure 16-20. In technical terms this is called *selective coordination*, and is presented in *Electrical Wiring—Commercial*.

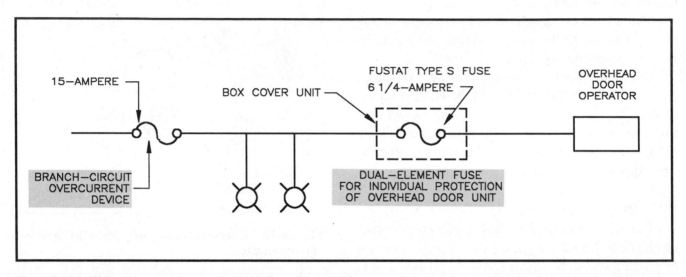

Fig. 16-20   This diagram shows the benefits of installing a small-ampere-rated dual element time-delay fuse that is sized to provide overload protection for the motor. If an electrical problem occurs at the overhead door operator, only the 6¼-ampere fuse will open. The 15-ampere branch circuit is not affected. All other loads connected to the 15-ampere circuit remain energized.

REVIEW

Note: Refer to the Code or plans where necessary.

1. What circuit supplies the garage? _____

2. What is the circuit rating? _____

3. How many GFCI-protected receptacle outlets are connected to the garage circuit?

   _____

4. a. How many cables enter the wall switch box located next to the side garage door?

   _____

   b. How many circuit conductors enter this box? _____

   c. How many equipment grounding conductors enter this box? _____

5. Show calculations on how to select a proper size box for question 4. What kind of box would you use? _____

   _____

   _____

   _____

   _____

   _____

6. a. How many lights are recommended for a one-car garage? _____ for a two-car garage? _____ for a three-car garage? _____

   b. Where are these lights to be located? _____
   _____

7. From how many points in the garage of this residence are the ceiling lights controlled? _____

8. GFCI breakers or GFCI receptacles are relatively inexpensive. How would *you* arrange the wiring of the garage circuit to make as economical installation as possible that complies with the Code? _____

   _____

   _____

   _____

   _____

   _____

9. The total estimated volt-ampere load of the garage circuit draws how many amperes? Show your calculations. _____

   _____

   _____

   _____

10. How high from the floor are the switches and receptacles to be mounted? _____

11. What type of cable feeds the post light? _____

12. All raceways, cables, and direct burial-type conductors require a "cover." Explain the term "cover." _____

   _____

   _____

   _____

13. In the spaces provided, fill in the cover (depth) for the following residential underground installations.

   a. Type UF. No other protection.   _____ inches

   b. Type UF below driveway.   _____ inches

   c. Rigid metal conduit under lawn.   _____ inches

   d. Rigid metal conduit under driveway.   _____ inches

   e. Electrical metallic tubing between house   _____ inches
       and detached garage.

   f. UF cable under lawn. Circuit is 120 volts,   _____ inches
       20 amperes, GFCI protected.

   g. Rigid nonmetallic conduit approved for direct
       burial. No other protection.   _____ inches

   h. Rigid nonmetallic conduit. Circuit is 120 volts, _____ inches
       20 amperes, GFCI protected.

   i. Rigid conduit passing over solid rock and covered _____ inches
       by 2 inches of concrete extending down to rock.

14. What section of the Code prohibits embedding Type UF cable in concrete? _____

_____

15. The following is a layout of the garage circuit. Using the suggested cable layout, make a complete wiring diagram of this circuit, using colored pens or pencils to indicate conductor insulation colors.

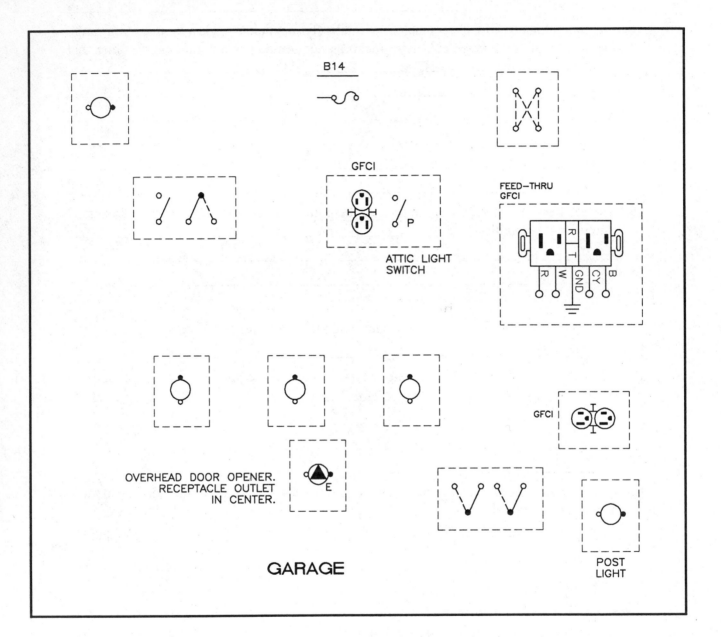

16. What type of motors are generally used for garage door openers? _____

17. How is the direction of a split-phase motor reversed? _____

_____

_____

18. Motor running overload protection is generally based on what percentage of the full-load running current of the motor? _____

19. Show the proper dual-element fuse size for a ½-horsepower, 115-volt, single-phase motor to provide running overload protection. See *Table 430-148*.

_____

_____

_____

20. When outdoor receptacles are installed where a cord will be plugged in permanently, such as in the case of decorative lighting, the cover must provide weatherproof integrity when the cord is plugged in. True or false? _____ Section _____ .

21. A homeowner purchased some single-conductor cable that was marked as suitable for direct burial in the ground to wire a post light. A separate equipment grounding conductor to the post light was not carried. Instead, the bottom of the metal post buried two feet into the ground was used as a ground for the post light. The inspector "red tagged" (turned down) the installation. Who is right? The homeowner or the electrical inspector? Explain.

_____

_____

_____

_____

UNIT 17

Recreation Room

OBJECTIVES

After studying this unit, the student will be able to

- understand three-wire (multiwire) branch-circuits.
- understand how to install lay-in fixtures.
- calculate watts loss and voltage drop in two-wire and three-wire circuits.
- understand the term *fixture whips*.
- understand the advantages of installing multiwire branch-circuits.
- understand problems that can be encountered on multiwire branch-circuits as a result of open neutrals.

RECREATION ROOM LIGHTING

The recreation room is well-lighted through the use of six lay-in 2' × 4' fluorescent fixtures. These fixtures are exactly the same size as two 2' × 2' ceiling tiles. The fixtures rest on top of the ceiling tee bars, figure 17-1.

Junction boxes are mounted above the dropped ceiling usually within 1 or 2 feet of the intended fixture location. Four to 6 feet of flexible conduit are installed between this junction box and the fixture. These are commonly called *fixture whips*. They contain the correct type and size of conductor suitable for the temperature ratings and load required by the Code for recessed fixtures. See unit 7 for a complete explanation of Code regulations for the installation of recessed fixtures. See figure 17-2.

Fluorescent fixtures of the type shown in figure 17-1 might bear a label stating "Recessed Fluorescent Fixture." The label might also state "Suspended Ceiling Fluorescent Fixture." Although they might look the same, there is a difference.

Recessed fluorescent fixtures are intended for installation in cavities in ceilings and walls and are to be wired according to *Section 410-64* of the *National Electrical Code.*® These fixtures may also be installed in suspended ceilings if they have the necessary mounting hardware.

Suspended fluorescent fixtures are intended only for installation in suspended ceilings where the acoustical tiles, lay-in panels, and suspended grid are not part of the actual building structure.

Underwriters Laboratories Standard 1570 covers recessed and suspended ceiling fixtures in detail.

The six recessed "lay-in" fluorescent fixtures in the recreation room are four-lamp fixtures with warm white energy-saving F40SPX30/RS lamps installed in them. The ballasts in these fixtures are high power factor, energy-saving ballasts. Five of the fixtures are controlled by a single-pole switch located at the bottom of the stairs. One fixture is controlled by the three-way switches that control the stairwell light fixture hung from

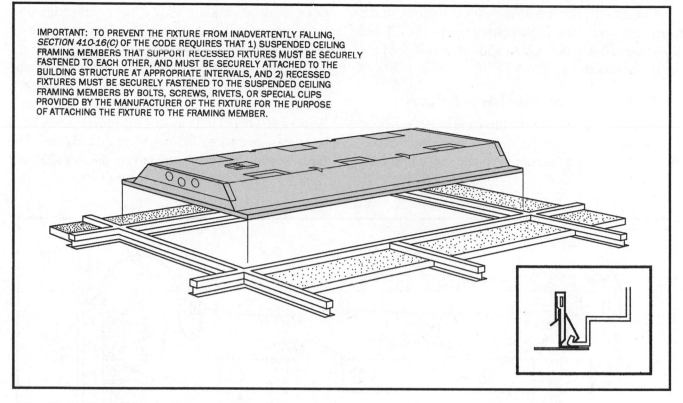

IMPORTANT: TO PREVENT THE FIXTURE FROM INADVERTENTLY FALLING, *SECTION 410-16(C)* OF THE CODE REQUIRES THAT 1) SUSPENDED CEILING FRAMING MEMBERS THAT SUPPORT RECESSED FIXTURES MUST BE SECURELY FASTENED TO EACH OTHER, AND MUST BE SECURELY ATTACHED TO THE BUILDING STRUCTURE AT APPROPRIATE INTERVALS, AND 2) RECESSED FIXTURES MUST BE SECURELY FASTENED TO THE SUSPENDED CEILING FRAMING MEMBERS BY BOLTS, SCREWS, RIVETS, OR SPECIAL CLIPS PROVIDED BY THE MANUFACTURER OF THE FIXTURE FOR THE PURPOSE OF ATTACHING THE FIXTURE TO THE FRAMING MEMBER.

Fig. 17-1   Typical lay-in fluorescent fixture commonly used in conjunction with dropped acoustical ceilings.

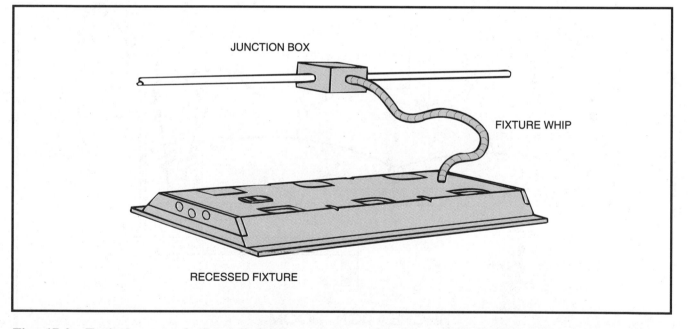

JUNCTION BOX

FIXTURE WHIP

RECESSED FIXTURE

Fig. 17-2   Typical recessed "lay-in" fluorescent fixture showing whip flexible connection to fixture. Conductors in whip must have temperature rating as required on label in fixture. For example, "For Supply Use 90°C conductors."

the ceiling above the stair landing midway up the stairs.

This was done so that the entire recreation room lighting would not be on continually whenever someone came down the stairs to go into the workshop. This is a practical energy-saving feature.

Circuit B12 feeds the fluorescent fixtures in the recreation room. See cable layout, figure 17-3. Table 17-1 summarizes the outlets and estimated load for the recreation room.

How to Connect Recessed Lay-In Fixtures

Figure 17-2 shows how this is done. A junction box is located above the ceiling near the fixture. Fixture whips are generally made up of ⅜-inch flexible metal conduit, either by the electrician or by a manufacturer that specializes in this type of manufacturing.

Section 350-10 permits ⅜-inch flexible metal conduit in lengths not to exceed 6 feet to make the connection between the junction box and the fixture. *Section 410-67(c)* tells us that the junction box shall be at least one foot (305 mm) from the fixture. The flex that contains the fixture tap conductors shall not

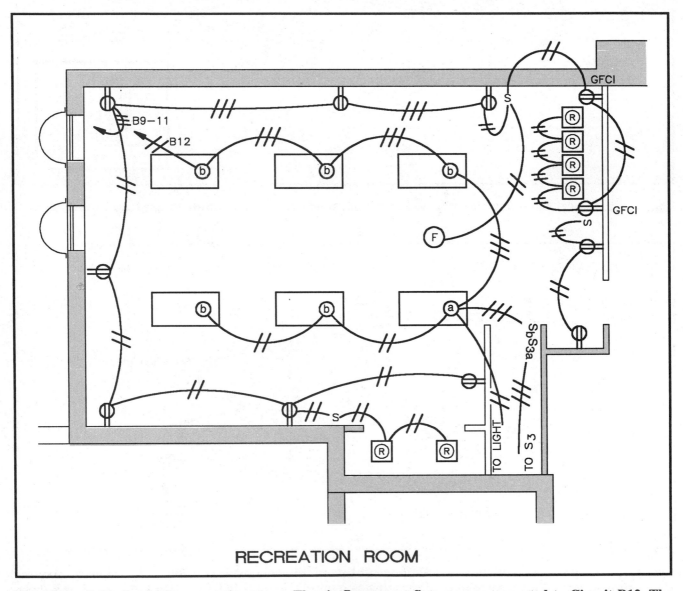

RECREATION ROOM

Fig. 17-3   Cable layout for recreation room. The six fluorescent fixtures are connected to Circuit B12. The receptacles located around the room are connected to Circuit B11. The four recessed fixtures over the bar plus the three receptacles located on the bar wall plus the receptacles located to the right of the steps as you come down the steps are connected to Circuit B9. Circuits B9 and B11 are three-wire circuits. Junction boxes are located above the dropped ceiling, and to one side of each fluorescent fixture. The electrical connections are made in these junction boxes. A flexible fixture "whip" connects between each junction box and each fixture. See figure 17-2.

DESCRIPTION	QUANTITY	WATTS	VOLT-AMPERES
Circuit Number B11			
Receptacles @ 120 W each	7	840	840
Closet recessed fixtures @ 75 W each	2	150	150
Exhaust fan (2.5 A × 120 V)	1	160	300
TOTALS	10	1150	1290
Circuit Number B12			
Fixture on stairway	1	100	100
Recessed fluorescent fixtures with four 40-W lamps each	6	960	1008
TOTALS	7	1060	1108
Circuit Number B9			
Receptacles @ 120 W each	4	480	480
Wet-bar recessed fixtures @ 75 W each	4	300	300
TOTALS	8	780	780

Table 17-1   Recreation room outlets and estimated load (three circuits).

be less than 4 feet (1.22 m) nor longer than 6 feet (1.83 m). The tap conductors must have a temperature rating as marked on the fixture.

Grounding of the fixture is done according to *Section 250-91(b), Exception 1*, through the metal of the ⅜-inch flexible metal conduit when:

1. The flex is not over 6 feet (1.83 m) long. See figure 4-33 when length exceeds 6 feet (1.83 m).

2. The branch-circuit overcurrent protection does not exceed 20 amperes.

3. The connectors for the flex are listed for grounding purposes. This refers to UL listing.

Flexible metal conduit must be supported within 12 inches (305 mm) of any outlet box, junction box, cabinet, or fitting according to *Section 350-18*. A "fixture whip" is exempt from this requirement per *Section 350-18, Exception No. 3*. Yet, many electricians do secure the flex within 12 inches of the outlet box so as not to pull the flex out of the connector while they are making up the electrical connections and installing the fixture in the lay-in ceiling.

Armored cable (BX) may also be used to connect recessed fixtures where the space above the

ceiling is accessible—either by "lift-out" ceiling tiles or gaining access by dropping down the fixture. See *Section 333-7, Exception No. 3*.

Refer to unit 7 for additional information regarding recessed, "lay-in" lighting fixtures.

RECEPTACLES AND WET BAR

The circuitry for the recreation room wall receptacles and the wet-bar lighting area introduces a new type of circuit. This is termed a *multiwire branch-circuit* or a *three-wire branch-circuit*.

In many cases the use of a multiwire branch-circuit can save money in that one three-wire branch-circuit will do the job of two two-wire branch-circuits.

Also if the loads are nearly balanced in a three-wire branch-circuit, the neutral conductor carries only the unbalanced current. This results in less voltage drop and watts loss for a three-wire branch-circuit as compared to similar loads connected to separate two-wire branch-circuits.

Figures 17-4 and 17-5 illustrate the benefits of a multiwire branch-circuit relative to watts loss and voltage drop. The example, for simplicity, is a purely resistive circuit, and loads are exactly equal.

The distance from the load to the source is 50 feet. The conductor size is No. 14 AWG solid uncoated copper. The resistance of each 50-foot length of No. 14 AWG is 0.154 ohm, as calculated from the data in *Table 8, Chapter 9* of the *National Electrical Code.*®

EXAMPLE 1:   TWO 2-WIRE BRANCH-CIRCUITS

Watts loss in each current-carrying conductor is

$$\text{Watts} = I^2R = 10 \times 10 \times 0.154 = 15.4 \text{ watts}$$

Watts loss in all four current-carrying conductors is

$$15.4 \times 4 = 61.6 \text{ watts}$$

Voltage drop in each current-carrying conductor is

$$E_d = IR$$
$$= 10 \times 0.154$$
$$= 1.54 \text{ volts}$$

E_d for both current-carrying conductors in each circuit:

$$2 \times 1.54 = 3.08 \text{ volts}$$

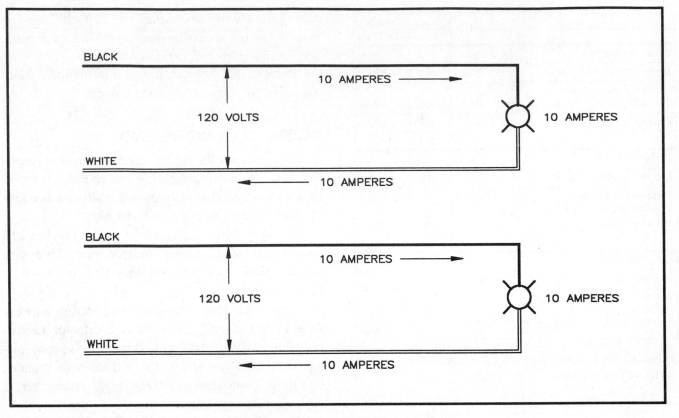

Fig. 17-4 Two 2-wire branch-circuits.

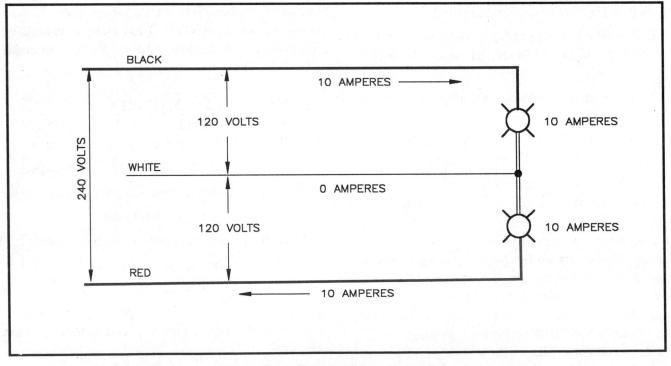

Fig. 17-5 A 3-wire branch-circuit.

Voltage available at load:

$$120 - 3.08 = 116.92 \text{ volts}$$

EXAMPLE 2:   ONE 3-WIRE BRANCH-CIRCUIT

Watts loss in each current-carrying conductor is

$$\text{Watts} = I^2R = 10 \times 10 \times 0.154 = 15.4 \text{ watts}$$

Watts loss in both current-carrying conductors is

$$15.4 \times 2 = 30.8 \text{ watts}$$

Voltage drop in each current-carrying conductor is

$$
\begin{aligned}
E_d &= IR \\
&= 10 \times 0.154 \\
&= 1.54 \text{ volts}
\end{aligned}
$$

E_d for both current-carrying conductors is:

$$2 \times 1.54 = 3.08 \text{ volts}$$

Voltage available at loads:

$$240 - 3.08 = 236.92 \text{ volts}$$

Voltage available at each load:

$$\frac{236.92}{2} = 118.46 \text{ volts}$$

It is readily apparent that a multiwire circuit can greatly reduce watts loss and voltage drop. This is the same reasoning that is used for the three-wire service-entrance conductors and the three-wire feeder to Panel B in this residence.

In the recreation room, a three-wire cable carrying Circuits B9 and B11 feeds from Panel B to the receptacle closest to the panel. This three-wire cable continues to each receptacle wall box along the same wall. At the third receptacle along the wall, the multiwire circuit splits, where the wet-bar Circuit B9 continues on to feed into the receptacle outlet to the left of the wet bar.

The white conductor of the three-wire cable is common to both the receptacle circuit and the wet-bar circuits. The black conductor feeds the receptacles. The red conductor is spliced straight through the boxes and feeds the wet bar area.

Care must be used when connecting a three-wire circuit to the panel. The black and red conductors *must* be connected to the opposite phases in the panel to prevent heavy overloading of the neutral grounded (white) conductor.

The neutral (grounded) conductor of the three-wire cable carries the unbalanced current. This current is the difference between the current in the black wire and the current in the red wire. For example, if one load is 12 amperes and the other load is 10 amperes, the neutral current is the difference between these loads, 2 amperes, figure 17-6.

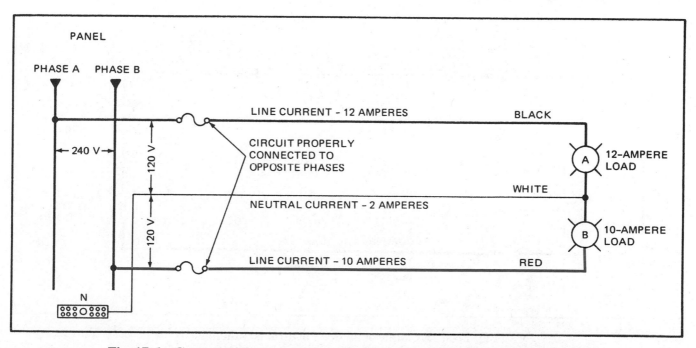

Fig. 17-6   Correct wiring connections for three-wire (multiwire) branch-circuit.

If the black and red conductors of the three-wire cable are connected to the same phase in Panel A, figure 17-7, the neutral conductor must carry the total current of both the red and black conductors rather than the unbalanced current. As a result, the neutral conductor will be overloaded. All single-phase, 120/240-volt panels are clearly marked to help prevent an error in phase wiring. The electrician must check all panels for the proper wiring diagrams.

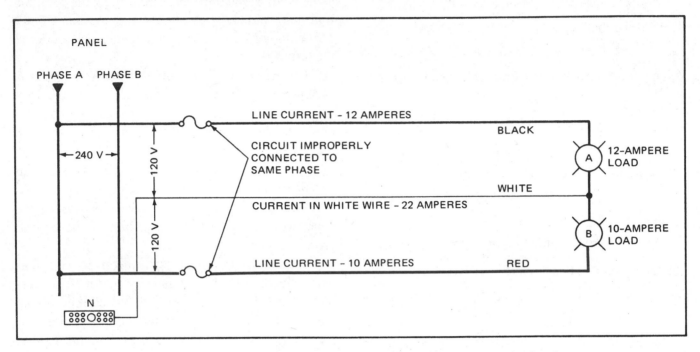

Fig. 17-7   Improperly connected three-wire (multiwire) branch-circuit.

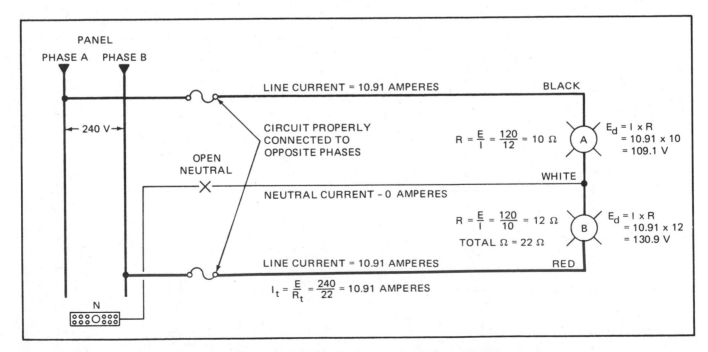

Fig. 17-8   Example of an open neutral conductor.

Figure 17-7 shows how an improperly connected three-wire branch-circuit results in an overloaded neutral conductor. If an open neutral occurs on a three-wire branch-circuit, some of the electrical appliances in operation may experience voltages higher than the rated voltage at the instant the neutral opens.

For example, figure 17-8 shows that for an open neutral condition, the voltage across load A decreases and the load across load B increases. If the load on each circuit changes, the voltage on each circuit also changes. According to Ohm's law, the voltage drop across any device in a series circuit is directly proportional to the resistance of that device. In other words, if load B has twice the resistance of load A, then load B will be subjected to twice the voltage of load A for an open neutral condition. To insure the proper connection, care must be used when splicing the conductors.

An example of what can occur should the neutral of a three-wire multiwire branch-circuit open is shown in figure 17-9. Trace the flow of current from phase A through the television set, then through the toaster, then back to phase B, thus completing the circuit. The following simple calculations show why the television set (or stereo or home computer) can be expected to burn up.

$$R_t = 8.45 + 80 = 88.45 \text{ ohms}$$

$$I = \frac{E}{R} = \frac{240}{88.45} = 2.71 \text{ amperes}$$

Voltage appearing across the toaster:

$$IR = 2.71 \times 8.45 = 22.9 \text{ volts}$$

Voltage appearing across the television:

$$IR = 2.71 \times 80 = 216.8 \text{ volts}$$

This example illustrates the problems that can arise with an open neutral on a three-wire, 120/240-volt multiwire branch-circuit.

The same problem can arise when the neutral of the utility company's incoming service-entrance conductors (underground or overhead) opens. The problem is minimized because the neutral of the service is solidly grounded to the metal water piping system within the building. However, there are cases

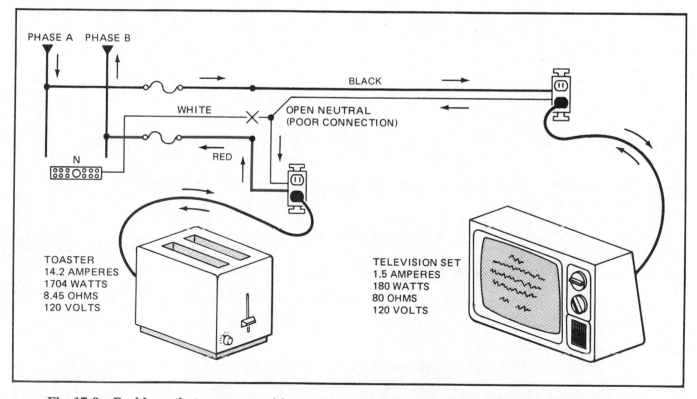

Fig. 17-9   Problems that can occur with an open neutral on a three-wire (multiwire) branch-circuit.

on record where poor service grounding has resulted in serious and expensive damage to appliances within the home because of an open neutral in the incoming service-entrance conductors. For example, poor service grounding results when relying on a driven iron pipe (that rusts in a short period of time) as the only means of obtaining the service equipment ground.

The integrity of using ground rods only is highly questionable, whether the rod is an iron pipe, a galvanized pipe, or a copper-clad rod. The Code prohibits the use of ground rods as the only means of obtaining a good service-entrance equipment ground.

See unit 28 for details on the grounding and bonding of service-entrance equipment.

In summary, three-wire branch-circuits are used occasionally in residential installations. The two hot conductors of three-wire cable are connected to opposite phases. One white grounded (neutral) conductor is common to both "hot" conductors.

The two receptacle outlets above the wet bar in the recreation room are GFCI protected, as indicated on the plans. ▶ This is required by *Section 210-8(a)(6)*, which states that receptacles serving countertop surfaces that are within 6 feet (1.83 m) of the outer edge of a wet bar sink must be GFCI protected. ◀

A single-pole switch to the right of the wet bar controls the four recessed fixtures above it. See unit 7 for details pertaining to recessed fixtures.

An exhaust fan of a type similar to the one installed in the laundry is installed in the ceiling of the recreation room to exhaust stale, stagnant, and smoky air from the room.

The basic switch, receptacle, and fixture connections are repetitive of most wiring situations discussed in previous units.

The wall boxes for the receptacles are selected based upon the measurements of the furring strips on the walls. Two-by-two furring strips will require the use of 4-inch square boxes with suitable raised covers. If the walls are furred with 2×4s, then possibly sectional device boxes could be used. Select a box that can contain the number of conductors, devices, and clamps to "meet Code," as discussed many times in this text. After all, the installation must be safe.

Probably the biggest factor in deciding what wiring method (cable or electrical metallic tubing) will be used to wire the recreation room is "what size and type of wall furring will be used?" See unit 4 for the in-depth discussion on the mechanical protection required for nonmetallic sheathed cables where the cables will be less than 1¼ inch from the edge of the framing members.

REVIEW

Note: Refer to the Code or plans where necessary.

1. What is the total current draw when all six fluorescent fixtures are turned on?

   _____

   _____

   _____

2. The junction box that will be installed above the dropped ceiling near the fluorescent fixtures closest to the stairway will have _____ (number of) No. 14 AWG conductors.

3. Why is it important that the "hot" conductor in a three-wire branch-circuit be properly connected to opposite phases in a panel? _____

_____

4. In the diagram, Load A is rated at 10 amperes, 120 volts. Load B is rated at 5 amperes, 120 volts.

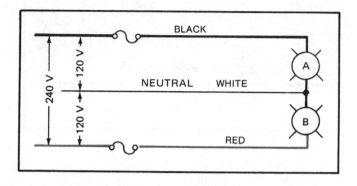

a. When connected to the three-wire branch-circuit as indicated, how much current will flow in the neutral? _____

b. If the neutral should open, to what voltage would each load be subjected, assuming both loads were operating at the time the neutral opened? Show all calculations.

5. Calculate the watts loss and voltage drop in each conductor in the following circuit. Show calculations.

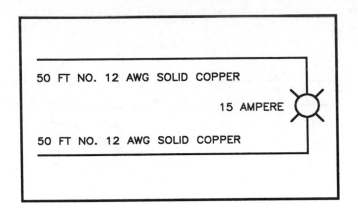

```
50 FT NO. 12 AWG SOLID COPPER

                                    15 AMPERE

50 FT NO. 12 AWG SOLID COPPER
```

6. Unless specifically designed, all recessed incandescent fixtures must be provided with factory-installed _____

7. If the fluorescent fixtures in the recreation room were to be mounted on the ceiling, what sort of marking would you look for on the label of the fixture? The ceiling is low-density cellulose fiberboard. _____

_____

8. What is the current draw of the recessed fixtures above the bar? _____

_____

_____

9. a. VA load for Circuit B9 is _____ volt-amperes.

   b. VA load for Circuit B11 is _____ volt-amperes.

   c. VA load for Circuit B12 is _____ volt-amperes.

10. Calculate the total current draw for Circuits B9, B11, and B12.

11. Complete the wiring diagram for the recreation room. Follow the suggested cable layout. Use colored pencils or pens to identify the various colors of the conductor insulation.

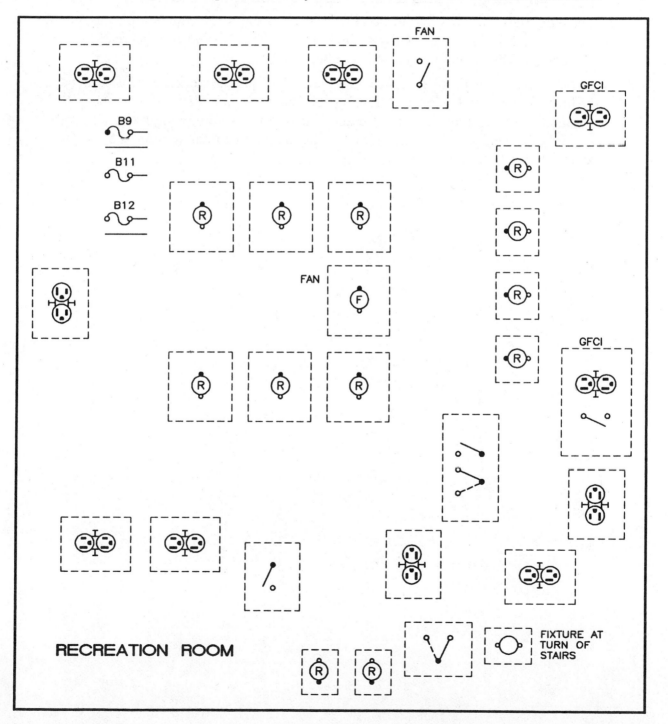

12. a. May a fluorescent fixture that is marked "Recessed Fluorescent Fixture" be installed in a suspended ceiling? _____

b. May a fluorescent fixture that is marked "Suspended Ceiling Fluorescent Fixture" be installed in a recessed cavity of a ceiling? _____

c. What part of the Code covers the installation of recessed fixtures? _____

_____

_____

d. What part of the Code covers the installation of suspended ceiling fixtures?

_____

_____

13. The *NEC®* requires GFCI protection for receptacles that serve countertop areas within 6 feet (1.83 m) of the edge of a wet bar sink, as in the case of the two recreation room receptacles. This requirement is found in Section _____ .

14. Lay-in fluorescent fixtures that rest on the metal framing members of a suspended ceiling cannot just lay on the framing members. *Section 410-16(c)* of the *NEC®* requires that (1) framing members used to support recessed lighting fixtures must be securely fastened to each other and to the building structure at appropriate intervals, and (2) the fixtures must _____

_____ .

UNIT 18

Lighting Branch-Circuit, Receptacle Circuits for Workshop

OBJECTIVES

After studying this unit, the student will be able to

- understand the meaning, use, and installation of multioutlet assemblies.
- understand where GFCI protection is required in basements.
- understand the Code requirements for conduit installation.
- select the proper outlet boxes to use for surface mounting.
- make conduit fill calculations based upon the number of conductors in the conduit.
- make use of derating and correction factors for determining conductor current-carrying capacity.

The workshop area is supplied by more than one circuit:

A13 Separate circuit for freezer

A17 Lighting

A18 Plug-in strip (multioutlet assembly)

A20 Two receptacles on window wall

The wiring method in the workshop is electrical metallic tubing (EMT). Figure 18-1 shows the conduit layout for the workshop.

A single-pole switch at the entry controls all five ceiling porcelain lampholders. However, note on the plans that three of these lampholders have pull-chains, which would allow the homeowner to turn these lampholders on and off as needed, figure 18-2. They would not have to be on all of the time, which would save energy.

In addition to supplying the lighting, Circuit A17 also feeds the smoke detectors, chime transformers, and ceiling exhaust fan. The current draw

of the smoke detectors and chime transformer is extremely small. Refer to Table 18-1, Workshop Outlet Count and Estimated Load for Lighting and Table 18-2, Outlet Count and Estimated Load for Three 20-Ampere Circuits.

Smoke detectors are discussed in detail in unit 26. Smoke detectors are *not* to be connected to GFCI-protected circuits.

DESCRIPTION	QUANTITY	WATTS	VOLT-AMPERES
Ceiling lights @ 100 W each	5	500	500
Fluorescent fixtures Two 40-W lamps each	2	160	200
Chime transformer	1	8	10
Exhaust fan	1	80	90
Smoke detectors @ ½ W each	4	2	2
TOTALS	13	750	802

Table 18-1 Workshop: outlet count and estimated load. (Lighting load) (Wire with conduit). Circuit A17.

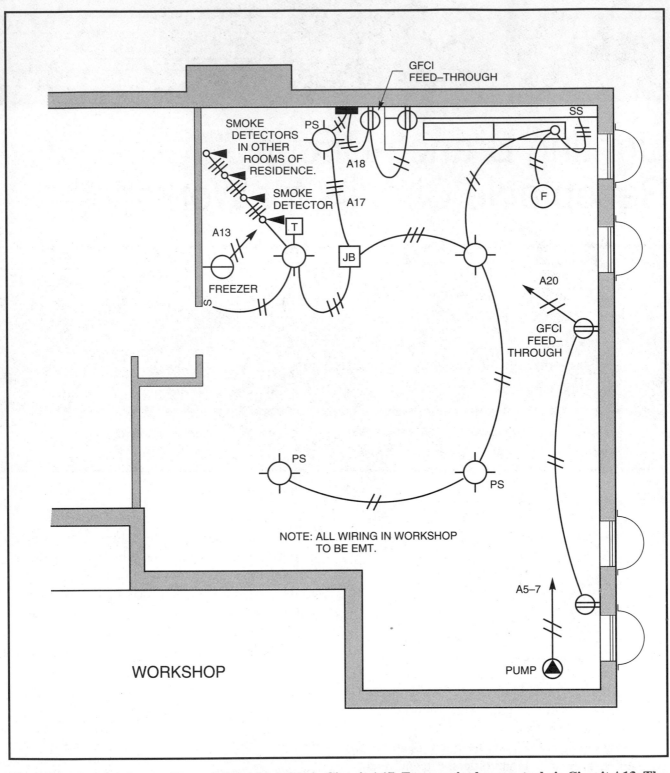

Fig. 18-1   Conduit layout for workshop. Lighting is Circuit A17. Freezer single receptacle is Circuit A13. The two receptacles on the window wall are connected to Circuit A20. Plug-in strip receptacles and receptacle to the right of Main Panel are connected to Circuit A18. All receptacles are GFCI protected except the single receptacle outlet serving the freezer. A junction box mounted on the ceiling near the fluorescent fixtures above the workbench will provide a convenient place to make up the necessary electrical connections.

WORKBENCH LIGHTING

Two two-lamp, 40-watt fluorescent fixtures are mounted above the workbench to reduce shadows over the work area. The electrical plans show that these fixtures are controlled by a single-pole wall switch. A junction box is mounted immediately above or adjacent to the fluorescent fixture so that the connections can be made readily.

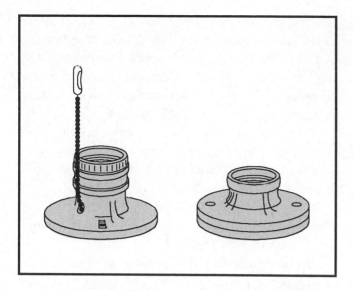

Fig. 18-2 This illustration shows porcelain lampholders of both the keyless and pull-chain types. These lampholders are also available with a receptacle outlet, which must be GFCI protected when installed in locations where GFCI protection is required by the Code.

DESCRIPTION	QUANTITY	WATTS	VOLT-AMPERES
20-A Circuit, Number A18			
Receptacles next to main panel	1	180	180
A six-receptacle plug-in multioutlet assembly at 1½ A per outlet (180 VA)	1	1080	1080
TOTALS	2	1260	1260
20-A Circuit, Number A20			
Receptacles on window wall @ 180 W each	2	360	360
TOTALS	2	360	360
20-A Circuit, Number A13			
Single receptacle for freezer (*not* GFCI)	1	—	696
TOTALS	1	1620	2316

Table 18-2 Workshop: outlet count and estimated load for three 20-A circuits A13, A18, A20.

RECEPTACLE OUTLETS

▶ *Sections 210-8(a)(4)* and *(5)* of the *National Electrical Code®* requires that all 125-volt, single-phase, 15- or 20-ampere receptacles installed in unfinished basements—such as workshop and storage areas that are not considered habitable, and crawl spaces that are at or below ground level—must be GFCI protected. ◀

Exceptions to this requirement are:

▶• Receptacles that are not readily accessible. ◀

▶• A single receptacle or a duplex receptacle for two appliances that occupy a dedicated space for each appliance, are cord- and plug-connected, and are not easily moved. ◀ Examples include freezers, refrigerators, water softeners, and central vacuum equipment.

• for the receptacle(s) installed for the laundry equipment as required by *Section 210-52(f)* and *Section 220-4(c).*

Most large freezers have an alarm or pilot light to indicate when the power to the freezer is lost.

See unit 6 for full discussion of ground-fault circuit interrupters.

Section 210-52(g) makes it mandatory to install at least one receptacle outlet in a basement. If the laundry area is in the basement, then a receptacle outlet must also be installed for the laundry equipment *(Section 210-52(f))* and be within 6 feet (1.83 m) of the intended location of the washer *(Section 210-50(c))* and must be a 20-ampere circuit.

The laundry receptacle outlet required in *Section 210-52(f)* is in addition to the basement receptacle required in *Section 210-52(g)*.

The circuit feeding the required laundry receptacle shall have no other outlets connected to it, *Section 220-4(c).*

When a single receptacle is installed on an individual branch-circuit, the receptacle rating must not be less than the rating of the branch-circuit, *Section 210-21(b)(1).*

CABLE INSTALLATION IN BASEMENTS

Unit 4 covers the installation of nonmetallic-sheathed cable (Romex) and armored cable (BX).

Some additional rules apply when nonmetallic-sheathed cable and armored cable are run exposed because of possible physical damage to the cables.

Armored cable, flexible metal conduit, or flexible cord (Type S, SJ, or equivalent) can be used to connect the fixture to the junction box. The fluorescent fixture must be grounded. Thus, any flexible cord used must contain a third conductor for grounding only.

In unfinished basements, nonmetallic-sheathed cables made up of conductors smaller than two No. 6 or three No. 8 must be run through holes bored in joists, or on running boards, *Section 336-12*. In unfinished basements, exposed runs of armored cable must closely follow the building surface or running boards to which it is fastened, *Section 333-11*. ▶ *Exception No. 2* to this requirement is for armored cable that is run on the undersides of joists in basements, attics, or crawl spaces, where supported on each joist and so located as not to be subject to physical damage. This exception is not permitted when using nonmetallic-sheathed cable. ◀ The local inspection authority usually interprets the meaning of "subject to physical damage."

Review units 4 and 15 for additional discussion regarding protection of cables installed in exposed areas of basements and attics.

CONDUIT INSTALLATION IN BASEMENTS

Where metal raceways are used in house wiring, in most cases the raceways will be electrical metallic tubing *(Article 348)*. In some instances, rigid metal conduit *(Article 347)* will be used for a mast service; refer to unit 28. The following text very briefly touches upon metal raceways, because most house wiring across the country is done with nonmetallic-sheathed cable that was covered in detail in unit 4.

Many local electrical codes do not permit cable wiring in unfinished basements other than where absolutely necessary. For example, a cable may be dropped into a basement from a switch at the head of the basement stairs. Transitions between cable and conduit also must be made. Usually this will be a junction box. The electrician must check the local codes before selecting conduit or cable for the installation, to prevent costly wiring errors.

Electrical utility companies can supply additional information on local regulations. The electrician should never assume that the *National Electrical Code®* is the recognized standard everywhere. Any city or state may pass electrical installation and licensing laws. In many cases these laws are more stringent than the *National Electrical Code®*

Types of electrical conduit are shown in figure 18-3.

Outlet boxes may be fastened to masonry walls with lead or plastic anchors, shields, concrete nails, or power-actuated studs.

Conduit may be fastened to masonry or wood surfaces using straps, such as the one-hole strap illustrated in figure 18-4.

The conduits on the workshop walls and on the ceiling are exposed to view. The electrician must install the exposed wiring in a neat and skillful manner while complying with the following practices:

- The conduit runs must be straight.

- Bends and offsets must be true.

- Vertical runs down the surfaces of the walls must be plumb.

- To assure that the conductor insulation will not be injured, be sure to ream the cut ends of rigid conduit and electrical metallic tubing to remove rough edges and burrs.

All conduits and boxes on the workshop ceiling are fastened to the underside of the wood joists, figure 18-5. Raceways shall be installed as a complete system between pull points before pulling in the wires, figure 18-5A.

If the electrician is able to get to the residence before the basement concrete floor is poured, he might wish to install conduit runs in or under the concrete floor from the main panel to the freezer outlet, the window wall receptacles, and to the water pump disconnect switch location. This can be a cost-saving (labor and material) benefit.

Metal raceways and metal cables (BX) are acceptable equipment grounding conductors,

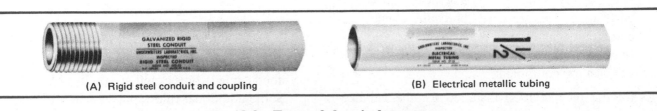

(A) Rigid steel conduit and coupling   (B) Electrical metallic tubing

Fig. 18-3 Types of electrical raceways.

Fig. 18-4   A one-hole conduit strap.

Section 250-91(b). The runs would have to be continuous between other effectively grounded boxes, panels, and/or equipment. When a nonmetallic wiring method (NM, NMC, UF, etc.) is installed where short sections of metal raceway are provided for mechanical protection—or for appearance—these short sections of raceway, if within reach of a grounded surface, may need to be grounded per *Section 250-56*.

Outlet Boxes for Use in Exposed Installations

Figure 18-6 illustrates outlet boxes and plaster rings that are used for concealed wiring, such as for the recreation room and first floor wiring in the residence. For surface wiring such as the workshop, various types and sizes of outlet boxes may be used. For example, a handy box, figure 18-7, could be used for the freezer receptacle. The remaining workshop boxes could be 4-inch square outlet boxes with raised covers, figure 18-8. The box size and type is determined by the maximum number of conductors contained in the box, *Article 370*.

▶ *Section 410-56(i)* prohibits fastening a receptacle outlet to a raised cover by one screw only unless the device, assembly, or box cover is listed and identified for the purpose. ◀ The listing and identifying for the purpose is a decision that is made by a recognized testing laboratory, such as Underwriters Laboratories. In the past, this was a common practice that resulted in loose receptacles when used over a period of time. Many short-circuits have been reported because of this. For the typical raised cover, as illustrated in figure 18-8, the duplex receptacle would be fastened to the cover by two small 6/32 screws and nuts.

Switch and outlet boxes are available with knockouts of the following sizes: ½ inch, ¾ inch, and 1 inch. The knockout size selected depends upon the size of the conduits entering the box. The conduit size is determined by the maximum percentage of fill of the total cross-sectional area of the conduit by the conductors pulled into the conduit.

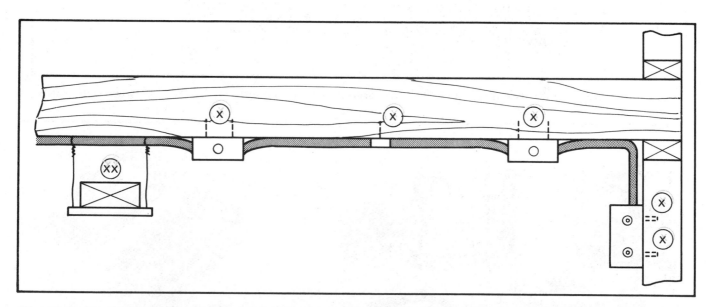

Fig. 18-5   Raceways, boxes, fittings, and cabinets must be securely fastened in place, points X, *Section 300-11(a)*. It is not permissible to hang other raceways, cables, or nonelectrical equipment from an electrical raceway, point XX, *Section 300-11(b)*. Refer to figure 24-20 for exception that allows low-voltage thermostat cables to be fastened to the conduit that feeds a furnace or other similar equipment. See unit 17, Recreation Room, for discussion regarding the use of suspended ceiling support wire for supporting equipment.

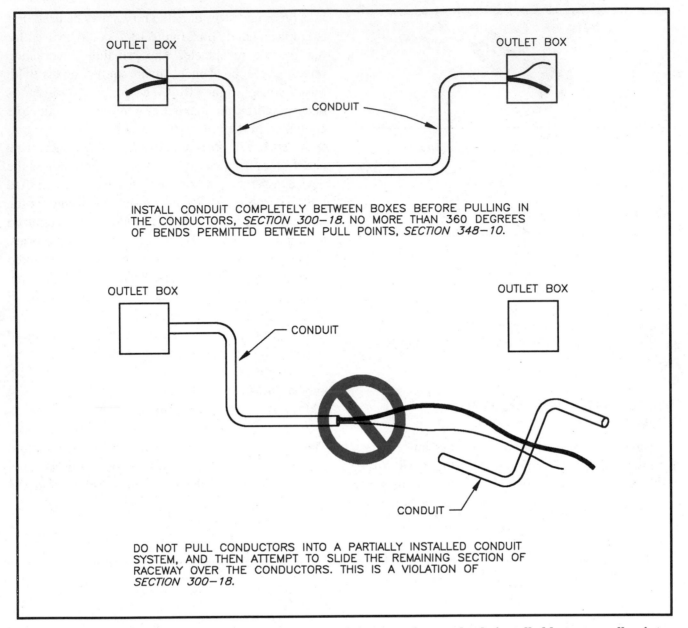

INSTALL CONDUIT COMPLETELY BETWEEN BOXES BEFORE PULLING IN
THE CONDUCTORS, *SECTION 300–18.* NO MORE THAN 360 DEGREES
OF BENDS PERMITTED BETWEEN PULL POINTS, *SECTION 348–10.*

DO NOT PULL CONDUCTORS INTO A PARTIALLY INSTALLED CONDUIT
SYSTEM, AND THEN ATTEMPT TO SLIDE THE REMAINING SECTION OF
RACEWAY OVER THE CONDUCTORS. THIS IS A VIOLATION OF
SECTION 300–18.

Fig. 18-5A   Don't pull wires in raceway until the conduit system is completely installed between pull points.

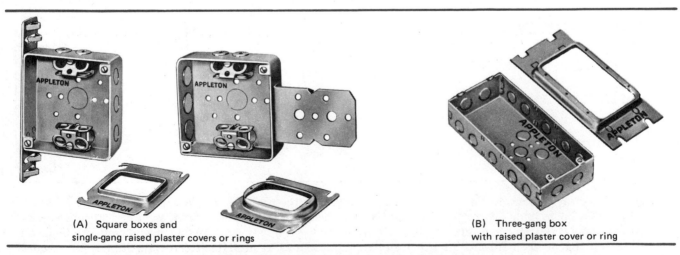

(A) Square boxes and
single-gang raised plaster covers or rings

(B)   Three-gang box
with raised plaster cover or ring

Fig. 18-6   Typical outlet boxes and raised covers.

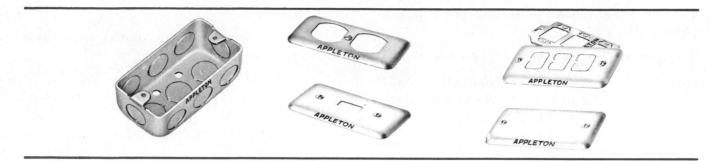

Fig. 18-7   Handy box and covers.

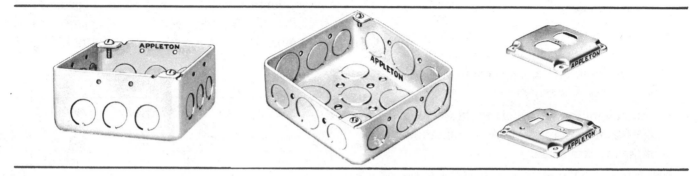

Fig. 18-8   Four-inch square boxes and raised covers.

Supporting equipment and/or raceways to the joists, walls, and so on, is accomplished with proper nails, screws, anchors, and straps. Be careful when trying to support other equipment from raceways as this is not always permitted. Refer to figure 18-6.

Conduit Fill Calculations

▶ An electrician must be able to select the proper size raceways based upon the number, size, and type of conductors to be installed in a particular type of raceway. To do this, the electrician must become familiar with the use of certain tables found in *Chapter 9* and *Appendix C* of the *National Electrical Code.*®

Percent Fill

The *National Electrical Code*® has established the following "percent fill" for conduit and tubing.

- one conductor . . . . 53% fill ⊙
- two conductors . . . . 31% fill ⊛
- three or more conductors . . . . . 40% fill ⊛

These percentages are taken from *Chapter 9, Table 1* of the *NEC.*®

Conduit Fill When All Conductors Are the Same Size and Type

- Determine how many conductors of one size and type are to be installed in the raceway.

- Refer to *Tables C2* through *C13A* in *Appendix C* of the *NEC*® for the type of raceway, and for the specific size and type of conductors.

- Select the proper size raceway directly from the table.

These tables are not difficult to use. Here are a few examples of conductor fill for *electric metallic tubing.* Look these up in the tables to confirm the accuracy of the examples.

- Not over five No. 10 THHN conductors may be pulled into a ½-inch EMT.

- Not over eight No. 12 THW conductors may be pulled into a ¾-inch EMT.

- Not over three No. 8 THHN conductors may be pulled into a ½-inch EMT.

- Not over six No. 8 THHW conductors may be pulled into a 1-inch EMT.

Conductor Fill When Conductors Are Different Sizes and/or Types

- Refer to *Chapter 9, Table 5* of the *NEC®*.

- Find the size and type of conductors used to determine the approximate cross-sectional area in square inches of the conductors.

- Refer to *Chapter 9, Table 8* for the square inch area of bare conductors.

- Total the square inch areas of all conductors.

- Refer to *Chapter 9, Table 4* for the type and size of raceway used.

- Determine the minimum size raceway based on the total square inch area of the conductors and the allowable square inch area fill for the particular type of raceway used.

- The sum must not exceed the allowable percentage fill of the cross-sectional square inch area of the raceway used.

For illustrative purposes, only the data for a few types of wires from *Table 5* are shown in figure 18-8A. The dimensions and areas for some electrical metallic tubing from *Table 4* are shown in figure 18-8B. You will need to refer to *Chapter 9* and *Appendix C* of the *NEC®* for the many other sizes and types of conductors and raceways.

EXAMPLE: If three No. 6 THW conductors and two No. 8 THW conductors are to be installed in one EMT, determine the proper size EMT.

SOLUTION:

1. Find the square inch area of the conductors in *Chapter 9, Table 5*.

Three No. 6 THW conductors	0.0726 in²
	0.0726 in²
	0.0726 in²
Two No. 8 THW conductors	0.0556 in²
	0.0556 in²
Total Area	0.3290 in²

2. In *Chapter 9, Table 4, NEC®* look at the fifth column (over 2 wires, 40% fill) for an EMT that has an area of 0.03290 square inch or more. Note that a ¾-inch EMT holds up to

TYPE	SIZE	APPROX. DIAM. INCHES	APPROX. AREA SQ. IN.
THHN	14	0.111	0.0097
THHN	12	0.130	0.0133
THHN	10	0.164	0.0211
THHN	8	0.216	0.0366
THHN	6	0.254	0.0507
THHN	3	0.352	0.0973
THW	6	0.304	0.0727
THW	8	0.266	0.0556

Fig. 18-8A   This sampling of typical conductors and their dimensions is taken from *Table 5, Chapter 9, NEC®*

Electrical Metallic Tubing					
Trade Size in Inches	Internal Diameter in Inches	Total Area 100% in Sq. In.	2 Wires 31% in Sq. In.	Over 2 Wires 40% in Sq. In.	1 Wire 53% in Sq. In.
1/2	0.622	0.304	0.094	0.122	0.161
3/4	0.824	0.533	0.165	0.213	0.283
1	1.049	0.864	0.268	0.346	0.458
1 1/4	1.380	1.496	0.464	0.598	0.793
1 1/2	1.610	2.036	0.631	0.814	1.079
2	2.067	3.356	1.040	1.342	1.778
2 1/2	2.731	5.858	1.816	2.343	3.105
3	3.356	8.846	2.742	3.538	4.688
3 1/2	3.834	11.545	3.579	4.618	6.119
4	4.334	14.753	4.573	5.901	7.819

Fig. 18-8B   Dimensional and percent fill data for electrical metallic tubing. This is one of the 12 charts found in *Chapter 9, Table 4, NEC®* for various types of raceways.

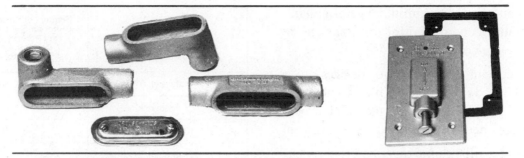

Fig. 18-9   A variety of common conduit bodies and covers.

0.213 square inch of conductor fill, and a 1-inch EMT holds up to 0.346 square inch of conductor fill. Therefore, 1-inch EMT is the minimum size for the combination of conductors in the example.

Many electrical ordinances require that a separate equipment grounding conductor be installed in all raceways, whether rigid or flexible, whether metallic or nonmetallic. This requirement resulted after inspectors found loose or missing set screws on set-screw type connectors and couplings. Corrosion of fittings also contributes to poor grounding as required by *Section 250-51* of the *NEC®* A separate equipment grounding conductor does take up space, and therefore must be counted when considering conduit fill.

EXAMPLE: A feeder to an electric furnace is fed with two No. 3 THHN conductors plus one equipment grounding conductor. This feeder is installed in EMT. The feeder overcurrent protection is 100 amperes. What is the minimum size EMT to be installed?

SOLUTION: According to *Table 250-95*, the minimum size equipment grounding conductor for a 100-ampere feeder is No. 8 AWG copper. *Table 5, Chapter 9, NEC®*:

Two No. 3 THHN	0.0973 in²
	0.0973 in²
One No. 8 THHN	0.0366 in²
Total Area	0.2312 in²

From *Table 4, Chapter 9, NEC®*:

40% fill for ¾-inch EMT = 0.213 in²
40% fill for 1-inch EMT = 0.346 in²

Therefore, install 1-inch electrical metallic tubing. Remember, *Section 300-18* of the Code states that a conduit system must be completely installed before pulling in the conductors. Refer to figure 18-6A. ◄

Conduit Bodies *(Section 370-16(c))*

Conduit bodies are used with conduit installations to provide an easy means to turn corners, terminate conduits, and mount switches and receptacles. They are also used to provide access to conductors, to provide space for splicing (when permitted), and to provide a means for pulling conductors, figure 18-9.

A conduit body:

• must have a cross-sectional area not less than twice the cross-sectional area of the largest conduit to which it is attached, figure 18-10(A). This is a requirement only when the conduit body contains No. 6 AWG conductors or smaller.

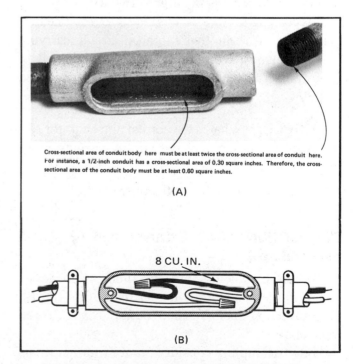

Cross-sectional area of conduit body here must be at least twice the cross-sectional area of conduit here. For instance, a 1/2-inch conduit has a cross-sectional area of 0.30 square inches. Therefore, the cross-sectional area of the conduit body must be at least 0.60 square inches.

(A)

8 CU. IN.

(B)

Fig. 18-10   Requirements for conduit bodies.

- may contain the maximum number of conductors the same as permitted for the size raceway attached to the conduit body.

- must not contain splices, taps, or devices unless the conduit body is marked with its cubic-inch capacity, figure 18-10(B), and the conduit body must be supported "in a rigid and secure manner."

- the conductor fill volume must be properly calculated according to *Table 370-16(b)*, which lists the free space that must be provided for each conductor within the conduit body. If the conduit body is to contain splices, taps, or devices, it must be supported rigidly and securely.

DERATING FACTORS

For more than three current-carrying wires in conduit or cable, refer to figure 18-11.

If a raceway is to contain more than three current-carrying conductors, the student must check *Note 8* to *Table 310-16*. *Note 8* states that for the conductors listed in the tables, the maximum allowable ampacity must be reduced when more than three current-carrying conductors are installed in a raceway. The reduction factors based on the number of conductors are noted in figure 18-11. When the conduit nipple does not exceed 24 inches, the reduction factor does not apply, figure 18-12.

According to *Note 10* to *Table 310-16*, a neutral conductor of a multiwire branch-circuit carrying only unbalanced currents shall not be counted in determining current-carrying capacities as provided for in *Note 8*. See figure 18-13.

EXAMPLE: What is the correct ampacity for No. 6 THHN when there are six current-carrying wires in the conduit?

$$75 \times 0.80 = 60 \text{ amperes}$$

CORRECTION FACTORS (Due to high temperatures)

When conductors are installed in locations where the temperature is higher than 30°C (86°F), the correction factors noted below *Table 310-16* must apply.

For example, consider that four current-carrying No. 3 THWN copper conductors are to be

Number of Current-Carrying Conductors	Percent of Values in Tables. If High Ambient Temperatures are Present, also apply Corrections Factors found Tables Below.
4 through 6	80
7 through 9	70
10 through 20	50
21 through 30	45
31 through 40	40
41 and above	35

Fig. 18-11   Derating factors necessary to determine conductor ampacity when there are more than 3 current-carrying conductors in a raceway or cable. See *Note 8* to *Table 310-16*.

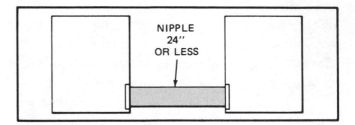

Fig. 18-12   The derating in figure 18-11 for more than three current-carrying conductors in a raceway is *not* required when the conduit nipple does not exceed 24 inches. See *Note 8*, *Exception No. 3* to *Table 310-16*, and *Note A3* to *Chapter 9* of the *NEC.*® Bundled cables also included in this requirement, figure 4-12.

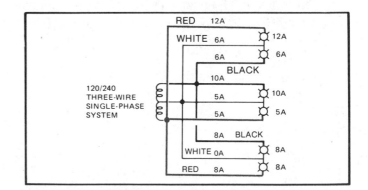

Fig. 18-13   A neutral conductor carrying the unbalanced currents from the other conductors need *not* be counted when determining the ampacity of the conductor, according to *Note 10* to *Table 310-16*. The Code would recognize this example as *two current-carrying conductors*. Thus, in the three multiwire circuits shown, the actual total is nine conductors, but only six current-carrying conductors are considered when the derating factor of figure 18-11 is applied.

installed in one raceway in an ambient temperature of approximately 90°F. "Ambient" temperature means "surrounding" temperature. This temperature is common in boiler rooms and is well exceeded in attics where the sun shines directly onto the roof.

The maximum ampacity for these conductors is determined as follows:

- The allowable ampacity of No. 3 THWN copper conductors from *Table 310-16* = 100 amperes.

- Apply the correction factor for 90°F found below *Table 310-16*. The first column shows Celsius; the last column shows Fahrenheit.

$$100 \times 0.94 = 94 \text{ amperes}$$

- Apply the derating factor for four current-carrying wires in one raceway.

$$94 \times 0.80 = 75.2 \text{ amperes}$$

- Therefore, 75.2 amperes is the new corrected ampacity for the example given.

Some authorities having jurisdiction in extremely hot areas of the country, such as the southwestern desert climates, will require that service-entrance conductors running up the side of a building or above a roof exposed to direct sunlight be "corrected" per the correction factors below *Table 310-16*. Check this out before proceeding with an installation where this situation might be encountered.

MAXIMUM SIZE OVERCURRENT PROTECTION

An important footnote below *Table 310-16* must be observed when installing No. 14, No. 12, and No. 10 AWG conductors. Notice in the table many obelisks (†) (also called daggers) referring to these sizes of conductors. As can be seen in figure 18-14, the maximum permitted size of the overcurrent protection is less than the ampacities listed in the table.

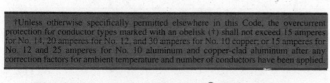

†Unless otherwise specifically permitted elsewhere in this Code, the overcurrent protection for conductor types marked with an obelisk (†) shall not exceed 15 amperes for No. 14, 20 amperes for No. 12, and 30 amperes for No. 10 copper; or 15 amperes for No. 12 and 25 amperes for No. 10 aluminum and copper-clad aluminum after any correction factors for ambient temperature and number of conductors have been applied.

Fig. 18-14   *Footnote* to NEC® *Table 310-16*.

OTHER CODE LIMITATIONS

We must also be aware of these other conductor loading limitations:

- *Section 210-19(a)*—Branch-circuit conductors shall have an ampacity not less than the maximum load to be served. This entire section discusses ranges, cooking appliances, and tap conductors.

- *Section 210-22*—This is the fundamental rule that the total load shall not exceed the branch-circuit rating, and shall not exceed the maximum computed load.

- *Section 210-22(a)*—This section refers to motor loads, *Article 430*. Size conductors for motors at 125% of motor full-load rating. For combination loads, take 125% of the largest motor's full-load current draw, plus the total current draw of all other loads.

▶• *Section 210-22(c)*—For branch-circuits in commercial or industrial installations where there are noncontinuous loads and continuous loads such as office, store, and plant lighting, the computed load must not be less than the noncontinuous load *plus* 125% of the continuous load. The minimum branch-circuit conductor ampacity, without applying any derating or correction factors, must not be less than the noncontinuous load *plus* 125% of the continuous load. ◀

▶• *Section 220-3(a)*—The wording in this section is much the same as in *Section 210-22(c)*. It states that the computed load shall not be less than the noncontinuous load *plus* 125% of the continuous load. The minimum branch-circuit conductor ampacity, without applying any derating or correction factors, must not be less than the noncontinuous load *plus* 125% of the continuous load. ◀

▶• *Section 220-10(b)*—For feeders in commercial or industrial installations where there are noncontinuous loads and continuous loads such as office, store, and plant lighting, the computed load must not be less than the noncontinuous load *plus* 125% of the continuous load. The minimum branch-circuit conductor ampacity, without applying any derating or correction factors, must not be less than the noncontinuous load *plus* 125% of the continuous load. ◀

- *Section 384-16(c)*—The total loading of any overcurrent device in a panel must not exceed 80% of its rating when the loads will continue for 3 hours or more.

- See *Articles 422 (Appliances)* and *424 (Fixed Electric Space Heating)* for specific loading limitations.

Note: In the above Code references, the term "continuous" is defined as "where the maximum current is expected to continue for 3 hours or more." Loads in homes are generally not considered to be continuous. Many loads in commercial and industrial installations can be considered to be continuous and must be computed as such.

EXAMPLES OF DERATING

To illustrate what has been learned so far about derating and maximum overcurrent protection, suppose that 8 THW copper current-carrying conductors are to be installed in the same conduit. The connected load will be 18 amperes. The size wire to use may be determined as follows:

- No. 14 THW copper wire
  a. The maximum overcurrent protection from figure 18-14 (*footnote* to *Table 310-16*) is 15 amperes.
  b. The ampacity of No. 14 THW copper wire from *Table 310-16* is 20 amperes.
  c. Apply the derating factor (8 current-carrying wires in the raceway) from figure 18-11 (*Note 8* to *Table 310-16*).
    $$20 \times 0.70 = 14 \text{ amperes}$$
    (derated ampacity)
    Thus, No. 14 THW wire is *too small* to supply the 18-ampere load.

- No. 12 THW copper wire
  a. The maximum overcurrent protection from figure 18-14 (*footnote* to *Table 310-16*) is 20 amperes.
  b. The ampacity of No. 12 THW copper wire from *Table 310-16* is 25 amperes.
  c. Apply the derating factor (8 current-carrying wires in the raceway) from figure 18-11 (*Note 8* to *Table 310-16*).
    $$25 \times 0.70 = 17.5 \text{ amperes}$$
    (derated ampacity)
    Thus, No. 12 THW is *too small* to supply the 18-ampere load.

- No. 10 THW copper wire
  a. The maximum overcurrent protection from figure 18-14 (*footnote* to *Table 310-16*) is 30 amperes.
  b. The ampacity of No. 10 THW copper wire from *Table 310-16* is 35 amperes.
  c. Apply the derating factor (8 current-carrying wires in the raceway) from figure 18-11 (*Note 8* to *Table 310-16*).
    $$35 \times 0.70 = 24.5 \text{ amperes}$$
    (derated ampacity)
    Thus, No. 10 THW copper wire is adequate to supply the 18-ampere load.

MULTIOUTLET ASSEMBLY

A multioutlet assembly has been installed above the workbench. For safety reasons, as well as from an adequate wiring viewpoint, it is recommended that any workshop receptacles be connected to a separate circuit. In the event of a malfunction in any power tool commonly used in the home workshop (saw, planer, lathe, or drill), only receptacle Circuit A18 is affected. The lighting in the workshop is not affected by a power outage on the receptacle circuit.

The GFCI feedthrough receptacle to the left of the workbench provides GFCI protection for the entire plug-in strip. See *Section 210-8(a)(5)*. See figure 18-15.

The installation of multioutlet assemblies must conform to the requirements of *Article 353* of the Code.

Load Considerations for Multioutlet Assemblies

Table 220-3(b) gives the general lighting loads in volt-amperes per square foot (0.093 m²) for types of occupancies. The *footnote* to this table mentions that *all* general-use receptacle outlets in one, two, and multifamily dwellings are to be considered as outlets for general lighting and as such have been included in the volt-amperes per square foot calculations. Therefore, no additional load need be added. This is further confirmed by the last sentence of *Exception No. 1* to *Section 220-3(c)*.

Because the multioutlet assembly, figure 18-16, has been provided in the workshop for the purpose of plugging-in portable tools, a separate Circuit A18 is provided. This circuit has been shown in the service-entrance calculations at 1500 volt-amperes, similar to the load requirements for small appliance circuits.

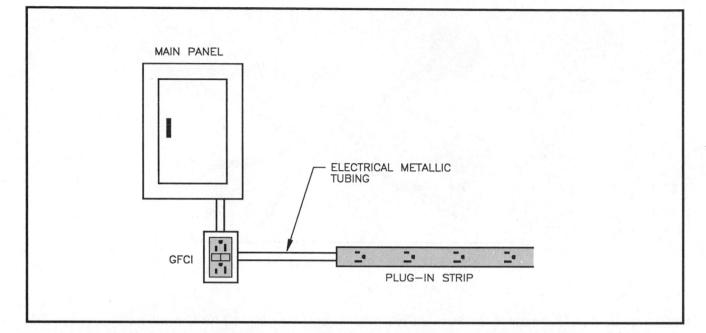

Fig. 18-15   Detail of how the workbench receptacle plug-in strip is connected by feeding through a GFCI feedthrough receptacle located below the Main Panel. This is Circuit A18.

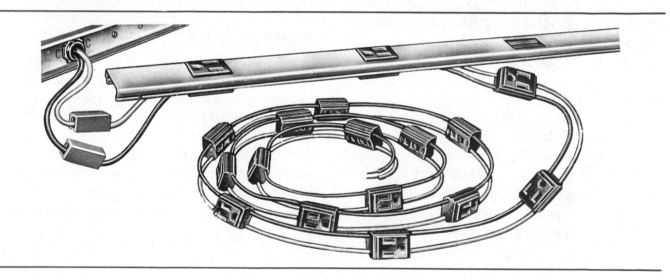

Fig. 18-16   A typical multioutlet assembly. Multioutlet assemblies are permitted to be run through (not within) a dry partition when no outlets are located within the partition, and all exposed portions of the assembly can have the cover removed.

Many types of receptacles are available for various sizes of multioutlet assemblies. For example, duplex, single-circuit grounding, split-circuit grounding, and duplex split-circuit receptacles can be used with multioutlet assemblies.

Wiring of the Multioutlet Assembly

The electrician first attaches the metal base of the multioutlet assembly to the wall or baseboard. The receptacles may be wired on the job. If the receptacles are factory prewired, they are snapped into the base, figure 18-17. Covers are then cut to the

Fig. 18-17   Prewired multioutlet assembly.

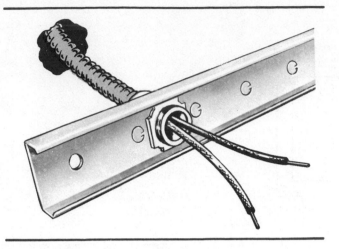

Fig. 18-18   Feed for multioutlet assembly.

lengths required to fill the spaces between the receptacles. Manufacturers of these assemblies can supply precut covers to be used when evenly spaced receptacles are installed. These receptacles are usually installed 12 inches (305 mm) or 18 inches (457 mm) apart.

Receptacles can also be installed by snapping them into the cover and then attaching the cover to the metal base.

In general, multioutlet installations are not difficult because of the large number of fittings and accessories available. Connectors, couplings, ground clamps, blank end fittings, elbows for turning corners, and end entrance fittings can all be used to simplify the installation.

Prewired multioutlet assemblies normally are wired with No. 12 AWG conductors and have either 15- or 20-ampere receptacles. The feed for a multioutlet assembly is shown in figure 18-18.

EMPTY CONDUITS

Although not truly part of the workshop wiring, Note 8 to the first-floor electrical plans indicates that two empty ½-inch EMT raceways are to be installed, running from the workshop to the attic. This is unique, but certainly welcomed when at some later date, additional wiring (telephone, television, etc.) might be required. The empty raceways will provide the necessary route from the basement to the attic, thus eliminating the need to fish through the wall partitions.

REVIEW

1. a. What circuit supplies the workshop lighting? _____

   b. What circuit supplies the plug-in strip over the workbench? _____

   c. What circuit supplies the freezer receptacle? _____

2. Approximately how much ½-inch EMT is used for connecting Circuits A13, A17, A18, and A20? _____

3. a. How is EMT fastened to masonry? _____

   _____

   b. The cut ends of rigid and electrical metallic tubing must be _____
      to prevent the insulation on conductors from being damaged.

4. What type of lighting fixtures are installed on the workshop ceiling? _____

   _____

   _____

5. A check of the plans indicates that two empty ½-inch thinwall conduits are installed
   between the basement and the attic. In your opinion do you feel this is a good idea?
   Briefly explain your thoughts. _____

   _____

   _____

   _____

   _____

   _____

   _____

6. To what circuit are the smoke detectors connected? _____

7. Is it a Code requirement to connect smoke detectors to a GFCI-protected circuit?

   _____

8. When a freezer is plugged into a receptacle, be sure that the receptacle is protected by
   a GFCI. Is this statement true or false? Explain. _____

   _____

   _____

   _____

   _____

   _____

   _____

9. A sump pump is plugged into a single receptacle outlet. This receptacle shall

   a. be GFCI protected

   b. not be GFCI protected

   Circle one.

10. When a 120-volt receptacle outlet is provided for the laundry equipment in a base-
    ment, what is the *minimum* number of additional receptacle outlets required? What
    Code reference? _____

    _____

11. Derating factors for conductor ampacities must always be taken into consideration
    when (a) _____ are contained in one raceway.
    Correction factors are applied when (b) _____ .

12. What is the current in the neutral conductor in this circuit at point ⊗?

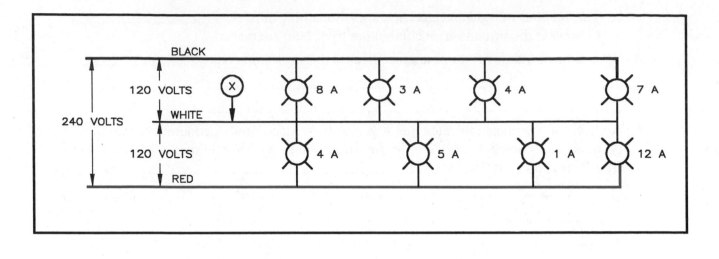

13. When a single receptacle is installed on an individual branch-circuit, the receptacle must have a rating _____ than the rating of the branch-circuit.

14. Calculate the total current draw of Circuit A17 if all lighting fixtures and the exhaust fan were turned on.

15. A conduit body must have a cross-sectional area not less than (two) (three) (four) times the cross-sectional area of the (largest) (smallest) conduit to which it is attached. Circle correct answers.

16. A conduit body may contain splices if marked with

    a. the UL logo.

    b. its cubic-inch area.

    c. the size of conduit entries.

    Circle one.

17. What size box would you use where Circuit A17 enters the junction box on the ceiling? _____

    _____

18. List the proper size of electrical metallic tubing for the following. Assume that it is new work and that THHN insulated conductors are used. See *Chapter 9, Table C2, NEC*®

    a. 3 No. 14 AWG _____   f. 4 No. 12 AWG _____

    b. 4 No. 14 AWG _____   g. 5 No. 12 AWG _____

    c. 5 No. 14 AWG _____   h. 6 No. 12 AWG _____

    d. 6 No. 14 AWG _____   i. 6 No. 10 AWG _____

    e. 3 No. 12 AWG _____   j. 4 No. 8 AWG _____

19. According to the Code, what size electrical metallic tubing is required for each of the following combinations of conductors? Assume that these are new installations. Show all calculations.

a. Three No. 14 AWG THHN, four No. 12 AWG THHN

b. Two No. 12 AWG TW, three No. 8 AWG TW

c. Three No. 3 THW, one No. 4 Bare

d. Three No. 3/0 THHN, two No. 8 THHN

20. a. When more than three current-carrying conductors are installed in one raceway, their allowable ampacities must be reduced according to _____ to *Table 310-16*.

b. Four No. 10 THW current-carrying conductors are installed in one raceway. The room temperature will not exceed 86°F. The connected load is nonmotor, noncontinuous. (1) What is the ampacity of these conductors before derating? (2) What is the maximum overcurrent protection permitted to protect these conductors? (3) What is the ampacity of these conductors after derating? _____

_____

_____

_____

_____

c. Why is it not necessary to count equipment grounding conductors and/or all of the "travelers" when installing 3-way and 4-way switches for the purposes of derating?

_____

_____

_____

21. What is the minimum number of receptacles required by the Code for basements?

_____

22. For laundry equipment in basements, what sort of electrical circuitry is required by the Code? _____

_____

_____

_____

23. How many receptacles are required to complete the multioutlet assembly above the workbench? _____

24. A Type NM cable carries a branch-circuit to the outside of a building, where it terminates in a metal weatherproof junction box. From this box, a short length of EMT is installed upward to another metal weatherproof junction box on which is attached a cluster of outdoor flood lights. The first box is located approximately 2 feet above the ground. To "meet Code," do the metal boxes and metal raceway have to be grounded? Yes _____ No _____ *Sections* _____ .

25. When raised covers are used for receptacles, does the Code permit fastening the receptacle to the cover with the one center 6/32 screw? Yes _____ No _____ *Section* _____ .

26. The following is a layout of the lighting circuits for the workshop. Using the conduit layout in figure 18-1, make a complete wiring diagram. Use colored pencils or pens to indicate conductors.

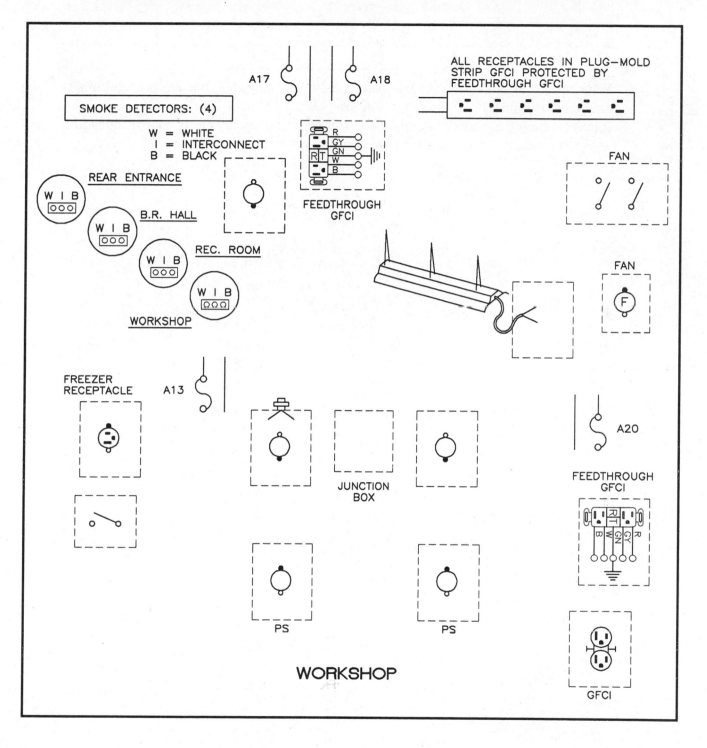

27. *Section 210-8(5)* of the *National Electrical Code*® requires that GFCI protection be provided for all receptacles in unfinished basements. There are some exceptions. What are these exceptions?

_____

_____

_____

28. Where loads are likely to be ON continuously, the *computed* load for branch-circuits and feeders must be figured at (100%) (125%) (150%) of the continuous load, (plus) (minus) the noncontinuous load. This requirement generally (does) (does not) apply to residential wiring. Circle the correct answers.

UNIT 19

Special-Purpose Outlets— Water Pump, Water Heater

OBJECTIVES

After studying this unit, the student will be able to

- list the requirements for deep-well jet pump and submersible pump installations.
- calculate (given the rating of the motor) the conductor size, conduit size, and overcurrent protection required for pump circuits.
- list some of the electrical circuits used for connecting and metering electric water heaters.
- describe the basic operation of the water heater.
- understand the meaning of the term "Time-of-Use."
- list the functions of the tank, the heating elements, and the thermostats of the water heater.
- understand "recovery" time for electric water heaters.
- make the proper electrical and grounding connections for water heaters, water pumps, and metal well casings.
- understand the hazards of possible scalding.
- discuss heat pump water heaters.

WATER PUMP CIRCUIT ▲B

All dwellings need a good water supply. City dwellings generally are connected to the city water system. In rural areas, where there is no public water supply, each dwelling has its own water supply system in the form of a well. The residence plans show that a deep-well jet pump is used to pump water from the well to the various plumbing outlets. The outlet for this jet pump is shown by the symbol ▲B. An electric motor drives the jet pump. A circuit must be installed for this system.

JET PUMP OPERATION

Figure 19-1 shows the major parts of a typical jet pump. The pump impeller wheel ① forces water down a drive pipe ② at a high velocity and pressure to a point just above the water level in the well casing. Some well casings may be driven to a depth of more than 100 feet (30.5 m) before striking water. Just above the water level, the drive pipe curves sharply upward and enters a larger vertical suction pipe ③. The drive pipe terminates in a small nozzle or jet ④ in the suction pipe. The water emerges from

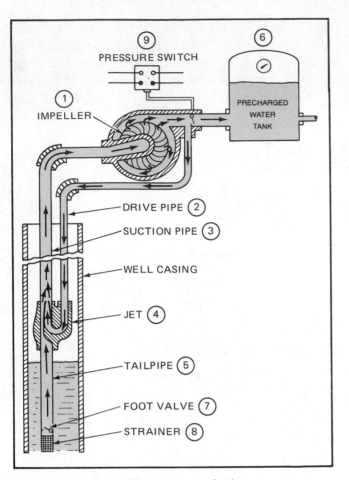

Fig. 19-1   Components of a jet pump.

the jet with great force and flows upward through the suction pipe.

Water rises in the suction pipe, drawn up through the tailpipe ⑤ by the action of the jet. The water rises to the pump inlet and passes through the impeller wheel of the pump. Part of the water is forced down through the drive pipe again. The remaining water passes through a check valve and enters the storage tank ⑥.

The tailpipe is submerged in the well water to a depth of about 10 feet (3.05 m). A foot valve ⑦ and strainer ⑧ are provided at the end of the tailpipe. The foot valve prevents water in the pumping equipment from draining back into the well when the pump is not operating.

When the pump is operating, the lower part of the storage tank fills with water and air is trapped in the upper part. As the water rises in the tank, the air is compressed. When the air pressure is 40 pounds per square inch, a pressure switch ⑨ disconnects the pump motor. The pressure switch is adjusted so that as the water is used and the air pressure

falls to 20 pounds per square inch, the pump restarts and fills the tank again. One pound (lb) = 0.4536 kilograms (kg). One kilogram (kg) = 2.2046 pounds (lb).

The Pump Motor

A jet pump for residential use may be driven by a 1-horsepower (hp) motor at 3400 r/min. A dual-voltage motor is used so that it can be connected to 120 or 240 volts. For a 1-hp motor, the higher voltage is preferred because the current at this voltage is half the value used by the lower voltage. It is recommended that a single-phase, capacitor-start motor be used for a residential jet pump. Such a motor is designed to produce a starting torque adequate for pumps with comparatively low starting current. In pumping operations, a high starting torque helps to overcome the back pressure within the tank and the weight of the water being lifted from the well.

The Pump Motor Circuit

Figure 19-2 is a diagram of the electrical circuit that operates the jet pump motor and pressure switch. This circuit originates at Main Service Panel A, Circuit A5-7. *Table 430-148* indicates that a 1-hp single-phase motor has a rating of 8 amperes at 240 volts. Using this information, the conductor

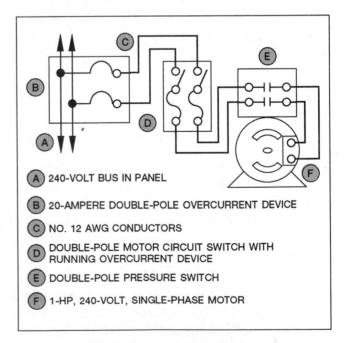

Ⓐ 240-VOLT BUS IN PANEL

Ⓑ 20-AMPERE DOUBLE-POLE OVERCURRENT DEVICE

Ⓒ NO. 12 AWG CONDUCTORS

Ⓓ DOUBLE-POLE MOTOR CIRCUIT SWITCH WITH RUNNING OVERCURRENT DEVICE

Ⓔ DOUBLE-POLE PRESSURE SWITCH

Ⓕ 1-HP, 240-VOLT, SINGLE-PHASE MOTOR

Fig. 19-2   The pump circuit.

size, conduit size, branch-circuit overcurrent protection, and running overcurrent protection can be determined for the motor.

Conductor Size *(Section 430-22)*

$$1.25 \times 8 = 10 \text{ amperes}$$

Table 310-16 of the Code indicates that the minimum conductor size permissible is No. 14 AWG. However, the **Schedule of Special Purpose Outlets** found in this text, which is part of the **Specifications** for the residence, indicates that No. 12 AWG conductors are to be used for the pump circuit. Installing larger size conductors can minimize voltage drop, and might be suitable at a later date should the need arise to install a larger capacity pump.

Conduit Size

Two No. 12 Type THHN conductors require ½-inch conduit, *Table C2, Chapter 9, NEC®*

Branch-Circuit Overcurrent Protection
(Section 430-52, Table 430-152)

The jet pump motor is a single-phase motor without code letters. Therefore, *Table 430-152*, line 1, can be used.

The motor's branch-circuit overcurrent device may be sized as follows *(Section 430-52* and *Table 430-152)*:

- NON-TIME-DELAY FUSES
  $8 \times 3 = 24$ amperes (use 25-ampere fuses)
  Maximum size: 400% (4.0)

- DUAL-ELEMENT, TIME-DELAY FUSES
  $8 \times 1.75 = 14$ amperes
  (use 15-ampere fuses)
  Maximum size: 225% (2.25)

- INSTANT TRIP BREAKERS
  $8 \times 8 = 64$ amperes setting
  Maximum setting: 1300% (13)

- INVERSE TIME BREAKERS
  $8 \times 2.5 = 20$-ampere setting
  Maximum setting: 400% for F.L.A. of
                  100 or less.
                  300% for F.L.A. of
                  over 100.

The motor branch-circuit overcurrent device must be capable of allowing the motor to start, *Section 430-52(a)*.

▶ Where the values for branch-circuit protective devices (fuses or breakers) as determined by *Table 430-152* do not correspond to the standard sizes, ratings, or settings as listed in *Section 240-6*, the next higher size, rating, or setting is permitted, *Section 430-52(c), Exception No. 1.*

If, after applying the values found in *Table 430-152* and after selecting the next higher standard size, rating, or setting of the branch-circuit overcurrent device, the motor still will not start, *Section 430-52(c), Exception No. 2* permits using an even larger size, rating, or setting. The maximum values (percentages) are shown in *Section 430-52(c)*. ◀

Whether or not a nuisance opening of a fuse, or tripping of a breaker will occur, can easily be predicted by referring to time/current curves for the intended fuse or breaker. This topic is covered briefly in unit 28.

Motor Running Overload Protection
(Sections 430-31 through *430-44)*

Overload protection is needed to keep the motor from dangerous overheating due to overloads or failure of the motor to start. *Sections 430-31* through *430-44* cover virtually every type and size of motor, starting characteristics, and duty (continuous or non-continuous) in use today. The overload protection for a motor might be thermal overloads, electronic sensing overload devices, built-in (inherent) thermal protection, or time-delay fuses.

Although there are exceptions, most of the common installations require motor overload protection not to exceed 125% of the motor's full load current draw. For the jet pump motor, this would be:

$$1.25 \times 8 = 10 \text{ amperes.}$$

Instructions furnished with a particular motor or motor operated appliance will probably indicate the maximum size overload protection and maximum size branch-circuit protection. These values must be followed.

Wiring for the Pump Motor Circuit

The water pump circuit is run in ½-inch EMT from Main Panel A to a disconnect switch and controller that are mounted on the wall next to the pump. The EMT may be run across the ceiling of the workshop or be embedded in the concrete floor.

Thermal overload devices (sometimes called *heaters*) are installed in the controller to provide running overcurrent protection. Two No. 12 conductors are connected to the double-pole, 20-ampere branch-circuit overcurrent protective device in Panel A to form Circuit A5-7. Dual-element, time-delay fuses may be installed in the disconnect switch. These fuses will serve the dual function of providing both branch-circuit overcurrent protection for the pump motor and back-up protection for the thermal overloads protecting the motor. These fuses may be rated at not more than 125% of the full-load running current of the motor.

A short length of flexible metal conduit is connected between the load side of the thermal overload switch and the pressure switch on the pump. This pressure switch is usually connected to the motor at the factory. In figure 19-2, a double-pole pressure switch is shown at Ⓔ. When the contacts of the pressure switch are open, all conductors feeding the motor are disconnected.

SUBMERSIBLE PUMP

A submersible pump, figure 19-3, consists of a centrifugal pump driven by an electric motor. The pump and the motor are contained in one housing, submersed below the permanent water level within the well casing. When running, the pump raises the water upward through the piping to the water tank. Proper pressure is maintained in the system by a pressure switch. The disconnect switch, pressure and limit switches, and controller are installed in a logical and convenient location near the water tank.

Most water pressure tanks have a precharged air chamber in an elastic bag (bladder) that separates the air from the water. This assures that air is not absorbed by the water. Water pressure tanks compress air to maintain delivery pressure in a usable range without cycling the pump at too fast a rate. When in direct contact with the water, air is gradually absorbed into the water, and the cycling rate of the pump will become too rapid for good pump life, unless the air is periodically replaced. With the elastic bag, the initial air charge is always maintained, so recharging is unnecessary.

Most pumps commonly called "two-wire pumps" contain all required starting and protection components and connect directly to the pressure switch—

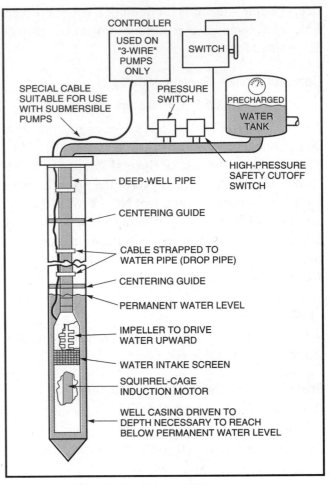

Fig. 19-3   Submersible pump.

with no aboveground controller. Generally, there are no moving electrical parts within the pump, such as the centrifugal starting switch found in a typical single-phase, split-phase induction motor.

Other pumps, referred to as "three-wire pumps," require an aboveground controller that contains any required components not in the motor, such as a starting relay, overload protection, starting and running capacitors, lightning arrester, and terminals for making the necessary electrical connections.

Rating data are on the pumps themselves, on the "three-wire" controllers, and on a separate nameplate furnished with each "two-wire" pump for aboveground identification.

The calculations for sizing the conductors, and the requirements for the disconnect switch, the motors' branch-circuit fuses, and the grounding connections are the same as those for the jet pump motor. The nameplate data and instructions furnished with the pump must be followed. The manufacturer's recommendations for conductor size, type, and length are extremely important if the pump is to operate properly.

See unit 16 for information about underground wiring.

Submersible pumps are covered by UL Standard 778.

Submersible Pump Cable

Power to the motor is supplied by a cable especially designed for use with submersible pumps. This cable is marked "submersible pump cable,"

and it is generally supplied with the pump and the pump's various other components. The cable is cut to the proper length to reach between the pump and its controller, figure 19-3, or between the pump and the well-cap, figure 19-4. When needed, the cable may be spliced according to the manufacturer's specifications.

Submersible water pump cable is "tag-marked" for use within the well casing for wiring deep-well water pumps where the cable is not subject to repetitive handling caused by frequent servicing of the pump units.

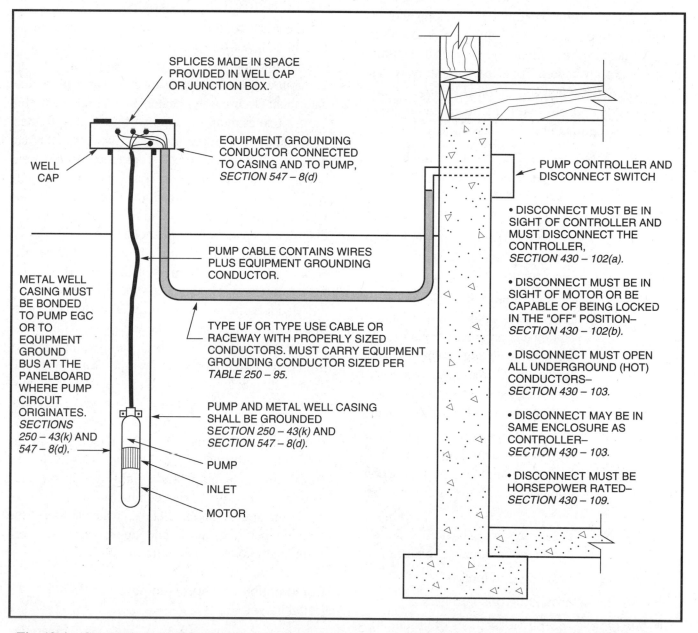

Fig. 19-4   Grounding requirements for submersible pump. *Section 250-43(k)* and *Section 547-8(d)* require that submersible pumps be grounded. In addition, *Section 547-8(d)* requires that the pump and well casing be bonded together. See text for details.

Submersible pump cable is *not* designed for direct burial in the ground unless it is marked Type USE or Type UF. Should it be required to run the pump's circuit underground for any distance, it is necessary to install underground feeder cable (UF), Type USE, or a raceway with suitable conductors, and then make up the necessary splices in a "listed" weatherproof junction box. See figure 19-4.

GROUNDING

Figure 19-4 illustrates how to provide proper grounding of a submersible water pump and the well casing. Proper grounding of the well casing and submersible pump motor will minimize or eliminate stray voltage problems that could occur if the pump motor is not grounded. The subject of stray voltage is discussed in the text *Agricultural Wiring*.

Key code sections that relate to grounding water pumping equipment are:

- *250-33* —metal enclosures for circuit conductors shall be grounded, other than short sections of metal enclosures that provide support or protection.

- *250-43(k)*—motor-operated water pumps, including submersible pumps, must be grounded

- *250-43(l)* —ground submersible pumps. Bond metal well casing to the pump's equipment grounding conductor, or run a separate equipment grounding conductor all the way back to the ground bus at the same panel that supplies the submersible pump circuit.

- *250-57(b)*— equipment fastened in place or connected by permanent wiring must have the equipment grounding conductor run with the circuit conductors.

- *250-91(c)*—the earth is not to be used as the sole equipment grounding conductor

- *430-142* —grounding of motors

- *430-144* —grounding of controllers

- *430-145* —acceptable methods of grounding

Because so much nonmetallic (PVC) water piping is used today, some electrical codes require

that a grounding electrode conductor be installed between the neutral bar of the main service disconnecting means and the metal well casing. This would satisfy the requirements of *Sections 250-80* (bonding), *250-81* (grounding electrode system), *250-83* (made and other electrodes), and *250-84* (resistance of made electrodes). This then is the service ground and is sized per *Table 250-94* of the Code. At the metal casing, the grounding electrode conductor is attached to a lug termination, or by means of exothermic (Cadweld) welding.

WATER HEATER CIRCUIT ⓐC

Electrical contractors many times are also in the plumbing, heating, and appliance business. They need to know more than how to do the electrical hook-up. The following text discusses electrical as well as other data about electric water heaters that will prove useful.

All homes require a supply of hot water. To meet this need, one or more automatic water heaters are generally installed as close as practical to the areas having the greatest need for hot water. Water piping carries the heated water from the water heater to the various plumbing fixtures and to appliances such as dishwashers and clothes washers.

For safety reasons, in addition to the regular temperature control thermostat that can be adjusted by the homeowner, the *National Electrical Code®* and Underwriters Laboratories Standards require that electric water heaters be equipped with a high temperature cut-off. This high temperature control limits the maximum water temperature to 190°F. This is 20°F below the 210°F temperature that causes the "temperature/pressure" relief valve to open. The high temperature cutoff is factory preset, and should never be tampered with or modified in the field. In conformance to *Section 422-14* of the *NEC®* the high temperature limiting control must disconnect all ungrounded conductors.

In most residential electric water heaters, the high temperature cutoff and upper thermostat are combined into one device, figure 19-5.

Combination Pressure/Temperature Safety Relief Valves

Look at figure 19-5. A pressure/temperature relief valve is an important safety device installed into an opening in the water heater tank within 6

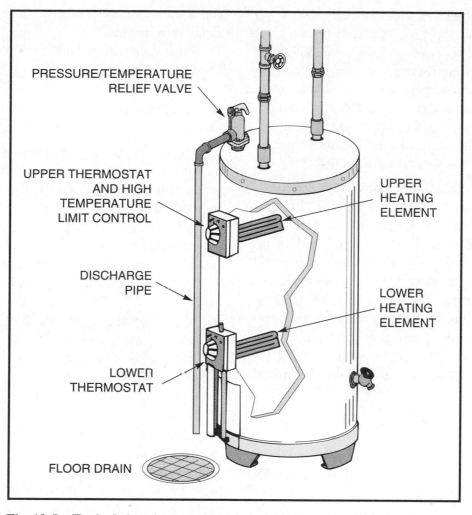

Fig. 19-5   Typical electric water heater showing location of heating elements and thermostats.

inches (152.4 mm) of the top. The opening is provided for and clearly marked by the manufacturer. Many water heater pressure/temperature relief valves are factory installed, although some models still require installation of the valve by the installing contractor. Proper discharge piping must be installed from the pressure/temperature relief valve downward to within 6 inches of the floor, preferably near a floor drain. Terminating the discharge pipe to within 6 inches of the floor is important to protect people from injury from scalding water or steam discharge should the relief valve activate for any reason. This also protects electrical equipment in and around the water heater from becoming soaked should the relief valve operate.

Pressure/temperature relief valves are installed for two reasons. Their pressure/temperature sensing features are interrelated.

• *Pressure:* Water expands when heated. Failure of the normal temperature thermostat and the high temperature limiting device to operate could result in pressures that exceed the rated working pressure of the water heater tank and the plumbing system components. The relief valve, properly installed and properly rated to local codes, will open to relieve the excess pressure in the tank to the predetermined setting marked on the relief valve. The rating is stamped on the valve's identification plate.

• *Temperature:* Should runaway water heating conditions occur within the water heater, temperatures far exceeding the atmospheric boiling point of water (212°C) can take place. Under high pressures, water will stay liquid at extremely high temperatures. For a typical municipal water supply pressure of 45 psi, water will not boil until it reaches approximately 290°F. At 290°F (143°C), opening a faucet, a broken pipe, or a ruptured water heater tank would allow the pressure to instantly drop to normal atmospheric pressure. The superheated water would immediately flash into steam, having about 1,600 times more volume than the liquid water occupied. This blast of steam can be catastrophic, and could result in severe burns or death.

Thus, the function of the temperature portion of the pressure/temperature relief valve is to open when a temperature of 210°F (98.7°C) occurs within the water heater tank. Properly rated and maintained, the pressure/temperature relief valve will allow the water to flow from the water heater through the discharge pipe at a rate faster than the heating process, thereby keeping the water temperature below the boiling point. Should the cold water supply be shut off, the relief valve will allow the steam to escape without undue build-up of temperature and pressure in the tank.

Sizes

Residential electric water heaters are typically available in sizes 6, 15, 20, 30, 40, 50 (52), 66, 75, 80 (82), 100, and 120 gallons.

Corrosion of Tank

To reduce corrosion of the steel tank, most water heater tanks are glass lined. Glass lining is a combination of silica and other minerals (rasorite, rutile, zircon, cobalt, nickel oxide, and fluxes). This special mix is heated, melted, cooled, crushed, then sprayed onto the interior steel surfaces and baked on at about 1,600°F. To further reduce corrosion, aluminum or magnesium anode rods are installed in special openings, or as part of the hot water outlet fitting, to allow for the flow of a protective current *from* the rod(s) *to* any exposed steel of the tank. This is commonly referred to as *Cathodic Protection*. Corrosion problems within the water heater tank are minimized as long as the aluminum or magnesium rod(s) remain in an active state. Anodes should be inspected at regular intervals to determine when replacement is necessary, usually when the rod(s) is reduced to about ⅓ of its original diameter, or when the rod's core wire is exposed. Further information can be obtained by contacting the manufacturer, and consulting the installation manual.

Some water heater tanks are fiberglass or plastic lined.

Heating Elements

The wattage ratings of electric water heaters can vary greatly, depending upon the size of the heater in gallons, the speed of recovery desired, local electric utility regulations, and codes. Typical wattage ratings are 1,500, 2,000, 2,500, 3,000, 3,800, 4,500, and 5,500 watts.

A resistance heating element might be dual rated. For example, the element might be marked 5,500 watts at 250 volts, and 3,800 watts at 208 volts.

Most residential-type water heaters are connected to 240 volts except the smaller 2-, 4-, and 6-gallon *point of use* sizes generally rated 1,500 watts at 120 volts. Commercial electric water heaters can be rated single-phase or three-phase, 208, 240, 277, or 480 volts.

Underwriters Laboratories requires that the power (wattage) input must not exceed 105% of the water heater's nameplate rating. All testing is done with a supply voltage equal to the heating element's *rated* voltage. Most heating elements may burn out prematurely if operated at voltages 5% higher than for which they are rated.

To help reduce the premature burnout of heating elements, some manufacturers will supply 250-volt heating elements, yet will mark the nameplate 240 volts with its corresponding wattage at 240 volts. This allows a safety factor if slightly higher than normal voltages are experienced.

Heating Element Construction

Heating elements generally contain a nickel-chrome (nichrome) resistance wire embedded in compacted powdered magnesium oxide to assure that no grounds occur between the wire and the sheath. The magnesium oxide is an excellent insulator of electricity, yet it effectively conducts the heat from the nichrome wire to the sheath.

The sheath can be made of tin-coated copper, stainless steel, or incolloy. The stainless steel and incolloy types can withstand higher operating

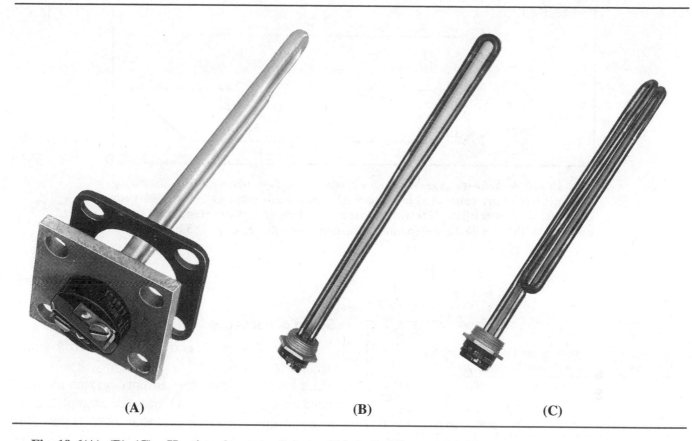

(A)   (B)   (C)

Fig. 19-6(A), (B), (C)   Heating elements. *Courtesy* Edwin L. Wiegand Division, Emerson Electric Company.

temperatures than the tin-coated copper elements. They are used in premium water heaters because of their resistance to lime scaling and dry firing burnout.

High watt density elements generally have a U-shaped tube, and give off a great amount of heat per square inch of surface. Figure 19-6(A) illustrates a flange-mounted high-density heating element. Figure 19-6(B) illustrates a high-density heating element that screws into a threaded opening in the tank. High-density elements are very susceptible to lime-scale burnout.

Low watt density elements, figure 19-6(C), generally have a double loop, like a U tube bent in half. They give off less heat per square inch, have a longer life, and are less noisy than high-density elements.

An option offered by at least one manufacturer of electric water heaters is a dual-wattage heating element. A dual-wattage heating element actually contains two heating elements in the one jacket. One element is rated 3800 watts and one element is rated 1700 watts. Three leads are brought out to the terminal block. The 3800-watt heating element is con-

nected initially, with the option of connecting the second 1700-watt heating element if the wiring to the water heater is capable of handling the higher total combined wattage of 5500 watts. To attain the higher wattage, a jumper (provided by the manufacturer) is connected between two terminals on the terminal block. See figure 19-6A.

Any electric heating element coated with lime scale may be noisy during the heating cycle. Limed up heating elements should be cleaned or replaced, following the manufacturer's instructions.

Do not energize the electrical supply to a water heater unless you are sure it is full of water. Most water heater heating elements are designed to operate only when submersed in water.

Residential electric water heaters are available with one or two heating elements. Figure 19-5 shows two heating elements. Single-element water heaters will have the heating element located near the bottom of the tank. If the water heater has two heating elements, the upper element will be located about two-thirds of the way up the tank.

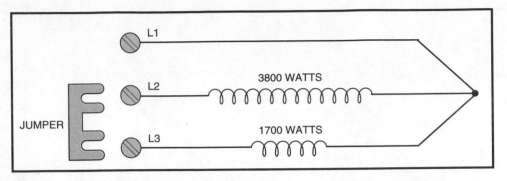

Fig. 19-6A A dual-wattage heating element. The two wires of the incoming 240-volt circuit are connected to L1 and L2. These leads connect to the 3800-watt element. To connect the 1700-watt element, the jumper is connected between L2 and L3. With both heating elements connected, the total wattage is 5500 watts.

Speed of Recovery

Speed of recovery is the time required to bring the water temperature to satisfactory levels in a given length of time. The current accepted industry standard is based on a 90°F rise. For example, referring to figure 19-7, we find that a 3,500-watt element can raise the water temperature 90°F of a little more than 16 gallons an hour. A 5,500-watt element can raise water temperature 90°F of about 25 gallons an hour. Speed of recovery is affected by the type and amount of insulation surrounding the tank, the supply voltage, and the temperature of the incoming cold water.

Figure 19-8 shows the approximate time it takes to run out of hot water using typical shower heads with various capacity electric water heaters. Regular residential-type shower heads deliver 5 to 6 gallons per minute. Water saver shower heads deliver 1 to 2 gallons per minute. This varies with water pressure. An easy way to determine a shower head's flow rate is to turn on the water, hold a large bucket of known capacity under it, and time how long it takes to fill the bucket.

Thermostats

Thermostats similar to the one illustrated in figure 19-9 turn the heating elements on and off, thereby maintaining the desired water temperature. In two-element water heaters, the top thermostat is an interlocking type sometimes referred to as a snap-over type. The bottom thermostat is a

Heating Element Wattage	Gallons per Hour Recovery for Indicated Temperature Rise					Equivalent NET BTU output
	60°F	70°F	80°F	90°F	100°F	
750	5.2	4.4	3.9	3.5	3.1	2,559
1,000	6.9	5.9	5.2	4.6	4.1	3,412
1,250	8.6	7.4	6.5	5.7	5.2	4,266
1,500	10.3	8.9	7.8	6.9	6.2	5,119
2,000	13.8	11.8	10.3	9.2	8.3	6,825
2,250	15.5	13.3	11.6	10.3	9.3	7,678
2,500	17.2	14.8	12.9	11.5	10.3	8,531
3,000	20.7	17.7	15.5	13.8	12.4	10,238
3,500	24.1	20.7	18.1	16.1	14.5	11,944
4,000	27.6	23.6	20.7	18.4	16.6	13,650
4,500	31.0	26.6	23.3	20.7	18.6	15,356
5,000	34.5	29.6	25.9	23.0	20.7	17,063
5,500	37.9	32.5	28.4	25.3	22.5	18,769
6,000	41.4	35.5	31.0	27.6	24.6	20,475

NOTES: If the incoming water temperature is 40°F (Northern states), use the 80° column to raise the water temperature to 120°F.

If the incoming water temperature is 60°F (Southern states), use the 60° column to raise the water temperature to 120°F.

For example, approximately how long would it initially take to raise the water temperature in a 42-gallon electric water heater to 120°F where the incoming water temperature is 40°F? The water heater has a 2,500-watt heating element.

Answer: 42 ÷ 12.9 = 3.26 hours (approximately 3 hours and 16 minutes)

Fig. 19-7   Electric water heater recovery rate per gallon of water for various wattages.

SHOWER HEAD(S) RATE IN GALLONS PER MINUTE	HOT WATER FLOW RATE IN GALLONS PER MINUTE BASED ON 120°F. STORED HOT WATER TO PROVIDE 105°F. MIXED WATER OUT OF SHOWER HEAD.	HOT WATER FLOW RATE IN GALLONS PER MINUTE BASED ON 120°F. STORED HOT WATER TO PROVIDE 105°F. MIXED WATER OUT OF SHOWER HEAD.	LENGTH OF CONTINUOUS SHOWER TIME — IN MINUTES — FOR TYPICAL SIZES OF RESIDENTIAL ELECTRIC WATER HEATERS. CHART BASED ON 70% DRAW EFFICIENCY. RECOVERY RATE IS NOT FIGURED IN BECAUSE IT IS NORMALLY NOT A FACTOR IN CONTINUOUS DRAW OF WATER SITUATIONS.											
			100 GALLON		80 GALLON		66 GALLON		50 GALLON		40 GALLON		30 GALLON	
	40°F. INCOMING WATER	60°F. INCOMING WATER	40°F. INCOMING WATER	60°F. INCOMING WATER	40°F. INCOMING WATER	60°F. INCOMING WATER	40°F. INCOMING WATER	60°F. INCOMING WATER	40°F. INCOMING WATER	60°F. INCOMING WATER	40°F. INCOMING WATER	60°F. INCOMING WATER	40°F. INCOMING WATER	60°F. INCOMING WATER
2.0	1.6	1.5	43	47	34	37	28	31	22	23	17	19	13	14
2.5	2.0	1.9	34	38	28	30	23	25	17	19	14	15	10	11
3.0	2.4	2.3	29	31	23	25	19	21	14	16	11	12	9	9
4.0	3.3	3.0	22	23	17	19	14	15	11	12	9	9	6	7
5.0	4.0	3.8	17	19	14	15	11	12	9	9	7	7	5	6
6.0	4.9	4.5	14	16	11	12	9	10	7	8	6	6	4	5
8.0	6.5	6.0	11	12	9	9	7	8	5	6	4	5	3	4
12.0	9.8	9.0	7	8	6	6	5	5	4	4	3	3	2	2

Example #1: A 40-gallon electric water heater has a 40°F incoming water supply. How many minutes can a 3 GPM shower head be used to obtain 105°F hot water?

Answer: 11 minutes

Example #2: A residence has an 80-gallon electric water heater. The incoming water supply is approximately 60°F. How many minutes can two 2 GPM shower heads be used at the same time to obtain 105°F hot water? Hint: Two 2 GPM shower heads are the same as one 4 GPM shower head.

Answer: 19 minutes

Fig. 19-8   This chart shows the approximate time in minutes it takes to run out of hot water when using various shower heads, different incoming water temperature, and different capacity water heaters.

Fig. 19-9   Typical electric water heater controls. The seven terminal control combines water temperature control plus high temperature limit control. This is the type commonly used to control the upper heating element as well as providing high temperature limiting feature. The two terminal control is the type used to control water temperature for the lower heating element. *Courtesy* THERM-O-DISC, Inc., subsidiary of Emerson Electric Company.

single-pole, ON-OFF type. Both thermostats have silver contacts, and have quick-make-quick break characteristics to prevent arcing. The wiring diagrams in figures 19-10 through 19-14 show that both heating elements cannot be energized at the same time.

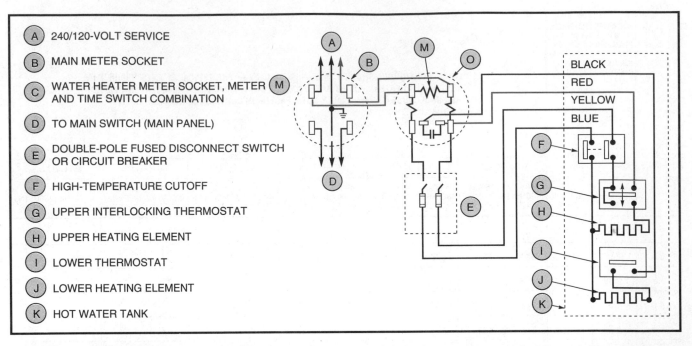

A 240/120-VOLT SERVICE

B MAIN METER SOCKET

C WATER HEATER METER SOCKET, METER AND TIME SWITCH COMBINATION

D TO MAIN SWITCH (MAIN PANEL)

E DOUBLE-POLE FUSED DISCONNECT SWITCH OR CIRCUIT BREAKER

F HIGH-TEMPERATURE CUTOFF

G UPPER INTERLOCKING THERMOSTAT

H UPPER HEATING ELEMENT

I LOWER THERMOSTAT

J LOWER HEATING ELEMENT

K HOT WATER TANK

BLACK
RED
YELLOW
BLUE

Fig. 19-10   Wiring for typical off-peak electric water heater circuit. A separate watt-hour meter/time clock controls the circuit to the water heater. The utility will turn the power off during certain peak periods of the day. This type of circuitry is called "time-of-day" program.

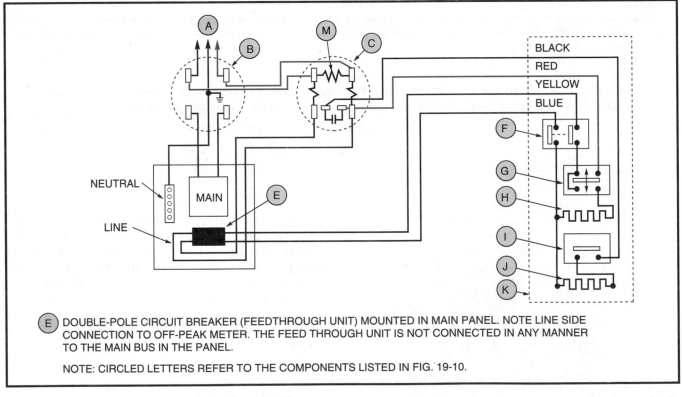

NEUTRAL

MAIN

LINE

BLACK
RED
YELLOW
BLUE

E DOUBLE-POLE CIRCUIT BREAKER (FEEDTHROUGH UNIT) MOUNTED IN MAIN PANEL. NOTE LINE SIDE CONNECTION TO OFF-PEAK METER. THE FEED THROUGH UNIT IS NOT CONNECTED IN ANY MANNER TO THE MAIN BUS IN THE PANEL.

NOTE: CIRCLED LETTERS REFER TO THE COMPONENTS LISTED IN FIG. 19-10.

Fig. 19-11   The water heater is supplied through a separate watt hour meter/time clock. From the load terminals of this separate meter, conductors are run through a "feed-through" circuit breaker in the main panel. From the "feed-through" circuit breaker, conductors are then run to the water heater. When the contacts of the meter/time clock are in the open position, the power to the water heater is "off."

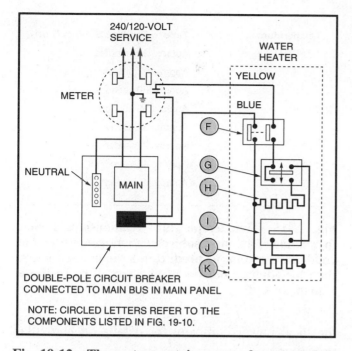

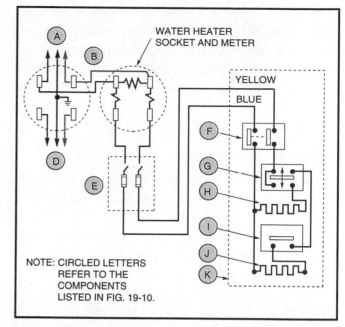

Fig. 19-12   The meter contains a set of contacts that can be controlled by the utility. The water heater circuit is fed from the main panel, through the contacts, then to the water heater.

Fig. 19-14   Twenty-four-hour water heater installation. The total energy demand for the water heater is fed through a separate watt hour meter at a lower kWh rate than the regular meter.

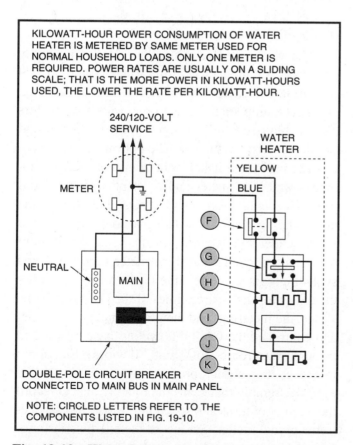

Fig. 19-13   Water heater connected to two-pole circuit in Main Panel.

Safety Note: Because water heater thermostats do not open **both** ungrounded conductors, a reading of 120 volts to ground will always be present at the terminals of the heating element on a 240-volt, single-phase system such as found in residential wiring. Should the heating element become grounded to the sheath, the element can continue to heat even when the thermostat is in the OFF position because of the presence of the 120 volts. Only the high temperature limit disconnects both conductors.

Basic Operation

Let's discuss a typical electric water installation, figure 19-12. When all of the water in the water heater is cold, the upper element heats the upper third of the tank. The lower thermostat is calling for power to the lower element, but the interlocking characteristic of the upper thermostat will not allow the bottom heating element to become energized. In other words, both elements cannot be on at the same time. This is referred to as *limited demand*.

When the water in the upper third of the tank reaches the temperature setting of the upper thermostat, the upper thermostat "snaps over," completing the circuit to the lower element. The lower element then begins to heat the lower two-thirds of the tank. Most of the time, it is the bottom heating element that keeps the water at the desired temperature. Should a large amount of water be used, the upper element comes on, resulting in the fast recovery of the water in the upper portion of the tank, which is where the hot water is drawn from the tank.

Scalding from Hot Water

Several states have laws regulating the maximum temperature settings for water heaters. To avoid scalding, this maximum setting is 120°F. See figure 19-15. Where such laws do not exist, the Consumer Product Safety Commission suggests a maximum temperature setting of 130°F. Water heater manufacturers post caution (warning, danger) labels on the appliance. Certain states do not accept the water heater thermostat as an acceptable water temperature control device, and require that showers and shower/bath combinations be provided with an automatic safety mixing device to prevent sudden unanticipated changes in water temperatures. Some manufacturers use nonadjustable upper thermostats, and set the adjustable thermostats at 120°F. Because 120°F is not hot enough to dissolve grease and to activate power detergents, most newer dishwashers will boost the water temperature in the dishwasher from 120°F to 140°–145°F.

Various Types of Electrical Connections

Because of the cumulative large power consumption of water heater loads, electric utilities across the country have been innovative in creating special lower rate structures (programs) for electric water heaters, air conditioners, and heat pumps. Electric utilities want to be able to shed some of this load during their peak periods. They will offer the homeowner choices of how the water heater, air conditioner, or heat pump is to be connected, and will provide financial incentives (lower rates) to those who are willing to use these restricted "time-of-day" connections. "Time-of-day," "time-of-use," and "off-peak" are terms used by different utilities. If homeowners have a large storage capacity electric water heater, they will have an adequate supply of hot water during those periods of time when the power to the water heater is OFF. If they cannot live with this, they have other options. Check with

Temperature	Time to Produce Serious Burn
120°F	Approx. 9½ minutes
125°F	Approx. 2 minutes
130°F	Approx. 30 seconds
135°F	Approx. 15 seconds
140°F	Approx. 5 seconds
145°F	Approx. 2½ seconds
150°F	Approx. 1⁸⁄₁₀ seconds
155°F	Approx. 1 second

Fig. 19-15   Time/temperature relationships related to scalding of an adult subjected to moving water. For children, the time to produce a serious burn is less than for adults. *Courtesy* of Shriners Burn Institute, Cincinnati Unit.

the local power company for details on particular programs.

The following text covers some of the methods in use around the country for hooking up electric water heaters and, in some cases, air-conditioning equipment and heat pumps.

Utility Controlled

The world of electronics has opened up many new ways for an electric utility to meter and control residential electric water heater, air-conditioning, and heat pump loads. Electronic watt-hour meters can be read remotely, programmed remotely, and can detect errors remotely. The options are endless.

The utility can install watt-hour meters that combine a watt-hour meter and a set of switching contacts. The contacts are switched ON and OFF by the utility during peak hours using a power line carrier signal. Older style versions of this type of watt-hour meter incorporated a watt-hour meter, time clock, and switching contacts. The time clock would be set by the utility for the desired times of the ON-OFF cycle. Today, the metering, and the programming for timing and control of the ON-OFF cycling is accomplished through electronic solid-state devices. Electronic meters contain a microprocessor that records, and an optical assembly that scans the rotating disc to record the number of turns and speed of the disc. These electronic digital meters can record the energy consumption during normal hours and during peak hours, allowing the electric utility to bill the homeowner at different rates for different periods of time. One type of residential single-phase watt-hour meter can register four different "time-of-use" rates. This type of meter is shown in figure 29-2.

One major utility programs their residential meters for "peak periods" from 9 A.M. to 10 P.M., Monday through Friday—except for holidays. Their "off-peak" periods are weekend days and all other hours. Programming of electronic meters can extend out for more than five years. The possibilities for programming electronic meters are endless. Older style mechanical meters used gears to record this data, and did not have the programming capabilities that electronic meters do.

The switching contacts in the meter are connected in *series* with the water heater, air-conditioner, or heat pump load that is to be controlled for the special "time-of-day" rates. Refer to figures 19-10, 19-11, and 19-12. The utility determines when they want to have the electric water heater ON or OFF.

Figure 19-10 shows a separate combination meter/time clock for the electric water heater load. The utility sets the clock to be "off" during their peak hours of high power consumption. Thus the term *off-peak metering*. Off-peak metering is oftentimes referred to as "time-of-use" or "time-of-day" metering. In exchange for lower electric rates, the homeowner runs the risk of being without hot water. Should homeowners run out of hot water during these off periods, they have no recourse but to wait until the power to the water heater comes on again. For this type of installation, large storage capacity water heaters are needed so as to be able to have hot water during these periods.

Figure 19-11 is electrically the same as figure 19-10, except that the water heater's separate disconnect switch is eliminated by using a "feed-through" circuit breaker in the main distribution panel.

Figure 19-12 illustrates a type of meter that contains an integral set of contacts, controlled by the utility via signals sent over the power lines. These meters are available with either 30-ampere or 50-ampere contacts, the contact rating selected to switch the load to be controlled. The electric utility determines when to shed the load. The homeowner must be willing to live with the fact that the power to the water heater, air-conditioner, or heat pump might be turned off during the peak periods.

In figure 19-16, the utility mounts a radio-controlled relay to the side of the meter socket. One major utility using this method makes use of ten different radio frequencies. When the utility needs to shed load, it can signal one or all of the relays to turn the water heater, air conditioner, or heat pump load OFF and ON. This might only be a few times each year, as opposed to time-clock programs that operate daily. This method is a simple one. One watt-hour meter registers the total power consumption. The homeowner signs up for this program. The program offers a financial incentive, and the homeowner knows up front that the utility might occasionally turn off the load during crucial peak periods.

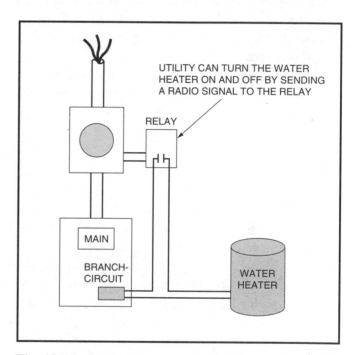

Fig. 19-16   In this diagram, the utility sends a radio signal to the relay, which in turn can "shed" the load of a residential water heater, air-conditioner, or heat pump. If the relay has two sets of contacts, four conductors would be required. Low-voltage relays are also used, in which case low-voltage conductors are run between the radio control unit and a relay in the appliance. For the control of an air-conditioner or heat pump, the radio control unit is connected "in series" with the room thermostat. This arrangement can be used with one watt hour meter using a sliding rate schedule, with a "time-of-day" control. Contact the electric utility for technical data regarding how they want the electrical connections to be made for their particular programs.

Customer Controlled

There are some rate schedules where customers can somewhat control their energy costs by doing their utmost to use electrical energy during off-peak hours. For example, they would not wash clothes, use the electric clothes dryer, take showers, or use the dishwasher during peak (premium) hours. The choice is up to homeowners. To accomplish this, watt-hour meters can be installed that have two sets of dials: one set showing the *total kilowatt hours*, and the second set showing *premium time kilowatt hours*. See figure 29-2. With this type of meter, the utility will have the total kilowatt hours used as well as the premium time kilowatt hours used. With this information, the utility bills customers at one rate for the energy used during peak hours, and at another rate for the difference between total kilowatt hours and premium time kilowatt hours. The electric utility will determine and define *premium time*.

Some homeowners have installed time-clocks on their water-heater circuit to shut-off during peak premium time hours to be sure they are not using energy during the high rate per kilowatt-hour periods.

Uncontrolled

The simplest uncontrolled method is shown in figure 19-13. There are no extra meters, switching contacts in the meter, extra disconnect switches, or controllable relays. The water heater circuit is connected to a double-pole branch-circuit breaker in the main distribution panel. The electrical power to the water heater is on 24 hours a day. This is the connection for the water heater in the residence in this text. Instead of separate "low-rate" metering that requires an additional watt-hour meter, the utility offers a sliding scale.

EXAMPLE:

METER READING 06-01-96	00532	
METER READING 05-01-96	00007	
TOTAL KILOWATT HOURS USED	00525	
BASIC MONTHLY ENERGY CHARGE		$11.24
1ST. 400 kWh × 0.10206		40.82
NEXT 125 kWh × 0.06690		8.36
AMOUNT DUE		$60.42

State, local, and applicable regulatory taxes would be added to the amount due.

Certain utilities will use the higher energy charge

during the June, July, August, and September summer months.

Another uncontrolled method is to use two watt-hour meters. See figure 19-14. The water heater has power 24 hours per day. One meter will register the normal lighting load at the regular residential energy charge. The second meter will register the energy consumed by the water heater, air conditioner, or heat pump at some lower energy rate.

Because each electric utility has its own unique time-of-use programs, the electrician must become familiar with these programs, and requirements for the electrical hook-up.

WATER HEATER LOAD DEMAND

The following Code sections apply to load demands placed on an electric water heater branch-circuit.

Section 422-4(a) requires that an appliance branch-circuit shall have a rating not less than the rating marked on the appliance.

Section 422-4(a), Exception No. 2 requires that a continuously loaded appliance must have a branch-circuit rating not less than 125% of the marked rating of the appliance. Most inspectors and electricians would not consider residential water heaters to be "continuous duty," as would be the case for most commercial electric water heaters.

Section 422-14(b) applies to all fixed storage water heaters having a capacity of 120 gallons (454.2 L) or less. Such heaters are required to have a branch-circuit rating not less than 125% of the nameplate rating of the water heater.

Disconnecting Means

Section 422-20 requires that the appliance must have a means of disconnect. This would normally be located in the panel (load center), or it could be provided by a separate disconnect switch. Although not normally done, *Section 422-25* of the Code does allow the unit switch that is an integral part of an appliance having an OFF position that disconnects all ungrounded conductors from the source to be considered the required disconnecting means. In many local codes, this is not acceptable. *Section 422-25* further requires that there be another means of disconnect for the appliance, which in the case of a one-family dwelling is the main service disconnect.

Section 422-26 requires that switches and circuit breakers used as the disconnecting means must be of the indicating type, meaning they must clearly show that they are in the ON or OFF position.

Section 422-21(b) requires that for permanently connected appliances rated greater than 300 volt-amperes or ⅛ horsepower, the disconnecting means must be within sight from the appliance, or be capable of being locked in the OPEN position. The electric water heater in the residence provides a good example of this Code requirement.

Branch-Circuit Overcurrent Protection

Section 422-28(e) states that for single non-motor-operated electrical appliances, the overcurrent protective device shall not exceed the protective device rating marked on the appliance. If the appliance has no such marking, then the overcurrent device is to be sized as follows:

- If appliance does not exceed 13.3 amperes— 20 amperes

- If appliance draws more than 13.3 amperes— 150% of the appliance rating

For a single non-motor-operated appliance, if the 150% sizing does not result in a standard size overcurrent device rating as listed in *Section 240-6*, then it is permitted to go to the next standard size. This is permitted in *Section 422-28(e), Exception.*

EXAMPLE: A water heater nameplate indicates 4500 watts, 240 volts. What is the maximum size fuse permitted by the Code?

SOLUTION:
$$\frac{4500}{240} = 18.75 \text{ amperes}$$

$$18.75 \times 1.5 = 28.125 \text{ amperes}$$

Therefore, the Code permits the use of a 30-ampere fuse.

Section 210-19(a) states that the branch-circuit conductors shall have an ampacity not less than the maximum load to be served.

Section 210-22 states that for any branch-circuit, the total load shall not exceed the rating of the branch-circuit.

The water heater for the residence is connected to Circuit A6-8. This is a 20-ampere, double-pole circuit breaker located in the main panel in the workshop. This circuit is straight 240 volts and does not require a neutral conductor.

Checking the Schedule of Special-purpose Outlets, we find that the electric water heater has two heating elements: a 2000-watt element and a 3000-watt element. Because of the thermostatic devices on the water heater, both elements cannot be energized at the same time. Therefore, the maximum load demand is

$$\text{amperes} = \frac{\text{watts}}{\text{volts}} = \frac{3000}{240} = 12.5 \text{ amperes}$$

For water heaters having two heating elements connected for nonsimultaneous operation, the nameplate on the water heater will be marked with the largest element's wattage at rated voltage. For water heaters having two heating elements connected for simultaneous operation, the nameplate on the water heater will be marked with the total wattage of both elements at rated voltage.

Conductor Size

Referring to *Table 310-16* of the Code, we find that the conductors will be No. 12 AWG. The specifications for the residence call for conductors to be Type THHN.

Conduit Size

Table C2, Chapter 9 of the Code indicates that ½-inch conduit (electrical metallic tubing in this residence for all exposed wiring in the basement) is required for two No. 12 THHN conductors. To simplify the installation, a short length (18 to 24 inches, or 457 to 610 mm) of ½-inch flexible metal conduit is attached to the ½-inch EMT and is connected to the knockout provided for the purpose on the water heater.

Cord/Plug Connections Not Permitted for Water Heaters

Sections 400-7(a)(8) and *422-8(c)* state the uses where flexible cords are permitted. A key code requirement, oftentimes violated, is that flexible cords shall be used only "where the fastening means and mechanical connections are specifically designed to permit ready removal for maintenance and repair, and the appliance is intended or identified for flexible cord connection." Certainly the plumbing does not allow for the "ready removal" of the water heater. The conductors in flexible cords cannot handle the high temperatures encountered on the water heater terminals. Check the instruction manual for proper installation methods.

Equipment Grounding

The water heater is grounded through the ½-inch conduit and the flexible metal conduit. Code references are *Sections 250-42, 250-57, 350-14*, and *422-16*. The subject of equipment grounding was covered in unit 4.

EFFECT OF VOLTAGE VARIATION

As previously stated, the heating elements in the heater in this residence are rated 240 volts. They will operate at a lower voltage with a reduction in wattage. If connected to voltages above their rating, they will have a very short life.

Ohm's law and the wattage formula show how the wattage and current depend upon the applied voltage.

$$R = \frac{E^2}{W} = \frac{240 \times 240}{3000} = 19.2 \text{ ohms}$$

$$I = \frac{E}{R} = \frac{240}{19.2} = 12.5 \text{ amperes}$$

If a different voltage is substituted, the wattage and current values change accordingly.

At 220 volts:

$$W = \frac{E^2}{R} = \frac{220 \times 220}{19.2} = 2521 \text{ watts}$$

$$I = \frac{W}{E} = \frac{2521}{220} = 11.46 \text{ amperes}$$

Another way to calculate the effect of voltage variance is:

$$\text{Correction Factor} = \frac{\text{Applied Voltage Squared}}{\text{Rated Voltage Squared}}$$

Using the previous example:

$$\text{Correction Factor} = \frac{220 \times 220}{240 \times 240} = \frac{48,400}{57,600} = 0.84$$

then, $3,000 \times 0.84 = 2,521$ watts

In resistive circuits, current is directly proportional to voltage and can be simply calculated using a ratio and proportion formula. For example, if a 240-volt heating element draws 12.7 amperes at rated voltage, the current draw can be determined at any other applied voltage, say 208 volts.

$$\frac{208}{240} = \frac{X}{12.7}$$

$$\frac{208 \times 12.7}{240} = 11.006 \text{ amperes}$$

To calculate this example if the applied voltage is 120 volts:

$$\frac{120}{240} = \frac{X}{12.7}$$

$$\frac{120 \times 12.7}{240} = 6.35 \text{ amperes}$$

Also, in a resistive circuit, wattage varies as the square of the current. Therefore, when the voltage on a heating element is doubled, the current also doubles and the wattage increases four times. When the voltage is reduced to one-half, the current is halved and the wattage is reduced to one-fourth.

Using the above formulae, it will be found that a 240-volt heating element when connected to 208 volts will have approximately three-quarters of the wattage rating than at rated voltage.

HEAT PUMP WATER HEATERS

With so much attention given to energy savings, heat pump water heaters have entered the scene.

Residential heat pump water heaters can save 50 to 60% over the energy consumption of resistance-type electric water heaters, depending upon energy rates. The heat energy transferred to the water is three to four times greater than the electrical energy used to operate the compressor, fan, and pump.

Heat pump water heaters work in conjunction with the normal resistance-type electric water heater. Located indoors near the electric water heater and properly interconnected, the heat pump water heater becomes the primary source for hot water. The resistance electric heating elements in the electric water heater are the back-up source should it be needed.

Heat pump water heaters remove low-grade heat from the surrounding air, but instead of transferring the heat outdoors as does a regular air-conditioning unit, it transfers the heat to the water. The unit shuts off when the water reaches 130°F. Moisture is removed from the air at approximately one pint per hour, reducing the need to run a dehumidifier. The exhausted cool air from the heat pump water heater is about 10°–15°F below room temperature. The cool air can be used for limited cooling purposes, or it can be ducted outdoors.

A typical residential heat pump water heater rating is 230 volts, 60 Hz, single-phase, 500 watts.

Heat pump water heaters operate with a hermetic refrigerant motor compressor. The electric hook-up is similar to a typical 240-volt, single-phase, residential air-conditioning unit. Article 440 of the *NEC®* applies. This is covered in unit 23.

Electric utilities are a good source to learn more about the economics of installing heat pump water heaters to save energy.

REVIEW

Note: Refer to the Code or the plans where necessary.

WATER PUMP CIRCUIT ⒶB

1. Does a jet pump have any electrical moving parts below the ground level? _____

2. Which is larger, the drive pipe or the suction pipe? _____

3. Where is the jet of the pump located? _____

4. What does the impeller wheel move? _____

5. Where does the water flow after leaving the impeller wheel?

    a. _____ b. _____

6. What prevents water from draining back into the pump from the tank? _____

7. What prevents water from draining back into the well from the equipment? _____

8. What is compressed in the water storage tank? _____

9. Explain the difference between a "two-wire" submersible pump and a "three-wire" submersible pump? _____

10. What is a common speed for jet pump motors? _____

11. Why is a 240-volt motor preferable to a 120-volt motor for use in this residence?

12. How many amperes does a 1-hp, 240-volt, single-phase motor draw? (See *Table 430-148.*) _____

13. What size are the conductors used for this circuit? _____

14. What is the branch-circuit protective device? _____

15. What furnishes the running protection for the pump motor? _____

16. What is the maximum ampere setting required for running protection of the 1-hp, 240-volt pump motor? _____

17. Submersible water pumps operate with the electrical motor and actual pump located (Circle correct answer.)
    a. above permanent water level
    b. below permanent water level
    c. half above and half below permanent water level

18. Because the controller contains the motor starting relay and the running and starting capacitors, the motor itself contains _____

19. What type of pump moves the water upward inside of the deep-well pipe? _____
    _____

20. Proper pressure of the submersible pump system is maintained by a _____
    _____

21. Fill in the data for a 16-ampere electric motor, single-phase, no Code letters.

    a. Branch-circuit protection, non-time-delay fuses.
       Normal size _____ A   Switch size _____ A
       Maximum size _____ A   Switch size _____ A

    b. Branch-circuit protection dual-element, time-delay fuses.
       Normal size _____ A   Switch size _____ A
       Maximum size _____ A   Switch size _____ A

    c. Branch-circuit protection instant-trip breaker.
       Normal setting _____ A
       Maximum setting _____ A

    d. Branch-circuit protection—inverse time breaker.
       Normal setting _____ A
       Maximum setting _____ A

    e. Branch-circuit conductor size. _____
       Ampacity _____ A

    f. Motor running protection using dual-element time-delay fuses.
       Maximum size _____ A

22. The *National Electrical Code*® is very specific in its requirement that submersible electric water pump motors be grounded. Where is this specific requirement found in the Code? _____
    _____

23. Does the *National Electrical Code*® allow submersible pump cable to be buried directly in the ground? _____
    _____

24. Must the disconnect switch for a submersible pump be located next to the well?
    _____
    _____

25. A metal well casing (shall) (shall not) be bonded to the pump's equipment grounding conductor or by grounding it with a separate equipment grounding conductor run all the way back to the same ground bus in the panel that supplies the pump circuit, *Section* _____ .

WATER HEATER CIRCUIT ⒶC

1. According to *Section 422-14* of the *NEC*® the high temperature limiting control must disconnect _____ of the ungrounded conductors. The high temperature control limits the maximum water temperature to _____°C.

2. A major hazard involved with water heaters is that they operate under whatever pressure the serving water utility supplies. Water stays liquid at temperatures higher than the normal boiling point of 212°F. If the high temperature limit control failed to operate for whatever reason, should a pipe burst or a faucet be opened, the pressure would instantly drop to normal atmospheric pressure, causing the superheated water to turn into _____ that could result in _____ . To prevent this from happening, water heaters are equipped with pressure/temperature relief valves.

3. Magnesium rods are installed inside the water tank to reduce _____ .

4. The heating elements in electric water heaters ar generally classified into two categories. These are _____ density and _____ density elements.

5. An 80-gallon electric water heater is energized for the first time. Approximately how many hours would it take to raise the water temperature 80°F (from 40°F to 120°F)? The water heater has a 3,000-watt heating element.

_____

_____

_____

6. What term is used by utilities when the water heater power consumption is measured using different rates during different periods of the day? _____

7. Explain how the electric utility in your area meters residential electric water heater loads.

_____

_____

_____

8. For residential water heaters, the Consumer Product Safety Commission suggests a maximum temperature setting of _____ °F. Some states have laws stating a maximum temperature setting of _____ °F.

9. An 80-gallon electric water heater has 60°F incoming water. How many minutes can a 3 gallon/minute showerhead be used to draw 105°F water? The water heater thermostat had been set at 120°F. _____

10. Approximately how long would it take to produce serious burns to an adult with 140°F water? _____

11. Two thermostats are generally used in an electric water heater.

    a. What is the location of each thermostat? _____

    _____

    b. What type of thermostat is used at each location? _____

    _____

12. a. How many heating elements are provided in the heater in the residence discussed in this text? _____

    b. Are these heating elements allowed to operate at the same time? _____

13. When does the lower heater operate? _____

    _____

14. The Code states that water heaters having a capacity of 120 gallons (454.2 L) or less shall be considered _____ duty and, as such, the circuit must have a rating of not less than _____ % of the rating of the water heater.

15. Why does the storage tank hold the heat so long? _____

16. The electric water heater in this residence is connected for "limited demand" so that only one heating element can be on at one time.

    a. What size wire is used to connect the water heater? _____

    b. What size overcurrent device is used? _____

17. a. If both elements of the water heater in this residence are energized at the same time, how much current will they draw? (Assume the elements are rated at 240 volts.)

    b. What size and Type THHN wire is required for the load of both elements? Show calculations.

18. a. How much current would the two elements in question 17 draw if connected to 220 volts?

    b. What is the wattage value at 220 volts? Show calculations for both.

19. A 240-volt heater is rated at 1,500 watts.
    a. What is its current rating at rated voltage? _____
    b. What current draw would occur if connected to a 120-volt supply?

    c. What is the wattage output at 120 volts?

20. For a single, non-motor-operated electrical appliance rated greater than 13.3 amperes that has no marking that would indicate the size of branch-circuit overcurrent device, what percentage of the nameplate rating is used to determine the maximum size branch-circuit overcurrent device? _____

21. A 7,000-watt resistance-type heating appliance is rated 240 volts. What is the maximum size fuse permitted to protect the branch-circuit supplying this appliance?

    _____

22. The Code requirements for standard resistance electric water heaters are found in *Article 422* of the Code. Because heat pump water heaters contain a hermetic refrigerant motor compressor, the Code requirements are found in *Article* _____ of the Code.

UNIT 20

Special-Purpose Outlets for Ranges, Counter-Mounted Cooking Unit ⏥G, and Wall-Mounted Oven ⏥F

OBJECTIVES

After studying this unit, the student will be able to

- interpret electrical plans to determine special installation requirements for counter-mounted cooking units, wall-mounted ovens, and freestanding ranges.

- understand the latest *National Electrical Code®* connection requirements for electric ranges, ovens, and counter-mounted cooking units.

- identify and understand the purpose of the National Electrical Manufacturers Association (NEMA) blade and slot configurations on 30-ampere and 50-ampere receptacles and cords.

- identify the terminals of 30-ampere and 50-ampere receptacles.

- understand when to use 3-wire and 4-wire receptacles and cords for electric range hook-ups.

- compute loads and select proper conductor sizes for electric ranges, wall-mounted ovens, and counter-mounted cooking units.

- properly ground electric ranges, wall-mounted ovens, and counter-mounted cooking units in conformance to the *National Electrical Code®* for both new installations and existing installations.

- describe a seven-heat control and a three-heat control for a dual-element heating unit for a cooktop.

- understand how infinite heat temperature controls operate.

- supply counter-mounted cooking units and wall-mounted ovens by connecting them to one feeder using the "tap" rule.

- understand how to install a feeder to a load center, then divide the feeder to individual circuits to supply the appliances.

The residence being discussed in this text has one built-in oven and one built-in cooktop. Checking the Schedule for Special-Purpose Outlets, we find that the built-in wall-mounted oven is supplied by Circuit B(6-8), and the built-in cooktop is supplied by Circuit B(2-4).

Running separate circuits to individual appliances is probably the most common method used to connect appliances. However, there might be a cost justification for running one larger circuit from the main panel to a junction box located in close proximity to both the oven and the cooktop. From this junction box, "taps" of smaller conductors are made to feed each appliance.

The following text explains the use of a load center and shows how to do the calculations for individual circuits, a single larger circuit with corresponding taps to the individual appliances, and a freestanding range. Which method to use will be governed by local codes and consideration of the cost of installation (time and material).

BASIC CIRCUIT REQUIREMENTS FOR RANGES, COUNTER-MOUNTED COOKING UNITS, AND WALL-MOUNTED OVENS

Section 210-19(b)

- the branch-circuit conductors must have an ampacity not less than the branch-circuit rating.

- the branch-circuit conductors' ampacity must not be less than the load being served.

- the branch-circuit rating must not be less than 40 amperes for ranges of 8¾ kW or more.

- for ranges 8¾ kW or more, the neutral may be 70% of the branch-circuit rating, but not smaller than No. 10.

- tap conductors connecting the ranges, ovens, and cooking unit to a 50-ampere branch-circuit must be rated at least 20 amperes, be adequate for the load, and must not be any longer than necessary for servicing the appliance.

Section 220-19

- The feeder demand load for household ranges, cook-tops, and ovens is computed by using the values found in *Table 220-19*. Also refer to the footnotes to *Table 220-19*.

Section 422-4(a), Exception No. 3

- Branch-circuit requirements for household cooking appliances shall be permitted to be in accordance with *Table 220-19*.

Section 422-28(b)

- for ranges with demand over 60 amperes (i.e., some large double-oven ranges), the power supply must be subdivided into two or more circuits each not over 50 amperes. This is done by the manufacturer of the appliance.

Section 422-28(e)

For non-motor-operated appliances, the rating of the overcurrent protection shall:

- not exceed the rated current marked on the appliance. Not all appliances are marked with the maximum size overcurrent device.

- not exceed 150% of the appliance rated current if not marked, and the appliance is over 13.3 kW.

- not exceed 20 amperes of the appliance rated current if not marked, and the appliance is rated 13.3 kW or less.

COUNTER-MOUNTED COOKING UNIT CIRCUIT ▲G

Counter-mounted cooking units are available in many styles. Such units may have two, three, or four surface heating elements. Some ranges have a smooth ceramic surface. The electric heating elements are in direct contact with the underside of the ceramic top. These units have the same controls as standard units and are wired according to the general Code guidelines for cooking units. When the elements of these ceramic-covered units are turned on, the ceramic above each element changes color slightly. Most of these cooking units have a faint design in the ceramic to locate the heating elements. The ceramic is rugged, attractive, and easy to clean.

In the residence in the plans, a separate circuit is provided for a standard, exposed-element, counter-mounted cooking unit. This unit is rated at 7450 watts and 120/240 volts. The ampere rating is:

$$I = \frac{W}{E} = \frac{7450}{240} = 31.04 \text{ amperes}$$

The cooking unit outlet is shown by the symbol ⬛G.

The demand load for individual wall-mounted oven circuits or individual counter-mounted cooking unit circuits must be calculated at the full nameplate rating (*Note 4* to *Table 220-19*).

According to *Section 210-19(b)*, the branch-circuit conductors supplying electric ranges, wall-mounted ovens, and counter-mounted cooking units must:

- have an ampacity not less than the rating of the branch-circuit, and

- not less than the maximum load to be served.

The two exceptions to these requirements are (1) for tap conductors and (2) for the permitted reduction in the neutral size to 70% of the branch-circuit rating but not smaller than No. 10 AWG, where the demand is 8¾ kW or greater as determined according to *Column A* of *Table 220-19*.

Installing the Cooking Unit

Table 310-16 shows that No. 8 THHN conductors may be used as the circuit conductors for the 31.04-ampere cooking unit installation. These circuit conductors are connected to Circuit B2-4. This is a two-pole, 40-ampere circuit in Panel B.

The electrician must obtain the roughing-in dimensions for the counter-mounted cooking unit so that the cable can be brought out of the wall at the proper height. In this way, the cable is hidden from view. To simplify both servicing and installation, a separable connector or a plug and receptacle combination may be used in the supply line, *Section 422-17*.

Cooktop manufacturers provide a junction box on their units. The electrician can terminate the circuit conductors of the appliance in this box and make the proper connections to the residence wiring. Instead of the junction box, the manufacturer may connect a short length of flexible conduit to the appliance. This conduit contains the conductors to which the electrician will connect the supply conductors. These connections must be made in a junction box furnished by the electrician or the manufacturer of the appliance.

In this residence, the cooktop range is installed in the island. See plans for details. The feed will be a three-conductor nonmetallic-sheathed cable that has three No. 8 AWG copper conductors plus an equipment grounding conductor. The cable runs out of the top of Panel B, across the face of the ceiling joists in the recreation room, then upward into the base of the island cabinet. Because the recreation room ceiling is a dropped ceiling, the cable is run to a point just below the intended location of the kitchen island, leaving ample cable length hanging in the basement. Later, when the kitchen cabinets and the island are installed, and other finishing touches are being done in the kitchen, the built-in appliances are set into place. Following this, the electrician then completes the final hookup of the built-in appliances.

Note on the plans that a receptacle outlet is to be installed on the side of the island containing the cooktop. This receptacle outlet is supplied by small appliance Circuit B16, figure 12-8. Also study unit 3 on "Countertops," and figure 3-6.

Temperature Effects on Conductors

All nonmetallic-sheathed cable has 90°C insulated conductors, with the ampacity based upon that of 60°C conductors, *Section 326-26*. Type THHN wire is also rated 90°C. In most instances, 90°C conductors are suitable for hooking up electric ranges. It is wise to check the installation instructions, nameplate, and any labels attached to the appliance to make sure that 90°C conductors are suitable.

Table 310-13 gives the conductor insulation temperature limitations and the accepted applications for conductors.

General Wiring Methods

Electric ranges, counter-mounted cooking units, wall-mounted ovens, and clothes dryers may be wired using any of the standard wiring methods, such as armored cable, nonmetallic-sheathed cable, electrical metallic tubing (EMT), flexible metal or nonmetallic conduit, or service-entrance cable.

Local electrical codes can contain information as to restrictions on the use of any of the wiring methods listed.

For residential electric ranges, wall-mounted ovens, and surface-mounted cooking units, receptacles and cord sets rated 30 amperes, 125/250 volts or 50 amperes, 125/250 volts are used, depending upon the ampere rating of the appliance. In the following discussion, 40-ampere and 45-ampere cord sets, using a 50-ampere plug cap on the end are also available. This becomes a cost issue to use a full 50-ampere rated cord set in all cases, or to match the cord ampere rating as close as possible to that of the appliance being connected.

Blade and Slot Configuration for Plug Caps and Receptacles

▶ Figure 20-1 shows the different configurations of receptacles commonly used for electric range hook-up.

- **30-ampere, 3-wire:** The L-shaped slot on the receptacle (NEMA 10-30R) and the L-shaped blade on the plug cap (NEMA 10-30P) are for the white neutral wire. Frames of electric ranges, ovens, and clothes dryers installed prior to adoption of the 1996 *National Electrical Code®* were permitted to be grounded to the neutral conductor as stated in *Sections 250-60 and 250-61*. The neutral connection had served a dual-purpose, that of the neutral and that of the appliance's required equipment ground.

- **30-ampere, 4-wire:** The L-shaped slot on the receptacle (NEMA 14-30R) and the L-shaped blade on the plug cap (NEMA 14-30P) are for the white neutral wire. The horseshoe-shaped

slot on the receptacle and the round or horseshoe shaped blade on the plug cap are for the equipment ground.

- **50-ampere, 3-wire:** The wide flat slot on the receptacle (NEMA 10-50R) and the matching wide blade on the plug cap (NEMA 10-50P) are for the white neutral wire. Frames of electric ranges, ovens, and clothes dryers installed prior to adoption of the 1996 *National Electrical Code®* were permitted to be grounded to the neutral conductor as stated in *Sections 250-60 and 250-61*. The neutral connection had served a dual-purpose, that of the neutral and that of the appliance's required equipment ground.

- **50-ampere, 4-wire:** The wide flat slot on the receptacle (NEMA 14-50R) and the matching wide blade on the plug cap (NEMA 14-50P) are for the white neutral wire. The horseshoe shaped slot on the receptacle and the round or horseshoe shaped blade on the plug cap are for the equipment ground. ◀

Cord Sets

A 30-ampere cord set contains:

- **3-wire:** three No. 10 AWG conductors. The attachment plug cap (NEMA 10-30P) is rated 30 amperes.

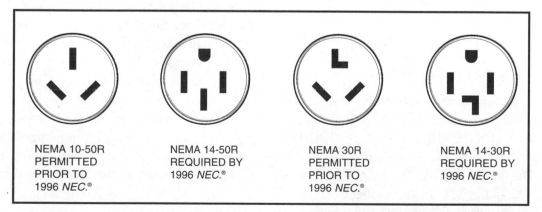

NEMA 10-50R
PERMITTED
PRIOR TO
1996 *NEC.®*

NEMA 14-50R
REQUIRED BY
1996 *NEC.®*

NEMA 30R
PERMITTED
PRIOR TO
1996 *NEC.®*

NEMA 14-30R
REQUIRED BY
1996 *NEC.®*

▶ **Fig. 20-1   Illustrations of 30-ampere and 50-ampere receptacles used for cord- and plug-connection of electric ranges, ovens, counter-mounted cooking units, and electric clothes dryers. Four-wire receptacles and four-wire cord sets are required for installations according to the 1996** *National Electrical Code.®* **Three-wire receptacles and three-wire cord sets were permitted prior to adoption of the 1996** *NEC.®* **See** *Sections* **250-60 and 250-61 of the** *National Electrical Code.®* ◀

- **4-wire:** four No. 10 AWG conductors. The attachment plug cap (NEMA 14-30P) is rated 30 amperes.

A 40-ampere cord set contains:

- **3-wire:** three No. 8 AWG conductors and one No. 10 AWG conductor. The attachment plug cap (NEMA 10-50P) is rated 50 amperes.
- **4-wire:** three No. 8 AWG conductors and one No. 10 AWG conductor. The attachment plug cap (NEMA 14-50P) is rated 50 amperes.

A 45-ampere cord set contains:

- **3-wire:** two No. 6 AWG conductors and one No. 8 AWG conductor. The attachment plug cap (NEMA 10-50P) is rated 50 amperes.
- **4-wire:** three No. 6 AWG conductors and one No. 8 AWG conductor. The attachment plug cap (NEMA 14-50P) is rated 50 amperes.

A 50-ampere cord set contains:

- **3-wire:** two No. 6 AWG, and one No. 8 AWG conductors. The attachment plug cap (NEMA 10-50P) is rated 50 amperes.
- **4-wire:** three No. 6 AWG, one No. 8 AWG conductors. The attachment plug cap (NEMA 14-50P) is rated 50 amperes.

Terminal Identification

Receptacle and cord terminals are marked:

- "X" and "Y" for the grounded conductors.
- "W" for the white grounded conductor. The "W" terminals will generally be whitish or silver (tinned) in color, in accordance with *Section 220-9* and *200-10* of the *NEC.*®
- "G" for the equipment grounding conductor. This terminal is green-colored hexagon shaped, and is marked "G", "GR", "GRN", or "GRND" in accordance with *Section 410-58(b)* of the *NEC.*® The grounding blade on the plug cap must be longer than the other blades so that its connection is made before the ungrounded conductors make connection in accordance with *Section 410-58(d)* of the *NEC.*®

Surface-mounted range outlets can be wired directly with nonmetallic-sheathed cable, armored cable, or conduit. Freestanding electric ranges might

require No. 6 or No. 8 AWG conductors, depending upon their kilowatt (kW) rating. Wall-mounted ovens, and surface-mounted cooking units might require No. 8, No. 10, or No. 12 AWG conductors, depending upon their kilowatt (kW) rating. Kilowatts are converted to amperes in a single-phase circuit by using the following formula:

$$\text{Amperes} = \frac{\text{kW} \times 1000}{\text{volts}}$$

A 50-ampere flush-mounted range receptacle is quite large. The range receptacle, plus the branch-circuit conductors, plus the equipment grounding conductor will probably require a 4-inch square, 1½ inch or 2⅛ inch deep outlet box with suitable single-gang or two-gang plaster ring. The correct box sizing is determined by applying the rules of *Section 370-16* and the tables located in this section.

GROUNDING FRAMES OF RANGES, WALL-MOUNTED OVENS, AND SURFACE-MOUNTED COOKING UNITS

New Installations Based on the 1996 *National Electrical Code*®

► For protection against electrical shock hazard, the metal frames of electric ranges, wall-mounted ovens, and surface-mounted cooking units must be grounded according to the rules found in *Sections 250-57* and *250-59*. An acceptable grounding means would be to connect the appliance using a metal raceway such as electrical metallic tubing (EMT), flexible metal conduit (Greenfield), armored cable (BX), or other means as listed in *Section 250-91(b)*. There are limitations on the use of flexible metal conduit as an equipment ground. See *Sections 250-91(b)* and *350-14*. This topic is covered in unit 4.

Proper connection of the separate equipment grounding conductor found in nonmetallic-sheathed cable is also an acceptable grounding means. The equipment grounding conductor found in nonmetallic-sheathed cable is sized according to *Table 250-95*.

Direct Connection: The metal raceway, armored cable, or nonmetallic-sheathed cable may be run directly to the junction box or knockout entry on the appliance. Be sure the supply conductor temperature ratings are suitable for direct connection to the appliance's terminal block. Supply conductor temperature requirements are marked on the appliance.

Receptacle/Cord Connection: The metal raceway or armored cable may also be run to a surface-mounted range receptacle or to a suitably sized flush-mounted outlet box in which a 4-wire, 30-ampere or 4-wire, 50-ampere receptacle is installed. The range, wall-mounted oven, or surface-mounted cooktop is then connected with a suitable 4-wire cord. Three-wire cords and three-wire receptacles are not permitted because the 1996 *NEC®* does not permit the frames to be grounded to the neutral of the circuit. Cord sets rated 40 amperes, 45 amperes, or 50 amperes are referred to as *range cords.* Cord sets rated 30 amperes are referred to as *dryer cords.* Freestanding electric ranges and surface-mounted cooking units will probably require 50-ampere receptacles. Wall-mounted ovens might only require 30-ampere receptacles. It all depends upon the appliance's ampere or kilowatt (kW) rating.

Type SE service-entrance cable may also be used provided *all* of the circuit conductors are insulated. This means that the cable must contain three insulated conductors plus one equipment grounding conductor. The equipment grounding conductor may be insulated (green) or bare. See *Section 338-3.* Type SE cable installed as interior wiring must be installed according to the installation requirements found in *Article 336* for nonmetallic-sheathed cable, *Section 338-4.*

The small copper bonding strap (link) that is furnished with electric ranges, wall-mounted ovens, and surface-mounted cooking units *must not* be connected between the neutral terminal and the metal frame of the appliance. This bonding strap (link) *must not* be used. It should be removed. ◄

It is interesting to note that the 4-wire cord/4-wire receptacle requirement for electric ranges, wall-mounted ovens, and surface-mounted cooking units in mobile homes has been in effect since 1965, when *Article 550* relating to mobile homes first appeared in the *NEC®*

Existing Installations

► Prior to the 1996 edition of the *National Electrical Code® Section 250-60* permitted the frames of electric ranges, wall-mounted ovens, surface-cooking units, and clothes dryers to be grounded to the neutral conductor. By way of *Tentative Interim Amendment No. 53*, this special permission was put into effect in July of 1942, and was supposedly an effort to conserve raw materials during World War II. In effect, this special permission allowed the neutral conductor to serve a dual purpose, that of the neutral conductor, and that of the equipment grounding conductor. This special permission remained in effect until the 1996 *National Electrical Code®*

Section 250-60 now applies *only* to existing branch-circuit installations. For those installations wired according to the rules of *Section 250-60* prior to the 1996 *NEC®* nothing need be changed. The required appliance grounding was accomplished through the use of the small copper bonding strap furnished by the appliance manufacturer that connected between the neutral terminal and the metal frame of the appliance. ◄

For existing installations:

- the supply circuit had to be 120/240 volt, single-phase, 3-wire or 208Y/120 volt derived from a 3-phase, 4-wire wye-connected system.

- the neutral had to be no smaller than No. 10 AWG copper or No. 8 AWG aluminum.

- the neutral conductor had to be insulated. The exception to this was for Type SE cable where the circuit originated at the main service, in which case the neutral conductor did not have to be insulated, only covered by the outer jacket of the cable.

- a metal outlet box used to hold a receptacle was permitted to be grounded to the neutral. See figure 15-6.

Surface Heating Elements for Cooking Units

The heating elements used in surface-type cooking units are manufactured in several steps. First, spiral-wound nichrome resistance wire is impacted in magnesium oxide (a white, chalklike powder). The wire is then encased in a nickel-steel alloy sheath, flattened under very high pressure, and formed into coils. The flattened surface of the coil makes good contact with the bottoms of cooking utensils. Thus, efficient heat transfer takes place between the elements and the utensils.

Surface unit heating elements with one-coil (one heating element) construction are used in conjunction with infinite heat switches that "pulse" the circuit to the heating element. Two-coil (two heating elements) surface unit heating elements are used with fixed-heat 3-, 5-, and 7-position switches. Three-coil (three heating elements) surface unit heating elements are used with 3-position switches.

Figures 20-2 and 20-3 illustrate typical electric range surface units.

TEMPERATURE CONTROL

Infinite-Position Temperature Controls

Modern electric ranges are equipped with "infinite"-position temperature knobs (controls). Newer switches have a rotating cam that provides an infinite number of heat positions. Older style heat controls connect the heating elements of a surface unit to 120 volts or 240 volts in series or in parallel to attain a certain number of specific heat levels.

The contacts open and close as a result of a heater-bimetal device that is an integral part of the rotary control. When the control is turned ON, a separate set of pilot light contacts closes and remains closed, while the contacts for the heating element open and close to attain and maintain the desired temperature of the surface heating unit, figure 20-4.

Inside these infinite position temperature controls are contacts that open and close repetitively. The circuit to the surface heating element is regulated by the ratio of the time the contacts are in a CLOSED position versus the OPEN position. This is termed *input percentage*.

Fig. 20-2 Typical 240-volt electric range surface heating element. This is the type of element used with infinite heat controls. *Courtesy* of Chromalox, Edwin L. Wiegand Division, Emerson Electric Company.

Fig. 20-3 Typical electric range surface unit of the type used with 7-position controls as discussed in this text. *Courtesy* of Chromalox, Edwin L. Wiegand Division, Emerson Electric Company.

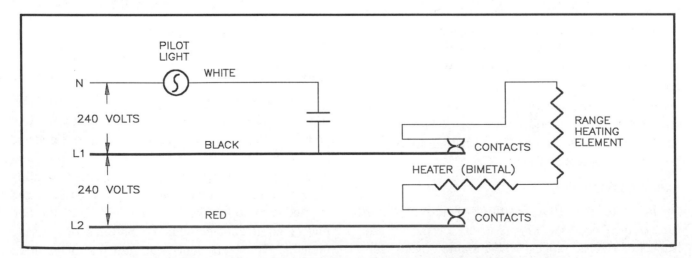

Fig. 20-4 Typical internal wiring of an infinite heat surface element control.

The contacts in a temperature control of this type have an expected life of over 250,000 automatic cycles. The knob rotation expected life exceeds 30,000 operations.

One-coil (single-element) surface units are used with infinite heat switches.

Fixed-Heat Temperature Controls

Still available are the older style rotary controls that have specific "indents" that lock the switch contacts into place as the knob is rotated to the various heat positions. Similar switching arrangements are found in push-button-type controls, where one button is pushed at a time. Pushing another button will raise and cancel the first button. Fixed-heat types of rotary and push-button switches, with or without pilot lights, are used in conjunction with two-coil surface unit heating elements. See figures 20-5 and 20-6. The different heat levels are obtained by switching between different voltages and coil circuitry. Their unique switching operation offers an excellent opportunity to make electrical calculations for series, parallel, and series/parallel circuits.

Seven-Heat Control for Two-Coil Heating Unit

One method of obtaining several levels of heat from a one-coil unit is to use a two-coil unit, figure 20-5. Figure 20-6A shows that seven levels of heat can be obtained from such a unit if a seven-point "fixed-heat" switch is used. Note that the two elements of the surface heating unit are both contained in the same steel jacket, and they are connected together at the inner end of the steel jacket. A return wire (not a resistance wire) is carried back to the terminal block. This arrangement means that the elements can be operated singly, together, or in series.

Figure 20-5 shows the heating element connected with element A as lead 2, element B as lead 1, and the

return wire as the common (C). Resistance readings can be taken between these leads as follows:

Test between C and 2	90 ohms (element A)
Test between C and 1 (single operation)	74 ohms (element B)
Test between 2 and 1 (elements in series)	164 ohms (elements A and B in series)
Test between C and 1–2 connected together (elements connected in parallel)	40.6 ohms (elements A and B in parallel)

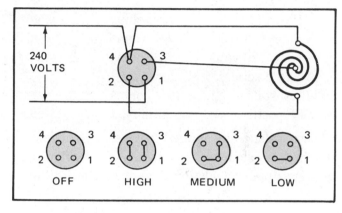

Fig. 20-6   Surface unit with three-heat series-parallel switch.

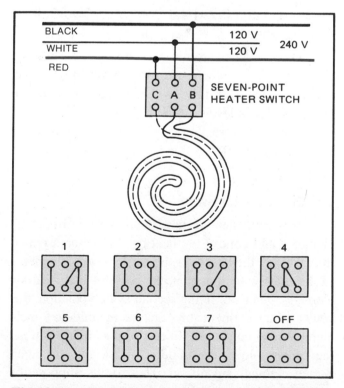

Fig. 20-6A   Surface heating element with seven levels of heat.

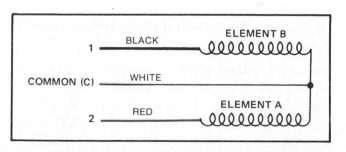

Fig. 20-5   Typical surface unit wiring.

Element B allows more current to flow than does element A because element B has 74 ohms of resistance as compared to 90 ohms of resistance for element A. As the current increases, the wattage also increases. The eight switch positions shown in figure 20-6A result in the wattages shown in the following list. For each switch position, the formula $W = E^2/R$ is used to find the wattage value.

1. A and B parallel,
    240 volts   40.6 ohms   1418 watts
2. B only, 240 volts   74   ohms   778 watts
3. A only, 240 volts   90   ohms   640 watts
4. A and B parallel,
    120 volts   40.6 ohms   ____ watts*
5. B only, 120 volts   74   ohms   ____ watts*
6. A only, 120 volts   90   ohms   ____ watts*
7. A and B series,
    120 volts   164   ohms   ____ watts*
8. OFF position   Infinity   0 watts

*These values are omitted because the student is required to calculate the values in the Review section. A sample calculation is completed as follows:

$$W = \frac{E^2}{R} = \frac{240 \times 240}{40.6} = \frac{57600}{40.6} = 1418 \text{ watts}$$

Three-Heat and Five-Heat Control for Two-Coil Heating Units

These switches are for use with two-coil heating elements. These switches function in much the same manner as the seven-heat switch, except for the fewer numbers of heat levels available.

A simple, three-heat series-parallel switch is shown in figure 20-6. A five-heat switch is not shown.

Coil Size Selector Switch

These switches are used with three-coil heating elements and control the heat by surface area rather than varying the heat over the entire heating element. In the first switch position, the center 4-inch heating element comes on. In the second switch position, the center 4-inch element stays on, and the middle 6-inch element also comes on. In the third switch position, the center 4-inch, the middle 6-inch, and the outer 8-inch element are all on. These units supposedly save energy because the heating can be focused on the contact area directly beneath a cooking utensil.

Automatic Sensor Control

Infinite heat is attained using a one-coil surface heating element has a sensor in the center of the surface unit. A special 12-volt control circuit is provided through a small transformer. The control knob is set to the desired temperature. A "responder" pulses the heating element off and on to maintain the set temperature. To operate properly, the cooking utensil must maintain good contact with the sensor. Use flat bottom utensils only!

High-Speed Control

Another type of heating element is the 120-volt flash or high-speed element. When connected to a special control switch, this type of element is connected briefly across 240 volts. In a few seconds, the element is returned automatically to the 120-volt supply for continued operation. A 1250-watt flash unit connected to 240 volts produces 5000 watts of heat in a matter of seconds. This brings the unit to maximum heat quickly. When the flash element is returned to the 120-volt supply, the unit continues to produce 1250 watts, the rated value.

To calculate the maximum wattage at 240 volts, assume that a 1250-watt heating element is rated at 120 volts. This unit has a resistance of:

$$R = \frac{E^2}{W} = \frac{120 \times 120}{1250} = \frac{14400}{1250} = 11.52 \text{ ohms}$$

When this 11.52-ohm element is connected to a 240-volt supply, the wattage is:

$$W = \frac{E^2}{R} = \frac{240 \times 240}{11.52} = \frac{57600}{11.52} = 5000 \text{ watts}$$

Note that the wattage quadruples when the supply voltage is doubled. Conversely, the wattage is reduced to one-fourth its original value when the supply voltage is reduced to one-half its original value.

Heat Generation by Surface Heating Elements

The surface heating elements generate a large amount of radiant heat. For example, a 1000-watt heating unit generates approximately 3412 Btu of heat per hour. Heat is measured in British thermal units (Btu). A *Btu* is defined as the amount of heat required to raise the temperature of one pound of water by one degree Fahrenheit.

WALL-MOUNTED OVEN CIRCUIT ⏃F

A separate circuit is provided for the wall-mounted oven. The circuit is connected to Circuit B(6-8), a two-pole, 30-ampere circuit in Panel B. The receptacle for the oven is shown by the symbol ⏃F.

The wall-mounted oven installed in the residence is rated at 6.6 kW or 6600 watts at 120/240 volts. The current rating is:

$$I = \frac{W}{E} = \frac{6600}{240} = 27.5 \text{ amperes}$$

Section 422-4(a), Exception No. 3 tells us that branch-circuits for household cooking units are to be determined according to *Table 220-19*.

Note 4 to *Table 220-19* states that the branch-circuit load for the wall-mounted oven must be based on the actual nameplate rating of the appliance.

Caution When Making Connections

After determining the current rating ($I = W/E$), refer to *Table 310-16*. We find that for 27.5 amperes, a No. 10 AWG copper TW, THW, or THHN may be used to supply this wall-mounted oven. As previously discussed, for nonmetallic-sheathed cable, even though the conductor insulation is 90°C, the allowable ampacity must be determined according to the 60°C column of *Table 310-16*. This is stated in *Section 336-26*.

When using armored cable, if the conductors are rated for 75°C as in Type ACTH, the ampacity is permitted to be based upon the 60°C or 75°C column of *Table 310-16*, depending upon the equipment and/or device terminal marking. If the conductors are rated for 90°C as in Type ACTHH, the ampacity would still have to be based upon the 60°C or 75°C column of *Table 310-16*, again depending upon the terminal marking. When armored cable is buried in insulation, it must have 90°C rated conductors, and its ampacity is based upon that of 60°C conductors, *Section 333-20*.

Recall in unit 4, under "Conductor Temperature Ratings," that, according to Underwriters Laboratories Standards, and *Section 110-14(c)* of the *NEC,*® when conductor sizes No. 14 through No. 1 AWG are installed, their allowable current-carrying capacity (ampacity) is to be that of the 60°C column of *Table 310-16* unless the terminals are marked with some other temperature rating, such as 75°C or 90°C.

That is why it is so important to carefully read the appliance manufacturer's installation data for recommended conductor size and type. Check particularly for the temperature rating.

Self-Cleaning Oven

Self-cleaning ovens are popular appliances because they automatically remove cooking spills from the inside of the oven. Lined with high temperature material, a self-cleaning oven can be set for a cleaning temperature that is much higher than baking and broiling temperatures. When the oven control is set to its "self-clean" position, a *sensor* located inside of the oven monitors the oven temperature to a preset value that approximates 800°F. The sensor is a resistor. The resistance of the sensor increases in proportion to the temperature inside the oven. This change in sensor resistance signals the electronic circuitry to provide the oven temperature needed for proper cleaning of the oven. At this temperature, all residue in the oven, such as grease and drippings, is burned until it is fine ash that can be removed easily. During the self-clean cycle, for safety reasons, a special latching mechanism on the oven door makes it impossible to open the oven door until a predetermined safe temperature is reached. On some self-cleaning electric ranges, the door locking lever also connects a *high temperature limit switch* into the circuit that protects against excessive high temperatures during the cleaning cycle.

Although the temperature and timing control features, and the internal associated circuitry found in modern freestanding and built-in electric ranges can be very complex, the actual branch-circuit sizing is the same for self-cleaning ranges and non-self-cleaning ranges. Branch-circuit calculations are discussed later in this unit.

Operation of the Oven

The oven heating elements are similar to the surface elements. The oven elements are mounted in removable frames. A thermostat controls the temperature of these heating units. Any oven temperature can be obtained by setting the thermostat knob to the proper point on the dial. The thermostat controls both the baking element and the broiling element. These elements can be used together to preheat the oven.

A combination clock and timer is used on most ovens. The timer can be preset so that the oven turns on automatically at a preset time, heats for a preset duration, and then turns off.

Microwave Oven

Modern microwave ovens, a result of electronic technology, have introduced many features never before possible. Such features include:

- control of time and temperature.
- programmable multiple-state settings.
- temperature probes for insertion into the food to assure desired temperature.
- ability to "hold" food temperatures following cooking cycle.
- touch-sensitive controls.
- programmable defrost settings with lower temperatures than those required for actual cooking or baking so that premature cooking does not occur.
- computer calculations to automatically increase the time of operation when recipe quantities are doubled, tripled, or more.
- delayed starting of oven for up to 12 hours.

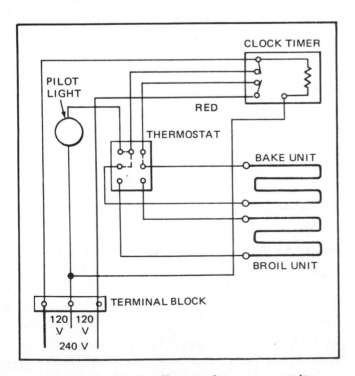

Fig. 20-7   Wiring diagram for an oven unit.

- preprogrammed temperature and time for recipes.

The electrician's primary concern is how to calculate the microwave oven load, and how to install the circuit for microwave appliances so as to conform to the Code. The same Code rules apply for both electrical and electronic microwave cooking appliances.

Ovens are insulated with fiberglass or polyurethane foam insulation placed within the oven walls to prevent excessive heat leakage.

Figure 20-7 shows a typical wiring diagram for a standard oven unit.

CIRCUIT *REQUIREMENTS* WHEN MORE THAN ONE WALL-MOUNTED OVEN AND COUNTER-MOUNTED COOKING UNIT ARE SUPPLIED BY ONE CIRCUIT

Note 4 to *Table 220-19* states that when a single branch-circuit supplies a counter-mounted cooking unit and not more than two wall-mounted ovens, all located in the same room, the calculation is made by adding up the nameplate ratings of the individual appliances, then treating this total as if it were one range.

See figure 20-8 for an example of a counter-mounted cooking unit and a wall-mounted oven connect to one branch-circuit. In accordance with *Section 210-19(b), Exception No. 1*, the tap conductors must:

- not be less than 20 amperes,
- be adequate for the appliance load to be served,
- not be longer than necessary for servicing the appliance.

The cost of installing one 50-ampere circuit might be higher than the cost of installing separate branch-circuits to each appliance because of the additional junction boxes, different size cables, cable connectors, conduit fittings, and wire connectors needed to complete the installation.

To compute the load demand for the counter-mounted cooking unit and the wall-mounted oven, the nameplate ratings are used.

Counter-mounted cooktop	7450 watts
Wall-mounted oven	6600 watts
Total	14050 watts

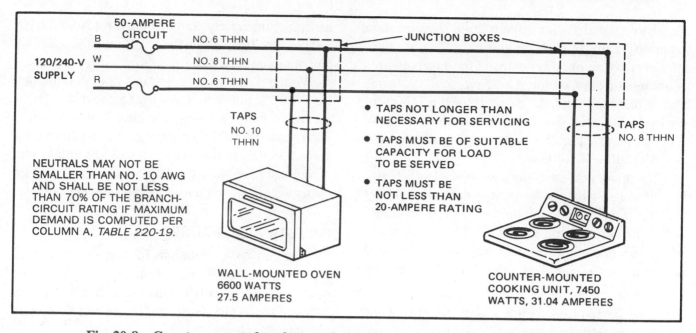

Fig. 20-8   Counter-mounted cooktop and wall-mounted oven connected to one circuit.

The total connected load of 14050 watts at 240 volts is

$$I = \frac{W}{E} = \frac{14050}{240} = 58.5 \text{ amperes}$$

According to *Note 1* of *Table 220-19*, for ranges rated over 12 kW but not over 27 kW, the demand load given in column A must be increased by 5% for each kilowatt or major fraction of a kilowatt in excess of 12 kW.

Assume that the combined load of the counter-mounted cooking unit and the wall-mounted oven is equivalent to the load of a freestanding range rated at 14050 watts. This value exceeds 12000 watts (12 kW) by 2050 watts, or roughly 2 kW.

The demand is determined as follows:

2 kW (kW over 12 kW) × 5% per kW = 10%
8 kW (from Column A of *Table 220-19*) × 0.10
    = 0.8 kW
Calculated demand = 8 kW + 0.8 kW = 8.8 kW or
                            = 8800 watts

and,

$$I = \frac{W}{E} = \frac{8800}{240} = 36.7 \text{ amperes}$$

Section 210-23(c) states that fixed cooking appliances that are fastened in place may be connected to 40- or 50-ampere circuits. If taps are to be made from 40- or 50- ampere circuits, the taps must

be able to carry the load to be served. In no case may the taps be less than 20 amperes, *Section 210-19(b)*.

For example, the 50-ampere circuit shown in figure 20-8 has No. 10 THHN and No. 8 THHN tap conductors. Each of these taps has an ampacity of more than 20 amperes. The branch-circuit requires No. 6 Type THHN conductors. *Section 210-19(b), Exception No. 2*, states that the load on the neutral conductor supplying the wall-mounted oven and the counter-mounted cooking unit may be based on 70% of the ampacity of the branch-circuit rating:

$$50 \times 0.7 = 35$$

According to *Table 310-16*, the neutral may be a No. 8 THHN conductor.

Watch out when selecting the conductors! Review "Caution When Making Connections" in the section on wall-mounted oven circuits in this unit.

USING A LOAD CENTER

In this residence, the built-in oven and range are very close to Subpanel B, in which case separate circuits are run to each appliance.

However, there are situations where the Main Panel and the built-in range components are far apart. In these instances, it might make economic sense to install a single subfeeder from the Main

Panel to a Load Center, figure 20-9, then create a separate circuit for each appliance.

The maximum power demand for the appliances is obtained from Column A of *Table 220-19*. It was shown that the calculated demand for the 7450-watt cooking unit and the 6600-watt oven is 8.8 kW (36.7 amperes). The two ungrounded conductors are No. 8 THHN wire (40 amperes).

The major portion of the load in an electric range is 240 volts. This results in low values of current flowing in the neutral conductor. *Section 210-19(b)* states that for electric ranges of 8¾ kW or more, where the maximum demand has been determined according to *Column A* of *Table 220-19*, the neutral conductor:

- may be reduced to not less than 70% of the branch-circuit rating, and

- shall not be smaller than No. 10AWG.

Thus, $40 \times 0.70 = 28$ amperes. This would be a No. 10 per *Table 310-16*.

In figure 20-9, a No. 8 THHN neutral conductor is shown because the current-carrying conductors in nonmetallic-sheathed cable are all of the same size. The equipment grounding conductor is No. 10. If the wiring method is electrical metallic tubing, there may be an economic advantage to install two No. 8 and one No. 10. *Table C2* in *Chapter 9* of the *NEC®* shows that three No. 8 THHN will fit into a ½-inch conduit or tubing. By computation three No. 8 THHN and one No. 10 THHN will also fit into a ½-inch conduit or tubing.

Watch out when selecting the conductors! Review "Caution When Making Connections" in the section on wall-mounted oven circuits in this unit.

The No. 8 THHN conductors are run from a 40-ampere circuit to the load center. The circuits supplying each appliance are protected at the load center according to their individual current ratings.

FREESTANDING RANGE

The demand calculations for a single, freestanding range are simple. For example, a single electric range is to be installed. This range has a rating of 14050 watts, which is exactly the same as the combined built-in units.

1. 14050 watts – 12000 watts = 2050 watts or 2 kW

2. According to *Note 1, Table 220-19*, a 5% increase for each kW in excess of 12 kW must be added to the load demand in Column A. 2 kW (kW over 12 kW) × 5% per KW = 10%

3. 8 kW (from Column A) × 0.10 = 0.8 kW

4. Calculated load = 8 kW + 0.8 kW = 8.8 kW

5. In amperes, the calculated load is:

$$I = \frac{W}{E} = \frac{8800}{240} = 36.7 \text{ amperes}$$

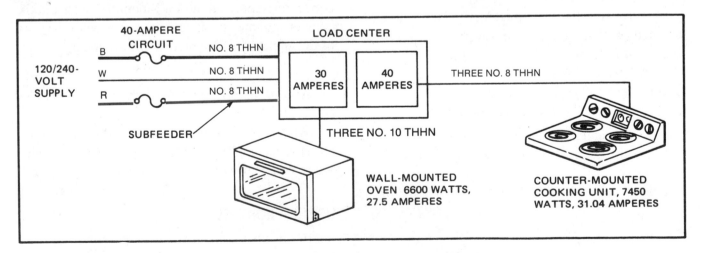

Fig. 20-9   Appliances connected to a load center.

No. 8 conductors could be used for the ungrounded conductors as they have an ampacity of 40 amperes, *Table 310-16*. The branch-circuit overcurrent device would be rated 40 amperes. The *rating* of a branch-circuit is based on the rating or setting of the overcurrent device protecting that branch-circuit, *Section 210-3*.

Section 210-19(b) states that for ranges of 8¾ kW or more, where the maximum demand had been computed using *Column A of Table 220-19*, the neutral may be reduced to not less than 70% of the branch-circuit rating, but never smaller than No. 10. The calculation is:

$$40 \times 0.70 = 28 \text{ amperes}$$

Checking ampacities in *Table 310-16*, we find that the neutral conductor for this range could be a No. 10, which has an ampacity of 30 amperes.

Using an 8/3 nonmetallic-sheathed cable, all current-carrying conductors are No. 8. The equipment grounding conductor is No. 10.

If a range is rated at not over 12 kW, then the maximum power demand is based on *Column A of Table 220-19*. For example, if the range is rated at 11.4 kW, the maximum demand is based on 8 kW:

$$I = \frac{W}{E} = \frac{8000}{240} = 33.3 \text{ amperes}$$

As in the previous example, two No. 8 THHN ungrounded conductors and one No. 10 THHN neutral conductor can be installed for this range. For ranges of 8¾ kW or higher rating, the minimum branch-circuit rating shall be 40 amperes, *Section 210-19(b)*.

Watch out when selecting the conductors! Review "Caution When Making Connections" in the section on wall-mounted oven circuits in this unit.

Disconnecting Means

The disconnecting means for an electric range may be a cord- and plug-connection located at the rear of the range that is accessible from the front by removing the lower drawer of the range, *Section 422-22(b)*.

Calculations When More Than One Electric Range, Surface-Mounted Cooking Unit, or Wall-Mounted Oven Is Connected to a Feeder or Service

▶ Although generally not an issue for wiring one- and two-family dwellings, the calculations for services and feeders of larger multifamily dwellings can be more complex where a number of household cooking units are to be installed.

Whenever electric ranges, surface-mounted cooking units, and/or wall-mounted ovens are involved, *Table 220-19* is used to determine the minimum ampacity requirements for the conductors. This table shows demand factors that are permitted to be applied to the nameplate rating of the appliance(s). This unit covers the computations for branch-circuits and feeders. Unit 29 covers the computations for services.

Table 220-19 has three columns. *Column A* is always used, with one exception. *Exception No. 3* shows that for household cooking units having a nameplate rating of more than 1¾ kW through 8¾ kW, *Columns B* and *C* are permitted to be used. *Column B* is for household cooking units rated less than 3½ kW. *Column C* is for household cooking units rated 3½ kW to 8¾ kW.

When using *Columns B* and *C*, apply the demand factors for the number of household cooking units whose kW ratings fall within that column. Then add the results of each column together. ◀

▶ **EXAMPLE:** Calculate the demand for a feeder that has four 3-kW wall-mounted ovens and four 6-kW counter-mounted cooktops connected to it.

Step 1: 4 ovens × 3 kW = 12 kW
12 kW × 66% = 7.92 kW

Step 2: 4 cooktops × 6 kW = 24 kW
24 kW × 50% = 12 kW

Step 3: 7.92 kW + 12 kW = 19.92 kW

Therefore, 19.92 kW is the demand load for the four ovens and four cooktops, even though the actual connected load is 36 kW. ◀

REVIEW

Note: Refer to the Code or the plans where necessary.

COUNTER-MOUNTED COOKING UNIT CIRCUIT ▲G

1. a. What circuit supplies the counter-mounted cooking unit in this residence? _____

    b. What is the rating of this circuit? _____

2. What three methods may be used to connect counter-mounted cooking units? _____

    _____

    _____

    _____

    _____

    _____

    _____

    _____

    _____

3. Is it permissible to use standard 60°C insulated conductors to connect all counter-mounted cooking units? Why? _____

    _____

    _____

4. What is the maximum operating temperature (in degrees Celsius) for a

    a. Type TW conductor? _____

    b. Type THW conductor? _____

    c. Type THHN conductor? _____

5. a. May service-entrance cable be used to connect counter-mounted cooking units? ___

    b. What Code section answers question (a)? _____

6. a. What section of the Code applies to grounding a counter-mounted cooking unit? ___

    _____

    b. What methods may be used to ground the counter-mounted cooking unit? _____

    _____

    _____

    _____

    _____

    _____

7. For electric ranges to have a computed demand of 8¾ kW or more, the neutral conductor can be reduced to _____% of the branch-circuit rating, but shall not be smaller than No. _____ AWG.

8. In the illustration of a typical seven-heat, eight-position switch, figure 20-6A, the wattage values for positions 4, 5, 6, and 7 are omitted. Calculate these values. Show all calculations.

   Position 4

   Position 5

   Position 6

   Position 7

9. A 120-volt flash or high-speed unit rated at 1000 watts produces _____ watts when connected briefly to the 240-volt source. Show all calculations.

10. When the voltage to an element is doubled, the wattage: (circle one)

    a. increases   b. decreases

11. By how much is the wattage in question 10 increased or decreased? (circle one)

    a. doubled   b. tripled   c. quadrupled   d. halved

12. One kilowatt equals _____ Btu per hour.

13. Older style electric ranges had heat control knobs that had "indents" that could be felt when rotating the knob to adjust temperature to three, five, or seven different heat positions. Most new electric ranges have control knobs that do not have "indents." This type of heat control is known as an _____ heat control.

WALL-MOUNTED OVEN CIRCUIT ▲F

1. To what circuit is the wall-mounted oven connected? _____

2. An oven is rated at 7.5 kW. This is equal to

   a. _____ watts.

   b. _____ amperes at 240 volts.

3. a. What section of the Code governs the grounding of a wall-mounted oven? _____

    _____

   b. By what methods may wall-mounted ovens be grounded? _____

    _____

    _____

    _____

    _____

    _____

4. What is the type and ampere rating of the overcurrent device protecting the wall-mounted oven in this residence? Hint: Refer to the *Schedule of Special Purpose Outlets* in the *Specifications*. _____

    _____

    _____

    _____

5. Approximately how many feet (meters) of cable are required to connect the oven in the residence? _____

    _____

6. When connecting a wall-mounted oven and a counter-mounted cooking unit to one feeder, how long are the taps to the individual appliances? _____

    _____

    _____

    _____

7. The branch-circuit load for a single wall-mounted self-cleaning oven or counter-mounted cooking unit shall be the _____ rating of the appliance.

8. A 6-kW counter-mounted cooking unit and a 4-kW wall-mounted oven are to be installed in a residence. Calculate the maximum demand according to *Column A, Table 220-19*. Show all calculations. Both appliances will be connected to the same branch-circuit.

9. The size of the neutral conductor supplying an electric range may be based on _____ % of the ampacity of the (ungrounded) conductor.

10. A freestanding electric range is rated 11.8 kW, 240 volts. For the following questions, base your answers on the 60°C column of *Table 310-16*.

    a. According to *Column A, Table 220-19*, what is the maximum demand?

    _____

    b. What is the correct rating for the branch-circuit overcurrent protection?

    _____

    _____

    _____

    c. What is the minimum size for the ungrounded conductors?

    _____

    _____

    _____

    d. What is the minimum size neutral conductor?

    _____

    _____

    _____

11. A double-oven electric range is rated at 18 kW, 240 volts. Calculate the maximum demand according to *Table 220-19*. Show all calculations.

12. For the range discussed in question 11:

    a. What size 60°C ungrounded conductors are required? _____

    b. What size 60°C neutral conductor is required? _____

    c. What size 75°C ungrounded conductors are required? The terminal block and lugs on the range, the branch-circuit breaker, and the panelboard are marked as suitable for 75°C wire. _____

    d. What size 75°C neutral conductor is required? Terminations the same as in c.

    _____

13. For ranges of 8¾ kW or higher rating, the minimum branch-circuit rating is _____ amperes.

14. When a separate circuit supplies a counter-mounted cooking unit, what Code reference tells us that it is "against Code" to apply the demand factors of *Table 220-19*, and requires us to compute the load based upon the appliance's actual nameplate rating? _____

15. A nonmetallic-sheathed cable is used to connect a wall-mounted oven. The insulated conductors are No. 10. What is the size of the equipment grounding conductor in this cable? _____

16. Match the following statements with the correct letter.

_____ A 30-ampere, 3-wire receptacle   (a) for the equipment ground
_____ A 50-ampere, 4-wire receptacle   (b) for the white neutral conductor
_____ The L-shaped blade on   (c) for the "hot" ungrounded conductors
        30-ampere plug cap   (d) NEMA 10-30R
_____ The horseshoe shaped slot   (e) NEMA 14-50R
        on a 50-ampere receptacle   (f) NEMA 14-50P
_____ A 50-ampere, 4-wire plug cap
_____ Terminals "X" and "Y"
_____ Terminal "W"
_____ Terminal "G"

UNIT 21

Special-Purpose Outlets—Food Waste Disposer ▲H, Dishwasher ▲I

OBJECTIVES

After studying this unit, the student will be able to

- install circuits for kitchen appliances such as a food waste disposer and a dishwasher.
- provide adequate running overcurrent protection for an appliance when such protection is not furnished by the manufacturer of the appliance.
- describe the difference in wiring for a food waste disposer with and without an integral ON-OFF switch.
- determine the maximum power demand of a dishwasher.
- make the proper grounding connections to the appliances.
- describe disconnecting means for kitchen appliances.

This unit discusses in detail the circuit requirements for food waste disposers and dishwashers as installed in residences. However, because there are many other types of appliances found in homes, the electrician needs to read carefully *Article 422* of the Code. *Article 422* covers all types of appliances in general, branch-circuit requirements, and installation requirements. It also discusses specific types of appliances.

CAUTION: Keep in mind that these 120-volt appliances are *not* permitted to be connected to the small appliance circuits provided for in the kitchen. The small appliance circuits are intended to be used for the numerous types of portable appliances used in the kitchen area.

FOOD WASTE DISPOSER ▲H

The food waste disposer outlet is shown on the plans by the symbol ▲H. The food waste disposer is rated at 7.2 amperes. This appliance is connected to Circuit B19, a separate 20-ampere, 120-volt branch-circuit in Panel B. To meet the specifications, No. 12 THHN conductors are used. The conductors terminate in a junction box provided on the disposer.

Overcurrent Protection

The food waste disposer is a motor-operated appliance. Food waste disposers normally are driven by a 1/4- or 1/3-hp, split-phase, 120-volt motor. Therefore, running overcurrent protection must be provided. Most food waste disposer manufacturers install a built-in thermal protector to meet Code requirements.

When running overcurrent protection is not built into the disposer, the electrician must install separate protection. Such protection must not exceed 125% of the full-load current rating of the motor. For example, a box-cover unit may be installed under the sink near the food waste disposer. A dual-element fuse of the proper size is then inserted into the box-cover unit. This unit also serves as the disconnecting means.

363

Disconnecting Means

All electrical appliances must be provided with some means of disconnecting the appliance, *Article 422, Part D*. The disconnecting means may be a separate ON-OFF switch, figure 21-1, a cord connec-tion, figure 21-2, or a branch-circuit switch or circuit breaker, within sight of the appliance or capable of being locked in the "OFF" position.

Some local codes require that food waste dispos-ers, dishwashers, and trash compactors be cord con-nected, as illustrated in figure 21-2, to make it easier to disconnect the unit, to replace it, to service it, and to reduce noise and vibration. The appliance shall be intended or identified for flexible cord connection, *Sections 400-7, 422-8(c)*, and *422-8(d)*.

Turning the Food Waste Disposer On and Off

Separate On-Off Switch. When a separate ON-OFF switch is used, a simple circuit arrangement can be made by running a two-wire supply cable to a switch box located next to the sink at a convenient location above the countertop. Not only is this con-venient, but it positions the switch out of the reach of children, figure 21-1. A second cable is run from the switch box to the junction box of the food waste disposer. A single-pole switch in the switch box pro-vides ON-OFF control of the disposer.

In the kitchen of this residence, the food waste disposer is controlled by a single-pole switch locat-ed to the right of the kitchen sink above the counter-top, in combination with the light switch and receptacle.

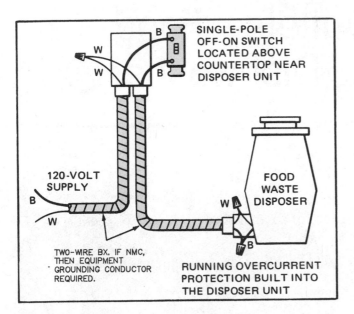

Fig. 21-1   Wiring for a food waste disposer operated by a separate switch located above countertop near the sink. Many local codes require this wall switch for all installations, to be sure that there is a safe means of easily disconnecting the appliance.

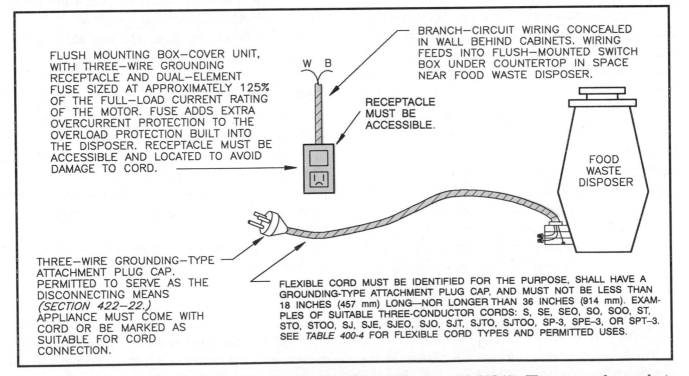

Fig. 21-2   Typical cord connection for a food waste disposer, *Section 422-8(d)(1).* **The same rules apply to built-in dishwashers and trash compactors, except the cord length must be 3 to 4 feet (0.914 to 1.22 m).**

Another possibility, but not as convenient, is to install the switch for the disposer inside the cabinet space directly under the sink near the food waste disposer.

Integral On-Off Switch. A food waste disposer may be equipped with an integral prewired control switch. This integral control starts and stops the disposer when the user twists the drain cover into place. An extra ON-OFF switch is not required with this type of control. To connect the disposer, the electrician runs the supply conductors directly to the junction box on the disposer and makes the proper connections, figure 21-3.

Disposer with Flow Switch in Cold Water Line. Some manufacturers of food waste disposers recommend that a *flow* switch be installed in the cold water line under the sink, figure 21-4. This switch is connected in series with the disposer motor. The flow switch prevents the disposer from operating until a predetermined quantity of water is flowing

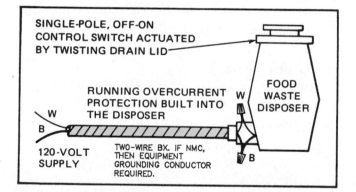

Fig. 21-3 Wiring for a food waste disposer with an integral ON/OFF switch.

through the disposer. The cold water helps prevent clogged drains by solidifying any grease in the disposer. Thus, the addition of a flow switch means that the disposer cannot be operated without water.

Figure 21-4 shows one method of installing a food waste disposer with a flow switch in the cold water line.

Grounding

Section 422-16 states that where required by *Article 250*, all exposed, noncurrent-carrying parts are to be grounded in the manner specified in *Article 250*.

The presence of any of the conditions outlined in *Sections 250-42* through *250-45* require that all electrical appliances in the dwelling be grounded. Food waste disposers can be grounded using any of the methods covered in the previous units, such as the metal armor of armored cable or the separate equipment grounding conductor of non-metallic-sheathed cable, *Sections 250-57(b)* and *250-91(b)*. NEVER, NEVER ground a food waste disposer to the grounded circuit conductor!

DISHWASHER

The dishwasher is supplied by a separate 20-ampere circuit connected to Circuit B5. No. 12 THHN conductors are used to connect the appliance. The dishwasher has a ⅓-hp motor rated at 7.2 amperes and 120 volts. The outlet for the dishwasher is shown by the symbol. During the drying cycle of the dishwasher, a thermostatically controlled, 1000-watt electric heating element turns on.

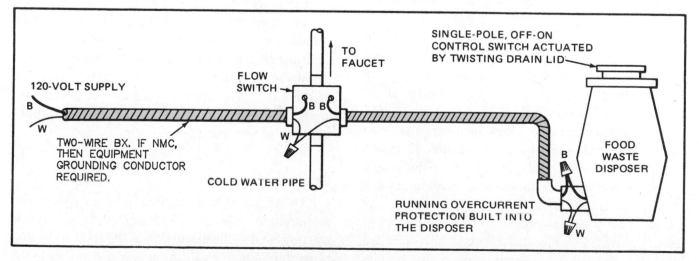

Fig. 21-4   Disposer with a flow switch in the cold water line.

Actual connected load		Circuit calculation	
Motor	7.20 A	$7.20 \times 1.25 =$	9.00 A
Heater	8.33 A		8.33 A
Total	15.53 A		17.33 A

For most dishwashers, the motor does not run during the drying cycle. Thus, the actual maximum demand on the branch-circuit would only be the larger of the two loads which in the previous example is the 1,000-watt heating element.

Water heater manufacturers ship their water heaters with the thermostat set as low as possible. When the water heater is installed, the thermostat must be set at recommended settings (such as 120°F). In fact, because of the problem of people being scalded, many municipalities and states have implemented code requirements that water heater thermostats not be set over 120°F. This temperature is OK for baths and showers, however, 120°F is not hot enough to properly clean dishes and glasses. The manufacturers of the dishwashers provide a feature that turns on the heating element for the final rinse cycle. This raises the water temperature from 120°F to the desired 140° to 145°F.

See unit 19 for a discussion on water heaters and the dangers of scalding with hot water of varying temperatures.

Wiring the Dishwasher

The dishwasher manufacturer supplies a terminal or junction box for the electrical hook-up. The electrician must make sure the supply conductors (cable or flex conduit) are brought in at the proper location. Usually, the electrician brings the supply conductors into the space under the sink, but the supply conductors could also be brought up from underneath or from behind the dishwasher. The manufacturer's installation manual provides all of the necessary information.

The dishwasher is a motor-operated appliance and requires motor overloaded protection. Overload protection is an integral part of the dishwasher's motor, and prevents the motor from burning out if it becomes overloaded, stalled, or overheated for any reason. The branch-circuit protection for the appliance is provided by the branch-circuit fuse or circuit breaker. Supplemental overcurrent protection is often provided by installing a box-cover "switch/fuse unit"

under the sink of a type similar to that shown in figure 16-19. For a cord/plug connection, the box-cover unit would be a "receptacle/fuse" type (fig. 21-2). Time-delay, dual-element Type S fuses sized at approximately 125% of the nameplate rating of the dishwasher would be installed in the box-cover unit. This also serves as the disconnecting means according to *Section 422-20*.

Common Feed for Dishwasher and Food Waste Disposer

There may be instances where the distance from the kitchen sink area to the electrical load center is very long, in which case it may be more economical to run one branch-circuit to feed the dishwasher and food waste disposer, instead of two separate circuits. In this case, the electrician must split the circuit in a junction box mounted in the cabinet space below the food waste disposer. If a box-cover unit is used and has the proper fuse sizes, then separate overcurrent protection and separate disconnecting means are available. The connections for this circuit are similar to those for the overhead garage door opener circuit described in figures 16-18, 16-19, and 16-20.

Both the food waste disposer and dishwasher in this residence are permitted to be supplied with one 20-ampere circuit because the circuit calculations are:

Food waste disposer
$$= 7.2 \times 1.25 = 9 \text{ amperes}$$
Heater in dishwasher
$$= 1000 \div 120 = 8.33 \text{ amperes}$$
Maximum calculated demand $= 17.33 \text{ amperes}$

For higher current or wattage rated appliances, it may be necessary to install a 30-ampere branch-circuit. Do not connect the garbage disposer or dishwasher to any of the small appliance branch-circuits in the kitchen. To do so would be a violation of the Code. The small appliance circuits are intended to supply cord-connected *portable* appliances only! See unit 12.

Grounding

The dishwasher is required to be grounded. The grounding of appliances is described fully in previous units and in the discussion of the food waste disposer. NEVER, NEVER ground a dishwasher to the grounded circuit conductor!

PORTABLE DISHWASHERS

In addition to built-in dishwashers, portable models are available. These portable units have one hose that is attached to the water faucet and a water drainage hose that hangs in the sink. The obvious location for a portable dishwasher is near the sink. Thus, the dishwasher probably will be plugged into the receptacle nearest the sink.

▶ *Section 210-8(a)(6)* requires that ALL 125-volt, single-phase, 15- and 20-ampere receptacles that serve countertop surfaces in kitchens must be GFCI protected. ◀

Portable dishwashers are supplied with a three-wire cord containing two circuit conductors and one equipment grounding conductor and a three-wire grounding-type attachment plug cap. If the three-wire plug cap is plugged into the three-wire grounding-type receptacle, the dishwasher is adequately grounded. Whenever there is a chance that the user may touch the appliance and a grounded surface at the same time, such as the water pipe or faucet, **the equipment must be grounded to reduce the shock hazard**. The exceptions to *Sections 250-45(c)* and *422-8(d)* permit the double insulation of appliances and portable tools instead of grounding. Double insulation means that the appliance or tool has two separate insulations between the hot conductor and the person using the device. Although refrigerators, cooking units, water heaters, and other large appliances are not double insulated, many portable hand-held appliances and tools have double insulation. All double-insulated tools and appliances must be marked by the manufacturer to indicate this feature.

CORD CONNECTION OF FIXED APPLIANCES

As previously discussed, food waste disposers, built-in dishwashers, and trash compactors in residential occupancies are permitted to be cord connected, figure 21-2.

The Code electrical installation requirements for dishwashers and trash compactors are very similar to the requirements for food waste disposers, except the cord is required to be 3 to 4 feet (0.914 to 1.22 m) long. The receptacle outlet should be located in the under-cabinet space adjacent to the appliance, which generally would be under the sink, allowing for easy access if the appliance needs to be disconnected. It is *not* recommended that the receptacle outlet be installed behind the appliance.

Receptacle outlets installed beneath the sink for the purpose of cord connecting a dishwasher, trash compactor, or food waste disposer are not required to be of the GFCI type, because these receptacles are not intended to serve the countertop area. Repeating what was stated at the beginning of this chapter, these receptacle outlets must not be connected to the required small appliance 20-ampere branch-circuits that serve the kitchen and/or dining room areas.

Some electrical inspectors feel that using the cord-connected method of connecting these appliances is better than direct connection (sometimes referred to as "hard-wired") because servicing of these appliances is generally done by appliance repair people.

Section 400-7 covers the uses permitted for flexible cords. Specifically, *Section 400-7(a)(8)* states that the appliance must be fastened in such a manner as to allow ready removal of the appliance for maintenance and repairs. The appliance must be intended or identified for flexible cord connection. The flexible cord must also be *identified for the purpose* per *Section 422-8(d)(1)* and *(2)*.

Simply stated, the above code requirements mean that field installing of "any type of cord using any type of connector" would be a violation of the *NEC®* and would invalidate the appliance's listing.

REVIEW

Note: Refer to the Code or the plans where necessary.

FOOD WASTE DISPOSER CIRCUIT ⒜H

1. How many amperes does the food waste disposer draw? _____

2. a. To what circuit is the food waste disposer connected? _____

   b. What size wire is used to connect the food waste disposer? _____

3. Means must be provided to disconnect the food waste disposer. The homeowner need not be involved in electrical connections when servicing the disposer if the disconnecting means is _____

_____

4. How is running overcurrent protection provided in most food waste disposer units?

_____

5. When running overcurrent protection is not provided by the manufacturer or if additional backup overcurrent protection is wanted, dual-element, time-delay fuses may be installed in a separate box-cover unit. These fuses are sized at not over _____ % of the full-load rating of the motor.

6. Why are flow switches sometimes installed on food waste disposers? _____

_____

7. What Code sections relate to the grounding of appliances? _____

_____

8. Do the plans show a wall switch for controlling the food waste disposer? _____

9. A separate circuit supplies the food waste disposer in this residence. How many feet (meters) of cable will be required to connect the disposer? _____

10. If a receptacle is installed underneath the sink for the purpose of plugging in a cord-connected food waste disposer, must the receptacle be GFCI protected? _____

DISHWASHER CIRCUIT Ⓐ₁

1. a. To what circuit is the dishwasher in this residence connected? _____

   b. What size wire is used to connect the dishwasher? _____

2. The motor on the dishwasher is (circle one)

   a. ¼ hp   b. ⅓ hp   c. ½ hp

3. The heating element is rated at (circle one)

   a. 750 watts   b. 1000 watts   c. 1250 watts

4. How many amperes at 120 volts do the following heating elements draw?

   a. 750 watts _____

   b. 1000 watts _____

   c. 1250 watts _____

5. How is the dishwasher in this residence grounded? _____

_____

_____

_____

_____

6. What type of cord is used on most portable dishwashers? _____

7. How is a portable dishwasher grounded? _____

_____

_____

8. What is meant by double insulation? _____

_____

_____

_____

9. Who is to furnish the dishwasher? _____

_____

10. What article of the Code specifically addresses electrical appliances? _____

UNIT 22

Special-Purpose Outlets for the Bathroom Ceiling Heat/Vent/Lights, ⊕J ⊕K, the Attic Fan ⊕L, and the Hydromassage Tub ⊕A

OBJECTIVES

After studying this unit, the student will be able to

- explain the operation and switching sequence of the heat/vent/light.
- describe the operation of a humidistat.
- install attic exhaust fans with humidistats, both with and without a relay.
- list the various methods of controlling exhaust fans.
- understand advantages of installing exhaust fans in residences.
- understand the electrical circuit and Code requirements for hydromassage bathtubs.
- understand grounding requirements for electrical equipment.

BATHROOM CEILING HEATER CIRCUITS ⊕K ⊕J

Both bathrooms contain a combination heater, light, and exhaust fan installed in the ceiling. The heat/vent/light is such a device and is shown in figure 22-1. The symbols ⊕K and ⊕J represent the outlets for these units.

Each heat/vent/light contains a heating element similar to the surface burners of electric ranges. The appliances also have a single-shaft motor with a blower wheel, a lamp with a diffusing lens, and a means of discharging air to the outside of the dwelling.

The heat/vent/lights specified for this residence are rated 1500 watts at 120 volts. The unit in the master bedroom bathroom ⊕J is connected to Circuit A12, a 20-ampere, 120-volt circuit. In the front bedroom bathroom, the unit is connected to Circuit A11, also a 20-ampere, 120-volt circuit.

The current rating of the 1500-watt unit is

$$I = \frac{W}{E} = \frac{1500}{120} = 12.5 \text{ amperes}$$

Article 424 of the *National Electrical Code®* covers fixed electric space heating. *Section 424-3(b)* requires that the branch-circuit conductors and the overcurrent protective device (fuses or circuit breakers) shall be not less than 125% of the appliances' ampere rating. Thus, 20-ampere, 120-volt branch-circuits rather than 15-ampere branch-circuits have been chosen to supply the electric ceiling heaters in the bathroom.

Wiring

In figure 22-2, we see the wiring diagram for the heat/vent/light. The 120-volt line is run from Panel A to the wall box. A raceway (flexible or EMT) is run

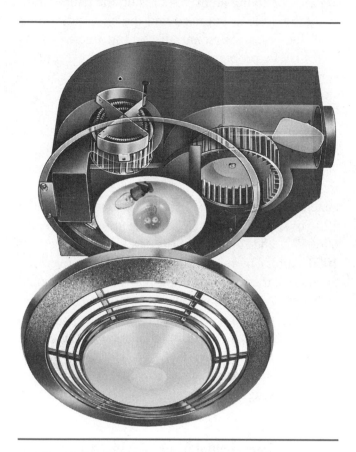

Fig. 22-1   Heat/vent/light. *Courtesy* of NuTone.

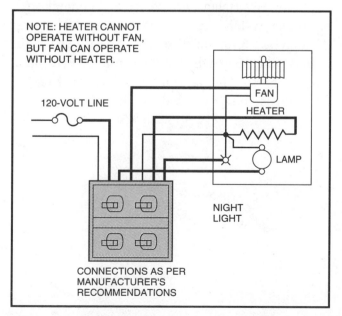

NOTE: HEATER CANNOT OPERATE WITHOUT FAN, BUT FAN CAN OPERATE WITHOUT HEATER.

120-VOLT LINE

FAN

HEATER

LAMP

NIGHT LIGHT

CONNECTIONS AS PER MANUFACTURER'S RECOMMENDATIONS

Fig. 22-2   Wiring for heat/vent/light.

between the wall box where the switches are located to the heat/vent/light unit. In this raceway, five wires are needed—one common white grounded circuit conductor, and one conductor each for the heating element, fan motor, 100-watt lamp, and the 7-watt night light. The four ungrounded switch leg wires may be any color except white, natural gray, or green, as discussed in *Sections 210-5* and *310-12* of the Code. Because the circuit rating is 20 amperes, No. 12 AWG conductors are installed. See the Schedule of Special Purpose Outlets contained in the specifications for this text.

Although done many times by nontrained electricians, it would be in violation of *Section 300-2* of the *NEC*® to run one 3-wire and one 2-wire cable between the wall box and heat/vent/light. *Section 300-2* states that "all conductors of the same circuit and, where used, the neutral and all equipment grounding conductors shall be contained within the same raceway, cable tray, trench, cable, or cord.

The switch location requires a two-gang opening, with the wall box being large enough to accommodate seven No. 12 AWG conductors, four switches, and cable clamps. Box sizing is covered in units 2 and 8.

Operation of the Heat/Vent/Light

The heat/vent/light is controlled by four switches, as shown in figure 22-2:

- one switch for the heater
- one switch for the light
- one switch for the exhaust fan
- one switch for the night light

When the heater switch is turned on, the heating element begins to give off heat. The heat activates a bimetallic coil attached to a damper section in the housing of the unit. The heat-sensitive coil expands until the damper closes the discharge opening of the exhaust fan. Air is taken in through the outer grille of the unit. The air is blown downward over the heating element. The blower wheel circulates the heated air back into the room area.

If the heater is turned off so that only the exhaust fan is on, the air is pulled into the unit and is exhausted to the outside of the house by the blower wheel.

Exhaust fans and fixtures are available with integral automatic moisture sensors and motion detectors. Step into a dark bathroom, and the light goes on. Step into and run the shower, and the exhaust fan turns on at a predetermined moisture level.

Grounding the Heat/Vent/Light

The heat/vent/light must be grounded. Equipment grounding is discussed elsewhere in this text. Do *not* ground the metal housing of the heat/vent/light (or any other piece of equipment for that matter) to the *grounded circuit conductor*. This is in violation of *Section 250-61(b),* and to do so could result in serious electrical shock hazard. *Section 250-61(b)* states that the grounded circuit conductor *shall not* be used to ground noncurrent-carrying metal parts of equipment *anywhere* beyond the main service disconnect. ▶ In the past, *Section 250-60* permitted the frames of ranges, wall-mounted ovens, counter-mounted cooking units, and clothes dryers to the grounded neutral conductor, but this is no longer permitted. ◀

ATTIC EXHAUST FAN CIRCUIT ⒶL

An exhaust fan is mounted in the hall ceiling between the master bedroom and the front bedroom, figure 22-3(A). When running, this fan removes hot, stagnant, humid, or smoky air from the dwelling. It will draw in fresh air through open windows. Such an exhaust fan can be used to lower the indoor temperature of the house as much as 10° to 20°F.

The exhaust fan in this residence is connected to Circuit A10, a 120-volt, 15-ampere circuit in the main panel utilizing No. 14 THHN wire.

Exhaust fans can also be installed in a gable of a house, figure 22-3(B), or in the roof.

On hot days, the air in the attic may reach a temperature of 150°F or more. Thus, it is desirable to provide a means of taking this attic air to the outside. The heat from the attic can radiate through the ceiling into the living areas. As a result, the temperature in the living area is raised and the air-conditioning load increases. Personal discomfort increases as well. These problems are minimized by properly vented exhaust fans, figure 22-4.

Fan Operation

Many sizes and types of attic exhaust fans are available. The attic exhaust fan in this residence has a ¼-horsepower direct-drive, permanent split capacitor motor. Direct-drive means that the fan blade is attached directly to the shaft of the motor. It is rated 120-volts, 60 Hz (60 cycles per second), 5.8 amperes, 696 VA, and operates at a maximum speed of 1050 r/min. The motor is protected against overload by internal thermal protection. The motor is mounted on rubber cushions for quiet operation.

When the fan is not running, the louver remains in the CLOSED position. When the fan is turned on, the louver opens. When the fan is turned off, the louver automatically closes to prevent the escape of air from the living quarters.

Fan Control

Typical residential exhaust fans can be controlled by a variety of switches.

Some electricians, architects, and home designers prefer to have fan switches (controls), figures 22-5 and 22-6, mounted 6 feet (1.83 m) above the floor so that they will not be confused with other wall switches. This is a matter of personal preference and should be verified with the owner.

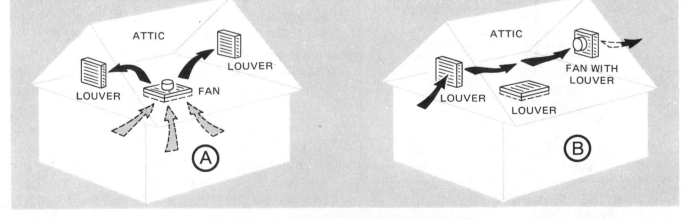

Fig. 22-3   Exhaust fan installation in an attic.

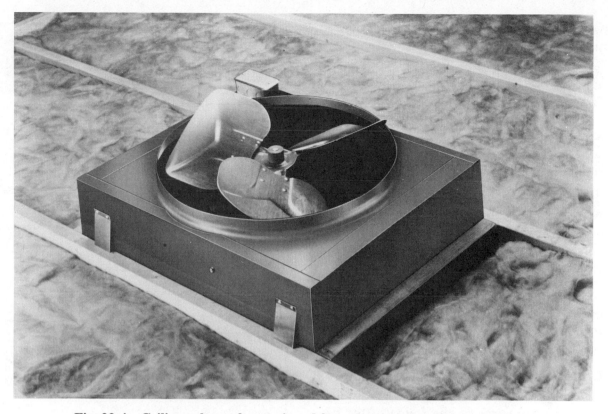

Fig. 22-4   Ceiling exhaust fan as viewed from the attic. *Courtesy* of NuTone.

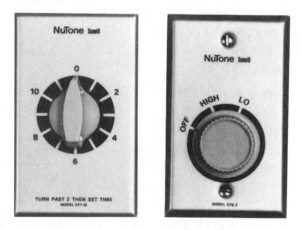

Fig. 22-5   Speed control switches. *Courtesy* of NuTone.

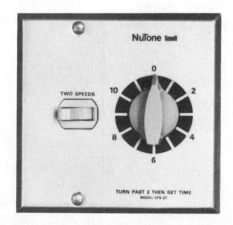

Fig. 22-6   Timer switch. *Courtesy* of NuTone.

Figure 22-7 shows some of the ways that an exhaust fan can be connected.

1. A simple ON-OFF switch. Refer to figure 22-7(A).

2. A speed control switch that allows multiple and/or infinite number of speeds. Refer to figure 22-7(B). (The exhaust fan in this residence is controlled by this type of control.) See also figure 22-5.

3. A timer switch that allows the user to select how long the exhaust fan is to run, up to 12 hours. Refer to figure 22-7(C). See also figure 22-6.

4. A humidity control switch that senses moisture build-ups. Refer to figure 22-7(D). See "Humidity Control" for further details on this type of switch.

5. Exhaust fans mounted into end gables or roofs of residences are available with an adjustable temperature control on the frame of the fan. This thermostat generally has a start range of 70° to 130°F, and will automatically stop at a temperature 10° below the start setting, thus providing totally automatic ON-OFF control of the exhaust fan. Refer to figure 22-7(E).

6. A high temperature automatic heat sensor that will shut off the fan motor when the temperature reaches 200°F. This is a safety feature so that the fan will not spread a fire. Connect in series with other control switches. Refer to figure 22-7(F).

7. A combination of controls, this circuit shows an exhaust fan controlled by an infinite-speed switch, a humidity control, and a high temperature heat sensor. Note that the speed control and the humidity control are connected in parallel so that either can start the fan. The high temperature heat sensor is in series so that it will shut off the power to the fan when it senses 200°F, even if the speed control or humidity control are in the ON position. Refer to figure 22-7(G).

Overload Protection for the Fan Motor

Several methods can be used to provide running overload protection for the ¼-hp attic fan motor. For

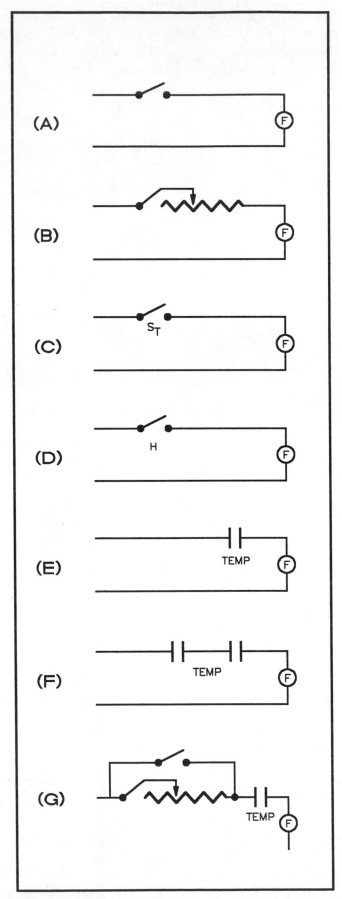

Fig. 22-7   Different options for the control of exhaust fans.

example, an overload device can be built into the motor or a combination overload tripping device and thermal element can be added to the switch assembly. Several manufacturers of motor controls also make switches with overcurrent devices. Such a switch may be installed in any standard flush switch box opening (2" × 3"). A pilot light may be used to indicate that the fan is running. Thus a two-gang switch box or 1½" × 4" square outlet box with a two-gang raised plaster cover may be used for both the switch and the pilot light.

NEC® Section 430-32(c) states that for automatically started motors, the running overload device must not exceed 125% of the full-load running current of the motor. *Table 430-148* shows that the exhaust fan motor has a full-load current rating of 5.8 amperes. Thus the overload device rating is 5.8 × 1.25 = 7.25 amperes.

Type S fuses may be installed for backup protection. These would be installed in the box-cover unit adjacent to the fan. Ratings of 6¼, 7, or 8 amperes would be a good choice, because this particular fan motor does have integral overload protection.

Branch-Circuit Short-Circuit Protection for the Fan Motor

For a full-load current rating of 5.8 amperes, *Table 430-152* shows that the rating of the branch circuit overcurrent device shall not exceed 20 amperes if fuses are used (5.8 × 3 = 17.4 amperes; the next standard fuse size is 20 amperes). The overcurrent device rating shall not exceed 15 amperes if circuit breakers are used (5.8 × 2.5 = 14.5 amperes; the next standard circuit breaker size is 15 amperes).

The exhaust fan is connected to Circuit A10, which is a 15-ampere, 120-volt circuit. This circuit does not feed any other loads. The circuit supplies the exhaust fan only.

Time-delay fuses sized at 115 to 125% of the full-load ampere rating of the motor provide both running overload protection and branch-circuit short-circuit protection. If the motor is equipped with inherent, built-in automatic reset thermal overload protection, then the time-delay fuse in the circuit provides secondary backup protection. Thus, we have double protection against possible motor burnout.

HUMIDITY CONTROL

An electrically heated dwelling can experience a problem with excess humidity due to the "tightness" of the house, because of the care taken in the installation of the insulation and vapor barriers. High humidity is uncomfortable. It promotes the growth of mold and the deterioration of fabrics and floor coverings. In addition, the framing members, wall panels, and plaster or drywall of a dwelling may deteriorate because of the humidity. Insulation must be kept dry or its efficiency decreases. A low humidity level can be maintained by automatically controlling the exhaust fan. One type of automatic control is the *humidistat*. This device starts the exhaust fan when the relative humidity reaches a high level. The fan exhausts air until the relative humidity drops to a comfortable level. The comfort level is about 50% relative humidity. Adjustable settings are from 0 to 90% relative humidity.

The electrician must check the maximum current and voltage ratings of the humidistat before the device is installed. Some humidistats are low-voltage devices and require a relay. Other humidistats are rated at line voltage and can be used to switch the motor directly (a relay is not required). However, a relay must be installed on the line-voltage humidistat if the connected load exceeds the maximum allowable current rating of the humidistat.

The switching mechanism of a humidistat is controlled by a nylon element, figure 22-8. This element is very sensitive to changes in humidity.

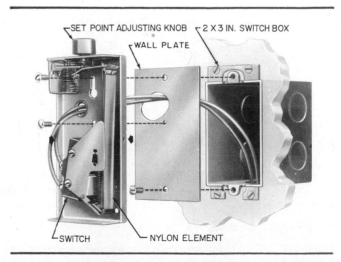

Fig. 22-8 Details of a humidity control used with an exhaust fan.

A bimetallic element cannot be used because it reacts to temperature changes only.

Wiring

The humidistat and relay in the dwelling are installed using two-wire No. 14 AWG cable. The cable runs from Circuit A10 to a 4-inch square, 1½-inch-deep outlet box located in the attic near the fan, figure 22-9(B). A box-cover unit is mounted on this outlet box. The box-cover unit serves as a disconnecting means within sight of the motor as required by *Sections 422-27* and *430-102*. Motor overload protection is provided by the dual-element time-delay fuses installed in the box-cover unit.

Mounted next to the box-cover unit is a relay that can carry the full-load current of the motor. Low-voltage thermostat wire runs from this relay to the humidistat. The humidistat is located in the hall between the bedrooms. The thermostat cable is a two-conductor cable because the humidistat has a single-pole, single-throw switching action. The line-voltage side of the relay provides single-pole switching to the motor. Normally the white grounded conductor is not broken.

Although the example in this residence utilizes a relay in order to use low-voltage wiring, it is certainly proper to install line-voltage wiring and use line-voltage thermostats and humidity controls. These are becoming more and more popular, with the switches actually being solid-state speed controllers permitting operation of the fan at an infinite number of speeds between high and low settings.

Grounding

As with most other appliances in this residence, the metal parts of the exhaust fan are to be grounded

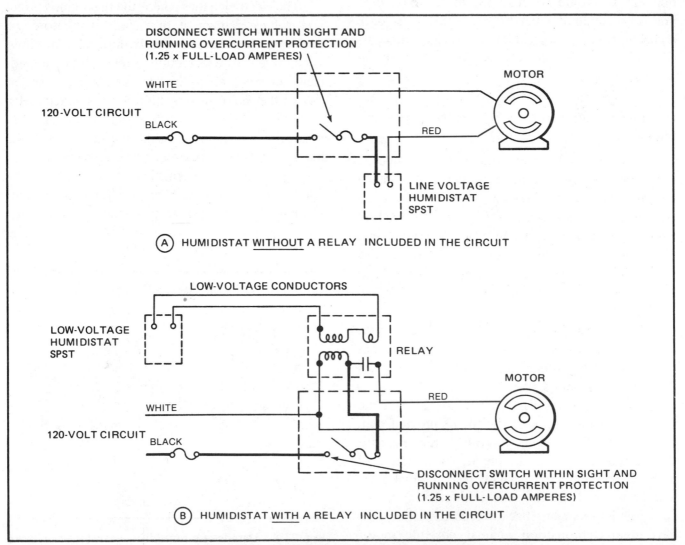

Fig. 22-9   Wiring for the humidistat control.

to the equipment grounding conductor that is part of the cable, or through the metal of the raceway if electrical metallic tubing is the wiring method used. Never ground equipment to the white grounded circuit conductor. This is prohibited in *Section 250-61(b)* of the Code.

APPLIANCE DISCONNECTING MEANS

To clean, adjust, maintain, or repair an appliance, it must be disconnected to prevent personal injury. *Sections 422-20* through *422-27* outline the basic methods for disconnecting fixed, portable, and stationary electrical appliances. Note that each appliance in the dwelling conforms to one or more of the disconnecting methods listed. The more important Code rules for appliance disconnects are as follows:

- each appliance must have a disconnecting means.

- the disconnecting means may be a separate disconnect switch if the appliance is permanently connected.

- the disconnecting means may be the branch-circuit switch or circuit breaker if the appliance is permanently connected and not over ⅛ hp or 300 volt-amperes.

- the disconnecting means may be an attachment plug cap if the appliance is cord connected.

- the disconnecting means may be the unit switch on the appliance only if other means for disconnection are also provided. In a single-family residence, the service disconnect serves as the other means.

- the disconnect must have a positive ON-OFF position.

- the disconnect must be capable of being locked in the "OFF" position or be within sight of a motor-driven appliance where the motor is more than ⅛ hp.

Read *Sections 422-20* through *422-27* for the details of appliance disconnect requirements.

HYDROMASSAGE TUB CIRCUIT ⒶA

The master bedroom is equipped with a hydromassage tub. A hydromassage bathtub is sometimes referred to as a whirlpool bath.

Section 680-4 of the Code defines a hydromassage bathtub as "a permanently installed bathtub equipped with a recirculating piping system, pump and associated equipment. It is designed so that it can accept, circulate, and discharge water upon each use."

In other words, fill—use—drain!

The significant difference between a hydromassage bathtub and a regular bathtub is the recirculating piping system and the electric pump that circulates the water. Both types of tubs are drained completely after each use.

Spas and hot tubs are intended to be filled, then used. They are *not* drained after each use because they have a filtering and heating system.

Another basic difference between a spa and a hot tub is that spas are constructed of manufactured material such as fiberglass, acrylics, plastics, or concrete, whereas hot tubs are made of wood, such as redwood, cypress, cedar, oak, or teak.

Electrical Connections

The hydromassage tub is fed with a separate Circuit A9, a 20-ampere, 120-volt circuit using No. 12 AWG THHN conductors.

The Schedule of Special-Purpose Outlets indicates that the hydromassage bathtub in this residence has a ½-hp motor that draws 10 amperes.

Section 680-70 requires that the circuit supplying a hydromassage bathtub must be GFCI protected as discussed in unit 6. ▶ *Section 680-70* also requires that all receptacles within 5 feet (1.52 m) or less of a hydromassage tub be GFCI protected. ◀

Section 680-71 tells us that lighting fixtures, switches, receptacles, and other electrical equipment located in the same room, and not directly associated with the hydromassage bathtub, shall be installed in accordance with the requirements of *Chapters 1* through *4* of the Code. This section recognizes that a hydromassage tub is used in much the same manner as a regular bathtub, and as such, does not introduce any additional hazards other than those of a regular bathtub.

Spas, hot tubs, and swimming pools do introduce additional electrical shock hazards, and these issues are covered in unit 30.

The hydromassage bathtub electrical control is prewired by the manufacturer. All that is necessary for the electrician to do is to run the separate

20-ampere, 120-volt GFCI-protected circuit to the end of the tub where the pump and control are located. Generally the manufacturer will supply a 3-foot length of watertight flexible conduit that contains one black, one white, and one green equipment grounding conductor, figure 22-10. Make sure that the equipment grounding conductor of the circuit is properly connected to the green grounding conductor of the hydromassage tub.

DO NOT CONNECT THE GREEN GROUNDING CONDUCTOR TO THE GROUNDED (WHITE) CIRCUIT CONDUCTOR!

Proper electrical connections are made in the junction box where the branch-circuit wiring is to be connected to the hydromassage wires. This junction box, because it contains splices, must be accessible.

The pump and power panel may also need servicing. Access may be from underneath or the end, whichever is convenient for the installation, figure 22-11.

Figure 22-12 is a photograph of a typical hydromassage tub.

Fig. 22-11   The basic roughing-in of a hydromassage bathtub is similar to that of a regular bathtub. The electrician runs a separate 20-ampere, 120-volt GFCI-protected circuit to the area where the pump and control are located. Check manufacturer's specifications for this data. An access panel from the end or from below is necessary to service the wiring, the pump, and the power panel.

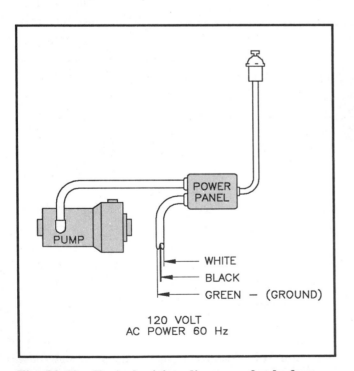

Fig. 22-10   Typical wiring diagram of a hydromassage tub showing the motor, power panel, and electrical supply leads.

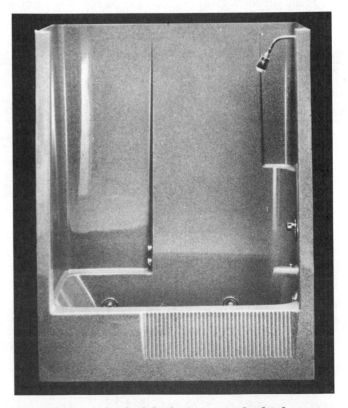

Fig. 22-12   A typical hydromassage bathtub, sometimes referred to as a whirlpool. *Courtesy* of the Kohler Company.

REVIEW

Note: Refer to the Code or the plans where necessary.

BATHROOM CEILING HEATER CIRCUITS Ⓐ﹨ Ⓐ﹈

1. What is the wattage rating of the heat/vent/light? _____

2. To what circuits are the heat/vent/lights connected? _____

3. Why is a 4-inch square, 2⅛-inch deep box with a two-gang raised plaster ring or similar deep box used for the switch assembly for the heat/vent/light? _____

   _____

   _____

   _____

   _____

4. a. How many wires are required to connect the wall switches and the heat/vent/light?

   _____

   b. What size wires are used? _____

5. Can the heating element be energized when the fan is not operating? _____

6. Can the fan be turned on without the heating element? _____

7. What device can be used to provide automatic control of the heating element and the fan of the heat/vent/light? _____

8. Where does the air enter the heat/vent/light? _____

9. Where does the air leave this unit? _____

10. Who is to furnish the heat/vent/light? _____

    _____

11. For a ceiling heater rated 1200 watts at 120 volts what is the current draw? _____

    _____

12. An electrician wired up a heat/vent/light as described in this unit. The wiring diagram showed that the electrician needed to run five conductors between the switches and the heat/vent/light. The electrician used one 12/3 nonmetallic-sheathed cable and one 12/2 nonmetallic-sheathed cable. Both cables had equipment grounding conductors. During the rough-in inspection, the electrical inspector cited "Non compliance with *Section 300-2.*" What was the problem? _____

    _____

    _____

ATTIC EXHAUST FAN CIRCUIT Ⓐ﹐

1. What is the purpose of the attic exhaust fan? _____

   _____

   _____

2. At what voltage does the fan operate? _____

3. What is the horsepower rating of the fan motor? _____

4. Is the fan direct or belt driven? _____

5. How is the fan controlled? _____

6. What is the setting of the running overcurrent device? _____

7. What is the rating of the running overcurrent protection if the motor is rated at 10 amperes? _____

8. What is the basic difference between a thermostat and a humidistat? _____

_____

_____

9. What size conductors are to be used for this circuit? _____

10. How many feet (meters) of cable are required to complete the wiring for the attic exhaust fan circuit? _____

11. May the metal frame of the fan be grounded to the grounded circuit conductor?

_____

12. What section of the Code prohibits grounding equipment to a grounding circuit conductor? _____

HYDROMASSAGE BATHTUB CIRCUIT ⓐA

1. What circuit supplies the hydromassage bathtub? _____

2. Which of the following statements "meets Code"?

   a. The circuit shall have GFCI protection.

   b. The circuit shall not have GFCI protection.

   Circle the correct statement.

3. What conductor size feeds the hydromassage tub? _____

4. What is the fundamental difference between a hydromassage bathtub and a spa?

_____

_____

5. What sections of the Code reference hydromassage bathtubs? _____

6. Must the metal parts of the pump and power panel of the hydromassage tub be grounded? _____

_____

_____

_____

_____

7. Is it permissible to connect the hydromassage tub's green equipment grounding conductor to the branch-circuit's grounded (white) conductor? _____

▶8. All receptacles within (5) (10) (15) feet of a hydromassage tub must be GFCI protected. Circle correct answer. ◀

UNIT 23

Special-Purpose Outlets—Electric Heating ⬤M, Air Conditioning ⬤N

OBJECTIVES

After studying this unit, the student will be able to

- list the advantages of electric heating.
- describe the components and operation of electric heating systems (baseboard, cable, furnace).
- describe thermostat control systems for electric heating units.
- install electric heaters with appropriate temperature control according to *National Electrical Code®* rules.
- discuss air conditioners, heat pumps, terminology, Code requirements, and electrical connections.
- explain how heating and cooling may be connected to the same circuit.

GENERAL DISCUSSION

There are many types of electric heat available for heating homes (e.g., heating cable, unit heaters, boilers, electric furnaces, duct heaters, baseboard heaters, heat pumps, and radiant heating panels). Unit 23 touches upon many of these types. The residence discussed in this text is heated by an electric furnace located in the workshop.

Detailed Code requirements are found in *Article 424* of the *National Electrical Code,® Fixed Electric Space Heating Equipment.*

This text cannot cover in detail the methods used to calculate heat loss and the wattage required to provide a comfortable level of heat in the building. For this residence, the total estimated wattage is 13,000 watts. Depending upon the location of the

residence (in the Northeast, Midwest, or South, for example), the heating load will vary.

Electric heating has gained wide acceptance when compared with other types of heating systems. It has a number of advantages. Electric heating is flexible when baseboard heating is used because each room can have its own thermostat. Thus, one room can be kept cool while an adjoining room is warm. This type of zone control for an electric, gas- or oil-fired central heating system is more complex and more expensive.

Electric heating is safer than heating with fuels. The system does not require storage space, tanks, or chimneys. Electric heating is quiet. Electric heat does not add or remove anything from the air. As a result, electric heat is cleaner. This type of heating is considered to be healthier than fuel heating systems that remove oxygen from the air. The only moving part of an electric baseboard heating system is the thermostat. This means that there is a minimum of maintenance.

If an electric heating system is used, adequate insulation must be provided. Proper insulation can keep electric bills to a minimum. Insulation also helps to keep the residence cool during the hot summer months. The cost of extra insulation is offset through the years by the decreased burden on the air-conditioning equipment. Energy conservation measures require the installation of proper and adequate insulation.

TYPES OF ELECTRIC HEATING SYSTEMS

Electric heating units are available in baseboard, wall-mounted, and floor-mounted styles. These units may or may not have built-in thermal overload protection. The type of unit to be installed depends on structural conditions and the purpose for which the room is to be used.

Another method of providing electric heating is to embed resistance-type cables in the plaster, or between two layers of drywall on the ceiling, *Sections 424-34* through *424-45*. Because premise wiring (nonmetallic-sheathed cable, etc.) will be located above these ceilings and will be subjected to the heat created by the heating cables, the Code does require the following:

- keep the wiring not less than 2 inches (50.8 mm) above the ceiling, and *also* reduce the ampacity of the wiring to the 50°C correction factors found at the bottom of *Table 310-16*, or
- keep the wiring above insulation that is at least 2 inches (50.8 mm) thick—in which case the correction factors do not have to be applied.

▶ For ease in identification, *Section 424-35* of the *National Electrical Code*® requires that the leads on heating cables be color coded as follows:

- 120 volts   yellow
- 208 volts   blue
- 240 volts   red
- 277 volts   brown
- 480 volts   orange ◀

Electric heat can also be supplied by an electric furnace and a duct system similar to the type used on conventional hot-air central heating systems. The heat is supplied by electric heating elements rather than the burning of fuel. Air conditioning, humidity control, air circulation, electronic air cleaning, and zone control can be provided on an electric furnace system.

SEPARATE CIRCUIT REQUIRED

Section 422-7 of the Code requires that central heating equipment must be supplied by a separate branch-circuit. This includes electric, gas, and oil central furnaces. It also includes heat pumps. The exception to *Section 422-7* is equipment directly associated to central heating equipment, such as humidifiers, electrostatic air cleaners, and so on, which is permitted to be connected to the same circuit.

The logic behind this "separate circuit" Code rule is that there is always a possibility that if the central heating equipment was to be connected to some other circuit, such as a lighting circuit, a fault on that circuit would shut off the power to the heating system. This could cause freezing of water pipes. Refer to figure 23-1.

Cord/Plug Connections Not Permitted for Furnaces

Sections 400-7(a)(8) and *422-8(c)* state the uses where flexible cords are permitted. A key code requirement, oftentimes violated, is that flexible cords shall be used only "where the fastening means and mechanical connections are specifically designed to permit ready removal for maintenance and repair, and the appliance is intended or identified for flexible cord connection." Certainly the gas piping and the shear size of a gas or electric furnace does not allow for the "ready removal" of the furnace. The conductors in flexible cords cannot handle the high temperatures encountered on the terminals of a furnace. Check the instruction manual for proper installation methods.

CONTROL OF ELECTRIC HEATING SYSTEMS

Line-voltage thermostats can be used to control the heat load for most baseboard electric heating systems, figure 23-2. Common ratings for line-voltage thermostats are 2500 watts, 3000 watts, and 5000 watts. Other ratings are available. The electrician must check the nameplate ratings of the heating

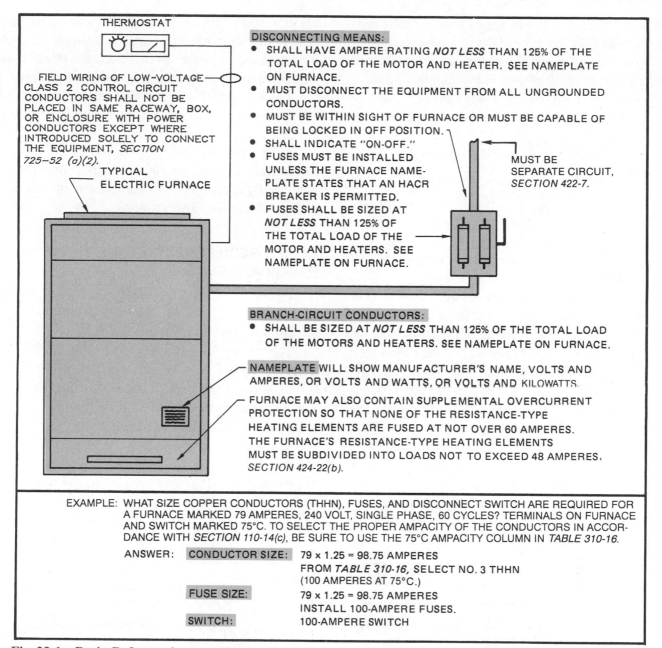

THERMOSTAT

FIELD WIRING OF LOW-VOLTAGE
CLASS 2 CONTROL CIRCUIT
CONDUCTORS SHALL NOT BE
PLACED IN SAME RACEWAY, BOX,
OR ENCLOSURE WITH POWER
CONDUCTORS EXCEPT WHERE
INTRODUCED SOLELY TO CONNECT
THE EQUIPMENT, *SECTION
725–52 (a)(2).*

TYPICAL
ELECTRIC FURNACE

DISCONNECTING MEANS:

- SHALL HAVE AMPERE RATING *NOT LESS* THAN 125% OF THE
  TOTAL LOAD OF THE MOTOR AND HEATER. SEE NAMEPLATE
  ON FURNACE.
- MUST DISCONNECT THE EQUIPMENT FROM ALL UNGROUNDED
  CONDUCTORS.
- MUST BE WITHIN SIGHT OF FURNACE OR MUST BE CAPABLE OF
  BEING LOCKED IN OFF POSITION.
- SHALL INDICATE "ON-OFF."
- FUSES MUST BE INSTALLED
  UNLESS THE FURNACE NAME-
  PLATE STATES THAT AN HACR
  BREAKER IS PERMITTED.
- FUSES SHALL BE SIZED AT
  NOT LESS THAN 125% OF
  THE TOTAL LOAD OF THE
  MOTOR AND HEATERS. SEE
  NAMEPLATE ON FURNACE.

MUST BE
SEPARATE CIRCUIT,
SECTION 422-7.

BRANCH-CIRCUIT CONDUCTORS:

- SHALL BE SIZED AT *NOT LESS* THAN 125% OF THE TOTAL LOAD
  OF THE MOTORS AND HEATERS. SEE NAMEPLATE ON FURNACE.

NAMEPLATE WILL SHOW MANUFACTURER'S NAME, VOLTS AND
AMPERES, OR VOLTS AND WATTS, OR VOLTS AND KILOWATTS.

FURNACE MAY ALSO CONTAIN SUPPLEMENTAL OVERCURRENT
PROTECTION SO THAT NONE OF THE RESISTANCE-TYPE
HEATING ELEMENTS ARE FUSED AT NOT OVER 60 AMPERES.
THE FURNACE'S RESISTANCE-TYPE HEATING ELEMENTS
MUST BE SUBDIVIDED INTO LOADS NOT TO EXCEED 48 AMPERES,
SECTION 424-22(b).

EXAMPLE: WHAT SIZE COPPER CONDUCTORS (THHN), FUSES, AND DISCONNECT SWITCH ARE REQUIRED FOR
A FURNACE MARKED 79 AMPERES, 240 VOLT, SINGLE PHASE, 60 CYCLES? TERMINALS ON FURNACE
AND SWITCH MARKED 75°C. TO SELECT THE PROPER AMPACITY OF THE CONDUCTORS IN ACCOR-
DANCE WITH *SECTION 110-14(c)*, BE SURE TO USE THE 75°C AMPACITY COLUMN IN *TABLE 310-16*.

ANSWER: CONDUCTOR SIZE: 79 × 1.25 = 98.75 AMPERES
FROM *TABLE 310-16,* SELECT NO. 3 THHN
(100 AMPERES AT 75°C.)

FUSE SIZE: 79 × 1.25 = 98.75 AMPERES
INSTALL 100-AMPERE FUSES.

SWITCH: 100-AMPERE SWITCH

Fig. 23-1 Basic Code requirements for an electric furnace. These "package" units have all of the internal components, such as limit switches, relays, motors, and contactors prewired. The electrician generally need only provide the branch-circuit supply and disconnecting means. See *Article 424* for additional Code information.

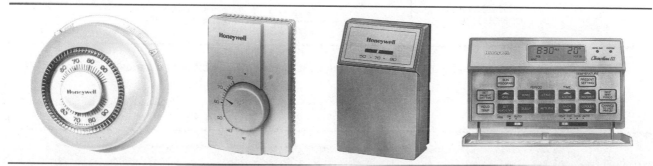

Fig. 23-2 Thermostats for electric heating systems. A thermostat or other switching device that is marked with an "off" position must open all ungrounded conductors of the circuit when turned to the "off" position. Typical mounting height is 52 inches to center. Mount perfectly level, particularly in the case of those thermostats that have a tipping vial of mercury that makes and breaks the circuit. *Courtesy* of Honeywell, Inc.

unit and the thermostat to insure that the total connected load does not exceed the thermostat rating.

The amperage limit of a thermostat is found using the wattage formula. For example, if a thermostat has a rating of 2500 watts at 120 volts, or 5000 watts at 240 volts, the current value is

$$I = \frac{W}{E} = \frac{2500}{120} = 20.8 \text{ amperes}$$

OR

$$I = \frac{5000}{240} = 20.8 \text{ amperes}$$

The total connected load for this thermostat must not exceed 20.8 amperes.

When the connected load is larger than the rating of a line-voltage thermostat, as in the case of an electric furnace, low-voltage thermostats may be used. In this case a relay must be connected between the thermostat and load. The relay is an integral component of the furnace. The low-voltage contacts of the relay are connected to the thermostat. The line-voltage contacts of the relay are used to switch the actual heater load. Low-voltage thermostat cable is run between the low-voltage terminals of the relay and the thermostat, figure 23-3.

When the connected load is larger than the maximum current rating of a thermostat and relay com-bination, the thermostat can be used to control a heavy-duty relay or contactor. An example of such a relay is the magnetic switch used for motor controls. The load side of the magnetic switch feeds a distribution panel containing as many 15- or 20-ampere circuits as needed for the connected load. A 40-ampere load may be divided into three 15-ampere circuits and still be controlled by one thermostat. The proper overcurrent protection is obtained by dividing the 40-ampere circuit into several circuits having lower ratings.

CIRCUIT REQUIREMENTS FOR BASEBOARD UNITS

Figure 23-4 shows typical baseboard electric heating units. These units are rated at 240 volts and are also available in 120-volt ratings.

The wiring for an individual baseboard heating unit or group of units is shown in figures 23-5 and 23-6. A two-wire cable (armored cable or non-metallic-sheathed cable with ground) would be run from a 240-volt, two-pole circuit in the main panel to the outlet box or switch box installed at the thermostat location. A second two-wire cable runs from the thermostat to the junction box on the heater unit. The proper connections are made in this junction box. Most heating unit manufacturers provide

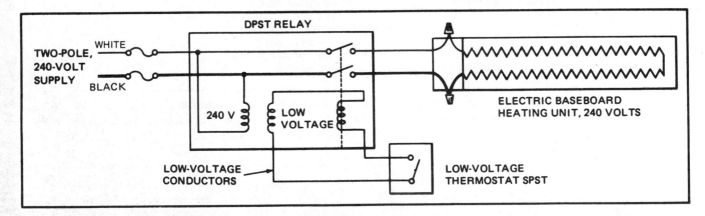

Fig. 23-3   Wiring for baseboard electric heating unit having a low-voltage thermostat and relay.

Fig. 23-4   Baseboard electric heating systems.

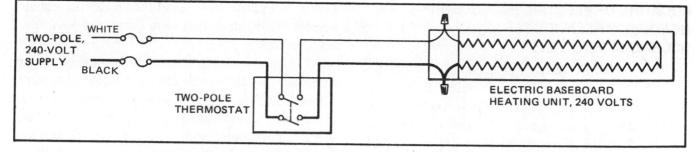

Fig. 23-5 Wiring for a single baseboard electric heating unit.

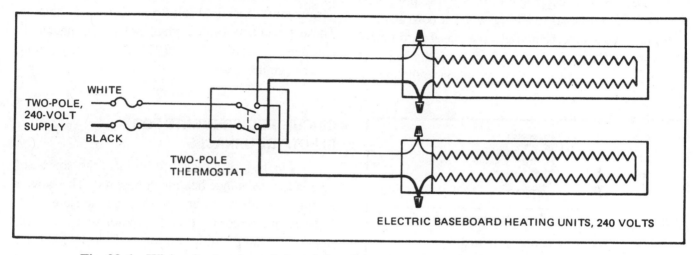

Fig. 23-6 Wiring for baseboard electric heating units at different locations within a room.

knockouts at the rear and on the bottom of the junction box. The supply conductors can be run through these knockouts. Most baseboard units also have a channel or wiring space running the full length of the unit, usually at the bottom. When two or more heating units are joined together, the conductors are run in this wiring channel. Most manufacturers indicate the type of wire required for these units because of conductor temperature limitations.

Supply conductors generally must be rated 90°C (194°F). This would be Type THHN conductors, or nonmetallic-sheathed cable or armored cable that has 90°C (194°F) conductors. The nameplate and/or instructions furnished with the baseboard electric heating unit provides this information.

A variety of fittings such as internal and external elbows (for turning corners) and blank sections are available from the manufacturer.

Most wall and baseboard heating units are available with built-in thermostats (figure 23-4). The supply cable for such a unit runs from the main panel to the junction box on the unit.

Some heaters have receptacle outlets. Underwriters Laboratories states that when receptacle sections are included with the other components of baseboard heating systems, they must be supplied separately using conventional wiring methods.

Manufacturers of electric baseboard heaters list "blank" sections ranging in length from 2 to 10 feet. These blank sections give the installer the flexibility needed to spread out the heater sections, and to install these blanks where wall receptacle outlets are encountered, figure 23-7.

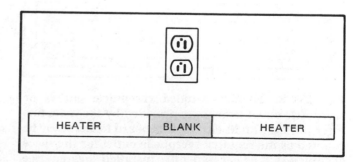

Fig. 23-7 Use of blank baseboard heating sections.

LOCATION OF ELECTRIC BASEBOARD HEATERS IN RELATION TO RECEPTACLE OUTLETS

Listed electric baseboard heaters shall not be installed below wall receptacle outlets unless the instructions furnished with the baseboard heaters indicate that they may be installed below receptacle outlets. See the Fine Print Note following the *Exception* to *Section 210-52(a)*. The reasoning for generally not allowing electric baseboard heaters to be installed below receptacle outlets is the possible fire and shock hazard resulting when a cord hangs over and touches the heated electric baseboard unit. The insulation on the cord can melt from the heat.

Electricians must pay close attention to the location of receptacle outlets and the electric baseboard heaters. They must study the plans and specifications carefully.

They may have to fine-tune the location of the receptacle outlets and the electric baseboard heaters, and/or install blank spacer sections, keeping in mind the Code requirements for spacing receptacle outlets and just how much (wattage and length) baseboard heating is required. See figures 23-8A and 23-8B.

The Code does permit factory-installed or factory-furnished receptacle outlet assemblies as part of permanently installed electric baseboard heaters. These receptacle outlets shall not be connected to the heater circuit, *Section 210-52(a), Exception*. Refer to figure 23-9.

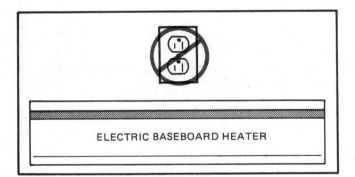

Fig. 23-8A CODE VIOLATION. Unless the instructions furnished with the electric baseboard heater specifically state that the unit is listed for installation below a receptacle outlet, then the above illustration is a violation of *Section 110-3(b)*. **In this type of installation, cords attached to the receptacle outlet would hang over the heater, creating a fire hazard . . . . and possible shock hazard should the insulation on the cord melt. See** *Exception* **and** *Fine Print Note* **to** *Section 210-52*.

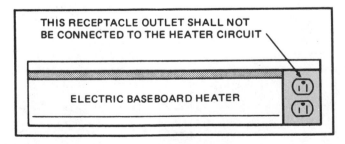

Fig. 23-8B Factory-installed receptacle outlets or receptacle outlet assemblies provided by the manufacturer for use with its electric baseboard heaters may be counted as the required receptacle outlet for the space occupied by a permanently installed heater. See *Exception* **to** *Section 210-52(a)*.

CIRCUIT REQUIREMENTS FOR ELECTRIC FURNACES

In the case of an electric furnace, the source of heat is the resistance heating element(s). The blower motor assembly, filter section, condensate coil, refrigerant line connection, fan motor speed control, high temperature limit controls, relays, and similar components are much the same as those found in gas furnaces.

In most installations, the supply conductors must be copper conductors having a minimum 90°C temperature rating, such as Type THHN. The instructions furnished with the unit will have information relative to supply conductor size and type.

Because proper supply voltage is critical to insure full output of the heating elements, voltage must be maintained at *not less* than 98% of the unit's rated voltage. Another way to state this is that the voltage drop shall not exceed 2% of the unit's rated voltage.

Many electric furnaces furnish the wattage (or kilowatts) output for more than one voltage level. Mathematically, for every 1% drop in voltage, there will be a 2% drop in wattage output. For instance,

Kilowatts	
@ 240 volts	@ 208 volts
12.0	9.0
15.0	11.3
20.0	15.0
25.0	18.8
30.0	22.5

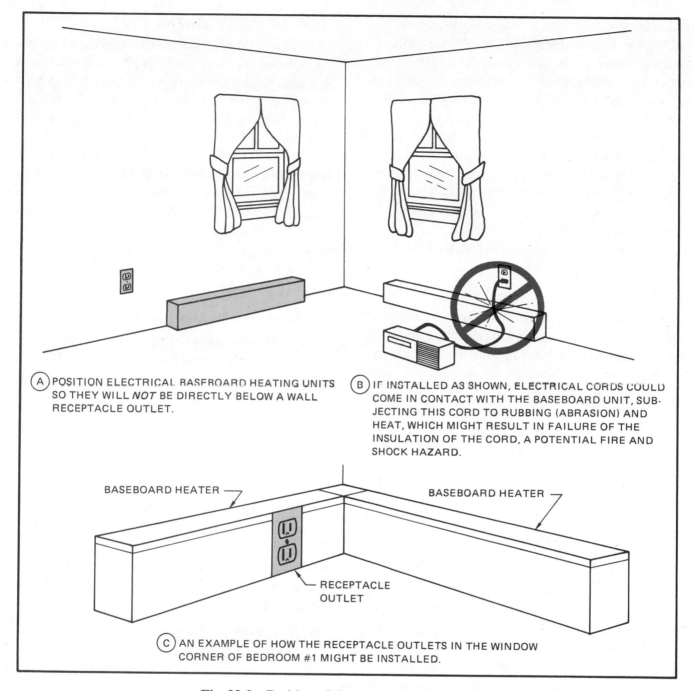

A) POSITION ELECTRICAL BASEBOARD HEATING UNITS SO THEY WILL *NOT* BE DIRECTLY BELOW A WALL RECEPTACLE OUTLET.

B) IF INSTALLED AS SHOWN, ELECTRICAL CORDS COULD COME IN CONTACT WITH THE BASEBOARD UNIT, SUBJECTING THIS CORD TO RUBBING (ABRASION) AND HEAT, WHICH MIGHT RESULT IN FAILURE OF THE INSULATION OF THE CORD, A POTENTIAL FIRE AND SHOCK HAZARD.

BASEBOARD HEATER

BASEBOARD HEATER

RECEPTACLE OUTLET

C) AN EXAMPLE OF HOW THE RECEPTACLE OUTLETS IN THE WINDOW CORNER OF BEDROOM #1 MIGHT BE INSTALLED.

Fig. 23-9   Position of electric baseboard heaters.

Therefore, the actual voltage at the supply terminals of the electric furnace or baseboard heating unit will determine the true wattage output of the heating elements.

The effect that different voltages have on electric heating elements is covered in unit 20.

The electric heating load used to determine the size of service-entrance conductors and feeders is computed at 100% of the total connected fixed space heating load, *Section 220-15*.

As previously discussed, the electric furnace must be supplied by an individual branch-circuit.

This is a requirement as referenced in *Section 422-7* of the Code. The "separate circuit" requirement pertains to any type of central heating, including oil, gas, electric, and heat pump systems.

HEAT PUMPS

A heat pump cools and dehumidifies on hot days and heats on cool days. To provide heat, the pump picks up heat from the outside air. This heat is added to the heat due to the compression of the refrigerant. The resultant heat is used to warm the air flowing through the unit. This warm air is then delivered to the inside of the building. The direction of flow of the refrigerant is reversed for the cooling operation.

Heat pumps are available as self-contained packages that provide both cooling and heating. In certain climates, the heat pump alone may not provide enough heat. To overcome this condition, the heat pump may utilize supplementary heating elements contained in the air-handling unit.

Heat pump water heaters were discussed in unit 19.

GROUNDING

Grounding of electrical baseboard heating units and other electrical equipment and appliances has been covered many times previously in this text and will not be repeated here. To repeat a caution, however, is in order. *Never* ground the equipment to the grounded (white) circuit conductor of the circuit.

MARKING THE CONDUCTORS OF CABLES

Two-wire cable contains one white and one black conductor. Because 240-volt circuits can supply electric baseboard heaters, it would appear that the use of two-wire cable is in violation of the Code. However, *Section 200-7* shows that this is not a Code violation.

According to *Section 200-7*, two-wire cable may be used for the 240-volt heaters if the white conductor is permanently reidentified by paint, colored tape, or other effective means. This step is necessary because it may be assumed that an unmarked white conductor is a grounded conductor having no voltage to ground. Actually, the white wire is connected to a hot phase and has 120 volts to ground. **Thus, a person can be subject to a harmful shock by touching this wire and the grounded baseboard heater (or any other grounded object) at the same time.** Most electrical inspectors accept black paint or tape as a means of changing the color of the white conductor. The white conductor must be made permanently reidentified at the electric heater terminals and at the panels where these cables originate. Reread the section on "Color Identification" in unit 5.

ROOM AIR CONDITIONERS

For homes that do not have central air conditioning, window or through-the-wall air conditioners may be installed. These types of room air conditioners are available in both 120-volt and 240-volt ratings. Because room air conditioners are plug-and-cord connected, the receptacle outlet and the circuit capacity must be selected and installed according to applicable Code regulations. The Code rules for air conditioning are found in *Article 440* of the *National Electrical Code.* The Code requirements for room air-conditioning units are found in *Sections 440-60* through *440-64.*

The basic Code rules for installing these units and their receptacle outlets are as follows:

- the air conditioners must be grounded.
- the air conditioners must be connected using a cord and attachment plug.
- the air-conditioner rating may not exceed 40 amperes at 250 volts, single phase.
- the rating of the branch-circuit overcurrent device must not exceed the branch-circuit conductor rating or the receptacle rating, whichever is less.
- the air-conditioner load shall not exceed 80% of the branch-circuit ampacity if no other loads are served.
- the air-conditioner load shall not exceed 50% of the branch-circuit ampacity if other loads are served.
- the attachment plug cap may serve as the disconnecting means.

RECEPTACLES FOR AIR CONDITIONERS

Figure 23-10 shows two types of receptacles that can be used for 240-volt installations. A combination receptacle is shown in figure 23-10(B). The lower portion of the receptacle is for 120-volt use and the upper portion is for 240-volt use. Note that the tandem and grounding slot arrangement for the 240-volt receptacle meets the requirements of *Section 210-7(f)*. This section states that receptacle outlets for different voltage levels must be noninterchangeable with each other.

CENTRAL HEATING AND AIR CONDITIONING

The residence discussed throughout this text has central electric heating and air conditioning consisting of an electric furnace and a central air-conditioning unit.

The wiring for central heating and cooling systems is shown in figure 23-11. Basic circuit requirements are shown in figure 23-12. Note that one feeder runs to the electric furnace and another feeder runs to the air conditioner or heat pump outside the dwelling. Low-voltage wiring is used between the inside and outside units to provide control of the systems. The low-voltage Class 2 circuit wiring must not be run in the same raceway as the power conductors, *Section 725-54(a)(1)*.

Refer to unit 24 for much more data relating to low-voltage circuitry.

"Time-of-Use" Lower Energy Rates

Many utilities offer lower energy rates for air-conditioner and/or heat pump loads through special programs, such as "time-of-use" usage and sliding scale energy rates. This is covered in unit 19.

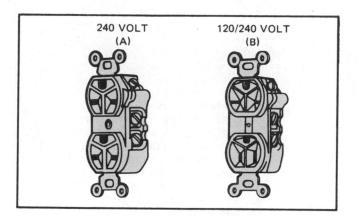

Fig. 23-10   Types of 240-volt receptacles.

UNDERSTANDING THE DATA FOUND ON AN HVAC NAMEPLATE

The letters HVAC mean **H**eating, **V**entilating, and **A**ir Conditioning.

Article 440 of the *NEC®* applies when HVAC equipment employs a hermetic refrigerant motor-compressor(s). *Article 440* is supplemental to the other articles of the *NEC®* and is needed because of the unique characteristics of hermetic refrigerant motor-compressors. The most common examples are air conditioners, heat pumps, and refrigeration equipment.

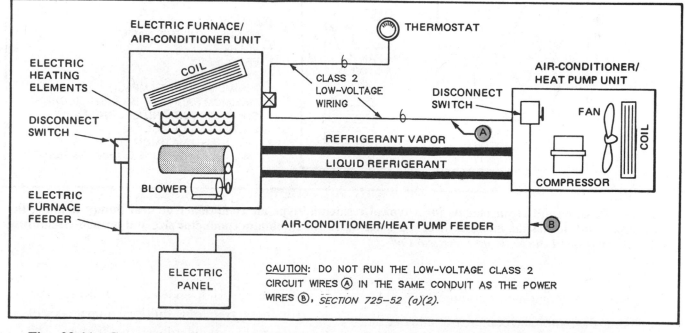

Fig. 23-11   Connection diagram showing typical electric furnace and air-conditioner/heat pump installation.

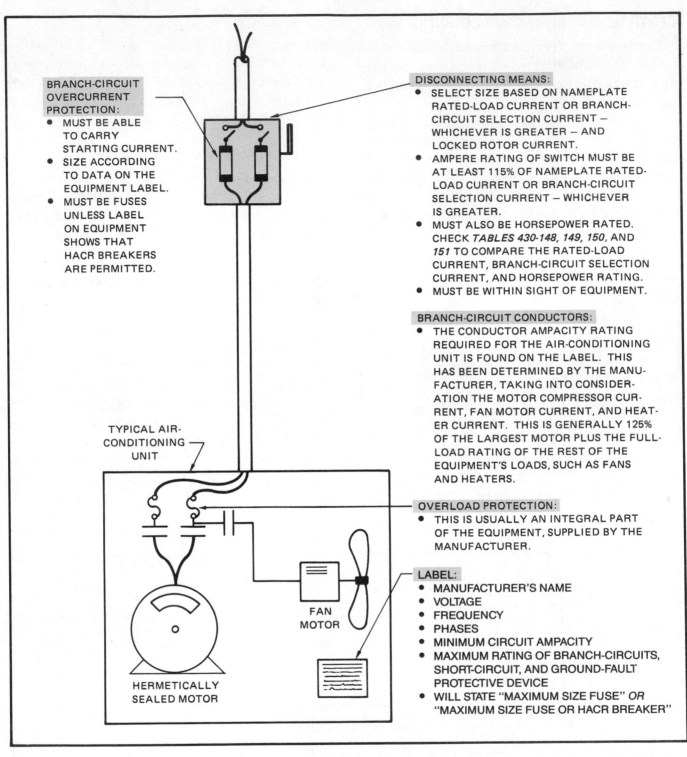

BRANCH-CIRCUIT OVERCURRENT PROTECTION:
- MUST BE ABLE TO CARRY STARTING CURRENT.
- SIZE ACCORDING TO DATA ON THE EQUIPMENT LABEL.
- MUST BE FUSES UNLESS LABEL ON EQUIPMENT SHOWS THAT HACR BREAKERS ARE PERMITTED.

DISCONNECTING MEANS:
- SELECT SIZE BASED ON NAMEPLATE RATED-LOAD CURRENT OR BRANCH-CIRCUIT SELECTION CURRENT — WHICHEVER IS GREATER — AND LOCKED ROTOR CURRENT.
- AMPERE RATING OF SWITCH MUST BE AT LEAST 115% OF NAMEPLATE RATED-LOAD CURRENT OR BRANCH-CIRCUIT SELECTION CURRENT — WHICHEVER IS GREATER.
- MUST ALSO BE HORSEPOWER RATED. CHECK *TABLES 430-148, 149, 150*, AND *151* TO COMPARE THE RATED-LOAD CURRENT, BRANCH-CIRCUIT SELECTION CURRENT, AND HORSEPOWER RATING.
- MUST BE WITHIN SIGHT OF EQUIPMENT.

BRANCH-CIRCUIT CONDUCTORS:
- THE CONDUCTOR AMPACITY RATING REQUIRED FOR THE AIR-CONDITIONING UNIT IS FOUND ON THE LABEL. THIS HAS BEEN DETERMINED BY THE MANU-FACTURER, TAKING INTO CONSIDER-ATION THE MOTOR COMPRESSOR CUR-RENT, FAN MOTOR CURRENT, AND HEAT-ER CURRENT. THIS IS GENERALLY 125% OF THE LARGEST MOTOR PLUS THE FULL-LOAD RATING OF THE REST OF THE EQUIPMENT'S LOADS, SUCH AS FANS AND HEATERS.

TYPICAL AIR-CONDITIONING UNIT

OVERLOAD PROTECTION:
- THIS IS USUALLY AN INTEGRAL PART OF THE EQUIPMENT, SUPPLIED BY THE MANUFACTURER.

LABEL:
- MANUFACTURER'S NAME
- VOLTAGE
- FREQUENCY
- PHASES
- MINIMUM CIRCUIT AMPACITY
- MAXIMUM RATING OF BRANCH-CIRCUITS, SHORT-CIRCUIT, AND GROUND-FAULT PROTECTIVE DEVICE
- WILL STATE "MAXIMUM SIZE FUSE" *OR* "MAXIMUM SIZE FUSE OR HACR BREAKER"

FAN MOTOR

HERMETICALLY SEALED MOTOR

Fig. 23-12   **Basic circuit requirements for a typical residential-type air conditioner or heat pump. Reading the label is important in that the manufacturer has determined the minimum conductor size and branch-circuit fuse size. See** *Article 440* **of the** *National Electrical Code.*®

Hermetic refrigerant motor-compressors:

- combine the motor and compressor into one unit
- have no external shaft
- operate in the refrigerant

- do not have horsepower and full-load current (often referred to as full-load amperes or FLA) ratings like standard electric motors. This is because as the compressor builds up pressure, the current increases, causing the windings to get hot. At the same time, the refrigerant gets colder, passes over the windings, and cools them. Because of this, a hermetically sealed motor can be "worked" much harder than a regular electric motor of the same size.

Special terminology is needed to understand how to properly make the electrical installation. These special terms are found in *Article 440* and in UL Standard 1995. An air-conditioning nameplate from Shiver Manufacturing Company will be used to help explain these special terms.

- *Rated-Load Current (RLC or RLA)*: Rated-load current is established by the hermetic motor-compressor manufacturer under actual operation at rated refrigerant pressure and temperature, rated voltage, and rated frequency. RLA is marked on the nameplate. In most instances, the marked RLA is at least equal to 64.1% of the hermetic refrigerant motor-compressor's maximum continuous current (MCC). In our example, the RLA is 17.8 amperes. See *Section 440-2*.

- *Maximum Continuous Current (MCC or MCA)*: This is the maximum current that the hermetic refrigerant motor-compressor can draw before damage occurs. Manufacturers know the damage capabilities of their equipment, and generally provide the necessary internal overcurrent protection that keeps the hermetic motor-compressor from burning out. This value is 156% of the marked RLA or branch-circuit selection current (BCSC). The overload protec-

tive system for the hermetic refrigerant motor-compressor must operate for currents in excess of the 156% value. Do not look for the MCC value on the nameplate of the end-use equipment. The MCC value might be marked on the motor-compressor. It is nice, but not necessary for the electrician to know the MCC value of the motor-compressor. See *Section 440-52*.

- *Branch-Circuit Selection Current (BCSC)*: The manufacturer of the HVAC end-use equipment might design better cooling and better heat dissipation into the equipment, in which case the hermetic refrigerant motor-compressor might be capable of being continuously "worked" harder than other equipment not so designed. "Working" the hermetic refrigerant motor-compressor harder will result in a higher current draw. This is safe insofar as the motor-compressor is concerned, but the conductors, disconnect switch, and branch-circuit overcurrent devices must also be capable of safely carrying this higher current draw. This higher value of current is marked on the nameplate using the term *Branch-Circuit Selection Current (BCSC)*.

Because the BCSC ampere value is always equal to or greater than the RLA ampere value, this BCSC value must be used instead of the RLA value when selecting conductors, disconnects, overcurrent devices, and other associated electrical equipment. In our example, the BCSC is 19.9 amperes. See *Sections 440-2* and *440-4(c)*.

SHIVER MANUFACTURING COMPANY

MODEL NO. XYZ

– ELECTRICAL RATINGS –

	VAC	HZ	PH	RLA	LRA	FLA
COMPRESSOR	230	60	1	17.8	107	—
OUTDOOR FAN MOTOR 1/4 HP	230	60	1	—	—	1.5

BRANCH CIRCUIT SELECTION CURRENT	19.9 AMPERES
MINIMUM CIRCUIT AMPACITY	26.4 AMPERES
MAXIMUM FUSE OR HACR TYPE BRKR.	45 AMPERES
OPERATING VOLTAGE RANGE:	197 MIN. 253 MAX.

- *Minimum Circuit Ampacity (MCA or MCC)*: This is one of two things (the other is MOP) the electrician needs to look for on the nameplate. MCA is the minimum circuit ampacity requirement to determine the conductor size and switch rating. MCA data is always marked on the nameplate of the end-use equipment, and is determined by the manufacturer of the end-use equipment as follows:

$$MCA = (RLA \text{ or } BCSC \times 1.25) + OTHER\ LOADS$$

Other loads would include condensing fans, electric heaters, coils, etc., that operate concurrently (at the same time).

In our example, the marked MCA is 26.4 amperes, calculated by the manufacturer of the end-use equipment as follows:

$$(19.9 \times 1.25) + 1.5 = 26.375 \text{ amperes}$$
(round off to 26.4)

Referring to *Table 310-16*, using the 60°C column, a No. 10 AWG copper conductor would be adequate for this air-conditioning unit.

See *Sections 440-33* and *440-35*.

A mistake that electricians often make is to multiply the MCA by 1.25 again. Certainly there is no hazard in doing this, but it might result in an unnecessary higher installation cost. For long runs, larger conductors would keep voltage drop to a minimum.

- *Type of Overcurrent Protection*: A typical nameplate will show the "maximum size fuse," or "maximum size fuse or HACR-type circuit breaker." If maximum size fuse is specified, then circuit breakers are not permitted. *Section 110-3(b)* of the Code states that "listed or labeled equipment shall be used or installed in accordance with any instructions included in the listing or labeling." Figure 23-17 illustrates the importance of using the proper type of overcurrent protective devices. In our example, the nameplate indicates "maximum fuse or HACR-type breaker 45 amperes."

In recent years, most 240-volt circuit breakers rated 100 amperes or less of the types used in residential wiring bear the letters HACR. But be very careful when connecting an air conditioner or heat pump to older style breakers that

were not tested to the HACR requirements and are not marked with the letters *HACR*. To make such a hook-up would be in violation of *Section 110-3(b)* of the *NEC.*®

See figures 23-13, 23-14, 23-15, and 23-17.

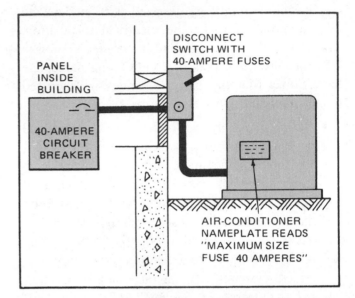

Fig. 23-13 This installation *conforms* to the Code, *Section 440-14*. The disconnect switch is within sight of the unit and contains the 40-ampere fuses called for on the air-conditioner nameplate as the branch-circuit protection.

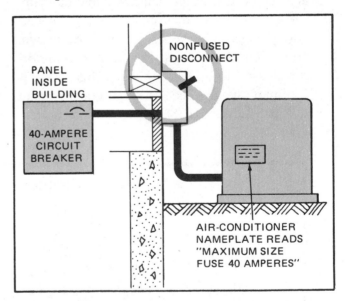

Fig. 23-14 This installation *violates* the Code, *Section 110-3(b)*. Although the disconnect is within sight of the air conditioner, it does not contain fuses. Note that the branch-circuit protection is provided by the 40-ampere circuit breaker inside the building. Note also that the nameplate requires that the branch-circuit protection be 40-ampere fuses maximum. If fused branch-circuit protection were provided at the panel inside the building, the installation would meet Code requirements.

• *Disconnecting Means to Be in Sight*: Section 440-14 of the Code requires that the disconnecting means must be within sight of the unit, and it must be readily accessible. See figures 23-13, 23-14, and 23-15. Mounting the disconnect on the side of the house instead of on the air condi-

tioner itself allows for easy replacement of the air-conditioning unit, should that become necessary. For cord-and-plug connected appliances, the cord and plug are considered to be the disconnecting means, *Section 440-63*.

• *Disconnecting Means Rating*: Figure 23-16 shows a fusible "pull-out" disconnect commonly used for residential air-conditioning and/or heat pump installations. The horsepower rating of the disconnecting means must be at least equal to the sum of all of the individual loads within the end-use equipment, at rated load conditions, and at locked rotor conditions, *Section 440-12(b)(1)*.

The ampere rating of the disconnecting means must be at least 115% of the sum of all of the individual loads within the end-use equipment, at rated load conditions, *Section 440-12(b)(2)*. Installing a disconnect switch that equals or exceeds the MOP value will in almost all cases be the correct choice. The MOP in our example is 45 amperes. Thus, a 60-ampere disconnect switch is the correct size. In cases where the MOP value is close to the ampere rating of the disconnect switch (i.e., 30, 60, 100, 200, etc.), *locked-rotor current* values must be considered.

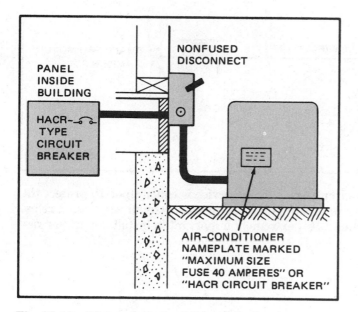

Fig. 23-15   This installation "Meets Code" because the circuit breaker in the panel is marked "HACR." This is permitted when the label on the breaker and the label on the air-conditioner bear the letters "HACR."

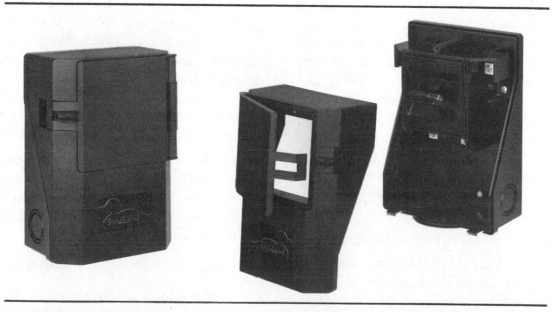

Fig. 23-16   One type of fuse-type disconnect for air-conditioning equipment. Available in 30-ampere, 240-volt and 60-ampere, 240-volt ratings. *Courtesy* of Midwest Electric Products, Inc.

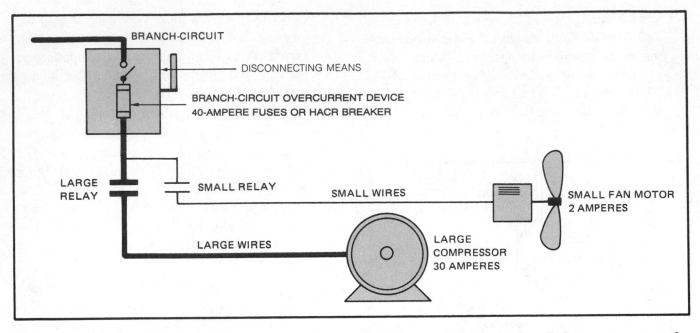

Fig. 23-17   In a typical air-conditioner unit, the branch-circuit overcurrent device is called upon to protect the large components (large wires, large relay, compressor) as well as the small components (small wire, small relay, fan motor) under short-circuit and/or ground fault situations. It is extremely important to install the proper *size* **and** *type* **of overcurrent device. Read the label!!!**

- *Locked-Rotor Current (LRA or LCR)*: This is the maximum current draw when the motor is in a LOCKED position. When the rotor is locked, it is not turning. The disconnect switch and controller (if used) must be capable of safely interrupting locked-rotor current. See *Sections 440-12* and *440-41.* In our example, the nameplate indicates that the hermetic refrigerant motor-compressor has a LRA of 107 amperes. The fan motor has a FLA of 1.5 amperes, and would have a LRA of approximately 6 times. According to *Section 440-12(b)(1)(b)*, all locked-rotor currents and other loads are added together when combined loads are involved. Therefore, the total locked-rotor current that the disconnect switch in our example would be called upon to interrupt is:

$$(1.5 \times 6) + 107 = 116 \text{ locked-rotor amperes}$$

▶ Checking *Table 430-151A* for the conversion of locked-rotor current to horsepower, we find that the locked-rotor current for a 230-volt, single-phase, 3-horsepower motor is 102 amperes, and the locked-rotor current for a 230-volt, single-

phase 5-horsepower motor is 168 amperes. ◀ Selecting a disconnect switch on the basis of the locked-rotor current in our example, the air-conditioning unit is considered to be a 5-horsepower unit, because 116 falls between 102 and 168. Manufacturers' technical literature indicates that a 30-ampere, 240-volt heavy-duty, single-phase disconnect switch has a 3-horsepower rating, and a 60-ampere, 240-volt heavy duty, single-phase disconnect switch has a 10-horsepower rating. Thus, a 60-ampere disconnect is selected. Note that this is the same size disconnect selected previously on the basis of MOP.

- *Maximum Overcurrent Protection (MOP)*: The MOP value is the maximum overcurrent protection. The MOP value is marked on the nameplate, and is determined by the manufacturer of the end-use equipment as follows:

$$(\text{RLA or BCSC} \times 2.25) + \text{OTHER LOADS}$$

Other loads would include condensing fans, electric heaters, coils, etc., that operate concurrently.

In our example, the marked MOP is 45 amperes, calculated by the manufacturer of the end-use equipment as follows:

$$(19.9 \times 2.25) + 1.5 = 46.275 \text{ amperes}$$
(round down to the next lower
standard size: 45 amperes)

See *Section 440-22*.

ENERGY RATINGS

Efficiency for air-conditioning units larger than 5 tons are marked with the term EER (energy efficiency ratio). Units 5 tons or less are marked with the term SEER (seasonal energy efficiency ratio).

These ratios are determined by dividing the appliance's cooling capacity (output) by the power input. Newer air-conditioning units have EER and SEER ratings ranging from 8.5 to 12. *The higher the EER or SEER rating, the more efficient the appliance.*

NONCOINCIDENT LOADS

Loads such as heating and air conditioning are not likely to operate at the same time. As an example, when calculating feeder sizes to an electric furnace and an air conditioner, only the larger load need be considered. See *Section 220-21* of the Code. This is discussed again in unit 29 when service-entrance calculations are presented.

REVIEW

Note: Refer to the Code or the plans where necessary.

ELECTRIC HEAT

1. a. What is the allowance in watts made for electric heat in this residence? _____
    b. What is the value in amperes of this load? _____
    _____

2. What are some of the advantages of electric heating? _____
    _____
    _____
    _____
    _____
    _____

3. List the different types of electric heating system installations. _____
    _____
    _____
    _____
    _____
    _____

4. There are two basic voltage classifications for thermostats. What are they? _____
    _____

5. What device is required when the total connected load exceeds the maximum rating of a thermostat? _____

6. The electric heat in this residence is provided by what type of equipment? _____
   _____

7. At what voltage does the electric furnace operate? _____

8. A certain type of control connects electric heating units to a 120-volt supply or a 240-volt supply, depending upon the amount of the temperature drop in a room. These controls are supplied from a 120/240-volt, three-wire, single-phase source. Assuming that this type of device controls a 240-volt, 2000-watt heating unit, what is the wattage produced when the control supplies 120 volts to the heating unit? Show all calculations.

9. What advantages does a 240-volt heating unit have over a 120-volt heating unit?

   _____
   _____
   _____
   _____
   _____
   _____
   _____
   _____
   _____

10. The white wire of a cable may be used to connect to a hot circuit conductor only if _____
    _____

11. Receptacle outlets furnished as part of a permanently installed electric baseboard heater, when not connected to the heater's branch-circuit, (may) (may not) be counted as the required receptacle outlet for the space occupied by the baseboard heater. Electric baseboard heaters (shall) (shall not) be installed beneath wall receptacle outlets unless the instructions furnished with the heaters indicate that it is OK to install the heater below a receptacle outlet. Circle correct answers.

12. The branch-circuit supplying a heater must be sized to at least _____ % of the heater's rating according to *Section* _____ .

13. Compute the current draw of the following electric furnaces. The furnaces are all rated 240 volts.

    a. 7.5 kW

    b. 15 kW

    c. 20 kW

14. For "ballpark" calculations, the wattage output of a 240-volt electric furnace connected to a 208-volt supply will be approximately 75% of the wattage output had the furnace been connected to a 240-volt supply. In question 13, calculate the wattage output of (a), (b), and (c). _____

_____

_____

_____

_____

15. A central electric furnace heating system is installed in a home. The circuit supplying this furnace:
   a. may be connected to a circuit that supplies other loads.
   b. shall be connected to a circuit that supplies other loads.
   c. shall be connected to a separate circuit.

16. What section of the Code provides the correct answer to question 15? _____

17. Electric heating cable embedded in plaster, or "sandwiched" between layers of dry wall, creates heat. Therefore, any nonmetallic-sheathed cable run above these ceilings and buried in insulation must have its current-carrying capacity derated according to the (40°C) (50°C) (60°C) correction factors found at the bottom of *Table 310-16* in the *NEC.* _____

AIR CONDITIONING

1. a. When calculating air-conditioner load requirements and electric heating load requirements, is it necessary to add the two loads together to determine the combined load on the system? _____

   b. Explain the answer to part (a). _____

   _____

   _____

2. The total load of an air conditioner shall not exceed what percentage of a separate branch-circuit? (circle one)

   a. 75%   b. 80%   c. 125%

3. The total load of an air conditioner shall not exceed what percentage of a branch circuit that also supplies lighting? (circle one)

   a. 50%   b. 75%   c. 80%

4. a. Must an air conditioner installed in a window opening be grounded if a person on the ground outside the building can touch the air conditioner? _____

   b. What Code section governs the answer to part (a)? _____

5. A 120-volt air conditioner draws 13 amperes. What size is the circuit to which the air conditioner will be connected? _____

6. What is the Code requirement for receptacles connected to circuits of different voltages and installed in one building? _____

7. When a central air-conditioning unit is installed and the label states "Maximum size fuse 50-amperes," is it permissible to connect the unit to a 50-ampere circuit breaker?

   _____

8. When the nameplate on an air-conditioning unit states "Maximum Size Fuse or Circuit Breaker," what type of circuit breaker must be used? _____

   _____

9. Briefly, what is the reasoning behind the UL requirement that air conditioners and other multimotor appliances be protected with fuses or special HACR circuit breakers?

   _____

   _____

   _____

   _____

   _____

   _____

   _____

10. What section of the *National Electrical Code*® prohibits the installing of Class 2 control circuit conductors in the same raceway as the power conductors? _____

   _____

11. Match the following terms with the statement that most closely defines the term. Enter the letters in the blank space.

   RLA _____   a) The maximum ampere rating of the overcurrent device.

   MCA _____   b) The current value to be used instead of the rated-load ampere.

   BCSC _____   c) The value used to determine the minimum ampere rating for the branch-circuit.

   MCC _____   d) A current value for the hermetic motor compressor that was determined by operating it at rated temperature, voltage, and frequency.

   MOP _____   e) The maximum current that the compressor can draw continuously without damage.

12. The disconnect for an air-conditioner or heat-pump must be installed _____

   _____ of the unit.

UNIT 24

Oil and Gas Heating Systems

OBJECTIVES

After studying this unit, the student will be able to

- understand the basics of gas and oil heating systems.
- understand some of the more important components of gas and oil heating systems.
- interpret basic schematic wiring diagrams.
- explain the principles of the thermocouple and the thermopile in a self-generating system.
- understand and apply the Code requirements for control-circuit wiring, *Article 725, NEC.*®

In unit 23, we discussed the circuit requirements for electric heating (furnace, baseboard, ceiling cable). This unit discusses typical residential oil and gas heating systems.

Heating a home with gas or oil is possible with:

- Warm air— fan-forced warm air or gravity (hot air rises naturally)

- Hot water—hot water is circulated through the piping system by a circulating pump. The radiators look very much like electric baseboard heating units. Hot water piping systems are not easily adaptable to central air-conditioning. Additional duct work would be required.

PRINCIPLES OF OPERATION

Most residential gas or oil heating systems operate as follows. A room thermostat is connected to the proper electrical terminals on the furnace or boiler, where the various controls and valves are inter-

connected in a manner that will provide safe and adequate heat to the home. Some typical wiring diagrams are shown in figures 24-1, 24-2, and 24-3.

Most residential gas and oil heating systems are "packaged units" in which the safety controls, valves, fan controls, limit switches, etc., are integral preassembled, prewired parts of the furnace or boiler.

For the electrician, the "field-wiring" generally consists of:

1. installing and connecting the wire to the thermostat, and

2. installing and connecting the power supply to the furnace or boiler.

Refer to figure 24-4.

MAJOR COMPONENTS

Thermostat, figure 24-5: Responds to temperature changes. Is connected to the controls on the furnace or boiler to turn the unit off and on, thereby maintaining the desired room temperature.

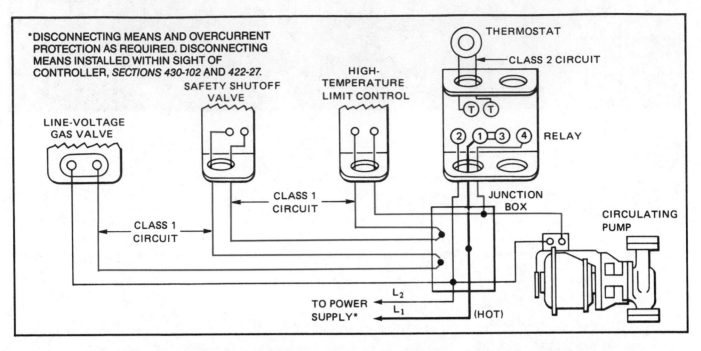

Fig. 24-1 Typical wiring diagram for a gas burner, forced hot water system.

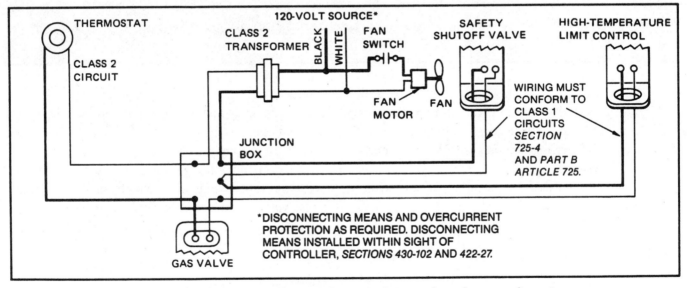

Fig. 24-2 Typical wiring diagram for a gas burner, forced warm air system.

A thermostat can have features such as day/night setback, a fan switch for continuous operation of the fan motor, a clock, and a thermometer. Some thermostats control both heating and air-conditioning functions.

For energy savings, thermostats of the types shown in figure 24-5 can maintain reduced temperatures at night and during the day if the home is unoccupied. Some thermostats feature an integral heat anticipator that allows the thermostat to maintain

specific comfort requirements to a more narrow range of temperature fluctuations. These heat anticipators are adjustable to match their setting with the current draw of the circuit controlled by the thermostat. Quality thermostats generally are accurate to 1½ degrees.

Thermostats in residences should be mounted approximately 52 inches above finish floor—and not mounted where influenced by drafts, concealed hot or cold water pipes or ducts, lighting fixtures,

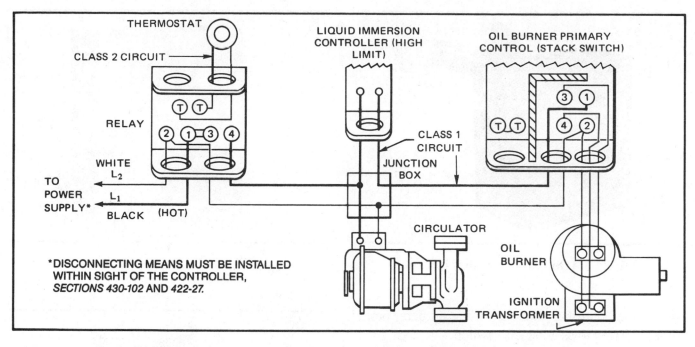

Fig. 24-3   Typical wiring diagram of an oil burner heating installation.

direct rays from the sun, radiation of heat from fireplaces, lamps, television sets, radios, or air currents from hot or cold air registers.

Primary Control, figure 24-6: Controls the oil burner motor, ignition, and oil valve in response to commands from the thermostat. If the flame fails to

Fig. 24-5   Photos of a typical thermostat that has a clock and day and night setback settings. Note the vials of mercury that "tip" to the right and to the left. When a vial "tips," the mercury rolls downward, opening and/or closing the circuit by bridging the gap between the electrical contacts inside the vial. To assure accuracy, the thermostat must be mounted perfectly level. The lever that points to the numbers .10, .20, .25, .30, etc., is to adjust the thermostat to match the current draw of the circuit that it controls. Instructions furnished with the thermostat explain this in detail. *Courtesy* of Honeywell Inc.

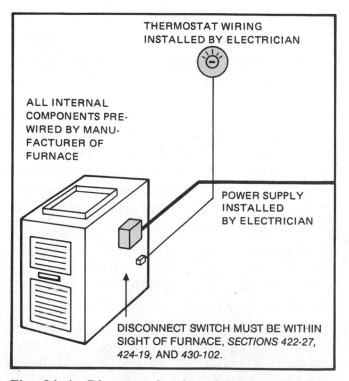

Fig. 24-4   Diagram showing thermostat wiring, power supply wiring, and disconnect switch.

Fig. 24-6   Oil burner primary control.

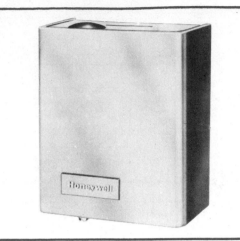

Fig. 24-7   Switching relay.

ignite, the controller shuts down the burner. Sometimes referred to as a "stack" switch.

This primary safety control is to an oil burning system what a main gas valve is to a gas burning system.

Primary controllers are available for both constant and intermittent ignition burners. Both types will recycle once if the flame goes out for any reason. If the flame fails to ignite, the controller trips "off" to safety. The control shown in figure 24-6 is designed for stack mounting. Other types of oil burner primary controls are mounted on the oil burner with the flame-sensing device mounted in the blast tube of the burner.

Switching Relay, figure 24-7:   Is the interconnect between the low-voltage wiring and line-voltage wiring. When a low-voltage thermostat "calls for heat," the relay "pulls in," closing the line-voltage contacts on the relay, which turns on the main power to the furnace or boiler. See figures 24-1 and 24-3. A built-in transformer provides low voltage for the thermostat circuit.

Liquid Immersion Control, figure 24-8:   Controls high water and circulating water temperatures. When acting as a "high-limit control," it will shut off the burner if water temperatures in the boiler reach dangerous levels. When acting as a "circulating water temperature control," it makes sure that the water being circulated through the piping system is at the desired temperature, not cold.

An immersion controller must be mounted in the boiler so that the sensing tube is either in direct con-

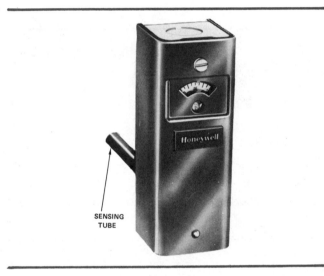

SENSING TUBE

Fig. 24-8   Liquid immersion controller.

tact with the water in the boiler or inserted into an immersion well that is in direct contact with the boiler water. The setting of the controller is usually in the range of 180 to 200°F. The setting can be changed by turning the adjusting screw on the face of the control. A stop prevents the temperature from being set any higher than 240°F.

Transformer, figure 24-9:   Is used where the line-voltage (120 volts) is reduced to low-voltage (24 volts), such as the low-voltage wiring used for thermostat wiring, and control-circuit wiring.

Low-Water Control: Is a device used to sense low water levels in a boiler. Burner cannot come on when water is too low. Similar in appearance to figure 24-8.

Fan Control, figure 24-10: Used to control the fan on forced warm-air systems. It will turn the fan off and on at some predetermined air temperature in the furnace's plenum chamber. Usually set to turn on between 120 and 140°F and to cut off at 90 to 100°F.

Limit Control, figure 24-10: When the air temperature in the plenum of the furnace reaches dangerously high levels (approximately 200°F), the limit control will shut off the burner. The fan, however, will continue to run. When the temperature in the plenum returns to safe levels, the contacts in the limit control will close, allowing the burner to resume normal operation. A limit control may also be called a limit switch. *NOTE:* Fan controllers and limit controllers are available as individual devices and are also available as combination devices.

Fan Motor: The electric motor that forces the warm air through the ducts of a heating system. The fan motor may be single-speed or multispeed. The fan itself may be belt driven or directly connected to the shaft of the motor. The motor is generally pro-vided with integral overload protection, but, in addition, may be protected with properly sized dual-element fuses installed in the disconnect switch for the furnace.

Water Circulating Pump, figure 24-11: Circulates hot water through the piping system. Is controlled by a liquid immersion control, figure 24-8.

Multizone control can be achieved by providing circulating pumps or zone valves (similar to figure 24-12, only electrically operated). Each circulating pump or zone valve would have its own thermostat, allowing each heating zone to be separately controlled. In multizone systems, the high-limit control is set to maintain the boiler water temperature at a certain value.

Flow Valve, figure 24-12: Prevents the gravity circulation of water through the piping system.

Solid-State Ignition: A feature on some gas furnaces, whereby the need for continuous burning of the pilot light is eliminated. The pilot light comes on only when the thermostat calls for heat. This is an energy-saving feature.

Main Gas Valve, figure 24-13: The valve that allows the gas to flow to the main burner of the furnace. It functions as a pressure regulator, a safety shutoff, and as the main gas valve. Should there be a "flame-out," or if the pilot light fails to operate, or in the event of a power failure, this control will shut off the flow of gas. It is the safety control of a gas system, just as a primary control is the safety control of an oil system.

Fig. 24-9   Transformer.

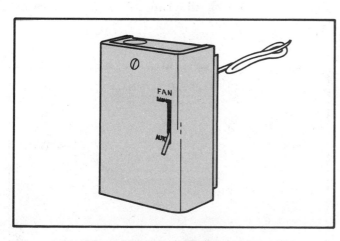

Fig. 24-10   Fan/limit control.

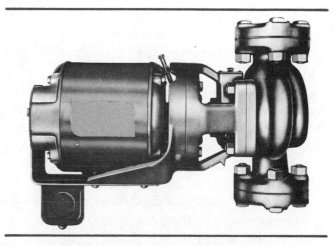

Fig. 24-11   Water circulator.

Fig. 24-12   Flow valve.

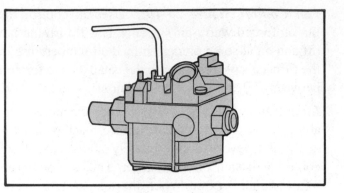

Fig. 24-13   Main gas valve.

SELF-GENERATING SYSTEM

Several manufacturers of gas burners provide self-generating systems. These systems do not require an outside power supply. Figure 24-14 shows the wiring diagram of a typical self-generating system. Note that the components are connected in series. The small amount of energy required by the gas valve is supplied by a thermopile. Many of these burners are prewired at the factory. Thus, the only wiring at the site is done by the electrician, who runs a low-voltage cable to the thermostat. This system is not affected by a power failure because there is no outside power source.

The self-generating system operates in the millivolt range. A special thermostat is required to function at this low voltage level. All other components of the system are matched to operate at the low voltage.

Thermocouple

Whenever two unlike metals are connected to form a circuit, an electric current flows if the junctions of the metals are at different temperatures, figure 24-15. As the temperature difference between the two junctions increases, the electric current increases. Thus, a thermocouple is formed by combining metals such as iron and copper, copper and iron constantan, copper-nickel alloy and chrome-iron alloy, platinum and platinum-rhodium alloy, or chromel and alumel. The types of metals used depend upon the temperatures involved.

In a gas burner system, the source of heat for the thermocouple is the pilot light. The cold junction of the thermocouple remains open and is connected to the pilot safety shutoff gas valve circuit.

A single thermocouple develops a voltage of about 25 to 30 dc millivolts. Thus, the circuit resistance must be kept low. The electrician should follow the manufacturer's recommendations regarding wire size. In addition, it is suggested that any connections other than those at screw terminals be soldered.

A very common use for thermocouples is in temperature-measuring instruments, where a chromel/alumel thermocouple generates approximately 40

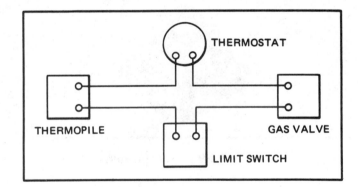

Fig. 24-14   Wiring diagram of a self-generating system.

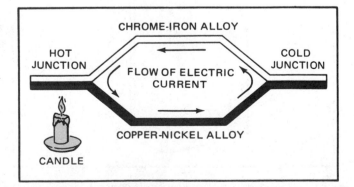

Fig. 24-15   Principle of a thermocouple.

microvolts per degree Celsius differential. The instrument is really an extremely sensitive voltmeter that is designed and calibrated to convert voltage to degrees.

Thermopile

When thermocouples are connected in series, a thermopile is formed, figure 24-16. The power output of the thermopile is greater than that of a single thermocouple; typically either 250 or 750 dc millivolts.

For example, 10 thermocouples (25 dc millivolts each) connected in series result in a 250 dc millivolt thermopile. Twenty six thermocouples (25 dc millivolts each) connected in series result in a 750 dc millivolt thermopile. For comparison, 750 millivolts is one-half the voltage of a 1½-volt "D" size flashlight battery. For both a thermocouple and a thermopile, there must be a temperature difference between the metal junctions. If the junctions marked A in figure 24-16 are heated and the junctions marked B remain cold, the resultant flow of current is shown by the arrows.

SUPPLY-CIRCUIT WIRING

The branch-circuit wiring must be adequate for the power requirements of the motors, fans, relays, pumps, ignition, transformers, and other power-consuming components of the furnace or boiler. The sizing of conductors, overcurrent protection, conduit fill, etc., has already been covered in other units of this text.

Note that *Section 422-7* of the *National Electrical Code®* requires that any central heating equipment (electric, gas, oil, heat pump) must be supplied by a separate branch-circuit. This has been discussed in unit 23, Special-purpose Outlets—Electric Heating, Air Conditioning.

CONTROL-CIRCUIT WIRING

▶ Installation requirements for motor control circuits are found in *Article 430, Part F* of the Code. Motor control circuit conductors must be protected according to *Section 430-72* and *Table 430-72(b)*. Transformers that supply motor control circuits must have overcurrent protection in accordance with *Article 450*, unless (1) the control transformer is an integral part of a motor controller, or (2) the control

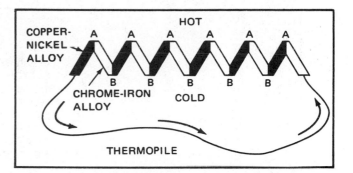

Fig. 24-16 Principle of a thermopile.

transformer primary current is 2 amperes or less and the overcurrent protection does not exceed 500% of the transformer's primary current rating, or (3) the control transformer supplies a Class 1, 2, or 3 circuit in which case the rules of *Article 725* apply.

Article 725 of the *National Electrical Code®* is rather lengthy and complex. It covers remote-control, signaling, and power-limited circuits that are *not* an integral part of a device or appliance. Wiring that leaves the device or appliance must conform to specific requirements of the *National Electrical Code®.* The internal wiring of a device or appliance must conform to a specific Underwriters Laboratories standard.

In part A of figure 24-17, the control circuitry is totally contained within the equipment. In this particular illustration, the equipment is a motor controller, covered by UL Standard 708. None of the internal wiring is governed by *Article 725* of the Code.

In part B of figure 24-17, the control wires leave the equipment. The external wiring is covered by the requirements of *Article 430, Part F* of the Code when the motor control circuit is connected to the load side of the motor's branch-circuit overcurrent protection. When the motor control circuit is supplied from other than being tapped to the load side of the motor's branch-circuit overcurrent protection, *Article 725* applies.

Class 1, 2, and 3 circuits offer alternative wiring methods regarding wire size, derating factors, overcurrent protection, insulation, wiring methods, and materials that "differentiate" these types of circuits from that of regular light and power wiring as covered in *Chapters 1 through 4* in the *National Electrical Code®.* Some key points for the three classes of circuits follow.

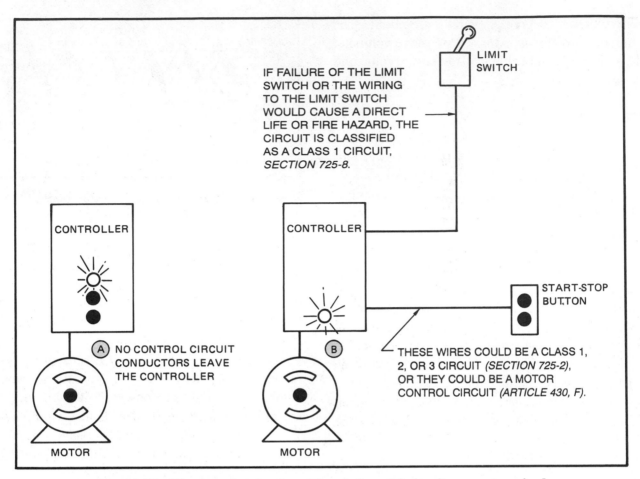

IF FAILURE OF THE LIMIT
SWITCH OR THE WIRING
TO THE LIMIT SWITCH
WOULD CAUSE A DIRECT
LIFE OR FIRE HAZARD, THE
CIRCUIT IS CLASSIFIED
AS A CLASS 1 CIRCUIT,
SECTION 725-8.

LIMIT
SWITCH

CONTROLLER

CONTROLLER

START-STOP
BUTTON

Ⓐ NO CONTROL CIRCUIT
CONDUCTORS LEAVE
THE CONTROLLER

Ⓑ

THESE WIRES COULD BE A CLASS 1,
2, OR 3 CIRCUIT *(SECTION 725-2),*
OR THEY COULD BE A MOTOR
CONTROL CIRCUIT *(ARTICLE 430, F).*

MOTOR

MOTOR

Fig. 24-17   Diagram showing how Class 1, 2, and 3 circuits are categorized.

Remote-control, signal, and power-limited circuits are classified as Class 1, Class 2, or Class 3, depending upon the voltage and power output capabilities of the circuit. These circuits are considered to be less hazardous than regular line-voltage light and power wiring because they are restricted to certain voltage maximums and power-output maximums.

Class 1 (There are two types of Class 1 circuits.)

- Class 1 circuits have their voltage and power limited to the values stipulated in *Section 525-21.*

- Class 1 *remote-control and signaling circuits* must not exceed 600 volts, *Section 725-21(b).*

- The conductors must be rated 600 volts, *Section 725-27(b).*

- Class 1 *power-limited circuits* must not exceed 30 volts and 1000 volt-amperes, *Section 725-21(a).*

- Class 1 circuits and power supply circuits may be run in the same raceway ONLY if the interconnected equipment is "functionally associated," *Section 725-26(b).* See figure 24-18.

- Class 1 wiring for safety control equipment, such as high limit and low water controls, must be installed in rigid or intermediate metal conduit, rigid nonmetallic conduit, electrical metallic tubing, Type MI, MC, or LWA cable, or be otherwise suitably protected from physical damage. A failure of the equipment, or a short-circuit or open circuit in the wiring would "introduce a direct fire or life hazard." See *Section 725-8(a)* and *(b).* See figure 24-18A.

- Class 1 circuits must have overcurrent protection sized to the ampacity of the conductor to be protected, *Section 725-23.* The overcurrent protection must be located where the conductors to be protected receive their supply, *Section 725-24.*

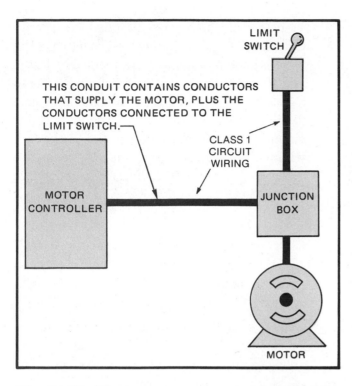

Fig. 24-18 The power conductors supplying the motor and the Class 1 circuit remote-control conductors to the limit switch and controller may be run in the same conduit when the interconnected equipment is "functionally associated," *Section 725-26(b)*. The insulation on the Class 1 circuit conductors must be insulated for the maximum voltage of any other conductors in the raceway. The minimum voltage rating for Class 1 conductors is 600 volts, *Section 725-27(b)*.

Class 2

In residential wiring, most of the low-voltage wiring comes under the Class 2 circuit category. In a home, some examples of Class 2 circuits are the thermostat wiring for furnaces and air-conditioners, burglar alarms, intercom systems, and the low-voltage wiring for chimes. Wiring diagrams for HVAC equipment clearly identifies where the line-voltage power wiring ends, and where the Class 2 wiring begins. Look at the data on the label of a chime transformer, or on the transformer and wiring diagram on HVAC equipment, and you will find reference to Class 2 circuitry. Fire alarm and smoke detectors come under the requirements of *Article 760* of the *NEC*.®

- Class 2 circuits are "power-limited," and are considered to be safe from fire hazard and from electric shock hazard.

- Class 2 wiring *does not* have to be in a conduit because a failure in the wiring, such as a short circuit or an open circuit in a thermostat cable, would not result in a direct fire or life hazard. The operation of the high-limit control or other safety control would prevent the occurrence of a fire or life hazard.

- Class 2 circuits *shall not* be run in the same raceway, cable, compartment, enclosure, outlet box, or fitting with electric light and power conductors,

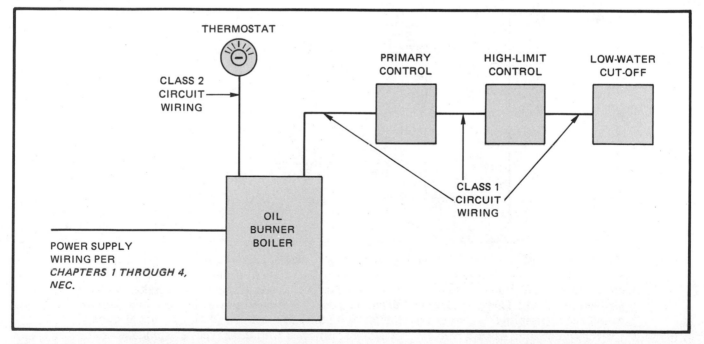

Fig. 24-18A Any failure in Class 1 circuit wiring would introduce a direct fire or life hazard. A failure in the Class 2 circuit wiring would *not* introduce a direct fire or life hazard. See *Section 725-8(a)* and *(b)*.

Section 725-54(a)(1). Even if the Class 2 conductors have the same 600-volt insulation as the power conductors, this is not permitted. See figure 24-19. A barrier inside of an enclosure that keeps the Class 2 wiring and the light and power conductors apart is permitted. Another execption is for installing Class 2 and Class 3 conductors in the same raceway, but only if the Class 2 conductors are insulated as required for Class 3 conductors, *Section 725-54(3)*. Another exception is where the conductors are "introduced" to connect the equipment, and are kept at least 0.25 inch (6.35 mm) apart. Appliances such as HVAC equipment will usually have a separate opening for the Class 2 wiring, and a separate opening (knockout) for the power wiring.

- Class 2 low-voltage control cable may be fastened to the conduit containing the power conductors that supply a furnace or similar equipment, *Sections 725-54(d)* and *300-11(b)*, *Exception No. 2*. See figure 24-20.

- Transformers that are intended to supply Class 2 and Class 3 circuits are listed by Underwriters Laboratories, and are marked "Class 2 Transformer" or "Class 3 Transformer." These transformers have "built-in" protection against the possibility of overheating should a short-circuit occur somewhere in the secondary wiring. These transformers may be connected to circuits having overcurrent protection not over 20 amperes, *Section 725-51*.

Class 3

- Class 3 circuits are power-limited and are considered safe from fire hazard, but could present an electrical shock hazard because of the higher voltage and current levels permitted, *Section 725-2*.

- Class 3 circuits are similar to Class 2 circuits, except that Class 3 circuits operate at higher voltage and current levels than Class 2 circuits.

- Class 3 cables have a voltage rating of not less than 300 volts, *Section 725-71(f)*.

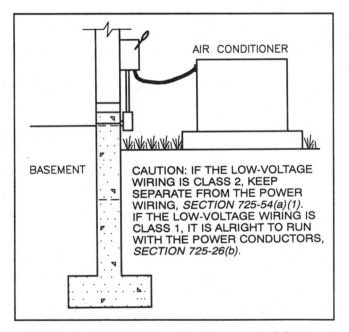

AIR CONDITIONER

BASEMENT

CAUTION: IF THE LOW-VOLTAGE WIRING IS CLASS 2, KEEP SEPARATE FROM THE POWER WIRING, *SECTION 725-54(a)(1)*. IF THE LOW-VOLTAGE WIRING IS CLASS 1, IT IS ALRIGHT TO RUN WITH THE POWER CONDUCTORS, *SECTION 725-26(b)*.

Fig. 24-19   Most residential air-conditioning units, furnaces, and heat-pumps have their low-voltage circuitry classified as Class 2. *Section 725-54(a)(1)* prohibits installing low-voltage Class 2 conductors in the same raceway as the power conductors ... even if the Class 2 conductors have 600-volt insulation. Therefore, the conduit running out of the basement wall in the above diagram would not be permitted to contain both the 240-volt power conductors and the 24-volt low-voltage Class 2 conductors. Always read the instructions furnished with these types of appliances to be sure your wiring is in compliance with these instructions, per *Section 110-3(b)* and the *NEC®* Review unit 4 for grounding methods.

Class 2 and Class 3 Power Source Requirements

Power source voltage limitations, current-limitations, and overcurrent protection requirements are found in *Chapter 9, Tables 11(a), 11(b), 12(a)*, and *12(c)* of the *National Electrical Code.®* Refer to the nameplate on Class 2 and Class 3 transformers, power supplies, and other equipment to confirm that they are U.L. "listed" for a particular class.

Cables for Class 2 and 3 Circuits

Low-voltage (single or multiple) conductors for Class 2 and Class 3 circuits are rated in a "hierarchy" based upon their ability to withstand fire. Their permitted uses and permitted substitutions between the different types are listed in *Tables 725-61* and *725-71* of the Code. These tables are presented in descending order relative to their fire resistance.

The low-cost Class 2 and Class 3 conductors generally installed in residences are Types CL2X and CL3X. The electrician will probably simply ask the electrical distributor for "bell wire," yet might really be asking for Type CL2X and not knowing exactly what to ask for.

Securing Class 1, 2, and 3 Cables

Class 1, 2, and 3 cables must be installed in a neat and workmanship manner, and shall be adequately supported by the *building structure* in such a manner that the cables will not be damaged by normal building use, *Section 725-7. Section 300-11(b)* states that "raceways shall not be used as a means of support for other raceways, cables, or nonelectric equipment." Taping, strapping, tie wrapping, or hanging Class 1, 2, and 3 cables to electrical raceways, piping, ducts, etc. is prohibited. But there is an exception to this rule.

Some leeway is found in the exception to *Section 725-54(d)*. The exception refers back to *Section 300-11(b), Exception No. 2*, which allows Class 2 circuit conductors to be supported by the raceway that contains the power conductors for the same equipment that the Class 2 circuit conductors are connected to. See figure 24-20.

Figure 24-21 illustrates the definitions of remote-control, signaling, and power-limited terminology. More discussion regarding remote-control and signaling systems is found in units 25, 26, and 27 of this text. ◄

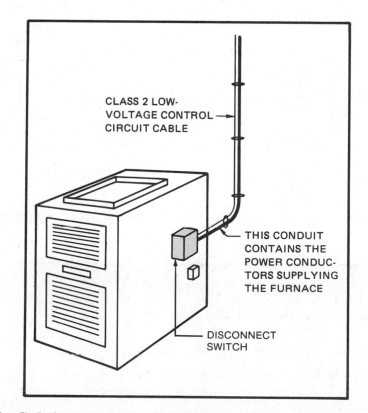

Fig. 24-20   The Code in *Section 300-11(b), Exception No. 2* and *725-54(d)*, allows Class 2 control circuit conductors (cables) to be supported by the raceway that contains the power conductors supplying electrical equipment such as the furnace shown.

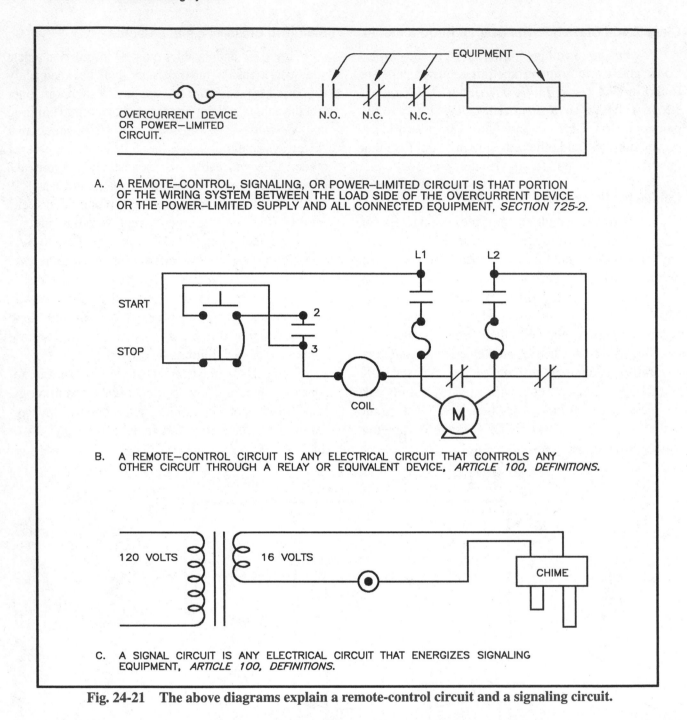

A.   A REMOTE-CONTROL, SIGNALING, OR POWER-LIMITED CIRCUIT IS THAT PORTION OF THE WIRING SYSTEM BETWEEN THE LOAD SIDE OF THE OVERCURRENT DEVICE OR THE POWER-LIMITED SUPPLY AND ALL CONNECTED EQUIPMENT, *SECTION 725-2.*

B.   A REMOTE-CONTROL CIRCUIT IS ANY ELECTRICAL CIRCUIT THAT CONTROLS ANY OTHER CIRCUIT THROUGH A RELAY OR EQUIVALENT DEVICE, *ARTICLE 100, DEFINITIONS.*

C.   A SIGNAL CIRCUIT IS ANY ELECTRICAL CIRCUIT THAT ENERGIZES SIGNALING EQUIPMENT, *ARTICLE 100, DEFINITIONS.*

Fig. 24-21   The above diagrams explain a remote-control circuit and a signaling circuit.

REVIEW

Note: Refer to the Code or the plans where necessary.

1. How is the residence in the text heated? _____

2. The Code requires that a disconnecting means must be _____

_____ of the furnace or boiler.

3. What device in an oil burning system will shut off the burner in the event of a flameout? _____

_____

4. Name the control that limits dangerous water temperatures in a boiler. _____

_____

5. Name the control that limits the temperature in the plenum of a furnace. _____

_____

6. In the wiring diagram in this unit, are the safety controls connected in series or parallel? _____

7. What is meant by the term "self-generating"? _____

_____

_____

_____

8. Because a self-generating unit is not connected to an outside source of power, the electrician need not use care in the selection of the conductor size or in making electrical connections. Is this statement true or false? Explain. _____

_____

_____

_____

_____

_____

_____

9. Explain what a thermocouple is and how it operates. _____

_____

_____

_____

_____

_____

10. If Class 2 low-voltage conductors are insulated for 32 volts, and the power conductors are insulated for 600 volts, does the Code permit pulling these conductors through the same raceway? _____

11. If Class 2 low-voltage conductors are used on a 24-volt system, but actually insulated for 600 volts (THHN for example) does the Code permit running these conductors together in the same raceway as the 600-volt power conductors? _____

_____

12. Under what conditions may Class 1 conductors and power conductors be installed in the same raceway? _____

_____

_____

13. The *National Electrical Code*® is very strict about securing anything to electrical raceways. Name the sections in the Code that prohibit this practice.

_____

_____

UNIT 25

Television, Telephone, and Low-Voltage Signal Systems

OBJECTIVES

After studying this unit, the student will be able to

- install television outlets, antennas, cables, and lead-in wires.
- list CATV installation requirements.
- describe the basic operation of satellite antennas.
- install telephone conductors, outlet boxes, and outlets.
- define what is meant by a signal circuit.
- describe the operation of a two-tone chime and a four-note chime.
- install a chime circuit with one main chime and one or more extension chimes.

INSTALLING THE WIRING FOR HOME TELEVISION

This unit touches upon the basics of installing outlets, cables, receivers, antennas, amplifiers, multiset couplers, and some of the *National Electrical Code®* rules for home television. Television is a highly technical and complex field. A trouble-free installation of a home system should be done by a competent television technician. These individuals receive training and certification through the Electronic Technicians Association in much the same way that electricians receive their training through apprenticeship and journeyman training programs, and learning about the *National Electrical Code®* via membership in the International Association of Electrical Inspectors organization. Some cities and states require that, when more than three television outlets are to be installed in a residence, the television technician making the installation be licensed and certified. Just as electricians can experience problems because of poor connections, terminations, and splices, poor reception problems can arise as a result of poor terminations. In most cases, poor crimping is the culprit. Electrical shock hazards are present when components are improperly grounded and bonded.

According to the plans and for this residence, television outlets are installed in the following rooms:

Front Bedroom	1
Kitchen/Nook	1
Laundry	1
Living Room	3
Master Bedroom	1
Recreation Room	3
Study/Bedroom	1
Workshop	1

Television outlets are usually installed in standard sectional switch (device) boxes, or 4-inch square 1½-inch deep outlet boxes with a single-gang plaster ring, figure 25-1. The boxes may be metallic or nonmetallic, and are roughed-in in the same manner as regular wiring. A box is installed at each point on the system where a television outlet is to be located. For new construction, shielded coaxial cables are installed concealed in the walls during the normal rough-in stage of construction. For remodel work, television cables can be concealed in the walls by "fishing" the cables through the walls.

Shielded 75-ohm coaxial cable is most often used to hook-up television sets to minimize interference and keep the color signal strong. There are different kinds of shielded coax cable. Double-shielded cable that has a 100% foil shield covered with a 40% or greater woven braid is recommended. This coax cable has a PVC outer jacket. The older style, flat 300-ohm twin lead cable can still be found on existing installations, but the reception might be poor. To improve reception, 300-ohm cables can be replaced with shielded 75-ohm coaxial cables.

For this residence, 12 television outlets are to be installed. For this many outlets, it is recommended that a shielded 75-ohm coaxial cable be run from each outlet location back to one central point, figure 25-2. This central point could be where the incoming CATV (Community Antenna Television) cable comes in, where the cable from an antenna comes in, or where the TV output of a satellite receiver is carried. There are so many possibilities that each installation must be analyzed to determine what is best for that particular installation. For rooftop or similar antenna mounting, the cable is run down the outside of the house, then into the basement, garage, or other convenient location where the proper connections are made. CATV companies generally run their incoming coaxial cables underground, again to some convenient point just inside of the house. Here, the technician can hook-up all of the coax cables using a

Fig. 25-1   Nonmetallic boxes and a nonmetallic raised cover.

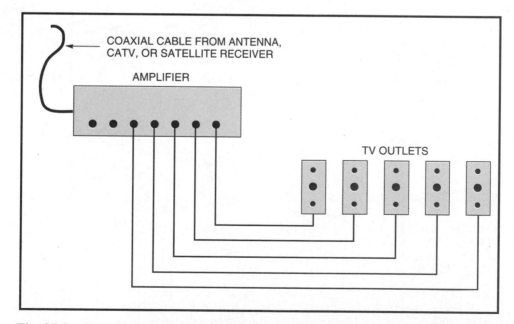

Fig. 25-2   Television master amplifier distribution system may be needed where many TV outlets are to be installed. This will minimize the signal loss. In simple installations, a multiset coupler can be used, figure 25-3.

multiset coupler, figure 25-3, or an amplifier, figure 25-2 to boost possible lost signals. An amplifier may not be necessary if *only* those coax cables that will be used are hooked up.

Cable television does not require an antenna on the house because the cable company has the proper antenna to receive signals from satellites. The cable company then distributes these signals throughout the community they have contractually agreed to serve. These contracts usually require the cable company to run their coaxial cable to a point just inside of the house. Inside the house, they will complete the

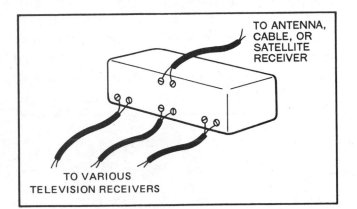

Fig. 25-3   Multiset coupler.

installation by furnishing cable boxes, controls, and the necessary wiring. In geographical areas where cable television is not available, antennas of the types shown in figure 25-4, or satellite "dishes," as illustrated in figure 25-6, must be installed.

Faceplates for the TV outlets are available in styles to match most types of electrical faceplates in the home.

Although combination two-gang faceplates can be used so that a 120-volt receptacle and a TV outlet can be mounted on one box, figure 25-5, most installers prefer to keep the television outlets completely separate from the 120-volt receptacles. Where a combination faceplate is used, the electrical wall box must have a metal barrier between the 120-volt wiring and the television outlet wiring, *Section 810-18(c)*. This barrier prevents line voltage interference with the television signal as well as keeping the two different voltages separate from each other for safety. More *National Electrical Code*® rules are discussed later in this unit.

Although shielded coaxial cable minimizes interference, it is good practice to keep the coaxial cables on one side of the stud space, and keep the light and power cables on the other side of the stud space.

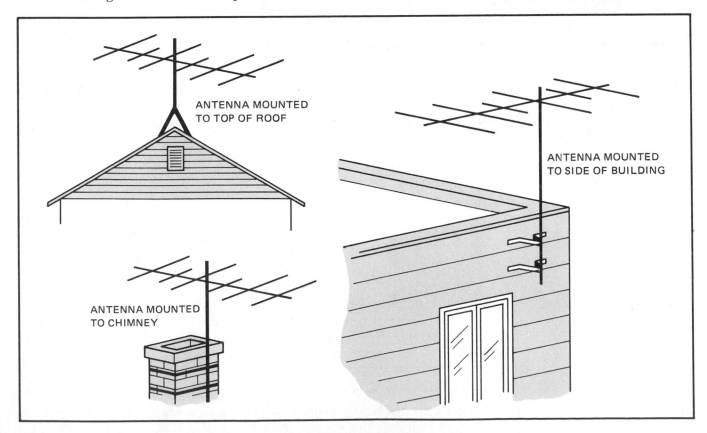

Fig. 25-4   Typical mounting of television antennas.

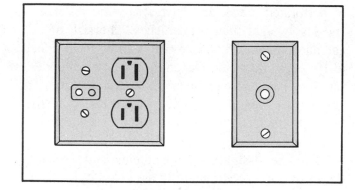

Fig. 25-5   Typical faceplates.

Code Rules for Cable Television (CATV)
(Article 820)

Cable television is more correctly called Community Antenna Television, hence the term CATV.

CATV systems are installed both overhead and underground in a community. Then, to supply an individual customer, the CATV company runs coaxial cables through the wall of a residence at some convenient point. Up to this point of entry *(Article 820, Part B)* and inside the building *(Article 820, Part E)*, the cable company must conform to the requirements of *Article 820* of the Code, plus local codes if applicable. Here in condensed form are some of the key rules.

1. The outer conductive shield of the coaxial cables must be grounded as close to the point of entry as possible, *Section 820-33*.

2. Coaxial cables shall not be run in the same conduits or box with electric light and power conductors, *Section 820-52(2)(b)*.

3. Do not support coaxial cables to raceways that contain electrical light and power conductors, *Section 820-52(e)*.

4. Where underground coaxial cables are buried, they must be separated by not less than 12 inches from a buried service-lateral—unless the service-lateral conductors are installed in a raceway, *Section 820-11(b)*.

5. The grounding conductor *(Article 820, Part D)*:
   a. Must be insulated.
   b. Must not be smaller than No. 14 copper or other corrosion-resistant conductive material.
   c. May be solid or stranded.
   d. Must be run in a line as straight as practicable.
   e. Shall be guarded from physical damage.
   f. Shall be connected to the nearest accessible location on one of the following:

      the building or structure grounding electrode *(Section 250-81)*

      the grounded interior water pipes *(Section 250-80(a))*

      the metallic power service raceway

      the service equipment enclosure

      the grounding electrode conductor or its metal enclosure

      the grounding conductor

      the grounding electrode of a building's disconnect when these are properly grounded.

   g. If *none* of *f* is available, then ground to any of the electrodes per *Section 250-81*, such as metal underground water pipes, metal frame of building, concrete encased electrode, or ground ring.

   h. If *none* of *f* or *g* is available, then ground to an effectively grounded metal structure or to any of the electrodes per *Section 250-83*, such as underground gas piping systems, driven rod or pipe, or a metal plate. If grounding to gas piping, make the connection between the gas meter and the street main.

   i. *CAUTION:* If any of the preceding results in one grounding electrode for the CATV shield grounding and another grounding electrode for the electrical system, bond the two electrodes together with a bonding jumper no smaller than No. 6 copper or its equal. Grounding and bonding the coaxial cable shield to the same grounding electrode as the main service ground minimizes the possibility that a difference of potential might exist between the shield and other grounded objects, such as a water pipe.

SATELLITE ANTENNAS

A satellite antenna is often referred to as a "dish," figure 25-6. A satellite has a parabolic shape that concentrates and reflects the signal beamed down from one of the many satellites located 22,247 miles above the equator. The orbit in which satellites travel is called the "Clarke" belt. See figure 25-9.

Satellite reception has made it possible for the homeowner to:

- use satellite reception only, in which case the homeowner would not receive local programs.

- use satellite reception *and* use CATV for local channels. By law, CATV companies must supply this basic service and cannot require the homeowner to sign up for premium programs.

- use satellite reception for all of the programs offered by the satellite company, *and* a roof-top or similar antenna for local channels.

Satellite antennas are available in solid or "see-through" types. See-through types are constructed of perforated metal, and have less wind resistance and less weight than solid dishes. The size of the antenna shown in figure 25-6(B) can be from 6 feet to 10 feet in diameter. Depending upon local codes, satellite antennas can be mounted on the roof, a chimney, a post, a pedestal, or on the side of the house. Some localities do not permit these large size antennas.

The line-of-sight from the antenna and the satellite must not be obstructed by trees, buildings, telephone poles, or other structures.

The latest satellite technology is the *digital satellite system* (DSS). A digital system allows the use of small antennas, approximately 18 inches in diameter, that can easily be mounted on the roof, the chimney, the side of the house, on a pipe, or on a pedestal, using the proper mounting hardware, figure 25-6(A). A digital system offers clearer pictures and clearer sound, greater choices of channel selection, and neater appearance insofar as the size of the antenna is concerned. The receiver of a digital system requires a connection to the telephone sys-

(A)   (B)

Fig. 25-6   Photo "A" shows a digital satellite system 18-inch antenna, a receiver, and a remote control. *Courtesy* **of Thomson Consumer Electronics. Photo "B" shows a large six-foot satellite antenna securely mounted to a post that is anchored in the ground.** *Courtesy* **of Winegard Company.**

tem for program billing and for updating the program access and to ensure continuous program service. Once each month the satellite company will call the receiver and download all of the information that has been stored. Any and all pay-per-view and other premium services used during that month will be recorded during this very brief downloading. The digital satellite receiver has a modular telephone phone jack on the back of the receiver for this connection.

The basic operation of a satellite system is for a transmitter on earth to beam an "uplink" signal to a satellite in space. Electronic devices on the satellite reamplify and convert this signal to a "downlink" signal, then retransmit the downlink signal back to earth, figure 25-7.

C-band satellites transmit at 3.7 to 4.2 GHz.

KU-band (low and medium power) transmit at 11.7 to 12.2 GHz.

KU-band (high power) transmit at 12.2 to 12.7 GHz. This high frequency makes it possible to use a small 18-inch antenna.

The term *gigahertz* (GHz) means billions of cycles per second.

The term *megahertz* (MHz) means millions of cycles per second.

On earth, the "downlink" signal is picked up by the "feedhorn" located on the front of a satellite antenna, figure 25-6.

A "feedhorn" is located at the exact focal point in relationship to the parabolic reflector to gather the maximum amount of signal. The feedhorn funnels the downlink satellite signal into a "low noise amplifier" where the signal is amplified, then fed to a "downconverter" where the signals are converted from high frequency to low frequency (70 MHz) that is proper for the receiver. All of these are mounted in a housing in front of the antenna.

The receiver is located inside the home and is connected to the "downconverter" on the dish with a 75-ohm coaxial cable. The receiver in turn is connected to the television set, figures 25-8A and 25-8B.

More and more satellite programmers are scrambling their uplink signals. The home satellite dish

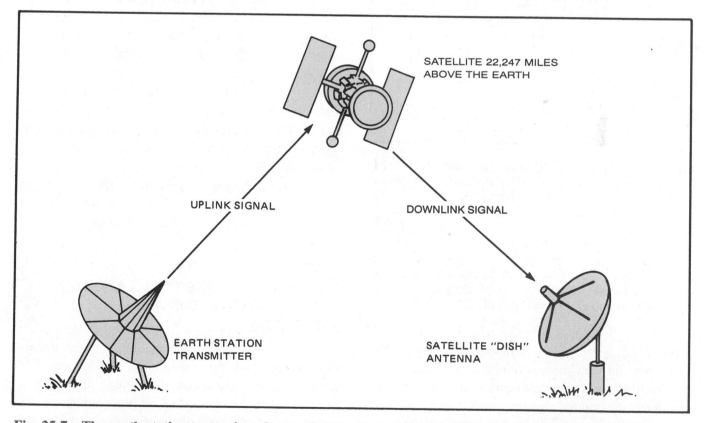

Fig. 25-7 The earth station transmitter beams the signal up to the satellite where the signal is amplified, converted, and beamed back to earth. The power requirements of the satellite are in the range of 5 watts and are provided by the use of solar cells.

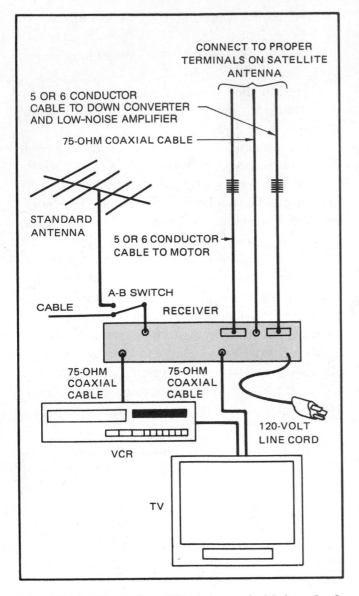

Fig. 25-8A   Typical satellite antenna/cable/standard antenna wiring. Connection to a standard antenna cable, CATV cable, or to a satellite cable enables the homeowner to watch one channel while recording another channel. The installation manual for the receiver usually has a number of different hook-up diagrams. In this diagram, the A-B switch allows a choice of connecting to a standard antenna or to the CATV cable. This hook-up can be desirable should the CATV or satellite reception fail.

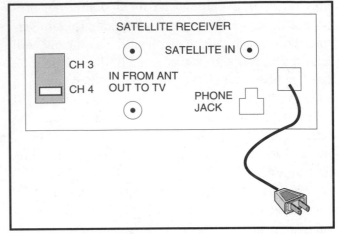

Fig. 25-8B   This diagram shows the terminals on the back of a digital satellite receiver. The connections are similar to those in figure 25-8A. Note the difference in that the digital satellite receiver has a modular telephone plug, a feature that uses a toll-free 800 number to update the access card inside the receiver to ensure continuous program service. The telephone can also handle program billing.

For the more complicated large dish installations, the cables that are installed between the receiver inside the home and the satellite antenna consist of color-coded conductors, making it easy to identify when doing the hook-up. The number of conductors necessary depends upon circuit designs of the receiver and the satellite. For the newer digital satellite systems, the hook-up is greatly simplified, using 75-ohm coaxial cable between the receiver and the antenna. Always follow the manufacturer's installation instructions.

All satellites in space are in the same orbit, figure 25-9. Figure 25-10 shows one method of installing a large satellite antenna to a metal post securely positioned in concrete.

CODE RULES FOR THE INSTALLATION OF ANTENNAS AND LEAD-IN WIRES
(ARTICLE 810)

Home television, video cassette recorders (VCRs), compact disc equipment, and AM and FM radios generally come complete with built-in antennas. For those locations in outlying areas on the fringe or out of reach of strong signals, it is quite common to install a separate antenna system.

Either indoor or outdoor antennas may be used with black and white and color televisions. The front

owner needs a special descrambling device to be able to decode and view these scrambled channels.

Small 18-inch antennas are easily focused following the manufacturer's instructions. Large antennas are more difficult to focus, and in some instances are motor driven. One manufacturer utilizes a reversible 36-volt dc motor, controllable from inside the home for fine-tune focusing of the antenna to the satellite.

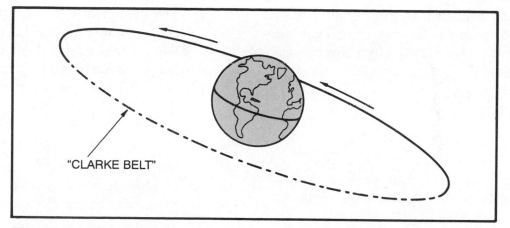

Fig. 25-9   All satellites travel around the earth in the same orbit called the "Clarke" belt. They appear to be stationary in space because they are rotating at the same speed at which the earth rotates. This is called "geosynchronous orbit (geostationary orbit)." The satellite receives the uplink signal from earth, amplifies the signal, and transmits it back to earth. The downlink signal is picked up by the satellite antenna.

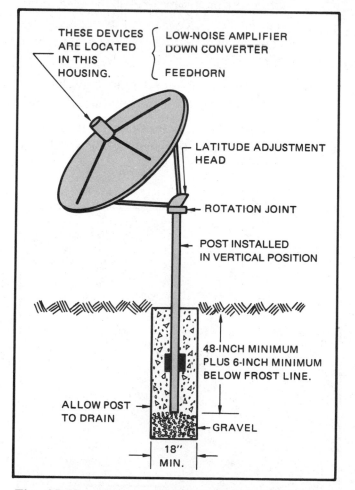

Fig. 25-10   Typical satellite antenna of the type installed in remote areas where reception is poor, or in areas served by CATV, but where the homeowner does not wish to subscribe and pay for the service of the CATV company. Always check with the local inspectors to be sure that all installation code rules are met.

of an outdoor antenna is aimed at the television transmitting station. When there is more than one transmitting station and they are located in different directions, a rotor is installed. A rotor turns the antenna on its mast so that it can face in the direction of each transmitter. The rotor is controlled from inside the building. The rotor controller's cord is plugged into a regular 120-volt receptacle to obtain power. A four-wire cable is usually installed between the rotor motor and the control unit. The wiring for a rotor may be installed during the roughing-in stage of construction, running the rotor's four-wire cable into a device wall box, allowing 5 or 6 feet of extra cable, then installing a regular single-gang switch plate when finishing.

Article 810 of the Code covers this subject. Although instructions are supplied with antennas, the following key points of the Code regarding the installation of antennas and lead-in wires should be followed.

1. Antennas and lead-in wires shall be securely supported.

2. Antennas and lead-in wires shall not be attached to the electric service mast.

3. Antennas and lead-in wires shall be kept away from all light and power conductors to avoid accidental contact with the light and power conductors.

4. Antennas and lead-in wires shall not be attached to any poles that carry light and power wires over 250 volts between conductors.

5. Lead-in wires shall be securely attached to the antenna.

6. Outdoor antennas and lead-in conductors shall not cross over light and power wires.

7. Lead-in conductors shall be kept at least 2 feet (610 mm) away from open light and power conductors.

8. Where practicable, antenna conductors shall not be run under open light and power conductors.

9. On the outside of a building:

   a. Position and fasten lead-in wires so they cannot swing closer than 2 feet (610 mm) to light and power wires having *NOT* over 250 volts between conductors; 10 feet (3.05 m) if *OVER* 250 volts between conductors.

   b. Keep lead-in conductors at least 6 feet (1.83 m) away from a lightning rod system, or bond together according to *Section 250-86.*

10. On the inside of a building:

   a. Keep antenna and lead-in wires at least 2 inches (51 mm) from other open wiring (as in old houses) unless the other wiring is in a raceway or cable.

   b. Keep lead-in wires out of electric boxes unless there is an effective, permanently installed barrier to separate the light and power wires from the lead-in wire.

11. Grounding: (figure 25-10A)

   a. The grounding wire must be copper, aluminum, copper-clad steel, bronze, or similar corrosion-resistant material.

   b. The grounding wire need not be insulated. It must be securely fastened in place; may be attached directly to a surface without the need for insulating supports; shall be protected from physical damage or be large enough to compensate for lack of protection; and shall be run in as straight a line as is practicable.

c. The grounding conductor shall be connected to the nearest accessible location on: the building or structure grounding electrode *(Section 250-81)*; or the grounded interior water pipe *(Section 250-80(a))*; or the metallic power service raceway; or the service equipment enclosure; or the grounding electrode conductor or its metal enclosure.

d. If none of *c* is available, then ground to any one of the electrodes per *Section 250-81*, such as metal underground water pipe, metal frame of building, concrete-encased electrode, or ground ring.

e. If neither *c* nor *d* are available, then ground to an effectively grounded metal structure or to any one of the electrodes per *Section 250-83*, such as underground gas piping system, driven rod or pipe, or a metal plate.

f. The grounding conductor may be run inside or outside of the building.

g. The grounding conductor shall not be smaller than No. 10 copper or No. 8 aluminum.

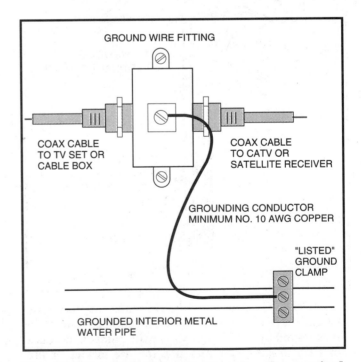

Fig. 25-10A Typical connections for the required grounding per *Section 810-21, National Electrical Code.*®

h. *CAUTION:* If any of the preceding rules results in one grounding electrode for the antenna and another grounding electrode for the electrical system, bond the two electrodes together with a bonding jumper not smaller than No. 6 copper or its equal.

The objective of grounding and bonding to the same grounding electrode as the main service ground is to reduce the possibility of having a difference of voltage potential between the two systems.

TELEPHONE WIRING (SYMBOL ▶|) *(ARTICLE 800)*

Since the deregulation of telephone companies, residential DO-IT-YOURSELF telephone wiring has become quite common. Following is an overview of residential telephone wiring.

The telephone company will install the service line to a residence and terminate at a protector device, figure 25-11. The protector protects the system from hazardous voltages. The protector may be mounted either outside or inside the home. Different telephone companies have different rules. Always check with the phone company before starting your installation.

The point where the telephone company ends and the homeowner's interior wiring begins is called the *demarcation point*, Figure 25-11. The preferred demarcation point is outside of the home, generally near the electric meter where proper grounding and bonding can be done. A *network interface* might be of the types depicted in figure 25-11, at **(A)** and **(B)**. In the past, interior telephone wiring in homes was installed by the telephone company. Today, the responsibility for roughing-in the interior wiring is usually left up to the electrician. Telephone wiring is *not* to be run in the same raceway with electrical wiring. In addition, it cannot be installed in the same outlet and junction boxes as electrical wiring.

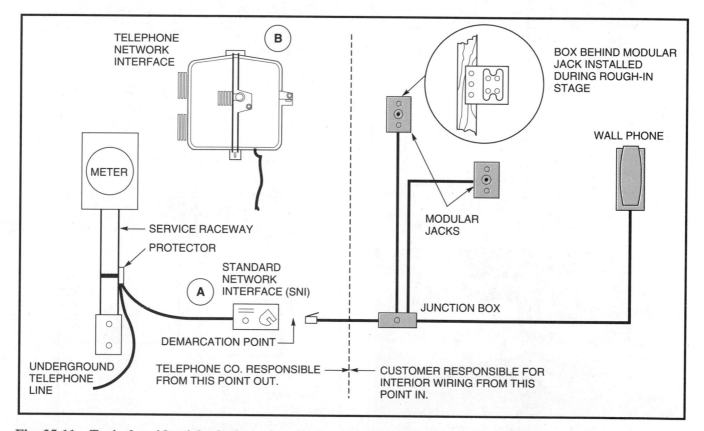

Fig. 25-11 **Typical residential telephone installation. The telephone company installs and connects the underground cable to the protector. In this illustration, the protector is mounted to the metal service-entrance raceway. This establishes the ground that offers protection against hazardous voltages such as a lightning surge on the incoming line. A cable is then run to a Standard Network Interface, referred to as SNI or just NI. These are shown in (A) and (B) in the diagram. The interior telephone wiring is then connected to the SNI, either with a modular plug as shown in (A), or "hard-wired" through a neoprene grommet on the bottom of (B). The SNI of the type depicted by (B) are preferred because they are weatherproof and can be mounted on the outside of the house.**

Whatever the case, the rules found in *Article 800* of the *National Electrical Code®* apply. Refer to figures 25-11, 25-12, and 25-13.

Article 800 of the Code covers telephone wiring as well as fire alarms and burglar alarms.

For a typical residence, the electrician will rough-in wall boxes wherever a telephone outlet is wanted. The boxes can be of the "single-gang" types illustrated in figures 25-1, 2-11, 2-12A, 2-18, and 2-23. Concealed telephone wiring in new homes is generally accomplished using multipair cables that are run from each telephone outlet to a common point, figure 25-12(A). The cable can also take the form of a "loop system," figure 25-13. A 3-pair, 6-conductor twisted cable is shown in figure 25-14.

There are some localities that require that these cables be run in a metal raceway, such as electrical metallic tubing (EMT).

The outer jacket is usually of a thermoplastic material. This outer jacket most often is a neutral color such as light gray or beige that blends in with decorator colors in the home. This is particularly important in existing homes where the telephone cable may have to be exposed. Cables, mounting boxes, junction boxes, terminal blocks, jacks, adaptors, faceplates, cords, hardware, plugs, etc., are all available through electrical distributors, builders' supply outlets, telephone stores, electronic stores, hardware stores, and similar wholesale and retail outlets. The selection of types of the preceding components is endless.

Telephone Conductors

The color coding of telephone cables is shown in figure 25-14. Figure 25-15 shows some of the many types of telephone cable available.

Cables are available in the following sizes:

2 pair – – – – – (4 conductors)
3 pair – – – – – (6 conductors)
6 pair – – – – – (12 conductors)
12 pair – – – – – (24 conductors)
25 pair – – – – – (50 conductors)
50 pair – – – – – (100 conductors)

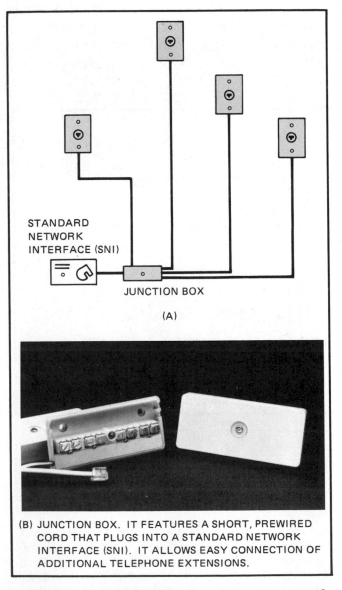

STANDARD
NETWORK
INTERFACE (SNI)

JUNCTION BOX

(A)

(B) JUNCTION BOX. IT FEATURES A SHORT, PREWIRED CORD THAT PLUGS INTO A STANDARD NETWORK INTERFACE (SNI). IT ALLOWS EASY CONNECTION OF ADDITIONAL TELEPHONE EXTENSIONS.

Fig. 25-12 Individual telephone cables run to each telephone outlet from the telephone company's connection point.

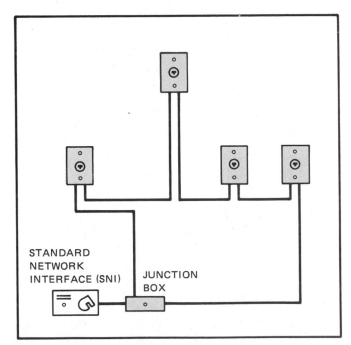

STANDARD
NETWORK
INTERFACE (SNI)

JUNCTION
BOX

Fig. 25-13 A complete "loop system" of the telephone cable. If something happens to one section of the cable, the circuit can be fed from the other direction.

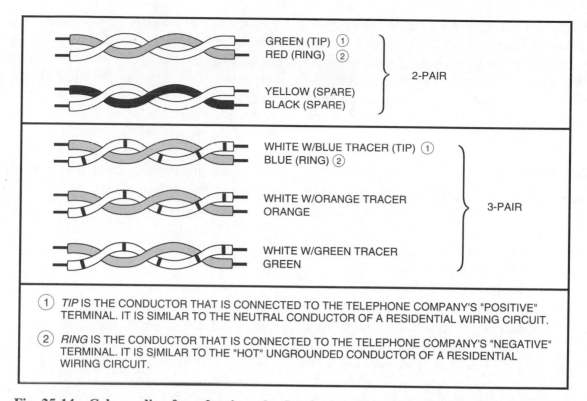

GREEN (TIP) ①
RED (RING) ②

YELLOW (SPARE)
BLACK (SPARE)

2-PAIR

WHITE W/BLUE TRACER (TIP) ①
BLUE (RING) ②

WHITE W/ORANGE TRACER
ORANGE

3-PAIR

WHITE W/GREEN TRACER
GREEN

① *TIP* IS THE CONDUCTOR THAT IS CONNECTED TO THE TELEPHONE COMPANY'S "POSITIVE" TERMINAL. IT IS SIMILAR TO THE NEUTRAL CONDUCTOR OF A RESIDENTIAL WIRING CIRCUIT.

② *RING* IS THE CONDUCTOR THAT IS CONNECTED TO THE TELEPHONE COMPANY'S "NEGATIVE" TERMINAL. IT IS SIMILAR TO THE "HOT" UNGROUNDED CONDUCTOR OF A RESIDENTIAL WIRING CIRCUIT.

Fig. 25-14   Color coding for a 2-pair and a 3-pair, 6-conductor twisted telephone cable. The color coding for 2-pair telephone cable stands alone. For 3 or more pair cables, the color coding becomes WHITE/BLUE, WHITE/ORANGE, WHITE/GREEN, WHITE/BROWN, WHITE/SLATE, RED/BLUE, RED/ORANGE, RED/GREEN, RED/BROWN, RED/SLATE, BLACK/BLUE, BLACK/ORANGE, BLACK/GREEN, BLACK/ BROWN, BLACK/SLATE, YELLOW/BLUE, YELLOW/ORANGE, YELLOW/GREEN, YELLOW/BROWN, YELLOW/ SLATE, VIOLET/BLUE, VIOLET/ORANGE, VIOLET/GREEN, VIOLET/BROWN, VIOLET/SLATE. In multi-pair cables, the additional pairs can be used for more telephones, fax's, security reporting, speaker phones, dialers, background music, etc.

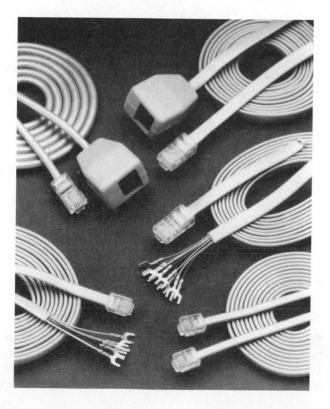

Fig. 25-15   Some of the many types of telephone cable available are illustrated. Note that for ease of installation, the modular plugs and terminals have been attached by the manufacturer of the cables. These cords are available in round and flat configurations, depending upon the number of conductors in the cable.

Telephone circuits require a separate pair of conductors from the telephone all the way back to the phone company's central switching center. To keep interference that could come from other electrical equipment to a minimum, such as electric motors and fluorescent fixtures, each pair of telephone wires (2-wires) is twisted.

Cross-Talk Problems

A common telephone interference problem in homes is "cross-talk." This problem is most often traced to 2-line telephones where flat cords are used. The signals traveling in one pair are picked up by the adjacent pair in the cord. This cross-talk problem is virtually eliminated by using cords that have conductors twisted at proper intervals. In figure 25-14, each pair of conductors is twisted, then all of the pairs are again twisted. Avoid the use of flat 2-line cords in lengths over 7 feet. Flat 1-line cords present no problem.

The recommended circuit lengths are:

No. 24 gauge – – – – – not over 200 feet

No. 22 gauge – – – – – not over 250 feet

Many telephone companies suggest that not more than five telephones be connected to one line because there may not be enough power to ring more than five ringers. Most newer telephones indicate on the label or on the carton the telephone REN (Ringer Equivalence Number). The REN refers to the older style electromechanical ringer which was considered to be 1-REN. Cordless telephones have a rating of 0.2 to 0.3 RENS. Electronic ringers use very little power. Therefore, many electronic phones can be connected to one line before running into problems. It is a good idea to call the local telephone company and ask if there are limitations concerning the number of phones that may be connected to one line.

EXAMPLE:

Five 1-REN phones = 5 RENS

Three 1-REN and one 2-REN phone = 5 RENS

Ten ½-REN phones = 5 RENS

If you are considering installing more than five telephones in a home, pay attention to the REN numbers. Possible line cut-off can occur when too many phones are connected to one line.

When several telephones are used in a residence, most telephone companies recommend that at least one of these telephones be permanently installed. This would probably be the wall phone in the kitchen. The remaining phones may remain portable. This means that they can be plugged into any of the phone jacks, figures 25-16 and 25-17. If all of the phones were portable, there might be a time when none of the phones were connected to phone jacks. As a result there would be no audible signal.

The electrical plans show that nine telephone outlets are provided as follows:

Front Bedroom	1
Kitchen/Nook	1
Laundry	1
Living Room	1
Master Bedroom	1
Rear Outdoor Patio	1
Recreation Room	1
Study/Bedroom	1
Workshop	1

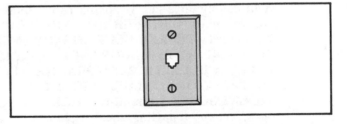

Fig. 25-16   Telephone modular jack.

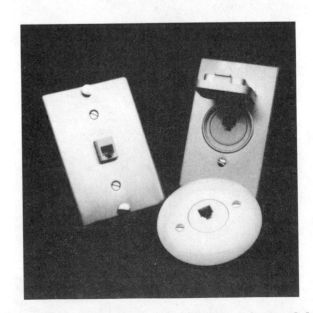

Fig. 25-17   Three styles of wall plates for modular telephone jacks: rectangular stainless steel, weatherproof for outdoor use, and circular. *Courtesy* of KEPtel, Armiger.

Installation of Telephone Conductors, *Sections 800-51* and *800-52*

Telephone conductors:

1. shall be Type CM or Type CMX (for dwellings only), listed for telephone installation as being resistant to the spread of fire.

2. shall be separated by at least 2 inches (51 mm) from light and power conductors unless the light and power conductors are in a raceway, or in nonmetallic-sheathed cable, Type AC cable, or Type UF cable.

3. shall not be placed in any conduit or boxes with electric light and power conductors unless the conductors are separated by a partition.

4. may be terminated in either metallic or nonmetallic boxes. Check with the local telephone company or local electrical inspector for specific requirements.

5. should not share the same stud space as the electrical branch-circuit wiring.*

6. should be kept at least 12 inches away from electrical branch-circuit wiring where the telephone cable is run parallel to the branch-circuit wiring. This guards against induction noise.*

7. should not share the same bored holes as electrical wiring, plumbing, or gas pipes.

8. should be kept away from hot water pipes, hot air ducts, and other heat sources that might harm the insulation.*

9. should be attached to the sides of joists and studs with insulated staples, being careful not to crush the cable.*

10. Do not support telephone conductors from raceways that contain light and power conductors.

*These are recommendations from residential telephone manufacturers' installation data.

Conduit

In some communities, electricians prefer to install regular device boxes at a telephone outlet location, then "stub" a ½-inch conduit to the basement or attic from this box. Then, at a later date, the telephone cable can be "fished" through the conduit.

Grounding *(Article 800, Part D)*

The telephone company will provide the proper grounding of their incoming cable sheath and protector, generally using wire not smaller than No. 14 AWG copper or equal. Or, they might clamp the protector to the grounded metal service raceway conduit to establish "ground." The grounding requirements are practically the same as for CATV and antennas, which were covered earlier in this unit.

Safety

Open circuit voltage between conductors of an idle pair of telephone conductors can range from 50 to 60 volts dc. The ringing voltage can reach 90 volts ac. Therefore, always work carefully with insulated tools and stay clear of bare terminals and grounded surfaces. Disconnect the interior telephone wiring if work must be done on the circuit, or take the phone off the hook, in which case the dc voltage level will drop and there should be no ac ringing voltage delivered.

SIGNAL SYSTEM (CHIMES)

A signal circuit is described in the *National Electrical Code®* as any electrical circuit that energizes signaling equipment, or one that is used to transmit certain types of signaling messages. Signaling equipment includes such devices as chimes, doorbells, buzzers, code-calling systems, and signal lights.

Door Chimes (Symbol CH)

Present-day dwellings often use chimes rather than bells or buzzers to announce that someone is at a door. A musical tone is sounded rather than a harsh ringing or buzzing sound. Chimes are available in single-note, two-note, eight-note (four-tube), and repeater tone styles. In a repeater tone chime, both notes sound as long as the push button is depressed. In an eight-note chime, contacts on a motor-driven cam are arranged in sequence to sound the notes of a simple melody when the chime button is pushed. This type of chime is usually installed in dwellings having three entrances. The chime can be connected so that the eight-note

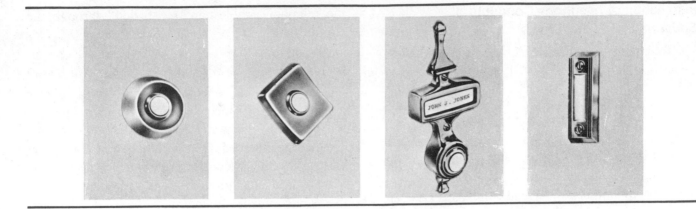

Fig. 25-18   Push buttons for door chimes. *Courtesy* Nutone, Inc.

melody sounds for the front door, two notes sound for the side door, and a single note sounds for the rear door. Chimes are also available with clocks and lights.

Electronic chimes may relay their chime tones through the various speakers of an intercom system.

When any chime is installed, the manufacturer's instructions must be followed.

The plans show that two chimes are installed in the residence. Two-note chimes are used. Each chime has two solenoids and two iron plungers. When one solenoid is energized, the iron plunger is

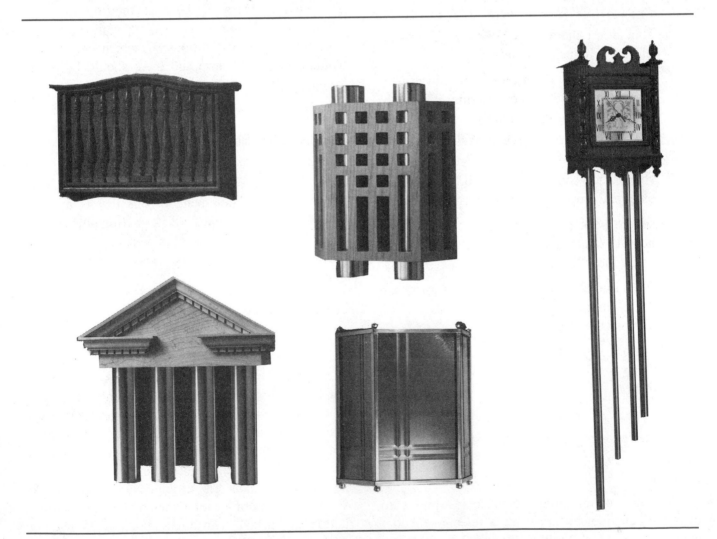

Fig. 25-19   Typical residential door chimes. *Courtesy* NuTone, Inc.

drawn into the opening of the solenoid. A plastic peg in the end of the plunger strikes one chime tone bar. When the solenoid is de-energized, spring action returns the plunger where it comes to rest against a soft felt pad so that it does not strike the other chime tone bar. Thus, a single chime tone sounds. As the second solenoid is energized, one chime tone bar is struck. When the second solenoid is de-energized, the plunger returns and strikes the second tone bar. A two-tone signal is produced. The plunger then comes to rest between the two tone bars. Generally, two notes indicate front door signaling, and one note indicates rear or side door signaling.

Figure 25-18 shows four push-button styles used for chimes. Many other styles are available. Figure 25-19 shows several types of typical residential-type wall-mounted chimes. The symbols used to indicate push buttons and audible signals on the plans are shown in figure 25-20.

Figures 25-21 and 25-22 show one way in which to provide proper backing for chimes. This is obviously done during the rough-in stages of the electrical installation before the walls are closed up.

Chimes for the Hearing Impaired

Chimes for the hearing impaired are available with accessory devices that can turn on a dedicated lamp at the same time the chime is sounded. Figure 25-22A shows the devices needed to accomplish this.

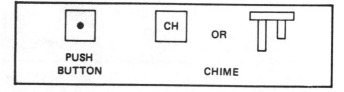

Fig. 25-20 Symbols for chimes and push buttons.

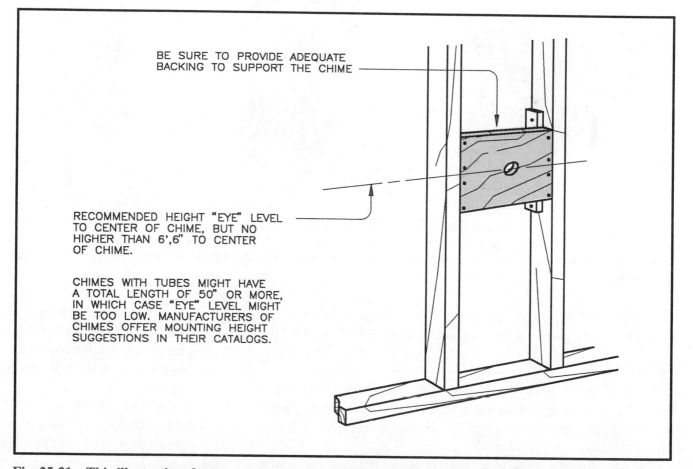

Fig. 25-21   This illustration shows one method of providing support for a surface-mounted chime. This is the rough-in stage, before the walls are closed.

Fig. 25-22   Chime transformer.

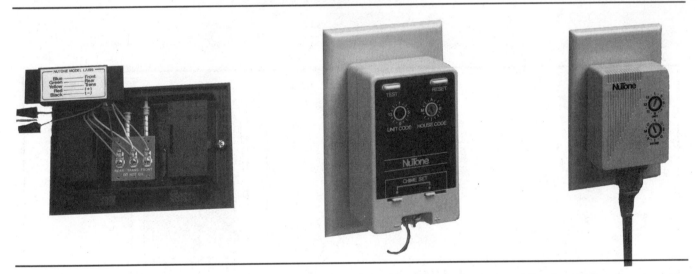

Fig. 25-22A   Accessory devices can be attached to new or existing chimes to provide visual signals when the chime is sounded. The first photo shows the module that is connected and mounted inside the chime. The second photo shows a transmitter that is plugged into a nearby receptacle. The third photo shows a receiver into which a lamp is plugged. *Courtesy* **NuTone, Inc.**

Figure 25-22B shows the wiring details. Figure 25-23 shows the roughing-in for a flush-mounted chime.

Transformers

Because of the low-voltage ratings and power limitations of chime transformers, the secondary wiring and the transformer itself are classified as Class 2 circuits.

The transformers used to operate door chimes have a greater capacity than the transformers used with bell and buzzer circuits, figure 25-22. The voltage output of chime transformers ranges between 10 and 24 volts. These transformers are rated from 5 to 30 volt-amperes (watts). Bell transformers have a voltage output range of 6 to 10 volts and a rating of 5 to 20 volt-amperes (watts).

Chime transformers used in dwellings are available with a 16-volt rating. Transformers that give a combination of voltages—such as 4, 8, 12, and 24 volts—also are used.

Underwriters Laboratories lists two types of chime transformers that are normally used for chime

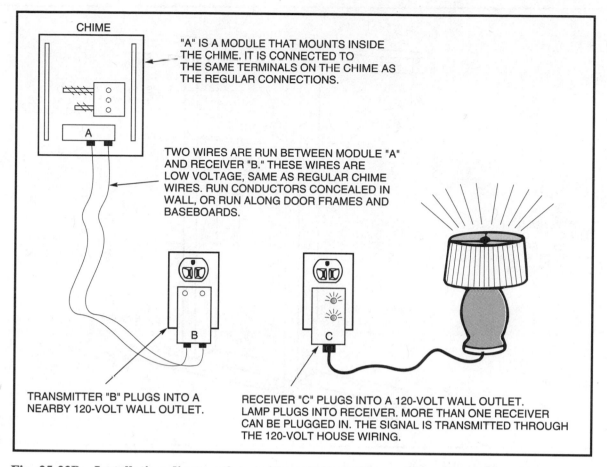

CHIME

"A" IS A MODULE THAT MOUNTS INSIDE THE CHIME. IT IS CONNECTED TO THE SAME TERMINALS ON THE CHIME AS THE REGULAR CONNECTIONS.

A

TWO WIRES ARE RUN BETWEEN MODULE "A" AND RECEIVER "B." THESE WIRES ARE LOW VOLTAGE, SAME AS REGULAR CHIME WIRES. RUN CONDUCTORS CONCEALED IN WALL, OR RUN ALONG DOOR FRAMES AND BASEBOARDS.

B

C

TRANSMITTER "B" PLUGS INTO A NEARBY 120-VOLT WALL OUTLET.

RECEIVER "C" PLUGS INTO A 120-VOLT WALL OUTLET. LAMP PLUGS INTO RECEIVER. MORE THAN ONE RECEIVER CAN BE PLUGGED IN. THE SIGNAL IS TRANSMITTED THROUGH THE 120-VOLT HOUSE WIRING.

Fig. 25-22B   Installation diagram for auxiliary devices for use with chimes that can help the hearing impaired. When the chime sounds, the signal is transmitted to the receiver, turning on the lamp that is plugged into the receiver. The "on" time is adjustable at "C". The lamp should be a "dedicated" lamp to light up only when the chime sounds.

wiring. The *inherently limited transformer* is designed to limit the short-circuit current to a maximum of 8 amperes. The *not inherently limited transformer* is rated at 100 volt-amperes or less. This transformer has an overcurrent protective device that limits the voltage and current. See *Section 725-41*. The open circuit voltage limitation of both types of transformers must not exceed 30 volts. Most transformers suitable for use in dwellings have a built-in thermal overload device. Whenever a short circuit occurs in the bell-wire circuit, the overload device opens and closes repeatedly until the short is cleared.

Additional Chimes

To extend a chime system to cover a larger area, a second or third chime may be added. In the residence, two chimes are used: one chime is mounted in the front hall and a second (extension) chime is mounted in the recreation room. The extension chime is wired in parallel to the first chime. The

wires are run from one chime terminal board to the terminal board of the other chime. The terminals are connected as follows: transformer to transformer, front to front, and rear to rear, figure 25-24.

When chimes are added to a chime circuit, a transformer with a higher wattage rating may be required to energize the greater number of solenoids being used at one time. Also, more wire is used in the circuit and the voltage drop and power loss are greater. This increase is the result of resistance in the wire between the chimes and the transformer. *Section 725-41(b)* does not allow transformers supplying Class 2 systems to be connected in parallel unless listed for interconnection.

If a buzzer (or bell) and a chime are connected to a single transformer and are used at the same time, the transformer will put out a fluctuating voltage. This condition does not allow either the buzzer or the chime to operate properly. The use of a transformer with a larger rating may solve this problem.

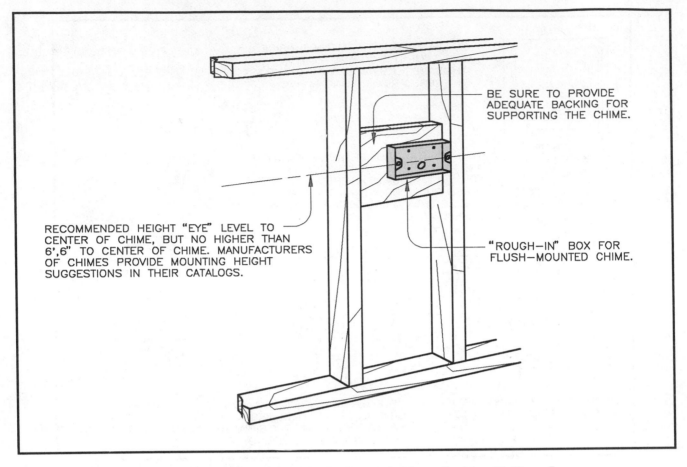

Fig. 25-23   **Roughing-in for a flush-mounted chime.** *Courtesy* **NuTone, Inc.**

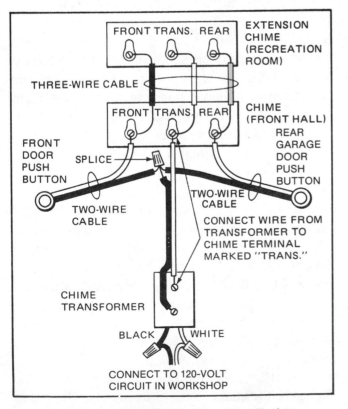

Fig. 25-24   **Circuit for chime installation.**

The wattage consumption of chimes varies with the manufacturer. Typical ratings are as follows:

TYPE OF CHIME	POWER CONSUMPTION
Standard two-note	10 watts
Repeating chime	10 watts
Internally lighted, two lamps	10 watts
Internally lighted, four lamps	15 watts
Combination chime and clock	15 watts
Motor-driven chime	15 watts
Electronic chime	15 watts

Transformers with ratings of 5, 10, 15, 20, and 30 watts (VA) are available. For a multiple chime installation, wattage ratings for the individual chimes are added. The total value is the minimum transformer rating needed to do the job properly. If there are still technical questions or problems, check the literature supplied by the chime and transformer manufacturers.

Wiring for Chime Installation

Bell Wire and Cable.   The wire used for low-voltage bell and chime circuits, and for connecting low-voltage thermostats, is called *bell wire, annunciator wire*, or *thermostat wire*.

These wires are generally made of copper and have a thermoplastic insulation suitable for use on 30 volts or less.

Because the current required for bell and chime circuits is rather small, No. 18 AWG conductors are quite often used. If the length of the circuit is very long, a larger conductor is suggested.

Multiconductor cables consist of two, three, or more single wires covered with a single protective insulation. This type of cable is often used for electrical installations because there is less danger of damage to individual wires and it gives a neat appearance to the wiring. The conductors within the cables are color coded to make circuit identification easy.

Bell wire and cable may be fastened directly to surfaces with insulated staples or cleats, or may be installed in the raceways. Be careful not to pierce or crush the wires. Do not pull bell wires or other low-voltage cables through the same holes in studs and joists that contain nonmetallic-sheathed cable, BX, conduit, or other piping because they cannot stand the physical abuse that they would be subjected to. The job requirements will determine how the conductors are to be attached. In the residence in the plans, the bell wire is run along the sides of the floor joists in the basement and on the sides of the studs in the walls.

Wiring Circuit.   The circuit shown in figure 25-24 is recommended for this chime installation because it provides a hot low-voltage circuit at the front hall location. However, it is not the only way in which these chimes may be connected.

Figure 25-24 shows that a two-wire cable runs from the transformer in the utility room to the front hall chime. A two-wire cable then runs from the chime to both the front and rear door push buttons. A three-wire cable also runs between the front hall chime and the recreation room chime. Because of the hot, low-voltage circuit at the front hall location, a chime with a built-in clock can be used. A four-conductor cable may be run to the extension chime so that the owner may install a clock-chime at this location also.

The plans show that the chime transformer is mounted on one of the ceiling boxes in the utility room. The line voltage connections are easy to make here. Some electricians prefer to mount the chime transformer on the top or side of the distribution panel or load center. The electrician decides where to mount the transformer after considering the factors of convenience, economy, and good wiring practice.

National Electrical Code® Rules for Signal Systems

Article 725 of the *National Electrical Code®* covers the requirements for low-voltage, Class 2 circuits. These were discussed in unit 24.

Repeating some of the most basic Code installation requirements:

- do *not* install Class 2 wiring in the same raceway or cable as light and power wiring.

- keep Class 2 bell wires at least 2 inches (50.8 mm) away from light or power wiring. This really pertains to old, open-knob, and tube wiring. Today, where light and power wiring is installed in Type NMC, Type AC, or in conduit, there is no problem.

- Class 2 wiring may not enter the same enclosure, unless there is a barrier in the enclosure that separates the Class 2 wiring from the light and power conductors. Examples of this barrier are illustrated in figure 27-12.

- Chime transformers are listed as Class 2 transformers. Thus, all of the wiring connected to the secondary of the transformer is Class 2 wiring.

- Do *not* run Class 2 cables through the same holes as light and power cables and conduits. There is too much chance of physical damage to the Class 2 cables.

- Because the insulation and the jacket of Class 2 cables and conductors are rather thin, be careful during installation to avoid damage. Secure these cables and conductors with care, using the proper type of staples. Staple "guns" are available that use rounded staples that nicely straddle the cable instead of flattening the cable, which might result in shorted out wires.

REVIEW

Note: Refer to the Code or the plans where necessary.

TELEVISION CIRCUIT

1. How many television outlets are installed in this residence? _____

2. What type of television cable is commonly used and recommended? _____
   _____

3. What determines the design of the faceplates used? _____
   _____

4. What must be provided when installing a television outlet and receptacle outlet in one wall box? _____
   _____
   _____
   _____
   _____

5. From a cost standpoint, which system is more economical to install: a master amplifier distribution system or a multiset coupler? Explain the basic differences between these two systems. _____
   _____
   _____
   _____
   _____
   _____
   _____
   _____

6. How many wires are in the cable used between a rotor and its controller? _____

7. Digital satellite systems use an antenna that is approximately (18 inches), (36 inches) (72 inches) in diameter. Circle the correct answer.

8. List the requirements for cable television inside the house. _____
   _____
   _____
   _____
   _____

9. What article of the Code references the requirements for cable community television installation? _____

10. It is generally understood that grounding and bonding together all metal parts of an electrical system and the metal shield of the cable television cable to the same grounding reference point in a residence will keep both systems at the same voltage level should a surge, such as lightning, occur. Therefore, if the incoming cable to a house that has been installed by the CATV cable company installer has the metal shield grounded to a driven ground rod, does this installation conform to the *National Electrical Code®*? _____

11. The basics of cable television are that a transmitter on earth sends an _____ signal to a _____ where the signal is reamplified and sent back to earth via the _____ signal. This signal is picked up by the _____ on the satellite antenna, which is sometimes referred to as _____ a. The feedhorn funnels the signal into a _____ _____ _____ , then on to a _____ , where the high-frequency signal is converted to a _____-frequency signal suitable for the television receiver.

12. All television satellites rotate above the earth in (the same orbit) (different orbits). Circle correct answer.

13. Television satellites are set in orbit (10,000) (18,000) (22,247) miles above the earth, which results in their rotating around the earth at (precisely the same) (different) rotational speed as the earth rotates. This is done so that the satellite "dish" can be focused on a specific satellite (once) (one time each month) (whenever the television set is used). Circle correct answer.

14. What section of the Code prohibits supporting coaxial cables from raceways that contain light or power conductors? Section _____

15. When hooking up one CATV cable and another cable from an outdoor antenna to a receiver that has only one antenna input terminal, an _____ switch is usually installed.

TELEPHONE SYSTEM

1. How many locations are provided for telephones in the residence? _____

2. At what height are the telephone outlets in this residence mounted? Give measurement to center.

_____

_____

_____

_____

_____

3. Sketch the symbol for a telephone outlet.

4. Is the telephone system regulated by the *National Electrical Code®*? _____

5. a. Who is to furnish the outlet boxes required at each telephone outlet? _____

   b. Who is to furnish the faceplates? _____

6. Who is to furnish the telephones? _____

7. Who does the actual installation of the telephone equipment? _____

8. How are the telephone cables concealed in this residence? _____

_____

_____

_____

9. The point where the telephone company's cable ends and the interior telephone wiring meet is called the _____ point. The device installed at this point is called a _____ .

10. What are the colors contained in a four-conductor telephone cable assembly and what are they used for? _____

_____

_____

_____

_____

_____

_____

11. Itemize the Code rules for the installation of telephone cables in a residence. _____

_____

_____

_____

12. If finger contact were made between the red conductor and green conductor at the instant a "ring" occurs, what shock voltage would be felt? _____

13. What section of the Code prohibits supporting telephone wires from raceways that contain light or power conductors? Section _____

14. The term *cross-talk* is used to define the hearing of the faint sound of voices in the background when you are using the telephone. Cross-talk can be reduced significantly by running (flat cables) (twisted pair cables) to the telephones. Circle the correct answer.

15. What section of the Code prohibits telephone cables from being installed in the same box or enclosure, or from being pulled into the same raceway as light and power circuits? _____

SIGNAL SYSTEM

1. What is a signal circuit? _____

_____

2. What style of chime is used in this residence? _____

3. a. How many solenoids are contained in a two-tone chime? _____

   b. What closes the circuit to the solenoid of a chime? _____

4. Explain briefly how two notes are sounded by depressing one push button (when two solenoids are provided). _____

_____

_____

_____

_____

_____

_____

5. a. Sketch the symbol for a push button. _____

   b. Sketch the symbol for a chime. _____

6. a. At what voltage do residence chimes generally operate? _____

   b. How is this voltage obtained? _____

7. What is the maximum volt-ampere rating of transformers supplying Class 2 systems?

_____

8. What two types of chime transformers for Class 2 systems are listed by Underwriters Laboratories? _____

_____

9. Is the extension chime connected with the front hall chime in series or in parallel?

_____

10. How many bell wires terminate at

   a. the transformer? _____

   b. the front hall chime? _____

   c. the extension chime? _____

   d. each push button? _____

11. a. What change in equipment may be necessary when more than one chime is connected to sound at the same time on one circuit? _____

_____

   b. Why? _____

_____

_____

_____

_____

12. What type of insulation is usually found on low-voltage wires? _____

13. What size wire is installed for signal systems of the type in this residence? _____

14. a. How many wires are run between the front hall chime and the extension chime in the recreation room? _____

   b. How many wires are required to provide a "hot" low-voltage circuit at the extension chime? _____

15. Why is it recommended that the low-voltage secondary of the transformer be run to the front hall chime location and separate two-wire cables be installed to each push button? _____

_____

16. a. Is it permissible to install low-voltage Class 2 systems in the same raceway or enclosure with light and proper wiring? _____

    b. What Code section covers this? _____

17. a. Where is the transformer in the residence mounted? _____

    b. To what circuit is the transformer connected? _____

18. a. How many feet of two-conductor bell wire cable are required? _____

    b. How many feet of three-conductor bell wire cable are required? _____

19. How many insulated staples are needed for the bell wire if it is stapled every 2 feet (610 mm)? _____

20. Should low-voltage bell wire be pulled through the same holes in studs and joists that also contain nonmetallic-sheathed cable? _____

21. What section of the Code prohibits supporting fire-protective signaling circuits from raceways that contain light or power conductors? *Section* _____ .

UNIT 26

Heat and Smoke Detectors, Security Systems

OBJECTIVES

After studying this unit, the student will be able to

- understand the *National Fire Protection Association Standard No. 72*, which refers to household fire warning equipment.
- name the two basic types of smoke detectors.
- discuss the location requirements for the installation of heat and smoke detectors for minimum acceptable levels of protection.
- discuss the location requirements for the installation of heat and smoke detectors for increased protection that exceed the minimum acceptable levels of protection.
- list the major components of typical residential smoke, heat, and security systems.
- discuss general *National Electrical Code* requirements for the installation of residential smoke, heat, and security systems.

THE IMPORTANCE OF HEAT, SMOKE, AND CARBON MONOXIDE DETECTORS

This unit will touch upon the basic requirements for protection against the hazards of fire, heat, smoke, and carbon monoxide.

Fire is the third leading cause of accidental death. Home fires account for the biggest share of these fatalities, most of which occur at night during sleeping hours.

Heat, smoke, and carbon monoxide detectors are installed in a residence to give the occupants early warning of the presence of fire or toxic fumes. Fires produce smoke and toxic gases that can overcome the occupants while they sleep. Most fatalities result from the inhalation of smoke and toxic gases, rather than from burns. Heavy smoke reduces visibility.

In nearly all home fires, detectable smoke precedes detectable levels of heat. Therefore, smoke detectors are considered to be the primary devices for protecting lives. Heat detectors should be installed in addition to smoke detectors, but should not replace them. A rather recent entry into the arena are carbon monoxide detectors, which sense dangerous carbon monoxide emitted from a malfunctioning furnace or other source. The *National Fire Protection Association* is developing a standard for carbon monoxide detection and protection. UL standard 2034 covers single and multiple station carbon monoxide detectors.

NATIONAL ELECTRICAL CODE®

The *National Electrical Code®* in *Article 760* covers the installation requirements for fire protective signaling systems that operate at 600 volts or less. Because the topic of fire protective signaling systems is so complex, the *NEC®* references a number of

National Fire Protection Association standards that contain information and recommendations not found in the *NEC.*® The following is a brief overview of some of the key points of *Article 760, NEC.*®

Fire protective signaling systems are classified into two groups:

1. Nonpower-limited circuits:
   - the conductors require overcurrent protection per the values found in *Section 310-15* (the same as for regular conductors). Maximum overcurrent protection for No. 18 AWG conductors is 7 amperes, and for No. 16 AWG it is 10 amperes.
   - derating the ampacity of the conductors required if the conductors carry continuous loads in excess of 10% of their ampacity.
   - shall not be installed in the same raceway as light and power conductors.
   - may be in the same enclosure, raceway, or cable *only* where connected to the same equipment.
   - only copper conductors permitted.
   ▶ • conductors larger than No. 16 AWG may be types listed in *Article 310*. No. 18 and No. 16 AWG conductors must be of the types listed for nonpower-limited signaling circuit use, Types NPLFP, NPLFR, NPLF. See *Section 760-31*. ◀
   - cables must be listed as resistant to the spread of fire. *Table 760-31(g)* shows the marking found on these cables.
   - cables may be installed on the surface or concealed, and shall be adequately supported and protected from physical damage.
   - in hoistways, cables must be installed in rigid metal conduit, rigid nonmetallic conduit, intermediate metal conduit, or electrical metallic tubing.
   - cables shall not be installed in ducts or plenums unless listed for that purpose.

2. Power-limited:
   - must have power output capabilities limited to the values specified in *Chapter 9, Tables 12(a)* and *12(b)*.
   - does not need separate overcurrent protection if the system itself inherently limits the power output, otherwise provide overcurrent protection that is not interchangeable with a higher ampere rating.
   - wiring on supply side (line side) to be installed according to nonpower-limited circuit requirements and in accordance with the normal wiring requirements found in *Chapter 3* of the *NEC.*®
   - wiring on load side to have copper conductors of the types listed in *Table 760-61(d)*, and need not be derated.
   - conductors and cables may be run on the surface or be concealed, shall be adequately supported and protected against physical damage.
   - conductors and cables shall not be supported by electrical raceways.
   - in hoistways, must be installed in rigid metal conduit, rigid nonmetallic conduit, intermediate metal conduit, or electrical metallic tubing.
   - conductors and cables shall be listed as suitable for the purpose, Types FPLP, FPLR, FPL.
   - *Table 760-61(d)* provides a table that shows the different types of cable that can be substituted for one another when necessary.
   - conductors and cables shall not be run in the same raceway, or be in the same enclosure as light and power wiring unless separated by a barrier or are in the enclosure to connect that particular piece of equipment.

For additional details of *National Electrical Code*® requirements for fire protective signaling systems, refer to *Article 760*.

NATIONAL FIRE PROTECTION ASSOCIATION (NFPA) STANDARD NO. 72

The *National Fire Protection Association (NFPA) Standard No. 72*, "National Fire Alarm Code," presents the minimum requirements for the proper selection, installation, operation, and maintenance of fire warning equipment. The standard also includes many recommendations that exceed the minimum requirements. Chapter 2 specifically covers household fire warning equipment.

Standard No. 72 defines a household fire warning system as "a system of devices that produces an audible alarm signal in the household for the purpose of notifying the occupants of the presence of a fire so they may evacuate the premises."

Fire warning devices commonly used in a residence are heat detectors, figure 26-1, and smoke detectors, figure 26-2.

The more elaborate systems connect to a central monitoring customer service center through a telephone line. These systems offer instant contact with police or fire departments, a panic button for emergency medical problems, low temperature detection, flood (high-water) detection, perimeter protection, interior motion detection, and other features. All detections are transmitted first to the company's customer service center, where personnel on duty monitor the system 24 hours a day, 7 days a week. They will verify that the signal is valid. After verification, they will contact the police department, the fire department, and individuals whose names appear on a previously agreed upon list that the homeowner

Fig. 26-1   Heat detector.

Fig. 26-2   Smoke detector.

prepared and submitted to the company's customer service center.

NFPA standards are *not* a law until adopted by a city, state, or other governmental body. Some NFPA standards are adopted in total, and others are adopted in part. Installers must be aware of all requirements in the locality in which they are doing work. Most communities require a special permit for the installation of fire warning equipment and security systems in homes, and require registration of the system with the police and/or fire departments. In most instances, fire protection requirements are found in the building codes of a community, and not necessarily spelled out in that community's electrical code. Remember, the *National Electrical Code®* sets forth the requirements of how to physically install a fire protective signaling system. It does *not* say where to install the detectors. The "where to install" information is found in NFPA 72, which is the standard adopted by most building officials.

A sufficient number of detectors, properly located, must be installed to provide adequate protection. These requirements and recommendations are discussed later.

TYPES OF SMOKE DETECTORS

Two common types of smoke detectors are the photoelectronic type and the ionization type. They usually contain an indication light to show that the unit is functioning properly. They also may have a test button that simulates smoke, so that when the button is pushed, the detector's smoke detecting ability as well as its circuitry and alarm are tested. Some systems are tested with a magnet.

Smoke detectors will not sense heat, flame, or gas.

Photoelectronic Type

The photoelectronic type of smoke detector has a light sensor that measures the amount of light in a chamber. When smoke is present, an alarm is sounded that indicates a reduction in light due to the obstruction of the smoke. This type of sensor detects smoke from burning materials that produce great quantities of smoke, such as furniture, mattresses, and rags. This type of detector is less effective for gasoline and alcohol fires, which do not produce heavy smoke.

Ionization Type

The ionization type of detector contains a low-level radioactive source that supplies particles that ionize the air in the detector's smoke chamber. Plates in this chamber are oppositely charged. Because the air is ionized, an extremely small amount of current (millionths of an ampere) flows between the plates. Smoke entering the chamber impedes the movement of the ions, reducing the current flow, which causes an alarm to go OFF.

The ionization type of detector is effective for detecting small amounts of smoke, as in gasoline and alcohol fires, which are fast-flaming.

TYPES OF HEAT DETECTORS

- **Fixed** temperature heat detectors are available that sense a specific **fixed** temperature, such as 135°F or 200°F. They shall have a temperature rating at least 25°F (14°C) above the normal temperature expected, but not to exceed 50°F (28°C) above the expected temperature.

- **Rate-of-Rise** heat detectors sense rapid changes in temperature (12 to 15 degrees per minute), such as those caused by flash fires.

- **Fixed/Rate of Rise** temperature detectors are available as a combination unit.

- **Fixed/Rate of Rise/Smoke** combination detectors are also available in one unit.

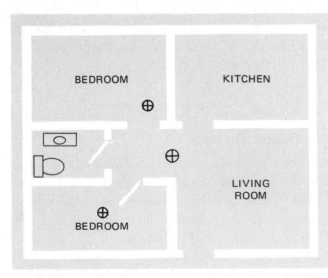

Fig. 26-3 Recommended location of heat or smoke detector *between* sleeping areas and rest of house. Locate outside of the bedroom, but near the bedrooms. ▶ In new construction, a smoke detector must be installed in each bedroom. ◀

INSTALLATION REQUIREMENTS

The following information includes specific recommendations for installing smoke and heat detectors in homes. Some of these requirements are illustrated in figures 26-3, 26-4, and 26-5. Complete data is found in *NFPA 72* and in the instructions furnished by the manufacturer of the equipment.

Individual detectors as well as complete systems can be purchased at electrical supply houses, home centers, hardware stores, and similar retail outlets. These are packaged with the necessary mounting hardware and detailed installation instructions.

Once installed, detectors should be checked periodically to make sure they are operating properly. Testing every six months is quite common. Refer to the manufacturer's recommendations.

Many companies specialize in installing complete fire and security systems. They also offer sys-

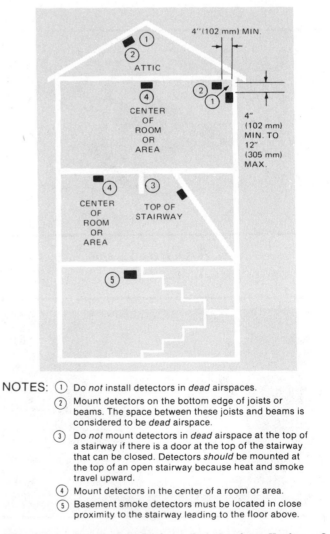

NOTES:
1. Do *not* install detectors in *dead* airspaces.
2. Mount detectors on the bottom edge of joists or beams. The space between these joists and beams is considered to be *dead* airspace.
3. Do *not* mount detectors in *dead* airspace at the top of a stairway if there is a door at the top of the stairway that can be closed. Detectors *should* be mounted at the top of an open stairway because heat and smoke travel upward.
4. Mount detectors in the center of a room or area.
5. Basement smoke detectors must be located in close proximity to the stairway leading to the floor above.

Fig. 26-4 Recommendations for the installation of heat and smoke detectors.

tem monitoring at a central office and can notify the police or fire department when the system gives the alarm. How elaborate the system should be is up to the homeowner.

Many local building codes do have requirements for the location, number, and type of smoke detectors to be installed in residential occupancies.

THE ABSOLUTE MINIMUM LEVEL OF PROTECTION

- DO install smoke detectors on each level of a residence.
- ►• DO install a smoke detector in each bedroom or sleeping area of a residence. ◄
- DO (in new construction) interconnect smoke detectors so that when one operates, all others will also operate. See figure 26-7.
- DO make sure that the detectors are listed by a recognized testing agency.

EXCEEDING MINIMUM LEVELS OF PROTECTION

The following are recommendations for attaining levels of protection that exceed the minimum level stated above and include guidelines for installing smoke and heat detectors in a home.

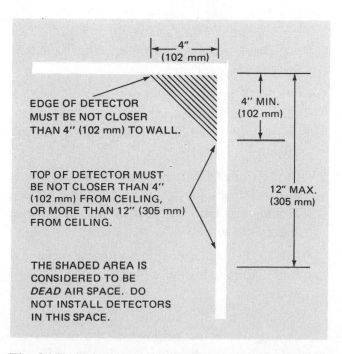

Fig. 26-5 Do not mount smoke or heat detectors in the *dead* airspace where the ceiling meets the wall.

- DO install smoke detectors for protection that *EXCEEDS* the minimum acceptable protection levels, in ALL rooms, basements, hallways, heated attached garages, and storage areas. Installing detectors in these locations will increase escape time, particularly if the room or area is separated by a door(s). In some instances, smoke detectors are installed in attics and crawl spaces. Check the manufacturer's recommendations for installing smoke detectors in attics, crawl spaces, and garages.

- DO consider special detectors that light up or vibrate when the occupants are hard of hearing.

- DO consider low temperature detectors (i.e, 45°F) that can detect low temperatures should the heating system fail. The damage caused by frozen water pipes bursting can be extremely costly.

INSTALLATION

- DO install smoke detectors on the ceiling, as close as possible to the center of the room or hallway.

- DO install smoke detectors at the high end of a room that has a sloped, gabled, or peaked ceiling where the rise is greater than 1 foot per 8 feet measured horizontally. Do not install in the dead air space, see figures 26-4 and 26-5.

- DO install HEAT detectors at the high end of a room that has a sloped, gabled, or peaked ceiling where the rise is greater than 1 foot per 8 feet measured horizontally. Mount them within 3 feet of the peak, but do not install in the dead air space, see figures 26-4 and 26-5.

- DO make sure that the path of rising smoke will reach the smoke detector when the detector is installed in a stairwell. This will usually be at the top of the stairway. The detector must not be located in a dead air space created by a closed door at the head of the stairway.

- DO install smoke detectors on a basement ceiling close to the stairway to the first floor.

- DO install smoke detectors in split-level homes. A smoke detector installed on the ceiling of an upper level can suffice for the protection of an adjacent lower level, provided the two levels are not separated by a door. Better protection is to install detectors for each level.

- DO install smoke detectors that are hard-wired directly to a 120 VAC source when installing detectors in NEW construction. In most cities this is mandatory. Dual-powered detectors having both 120 VAC and battery power provide the best protection. In existing homes, battery-powered detectors are the most common, but 120 VAC detectors or dual-powered detectors can be installed. Remember, battery-powered units will not operate with dead batteries. AC powered units will not operate when the power supply is off.

- DO install interconnected detectors when required by local codes. When one goes off, all other detectors also are triggered.

- DO always consider the fact that doors, beams, joists, walls, partitions, and similar obstructions will interfere with the flow of heat and smoke and, in most cases, create new areas needing additional smoke and heat detectors.

- DO consider that the accepted maximum distance between HEAT detectors mounted on flat ceilings is 50 feet (15 m), and from detector to wall 25 feet (7.6 m). This information is explained in detail by *NFPA 72*. Where beams, joists, etc. will obstruct the flow of heat, the 50-foot distance is reduced to 25 feet (7.6 m), and the 25-foot distance is reduced to 12½ feet.

- DO NOT install smoke or heat detectors in the dead air space at the top of a stairway that can be closed off by a door.

- DO NOT place the edge of a ceiling-mounted smoke or heat detector closer than 4 inches (102 mm) from the wall.

- DO NOT place the top edge of a wall-mounted smoke or heat detector closer than 4 inches (102 mm) from the ceiling and farther than 12 inches (305 mm) down from the ceiling. Some manufacturers recommend not farther down than 6 inches.

- DO NOT install smoke or heat detectors where the ceiling meets the wall because this is considered dead air space where smoke and heat may not reach the detector.

- DO NOT connect smoke or heat detectors to wiring that is controlled by a wall switch.

- DO NOT connect smoke or heat detectors to a circuit that is protected by a ground-fault circuit interrupter (GFCI).

- DO NOT install smoke or heat detectors where the relative humidity exceeds 85%, such as in bathrooms with showers, laundry areas, or other areas where large amounts of visible water vapor collect. Check the manufacturer's instructions.

- DO NOT install smoke or heat detectors in front of air ducts, air conditioners, or any high-draft areas where the moving air will keep the smoke or heat from entering the detector.

- DO NOT install smoke detectors in kitchens where the accumulation of household smoke can result in setting off an alarm, even though there is no real hazard. The person in the kitchen will know why the alarm sounded, but other people in the house may panic. This problem exists in multifamily dwelling units, where unwanted triggering of the alarm in one dwelling unit might cause panic for people in the other units. The photoelectronic type *may* be installed in kitchens, but must *not* be installed directly over the range or cooking appliance. A better choice in a kitchen would be to install a heat detector. Consult the manufacturer's recommendations.

- DO NOT install smoke detectors where the temperature can fall below 32°F or rise above 120°F unless the detector is specifically identified for this application.

- DO NOT install smoke detectors in garages where vehicle exhaust might set off the detector. Instead of a smoke detector, install a heat detector.

- DO NOT install smoke detectors in airstreams that will pass air originating at the kitchen cooking appliances across the detector. False alarms will result.

- DO NOT install smoke detectors near fluorescent fixtures. Install at least 6 feet (1.8 m) away.

- DO NOT install smoke or heat detectors on a ceiling where the ceiling will be excessively cold or hot. Smoke and heat will have difficulty reaching the detectors. This could be the case in older homes that are not insulated—or are poorly insulated. Instead, mount the detectors on the side wall. See figure 26-5.

- DO NOT install smoke or heat detectors on an outside wall that is not insulated, or poorly insulated. Instead, mount the detectors on an inside wall.

- DO NOT install smoke or heat detectors on a ceiling that employs radiant heating.

MANUFACTURERS' REQUIREMENTS

Here are a few of the responsibilities of the manufacturer of smoke and heat detectors. Complete data is found in *NFPA Standard No. 72.*

- The power supply must be capable of operating the signal for at least 4 minutes continuously.

- Battery-powered units must be capable of a low battery warning "beep" of at least one beep per minute for seven consecutive days.

- Direct-connected 120V ac detectors must have a visible indicator that shows "power-on."

- Detectors must not signal when a power loss occurs or when power is restored.

Wiring Requirements

Direct-Connected Units.   These units are connected to a 120-volt circuit. The black and white wires on the unit are spliced to the black and white wires of the 120-volt circuit, figure 26-6.

Feedthrough (Tandem) Units.   These units may be connected in tandem up to 10 units. Any one of these detectors can sense smoke, then send a signal to the remaining detectors, setting them OFF so that all detectors sound an alarm. These units contain three wires: one black, one white, and one yellow, figure 26-7. The yellow wire is the interconnecting wire.

Cord-Connected Units.   These units operate in the same way as direct-connected units, except that they are plugged into a wall outlet that is not controlled by a switch. These units are mounted to the wall on a bracket or screws. A clip on the cord is attached to the wall outlet with the same screw that fastens the faceplate to the outlet, figure 26-8. This prevents the cord from being pulled out, rendering the unit inoperative.

Battery-Operated Units.   These units require no wiring; they are simply mounted in the desired locations. Batteries last for about a year. Battery-operated units should be tested periodically, as recommended by the manufacturer.

COMBINATION DIRECT/BATTERY/FEED-THROUGH DETECTORS

These units are connected directly to a 120V ac circuit (not GFCI protected) and are equipped with a 9-volt dc battery, figure 26-9. Thus, in the event of a power outage, these detectors will continue to

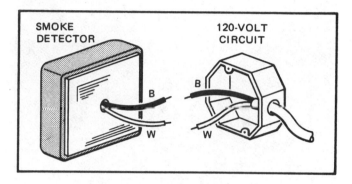

Fig. 26-6   Wiring of direct-connected detector units.

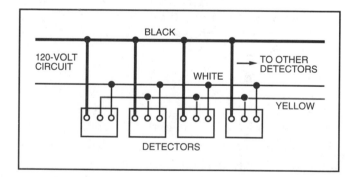

Fig. 26-7   Wiring of feedthrough (tandem) detector units. These detectors are "interconnected."

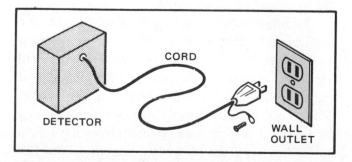

Fig. 26-8   Cord-connected detector units are plugged into wall outlet.

Fig. 26-9   Smoke detector. Operates on 120 volts ac. However, if power fails, the detector continues to operate on the 9-volt battery. *Courtesy* **of BRK Electronics.**

SECURITY SYSTEMS

It is beyond the scope of this text to cover each and every type of residential security system. Instead, we will focus on some of the features available from the manufacturers of these systems.

Installed security systems can range from the most simple system to very complex systems. Features and options include intruder detectors for doors and windows, motion detectors, infrared detectors, and under-the-carpet floor mat detectors for "space" protection, inside and outdoor horns, bells, electronic buzzers, and strobe lights to provide audio (sound) as well as visual detection, and telephone interconnection to preselected telephone numbers such as the police or fire departments, figures 26-10 through 26-19.

The decision about how complex an individual residential security system should be generally begins with a meeting between the homeowner and the electrician prior to the actual installation. Most electricians will become familiar with a particular manufacturer's selection of systems. Figure 26-20 shows the range of devices available for such a system.

These systems are generally on display at electrical distributors and home centers, which offer the homeowner an opportunity to see a security system "in action."

operate on the 9-volt dc battery. The ability to interconnect a number of detectors offers "whole house" protection, because when one unit senses smoke, all the other interconnected units will sound an alarm.

Read carefully the application and installation manuals that are prepared by the manufacturers of the smoke and heat detectors.

In this residence we find four smoke detectors connected to Circuit A17. These units are hardwired directly to the 120-volt ac branch-circuit. They are interconnected feedthrough units that will set off signaling in all units should one unit trigger. These interconnected smoke detectors are also powered by a 9-volt dc battery that allows the unit to operate in the event of a power failure in the 120-volt ac supply.

Fig. 26-10   Intruder alarm system. *Courtesy* **of NuTone, Inc.**

Fig. 26-11   Infrared detector. *Courtesy* **of NuTone, Inc.**

Fig. 26-12 Glass-break detector. *Courtesy* of NuTone, Inc.

Fig. 26-13 Security alarm keyboard. *Courtesy* of NuTone, Inc.

Fig. 26-14 Outdoor remote switch. *Courtesy* of NuTone, Inc.

Fig. 26-15 Indoor remote switch. *Courtesy* of NuTone, Inc.

Fig. 26-16 Indoor horn alarm. *Courtesy* of NuTone, Inc.

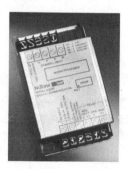

Fig. 26-17 Indoor/outdoor alarm. *Courtesy* of NuTone, Inc.

Fig. 26-18 Automatic digital communicator. *Courtesy* of NuTone, Inc.

Fig. 26-19 Telephone dialer. *Courtesy* of NuTone, Inc.

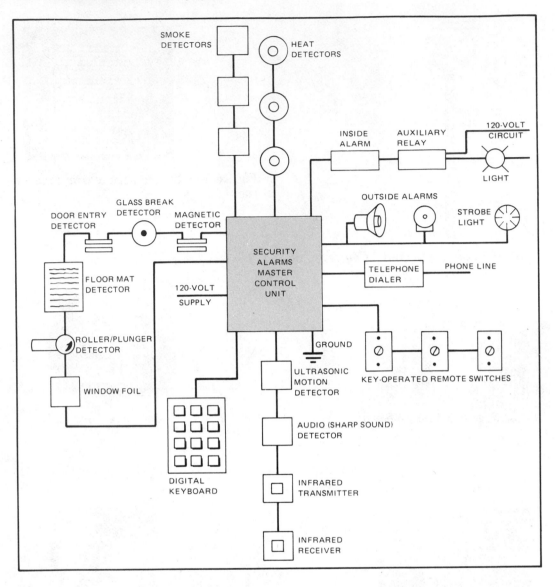

Fig. 26-20   Diagram of typical residential security system showing some of the devices available. Complete wiring and installation instructions are included with these systems. Check local code requirements in addition to following the detailed instructions furnished with the system. Most of the interconnecting conductors are No. 18 AWG.

The wiring of a security system consists of small, easy to install, low-voltage, multiconductor cables made up of No. 18 AWG conductors. The actual installation of these conductors should be done after the regular house wiring is completed to prevent damage to these smaller cables. Usually the wiring can be done at the same time as the chime wiring is being installed. Security system wiring comes under the scope of *Article 725* of the Code.

When wiring detectors such as door entry, glass break, floor mat, and window foil detectors, circuits are electrically connected in series so that if any part

of the circuit is opened, the security system will detect the open circuit. These circuits are generally referred to as "closed" or "closed loop." Alarms, horns, and other signaling devices are connected in parallel, because they will all signal at the same time when the security system is set off. Heat detectors and smoke detectors are generally connected in parallel because all of these devices will "close" the circuit to the security master control unit, setting off the alarms.

The instructions furnished with all security systems cover the installation requirements in detail, alerting the installer to Code regulations, clearances, suggested locations, and mounting heights of the systems components.

Always check with the local electrical inspector to determine if there are any special requirements in your locality relative to the installation of security systems.

Do not support power-limited fire-protective signaling circuit conductors from raceways that contain light and power conductors, *Section 760-54(c)*.

REVIEW

Note: Refer to the Code or the plans where necessary.

1. What article in the *National Electrical Code*® covers fire protective signaling systems?

   _____

2. Is it permissible to connect smoke and fire detectors with the same kind of low-voltage wire that is used to hook-up door chimes, or must cables that are listed for the purpose be used? _____

   _____

3. Find in *Article 760* the section that prohibits electrical raceways from being used to support the cables of a fire protective signaling system installation. *Section* _____

4. Find in *Article 760* the sections that prohibit fire protective system wires and cables from being installed in the same raceway as light and power conductors. *Section*

   _____

5. If it is necessary to install fire protective system cables through a duct, the cables (circle the correct answer):

   a. must be listed for the purpose.

   b. must be at least No. 14 AWG.

   c. are not permitted to be run through ducts.

6. What is the name and number of the NFPA standard that covers smoke and heat detectors for homes? _____

7. Has the standard that covers smoke and heat detectors for homes been adopted by the community in which you live? _____

8. a. Name the two basic types of heat detectors. _____

   _____

   b. Name the two basic types of smoke detectors. _____

   _____

9. Why is it important to mount a smoke or heat detector on the ceiling not closer than 4 inches (102 mm) from a wall? _____

   _____

   _____

   _____

10. In new residential construction, a smoke detector must be installed (circle the correct answer):

   a. in each bedroom or sleeping area.

   b. at the top of a basement stairway that has a closed door at the top.

   c. in a garage that is subject to subzero temperatures.

11. Always follow the installation instructions of (circle the correct answer):

   a. the manufacturer of the system.

   b. your neighbor.

   c. the man at the hardware store.

12. Smoke detectors, heat detectors, and security system wires and cables are much smaller than regular house wiring. Would it be better to install these wires (before) or (after) installing the regular house wiring to avoid possible damage to the smaller cables? Circle the correct answer.

13. Fire protection system conductors are generally No. _____ AWG.

14. When more than one smoke detector is installed in a new home, these units must be

   a. connected to separate 120-volt circuits.

   b. connected to one 120-volt circuit.

   c. interconnected so that if one detector is set off, the others would also sound a warning.

   Circle the correct answer(s).

15. It is advisable to connect smoke and heat detectors to circuits that are protected by GFCI devices.   TRUE   FALSE

16. Because a smoke detector might need servicing or cleaning, be sure to connect it in such a manner that it is controlled by a wall switch.   TRUE   FALSE

17. Smoke detectors of the type installed in homes must be able to sound a continuous alarm when set off for at least

   a. 4 minutes

   b. 30 minutes

   c. 60 minutes

UNIT 27

Remote-Control Systems—Low Voltage

OBJECTIVES

After studying this unit, the student will be able to

- explain the operation of a low-voltage, remote-control system for lighting circuits.
- interpret the wiring diagrams of various types of low-voltage, remote-control systems.
- install a low-voltage, remote-control system to comply with the requirements of the *National Electrical Code*®

The general lighting circuits described in this text are wired using standard methods. Another way of installing lighting circuits is to use a remote-control or low-voltage switching system. This system provides very flexible switching.

The remote-control wiring discussed in this unit is one manufacturer's system. In general, it is difficult to mix components of one manufacturer with those of another.

The following text is an overview of how typical remote-control, low-voltage systems operate. There are differences in the form, fit, and function of systems among the many manufacturers of these systems. Electrical wholesalers and distributors usually carry one or two brands of these systems. Once a decision has been reached to install a remote-control, low-voltage system, it is best to stay with one manufacturer's components.

Another type of full remote-control system, plus many added features is covered in unit 31, The Smart House.

A remote-control wiring system uses controlling devices (such as relays) at the equipment to be controlled. An auxiliary means is provided to operate the controlling device. A room thermostat in a residence is an example of a low-voltage remote-control system. For example, when calling for cooling, the low-voltage contacts in the thermostat "pull-in" a relay. This closes the line-voltage contacts in the relay, turning on the air conditioner. When the thermostat is satisfied, the low-voltage contacts in the thermostat open, dropping out the relay. The line-voltage contacts in the relay open, turning off the air conditioner.

A low-voltage, remote-control system can be installed in a home for the control of lighting and receptacle outlets. The system is generally more expensive than conventional wiring, but offers tremendous flexibility for the control of lighting and receptacle outlets. This type of installation lends itself to large upscale residences as well as commercial and industrial buildings.

A remote-control system for lighting circuits consists of low-voltage cable, low-voltage relays, a transformer(s), low-voltage switches, and master controls for the more elaborate systems. Because the system operates on 24 volts, the shock hazard is less than for conventional 120-volt circuitry.

REMOTE-CONTROL SWITCHES

The low-voltage, remote-control system uses special low-voltage switches that are normally open, single-pole, double-throw momentary contact switches. These switches are smaller than conventional switches and mount on yokes provided with the switches. They are also available in the same configuration as the interchangeable line of switches. They may have color-coded lead wires, screw terminals, or quick-connect terminals. Figure 27-1 shows a wiring diagram of a low-voltage switch. The white lead of the

switch is connected to the transformer. The red lead of the switch is connected to the red lead of the relay (on) and the black lead is connected to the black lead of the relay (off).

Pilot light switches have a fourth wire, which is yellow. The pilot light is energized through the yellow wire returning from a relay that has the necessary auxiliary contacts.

Figure 27-2 shows various types of low-voltage, remote-control switches.

Figure 27-3 shows a master selector switch module that has a combined total of eight switches, pilot lights, or blanks for future use. In the residence, for example, master selector switch modules could be located in one of the bedrooms, the kitchen, the living room, or any other location. If one or more of these switches is used to control a master group control or master sequencer, figure 27-4, the number of lighting and receptacle outlets to be controlled is endless.

For example, one switch in a cluster of eight switches could be hooked up to control a group controller that turns on all of the outside lights should the homeowner hear prowlers. Using a master is faster and easier than operating a number of individual switches. For more than eight switches in one location, additional master modules can be installed in a group.

Low-voltage switches may have pilot lights that indicate when a circuit is on, or locator lights for ease in finding the switch in the dark. Key-operated switches are shown in figure 27-2.

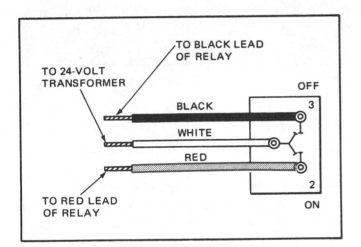

Fig. 27-1   Wiring diagram of a low-voltage switch.

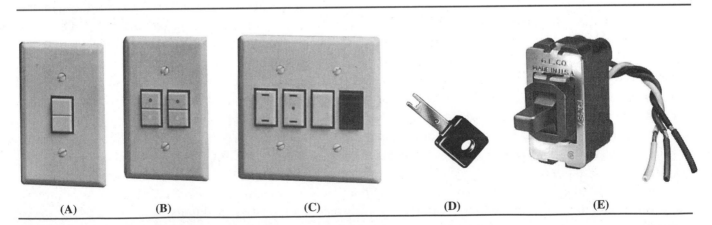

Fig. 27-2   Types of low-voltage switches. From left to right: (A) switch with location light, (B) two unlighted switches, (C) unlighted key operated switch, pilot light key operated switch, blank filler, pilot light, (D) a key, and (E) switch sized the same as the interchangeable line of switches shown in figure 2-22. *Courtesy* **General Electric Company, Wiring Device Department.**

These low-voltage switches are mounted in switch (device) boxes, on standard raised plaster rings selected for the thickness of the finished wall, or on brackets supplied by the manufacturer of the system. Single-gang mounting can accommodate one or two devices. Two-gang mounting can accommodate three or four devices. Master switch modules that can accommodate one through eight devices, as shown in figure 27-3, require a 4$\frac{11}{16}$-inch square box.

Installing switch boxes (single-gang, two-gang, or 4-inch square with suitable plaster ring) at switch locations provides the room needed to make splices and connections, rather than merely pushing the wires back into the insulation or vapor barrier if plaster rings alone were used. It is a matter of choice— and it might be a code issue in your community. Check this out before beginning an installation.

LOW-VOLTAGE RELAYS

Figure 27-5 shows a low-voltage relay. Figure 27-6 shows the internal wiring of the relay. The relay operates on 24 volts, supplied by a transformer, figure 27-7. The relay opens and closes the 120-volt power contacts. Relays with four low-voltage leads have a separate set of low-voltage "pilot" contacts connected to the relay's 24-volt supply. Relays with five low-voltage leads have the separate set of low-voltage "pilot" contacts connected by a 24-volt circuit independent of the relay's 24-volt supply. The four and five conductor relays are used to control other circuits, relays, master controls, pilot lights, etc.

The relay illustrated in figure 27-5 is rated for 1 hp, 20 amperes, 125 and 277 volts. This type of

Fig. 27-3   Illustration of a Master Selector Switch that has up to eight switches, pilot lights, or blank spaces. Note the key slot. This is a key operated lockout switch that can be connected to deactivate one or all of the switches in the plate. *Courtesy* **General Electric Company, Wiring Device Department.**

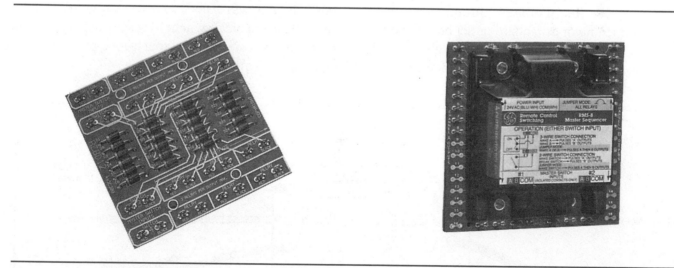

Fig. 27-4   Where simultaneous control of more than one relay is needed, a Master Group Control can be installed. This provides group control, but still permits control by other individual switches. This Master Group Control has four master switch inputs. Each master switch input is connected to a bank of three relay outputs. This type of Master Group Control is more than adequate for most residential applications. For very involved installations, a Master Sequencer that contains a microprocessor can be installed. A Master Sequencer provides virtually unlimited amount of control. *Courtesy* **General Electric Company, Wiring Device Department.**

Fig. 27-5   A low-voltage relay. It has three leads: Red (on), black (off), and blue (common). The 3-wire relay is the basic type. Relays for pilot control of other relays, master controls, and lights contain an extra set of internal contacts. These relays will have four leads, if the pilot contacts are fed by the same circuit as the relay. Relays will have five leads if the pilot contacts are totally isolated from the circuit that supplies the relay, which allows the pilot contacts to control some other low-voltage circuit. The relay shown requires splicing of the leads. Relays are available that have quick-connect terminals, making it easier to make up the connections where many relays are mounted in a control cabinet. *Courtesy* General Electric Company, Wiring Device Department.

relay can be used to control alternating-current loads, including tungsten filament and fluorescent loads, up to the full ampere rating of the relay.

Relays can be mounted through a ½-inch knockout opening of a standard outlet box. For quiet operation, they may be mounted through a ¾-inch knockout using a rubber grommet. When the relay is mounted from the inside of the outlet box through the knockout, the two 12-volt leads of the relay remain inside the box. The 24-volt leads are kept outside the box. See figures 27-9 and 27-10.

For complex systems, component cabinets are used. See figure 27-8A. These cabinets will contain conductor termination strips, relays, transformer, master switch controls, and master sequencers. These cabinets can be flush or surface mounted, and are generally installed near the fuse/circuit breaker distribution panel. A barrier in the box separates the 120-volt and 24-volt wiring.

Basically, the relay is a double solenoid. Touching the ON position of a switch momentarily energizes the coil connected to the red and blue leads. The iron core tries to center itself within coil. In figure 27-6, the core moves to the left. This mechanically latches the 120-volt contacts in the ON position.

Touching the OFF position of the switch momentarily energizes the coil connected to the black and blue leads. The iron core again tries to center itself within that particular coil. In figure 27-6, the core moves to the right. This mechanically latches the 120-volt contacts in the OFF position.

For complex installations, control of relays is also possible using photo cells and touch-tone telephones. These are used in combination with master controls and sequencers, and can be connected to override the manual switches for selected circuits. In other words, you can use a touch-tone phone to turn lights on and off.

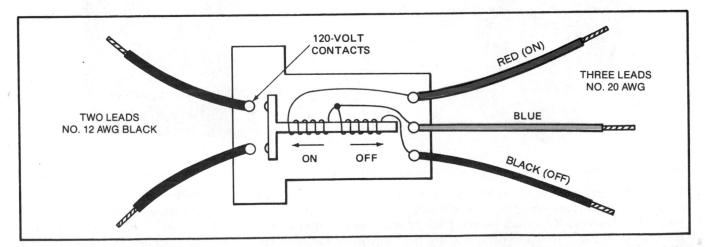

Fig. 27-6   A cut-a-way view of a low-voltage relay showing the internal mechanism as well as the external line voltage and low voltage leads.

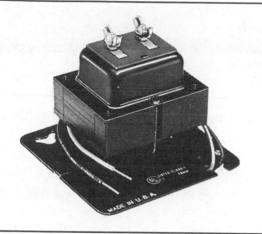

Fig. 27-7 Transformer for remote-control systems. *Courtesy* General Electric Company, Wiring Device Department.

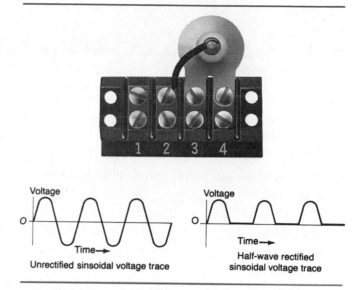

Fig. 27-7A Illustration of a silicon rectifier that converts an alternating current full sine wave to a half-wave. Although not absolutely required, they are recommended for use in low-voltage systems to extend the life of relays, pilot lamps, and locator lamps. The diagram in figure 27-10 shows a rectifier connected to the secondary of the transformer. *Courtesy* General Electric Company, Wiring Device Department.

TRANSFORMERS

A low-voltage, remote-control system typically requires just one transformer, figure 27-7. The transformer steps down the lighting circuit voltage from 120 volts to 24 volts for the low-voltage control circuit. The transformer rating is no larger than 100 volt-amperes. These transformers are listed as Class 2 transformers. When the transformer is overloaded, its output voltage decreases and there is less current output. In other words, the transformer has energy-limiting characteristics that counteract any overload. A specially designed core limits the amount of electrical energy that can be delivered at the secondary or output terminals. Some transformers have an internal thermal breaker that opens the primary circuit to protect the transformer from overheating. As soon as it cools down, the thermal breaker resets itself and the transformer is automatically reconnected to the line.

Rarely is more than one switch pressed at the same time in a low-voltage, remote-control system. Although many relays may be connected to the transformer, it is the same as if there were only one. The transformer seems to be underrated for the connected load. Actually, it can control many relays. The manufacturer will specify the maximum number of relays that a transformer can supply. One manufacturer limits the number of simultaneously operated relays to five. When the number of relays to be operated at the same time exceeds the transformer's recommended limit, master controls are used.

Some manufacturers of remote-control systems suggest that a rectifier be added to the secondary circuit, figure 27-7A. This device changes the alternating-current supply to direct or pulsating direct current, depending upon whether a half-wave or a full-wave rectifier is used. If a relay is energized for a long time, the alternating-current supply will produce eddy currents inside the laminations. As a result, there is a temperature buildup that may damage the relay. Recall that this type of relay is meant to provide momentary contact operation only. Direct current or pulsating direct current does not produce as much heat in this type of relay as does alternating current.

CONDUCTORS

No. 20 AWG stranded copper conductors are generally used for low-voltage, remote-control systems, figure 27-8. To minimize voltage drop, larger conductors may be required for long runs, or where a number of relays are operated at the same time. The equipment manufacturers' technical manuals and installation manuals must be consulted to be

Fig. 27-8   Illustration of typical cable used for remote-control wiring. *Courtesy* **General Electric Company, Wiring Device Department.**

sure the completed installation will perform as intended.

The thermoplastic insulation on the individual conductors is color coded for ease in identification. The conductors are protected by an outer jacket. Cables come in:

- 3-conductor cable (red/black/white) for connecting switches.

- 4-conductor cable (red, black, yellow, white) for connecting switches with pilot and locator lights.

- 4-conductor cable (red/black/blue/white) for connecting sensors.

- 5-conductor cable (red/black/yellow/yellow/ blue) for connecting relays that have auxiliary contacts.

- 22-conductor cable, for connecting remote groups of six relays, has six sets of 3-conductor (red/black/yellow), two whites, a blue, and a yellow.

- 25-conductor cable, for connecting eight-gang master switch modules, has eight sets of 3-conductor (red/black/yellow) plus a No. 18 AWG white common conductor.

Individual color-coded No. 18 AWG conductors for miscellaneous connections should be black, red, blue, blue/white, and white (common) to correspond to the color coding of the multiconductor cable mentioned earlier.

The cables discussed here are suitable for use in Class 2 circuits. See *Table 725-61* of the *National Electrical Code®* for cable markings and their permitted uses. Cables marked CL2 are for general-purpose use. Cables marked CL2P are Class 2 plenum cables that can be used in air-handling plenums. Cables marked CL2X are limited to residential installations. These different markings are related to the cables' fire resistance capabilities.

Later in this unit you will learn that Class 2 conductors are not permitted to be run together or mixed with 120-volt wiring.

Because the jacket on low-voltage cables is rather thin, the electrician must be careful not to damage the jacket during the rough-in stages of the installation. *Do not* install low-voltage wiring through the same holes that the power and lighting 120-volt cables or conduits are installed. Physical damage to the jacket and individual conductors of the low-voltage cables is possible. The possibility of short-circuits can result in relay malfunction or failure of other system components.

INSTALLATION PROCEDURE

The installation of a remote-control system in new construction begins with the roughing-in of the 120-volt system. All of the switch legs or switch loops are omitted for the system. In other words, the 120-volt supply conductors are brought directly to each outlet or switch box that will have a switched lighting fixture or switched outlet attached to it. Although the various devices used may be numbered or color coded in different ways, the electrician must remember that only three low-voltage conductors are required between the relay and the switch. In addition, two low-voltage conductors must be run from the transformer to either the relay or the switch to carry the low-voltage supply. These wires are spliced like color to like color. The wiring is then completed by slipping the relay through the ½-inch knockout in the outlet or switch box. The 120-volt connections are made in the standard way.

The low-voltage cables are stapled in place in new construction. When existing buildings are rewired, the low-voltage cables may be fished through the walls and ceiling or run behind moldings and baseboards.

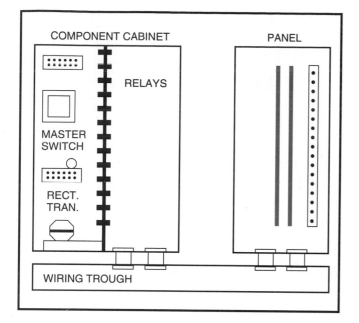

Fig. 27-8A   For complex low-voltage, remote-control installations, a component cabinet is installed near the electrical distribution panel. This cabinet provides adequate space and hardware for the proper mounting of relays, transformer, master switch(es), rectifier. Terminal strips are provided for ease in making up the conductor connections and terminations. The wiring trough ties the component cabinet and the electrical distribution panel together for routing of the 120-volt conductors. An equally satisfactory installation would have been to install nipples between the cabinet and the panel.

For some installations, many or all of the relays should be placed at one location using a component cabinet, figure 27-8A. The 120-volt connections are done on one side of the barrier; the 24-volt connections are made on the other side of the barrier. Thus, the 120-volt wiring is separated from the 24-volt wiring in conformance to *Section 725-54(a)(1)* of the *National Electrical Code.*®

One method of installing remote-control, low-voltage switches is shown in figure 27-9. Multipoint control of the light results when the low-voltage, remote-control switches are connected in parallel.

Figure 27-10 shows one method of connecting a master selector module with individual low-voltage switches. The lighting outlets can be placed in different parts of the building and can be connected to operate 120-volt circuits. These outlets are then controlled by their own individual low-voltage switches or the master selector control. For simplicity, only three switches are shown connected. This particular master module can have as many as eight switches, pilot lights, or blank spaces.

In this residence, an eight-switch module could be installed in the master bedroom to control the weatherproof receptacle(s) on the front porch and on the outside wall of the front bedroom, the front entrance and garage lights (outside front), post light

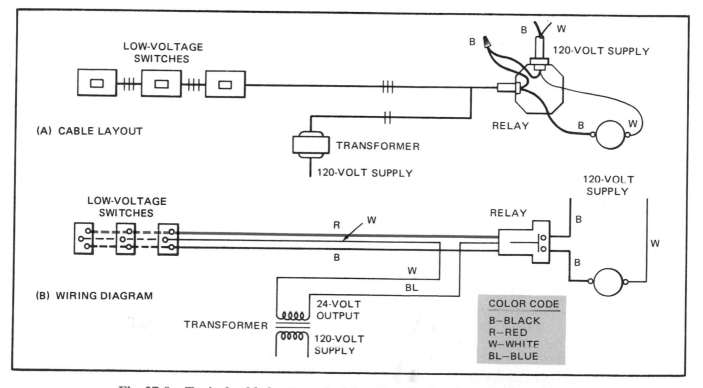

Fig. 27-9   Typical cable layout and wiring diagram for low-voltage switches.

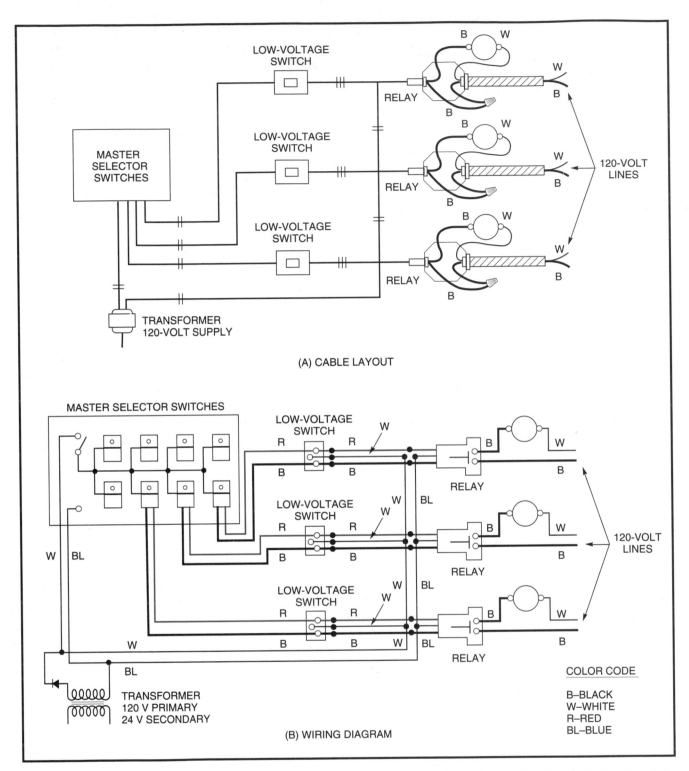

Fig. 27-10 Cable layout and wiring diagram for a low-voltage remote-control system that has individual switches as well as a master selector module. The master module has eight switches. For simplicity, the diagram shows only three switches connected. Note that the individual switches and the corresponding switches in the master module are connected in parallel. The blue conductor is carried to the master module in the event that pilot light or locator switches are installed. For a more elaborate system, one or more of the switches in the master module could control a master group control or a master sequencer, as illustrated in figure 27-4.

(front), rear yard floods, side garage door light, and other lighting in the home that would be desirable to turn on from a single location. Eight-switch modules could be installed in other areas in the home. If eight switches in one location is not enough, then a master sequencer can be installed, figures 27-4 and 27-8A. The possibilities of remote control are endless, all dependent upon the complexity needed to fulfill the homeowner's needs. Cost can become the deciding factor.

NATIONAL ELECTRICAL CODE®

The low-voltage, remote-control system is not subject to the same Code restrictions as the standard 120-volt system. The low-voltage (24-volt) portion of the remote-control system is a Class 2 system and is regulated by *NEC® Article Part C*. See figure 27-11. UL sets 30 volts RMS as the maximum secondary voltage for Class 2 transformers.

For a remote-control system, all wiring on the supply side of the transformer (120 volts) must conform to the wiring methods given in *Chapter 3* of the *National Electrical Code,® Section 725-51.* All wiring on the load side of the transformer (24 volts) must conform to *Section 725-52.*

The Class 2 or Class 3 transformer for a remote-control, low-voltage system is designed so that the power output is limited to meet the requirements of the Code, *Table 725-41.* The power output can be limited by a combination of the power source and the overcurrent protection. The maximum nameplate rating of the transformer is 100 volt-amperes.

Conductors with thermoplastic insulation are normally used for a low-voltage system. However, the Code does not require thermoplastic insulation.

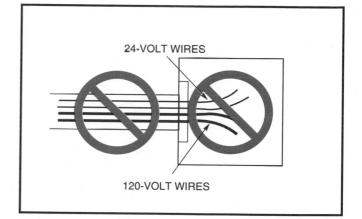

Fig. 27-11   Low-voltage Class 2 wiring *shall not* **be run in the same raceway or be in the same outlet box or enclosure with light and power circuits unless separated by a barrier,** *Section 725-52(a)(2).* **In figure 27-8A, the barrier serves as a mounting means for the relays, keeping the 120-volt wiring to the right of the barrier, and the low-voltage Class 2 wiring to the left of the barrier.**

Conductors for Class 2 systems shall not be run with regular light and power wiring, *Section 725-54(a).* Class 2 wiring may enter the same compartment or enclosure as power conductors if separated by a barrier or if introduced solely to connect the equipment, *Section 725-54(a)(1), Exceptions 1 and 2.*

REVIEW

Note: Refer to the Code or the plans where necessary.

1. What is the approximate voltage used on low-voltage remote-control systems? _____
   _____

2. What are the advantages of a low-voltage system? _____
   _____
   _____
   _____
   _____
   _____

3. What type of switch is used to control the relays of a low-voltage system? _____
   _____

4. Are switch boxes recommended at all low-voltage switch locations? Explain. _____
   _____

5. Are three-way and four-way switches used in remote-control circuits? Explain your answer. _____

_____

_____

_____

_____

_____

6. What is a master selector switch module? _____

_____

_____

_____

_____

_____

7. What is a master group control? _____

_____

_____

_____

_____

_____

8. Regarding the transformer for a low-voltage, remote-control system, what must be taken into consideration? _____

_____

_____

_____

_____

9. Explain briefly the operating principle of a low-voltage relay. _____

_____

_____

_____

_____

_____

10. What is the electrical rating of relays discussed in this unit? _____

_____

11. What will happen if the common lead from the transformer is connected to the red lead of the relay instead of the blue lead, and the blue and black leads are connected to the switch leads? _____

_____

_____

_____

_____

12. a. Is it permitted to run low-voltage conductors and line-voltage (120-volt) conductors in the same raceway or into the same box? Explain your answer. _____

_____

_____

_____

 b. Low-voltage, remote-control wires and cables (should) (should not) be installed through the same bored holes that contain nonmetallic-sheathed cables, armored cables, electrical conduits, or water pipes. Circle the correct answer.

13. The maximum nameplate rating of Class 2 transformers is _____ volt-amperes.

14. Because the insulation on low-voltage wire is much thinner than the insulation on regular building wire, the electrician should _____ when installing the conductors to avoid _____ the insulation.

15. a. What is the purpose of a rectifier? _____

_____

 b. Must rectifiers be used on low-voltage systems? _____

_____

16. What type of conductors are generally used for a low-voltage remote-control system?

_____

_____

_____

_____

_____

17. a. Why are color-coded wires and cable used in low-voltage control systems? _____

_____

_____

 b. What is the color coding for three-conductor low-voltage cable that is used for connecting switches? _____

_____

18. The type of low-voltage wiring discussed in this unit is classified as (Class 1) (Class 2). Circle one.

19. When overcurrent protection is not provided in the secondary circuits of the transformers described in this unit, the transformers must _____ their power output.

20. Conductors supplying transformers are governed by what part of the Code? _____

21. Complete all line voltage and low-voltage connections in the following diagram. Use colored pencils to indicate conductors. Show low-voltage connections with a dot. Switch No. 1 controls lamps A and C, switch No. 2 controls lamp B, and switch No. 3 controls the bottom of each convenience receptacle. The top of each receptacle is to be hot at all times. The line-voltage wiring is in armored cable.

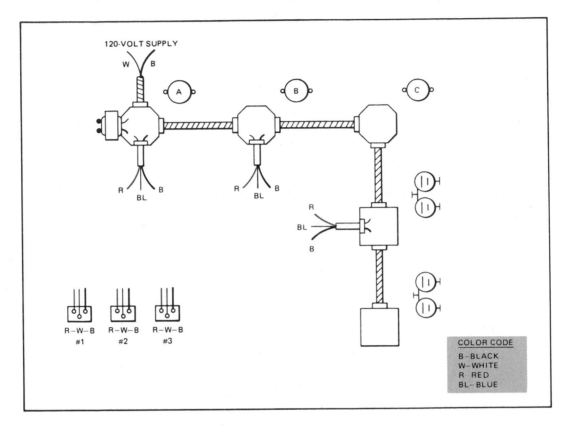

UNIT 28

Service-Entrance Equipment

OBJECTIVES

After studying this unit, the student will be able to

- define electrical service, overhead service, service drop, and underground service.
- list the various Code sections covering the installation of a mast-type overhead service and an underground service.
- discuss the Code requirements for disconnecting the electrical service using a main panel and load centers.
- discuss the grounding of interior alternating-current systems and the bonding of all service-entrance equipment.
- discuss the "UFER" ground method.
- describe the various types of fuses.
- select the proper fuse for a particular installation.
- explain the operation of fuses and circuit breakers.
- explain the term *interrupting rating*.
- determine available short-circuit current using a simple formula.

An electric service is required for all buildings containing an electrical system and receiving electrical energy from a utility company. The *National Electrical Code®* describes the term *service* as the conductors required to deliver energy from the electrical supply source to the wiring system of the premises. The utility company generally must be contacted to determine where they want the meter to be located.

OVERHEAD SERVICE

The Code defines a *service drop* as the overhead service conductors, including splices, if any, which are connected from an outdoor support to the service-entrance conductors at the structure.

The overhead service includes all of the service equipment and installation means from the attachment of the service-drop wires on the outside of the building to the point where the circuits or feeders are tapped to supply specific loads or load centers. In general, watt-hour meters are located on the exterior of a building. Local codes may permit the watt-hour meter to be mounted inside the building. In some cases, the entire service-entrance equipment may be mounted outside the building. This includes the watt-hour meter and the disconnecting means. Figure 28-1 illustrates the Code terms for the various components of a service entrance. Figure 28-1A illustrates the service entrance, main panel, sub-panel, and grounding for this service.

MAST-TYPE SERVICE ENTRANCE

The mast service, figure 28-1, is a commonly used method of installing a service entrance. The mast service is often used on buildings with low roofs, such as ranch-style dwellings, to insure adequate clearance between the ground and the lowest service conductor.

The service raceway is run through the roof, as shown in figure 28-1, using a roof flashing and neoprene seal fitting, as illustrated in figure 28-2(A). This fitting keeps water from seeping in where the raceway penetrates the roof. The conduit is securely fastened to comply with code requirements. Many types of fittings are readily available at electrical supply houses for securely fastening raceways and other electrical equipment to wood, brick, masonry, and other surfaces. Figure 28-2(B) shows a typical through-the-wall fitting.

Clearance Requirements for Mast Installations

Several factors determine the maximum length of conduit that can be installed between the roof support and the point where the service-drop conductors are attached. These factors are the service-drop length, the system voltage, the roof pitch, and the size and type of conduit (aluminum or steel), figure 28-3.

Service Mast as Support *(Section 230-28)*

The bending force on the conduit increases with an increase in the distance between the roof support and the point where the service-drop conductors are attached. The pulling force of a service drop on a mast service conduit increases as the length of the service drop increases. As the length of the service drop decreases, the pulling force on the mast service conduit decreases.

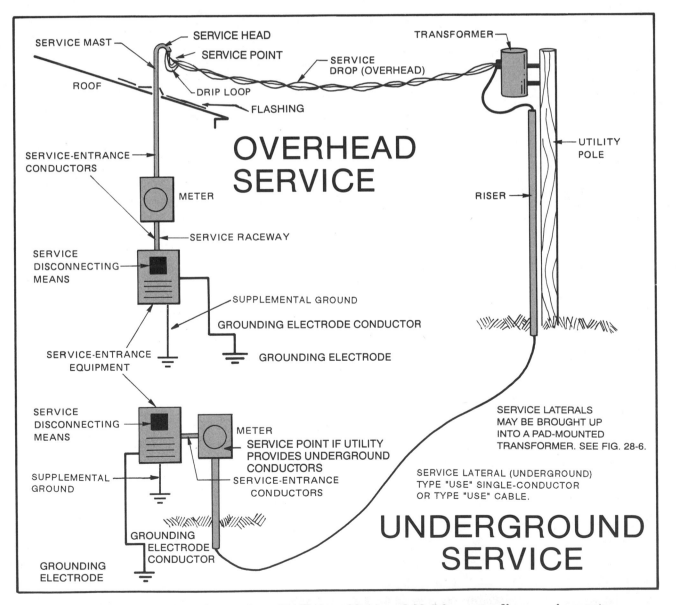

Fig. 28-1   Code terms for services. See figures 28-1A and 28-5 for grounding requirements.

If extra support is not to be provided, the mast service conduit must be at least 2 inches in diameter. This size prevents the conduit from bending due to the strain of the service-drop conductors.

If extra support is needed, it is usually in the form of a guy wire attached to a mast fitting (figure 28-2(C)) and to a roof fitting (figure 28-2(D)). Again, many support fittings and devices are available at electrical supply houses.

When service-entrance cable is used, a fitting of the type shown in figure 28-2(E) can be used to protect the cable from physical damage where it penetrates the outside wall of the house.

▶ When used as a service mast, rigid metal conduit that passes through a roof *does not* require secure fastening within 3 feet (914 mm) of the service head, *Section 346-12, Exception No. 3.* ◀ The roof mast kit provides adequate support when properly installed. A guy wire might be necessary if the service-drop conductors are long or heavy.

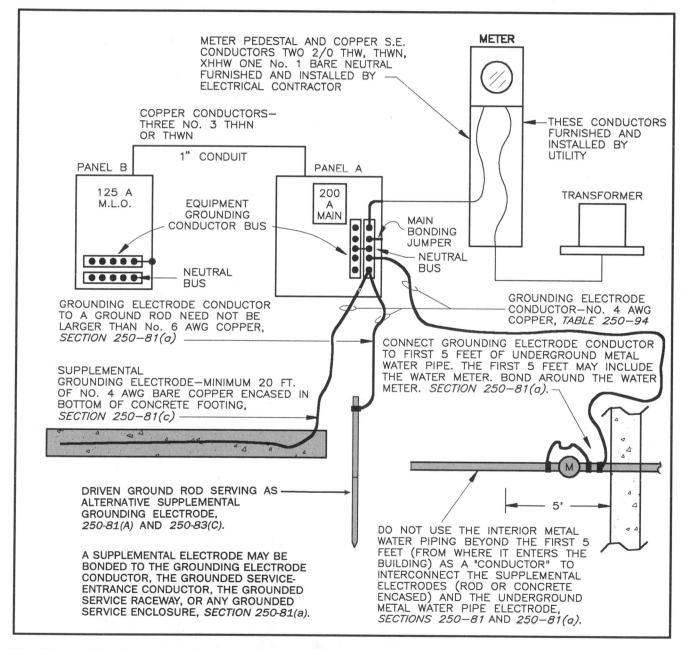

Fig. 28-1A   This illustration shows the service-entrance, main panel, subpanel, and grounding for the service in this residence. Note that because the integrity of underground water pipes is questionable, the Code in *Section 250-81(a)* requires a supplemental grounding electrode. Shown are two common types of supplemental grounding electrodes: a ground rod or a concrete-encased grounding electrode. In many instances, the concrete-encased grounding electrode is better than the water pipe ground, particularly since the advent of non-metallic (PVC) water piping systems.

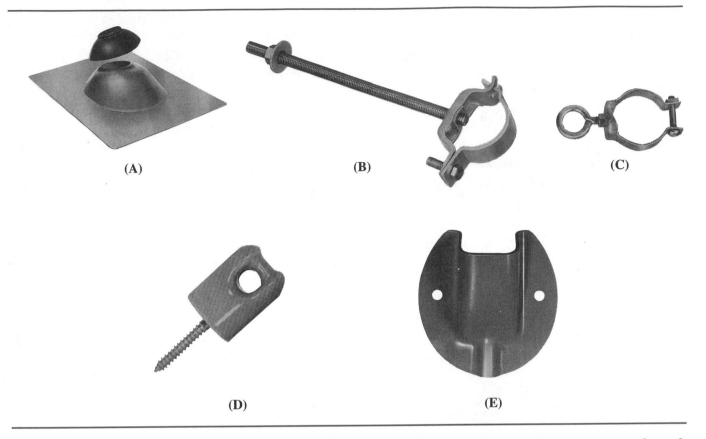

Fig. 28-2   Illustrations of (A) a neoprene seal with metal flashing to use where the service raceway comes through the roof, (B) a through-the-wall support for securing the service raceway, (C) a clamp that fastens to the service raceway to which a guy wire can be attached for support of the service riser, and (D) an insulator that can be screwed through the roof into a roof truss or rafter. A guy wire is attached to it and to the clamp (C). Commonly referred to as a "sill plate," (E) is used to protect service-entrance cable where it enters the building.

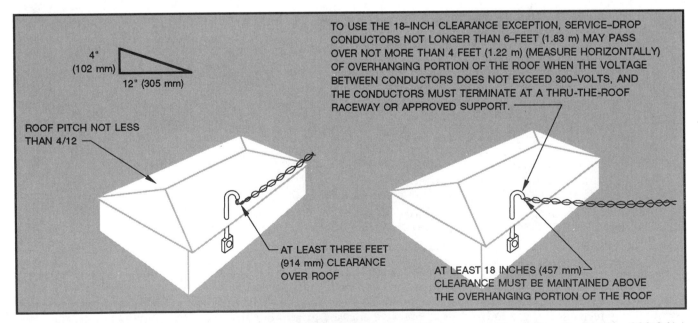

Fig. 28-3   Clearance requirements for service-drop conductors passing over residential roofs, *Section 230-24(a)*, where voltage between conductors does not exceed 300 volts.

▶ Intermediate metal conduit is also permitted to be used as a service mast. It *does not* require secure fastening within 3 feet (914 mm) of the service head. To use IMC as a service mast requires special permission from the electrical inspector, referred to in *Section 90-4* of the *NEC®* as the *Authority Having Jurisdiction*. See *Section 345-12, Exception No. 3.* ◀

▶ Only power service-drop conductors are permitted to be attached to and supported by the service mast, *Section 230-28.* Refer to figure 28-3A. ◀

All fittings used with raceway-type service masts must be identified for use with service masts, *Section 230-28.*

Consult the utility company and electrical inspection authority for information relating to their specific requirements for clearances, support, and so on, for service masts.

The *NEC®* rules for insulation and clearances apply to the service-drop and service-entrance conductors. For example, the service conductors must be insulated, except where the voltage to ground does not exceed 300 volts. In this case, the grounded neutral conductors are not required to have insulation.

Section 230-24(a) gives clearance allowances for the service drop passing over the roof of a dwelling, figure 28-3.

The installation requirements for a typical service entrance are shown in figures 28-4 and 28-5. Figure 28-4 shows the required clearances above the ground. The wiring connections, grounding requirements and Code references are given in figure 28-5.

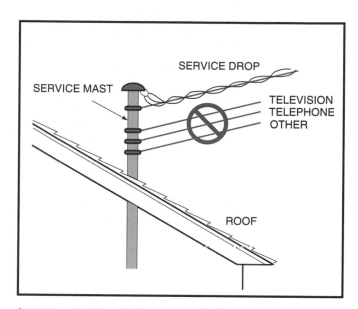

▶ **Fig. 28-3A   Only power service-drop conductors are permitted to be attached to and supported by the service mast. Do *not* attach or support television cables, telephone cables, or anything else to the mast, *Section 230-28.* ◀**

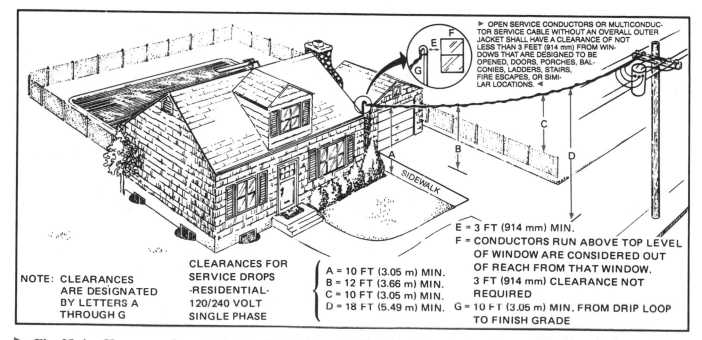

NOTE: CLEARANCES ARE DESIGNATED BY LETTERS A THROUGH G

CLEARANCES FOR SERVICE DROPS -RESIDENTIAL- 120/240 VOLT SINGLE PHASE

A = 10 FT (3.05 m) MIN.
B = 12 FT (3.66 m) MIN.
C = 10 FT (3.05 m) MIN.
D = 18 FT (5.49 m) MIN.

▶ OPEN SERVICE CONDUCTORS OR MULTICONDUCTOR SERVICE CABLE WITHOUT AN OVERALL OUTER JACKET SHALL HAVE A CLEARANCE OF NOT LESS THAN 3 FEET (914 mm) FROM WINDOWS THAT ARE DESIGNED TO BE OPENED, DOORS, PORCHES, BALCONIES, LADDERS, STAIRS, FIRE ESCAPES, OR SIMILAR LOCATIONS. ◀

E = 3 FT (914 mm) MIN.
F = CONDUCTORS RUN ABOVE TOP LEVEL OF WINDOW ARE CONSIDERED OUT OF REACH FROM THAT WINDOW. 3 FT (914 mm) CLEARANCE NOT REQUIRED
G = 10 FT (3.05 m) MIN. FROM DRIP LOOP TO FINISH GRADE

▶ **Fig. 28-4   Clearances for a typical residential service in accordance with *Sections 230-24(b)* and *(c)* and *230-26*. Clearances (A) and (C) of 10 feet permitted only if the service-entrance drop conductors are supported on and cabled together with a grounded bare messenger wire where the voltage to ground *does not* exceed 150 volts. This 10-foot clearance must also be maintained from the lowest point of the drip-loop. ◀ Clearance (B) of 12 feet permitted over residential property and driveways where the voltage to ground *does not* exceed 300 volts.**

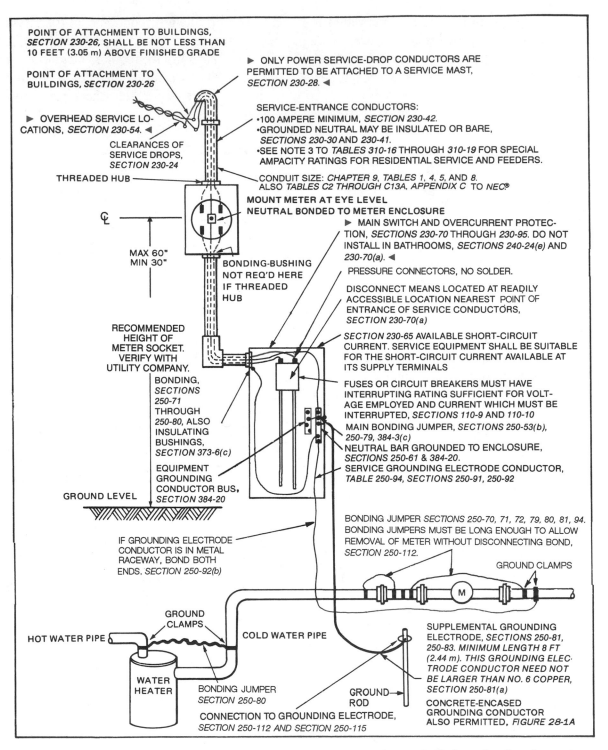

POINT OF ATTACHMENT TO BUILDINGS, *SECTION 230-26,* SHALL BE NOT LESS THAN 10 FEET (3.05 m) ABOVE FINISHED GRADE

POINT OF ATTACHMENT TO BUILDINGS, *SECTION 230-26*

▶ OVERHEAD SERVICE LO-CATIONS, *SECTION 230-54.* ◀

CLEARANCES OF SERVICE DROPS, *SECTION 230-24*

THREADED HUB

▶ ONLY POWER SERVICE-DROP CONDUCTORS ARE PERMITTED TO BE ATTACHED TO A SERVICE MAST, *SECTION 230-28.* ◀

SERVICE-ENTRANCE CONDUCTORS:
•100 AMPERE MINIMUM, *SECTION 230-42.*
•GROUNDED NEUTRAL MAY BE INSULATED OR BARE, *SECTIONS 230-30* AND *230-41.*
•SEE NOTE 3 TO *TABLES 310-16* THROUGH *310-19* FOR SPECIAL AMPACITY RATINGS FOR RESIDENTIAL SERVICE AND FEEDERS.

CONDUIT SIZE: *CHAPTER 9, TABLES 1, 4, 5,* AND *8.* ALSO *TABLES C2* THROUGH *C13A, APPENDIX C* TO NEC®

MOUNT METER AT EYE LEVEL NEUTRAL BONDED TO METER ENCLOSURE

C_L

MAX 60" MIN 30"

▶ MAIN SWITCH AND OVERCURRENT PROTEC-TION, *SECTIONS 230-70* THROUGH *230-95.* DO NOT INSTALL IN BATHROOMS, *SECTIONS 240-24(e)* AND *230-70(a).* ◀

BONDING-BUSHING NOT REQ'D HERE IF THREADED HUB

PRESSURE CONNECTORS, NO SOLDER.

DISCONNECT MEANS LOCATED AT READILY ACCESSIBLE LOCATION NEAREST POINT OF ENTRANCE OF SERVICE CONDUCTORS, *SECTION 230-70(a)*

RECOMMENDED HEIGHT OF METER SOCKET. VERIFY WITH UTILITY COMPANY.

SECTION 230-65 AVAILABLE SHORT-CIRCUIT CURRENT. SERVICE EQUIPMENT SHALL BE SUITABLE FOR THE SHORT-CIRCUIT CURRENT AVAILABLE AT ITS SUPPLY TERMINALS

BONDING, *SECTIONS 250-71* THROUGH *250-80,* ALSO INSULATING BUSHINGS, *SECTION 373-6(c)*

FUSES OR CIRCUIT BREAKERS MUST HAVE INTERRUPTING RATING SUFFICIENT FOR VOLT-AGE EMPLOYED AND CURRENT WHICH MUST BE INTERRUPTED, *SECTIONS 110-9* AND *110-10*

MAIN BONDING JUMPER, *SECTIONS 250-53(b), 250-79, 384-3(c)*

EQUIPMENT GROUNDING CONDUCTOR BUS, *SECTION 384-20*

NEUTRAL BAR GROUNDED TO ENCLOSURE, *SECTIONS 250-61* & *384-20.*

SERVICE GROUNDING ELECTRODE CONDUCTOR, *TABLE 250-94, SECTIONS 250-91, 250-92*

GROUND LEVEL

IF GROUNDING ELECTRODE CONDUCTOR IS IN METAL RACEWAY, BOND BOTH ENDS. *SECTION 250-92(b)*

BONDING JUMPER *SECTIONS 250-70, 71, 72, 79, 80, 81, 94.* BONDING JUMPERS MUST BE LONG ENOUGH TO ALLOW REMOVAL OF METER WITHOUT DISCONNECTING BOND, *SECTION 250-112.*

GROUND CLAMPS

M

GROUND CLAMPS

HOT WATER PIPE

COLD WATER PIPE

WATER HEATER

BONDING JUMPER *SECTION 250-80*

CONNECTION TO GROUNDING ELECTRODE, *SECTION 250-112* AND *SECTION 250-115*

GROUND ROD

SUPPLEMENTAL GROUNDING ELECTRODE, *SECTIONS 250-81, 250-83.* MINIMUM LENGTH 8 FT (2.44 m). THIS GROUNDING ELEC-TRODE CONDUCTOR NEED NOT BE LARGER THAN NO. 6 COPPER, *SECTION 250-81(a)*

CONCRETE-ENCASED GROUNDING CONDUCTOR ALSO PERMITTED, *FIGURE 28-1A*

Fig. 28-5   The wiring of a typical service-entrance installation and Code rules for the system grounding. Refer also to figures 28-1A and 28-19.

UNDERGROUND SERVICE

The underground service means the cable installed underground from the point of connection to the system provided by the utility company.

New residential developments often include underground installations of the high-voltage elec-

trical systems. The conductors in these distribution systems end in the bases of pad-mounted trans-formers, figure 28-6. These transformers are placed at the rear lot line or in other inconspicuous loca-tions in the development. The transformer and pri-mary high-voltage conductors are installed by, and are the responsibility of, the utility company.

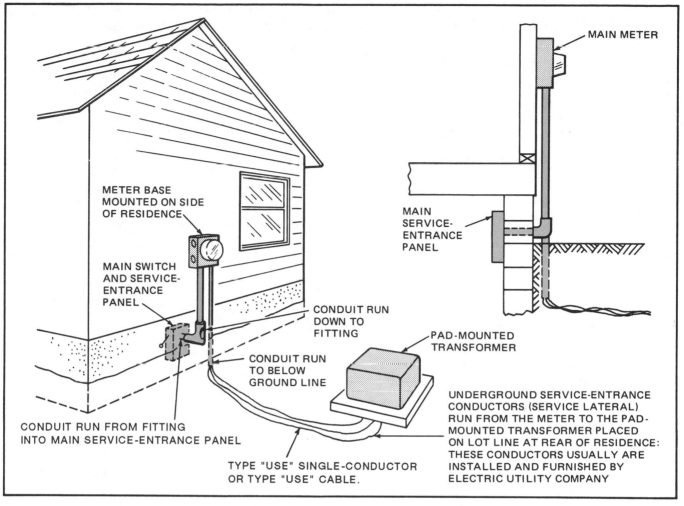

Fig. 28-6   Underground service.

The conductors installed between the pad-mounted transformers and the meter are called *service lateral conductors*. Normally, the electrical utility supplies and installs the service laterals. Figure 28-6 shows a typical underground installation.

The wiring from the external meter to the main service equipment is the same as the wiring for a service connected from overhead lines, figure 28-5. Some local codes may require conduit to be installed underground from the pole to the service-entrance equipment. *NEC®* requirements for underground services are given in *Sections 230-30, 230-31* and *230-32*. The underground conductors must be suitable for direct burial in the earth, Type USE single-conductor or Type USE cable containing more than one conductor. See *Article 338*.

If the electric utility installs the underground service conductors, the work must comply with the rules established by the utility. Utilities follow the National Electrical Safety Code, NESC. These rules may not be the same as those given in *NEC® Section 90-2*.

When the underground conductors are installed by the electrician, *Section 230-49* applies. This section deals with the protection of conductors against physical damage. *Section 230-49* also refers you to *Section 300-5*, which covers all situations involving underground wiring, such as the sealing of raceways where they enter a building, figure 28-15.

MAIN SERVICE DISCONNECT LOCATION

The main service disconnecting means shall be installed at a readily accessible location outside or inside of the building so that the service conductors inside the building are as short as possible, *Section 230-70(a)*. See figure 28-7.

If for some reason it is necessary to locate service-entrance equipment some distance inside the building, *Section 230-6* considers service-entrance conductors to be outside of a building if the service-entrance conductors are:

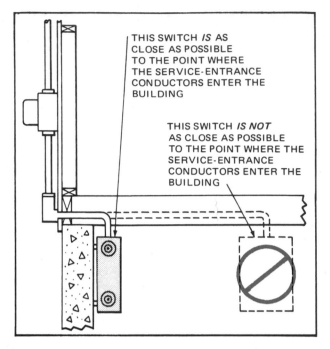

► **Fig. 28-7 Main service disconnect location. Do** *not* **install in a bathroom. Refer to** *Sections 230-70(a)* **and** *240-24(e).* ◄

• installed under not less than 2 inches (50.8 mm) of concrete beneath the building

or

• installed within the building in a raceway encased in at least 2 inches (50.8 mm) of concrete or brick.

Panelboard Directory

Section 384-13 states that all panels must have a "legibly identified" circuit directory indicating what the circuits are for. *Section 110-22* also requires that disconnect switches be "legibly marked" as to what the disconnect is for. The directory is furnished and attached to the inside of the cover by the panelboard manufacturer. Figure 28-8 shows two typical 240/120-volt, single-phase, 3-wire main panelboards. Figure 28-9 shows the circuit identification for Main Service Panel A. Figure 28-12 shows the circuit identification for Panel B. Many times, the electrician will find the directory furnished by the

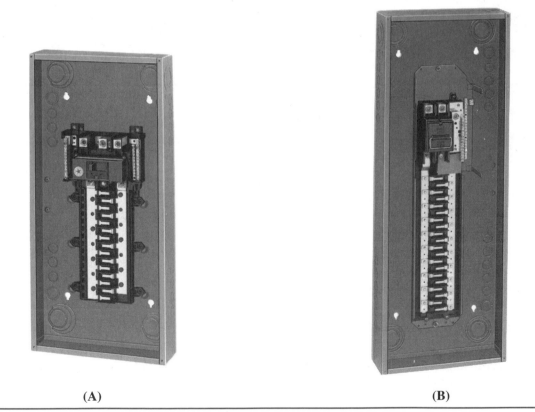

(A)   (B)

Fig. 28-8   Typical 240/120-volt, single-phase, 3-wire main panelboards, sometimes referred to as load centers. The panel on the left (A) has a main fusible pullout disconnect suitable for use with 200-ampere Class T fuses. The panel to the right (B) has a main circuit breaker. The branch circuit-breakers are not shown. This is the type of panel that could be used as Main Panel A in the residence discussed in this text. Many panels have a separate terminal bus on which to terminate the equipment grounding conductors of nonmetallic-sheathed cable. See figure 28-10. When used as service-equipment, the panel must be listed as suitable for use as service-equipment, *Section 230-70. Courtesy* **Square D Company,** *Group Schneider.*

MAIN SERVICE PANEL "A"

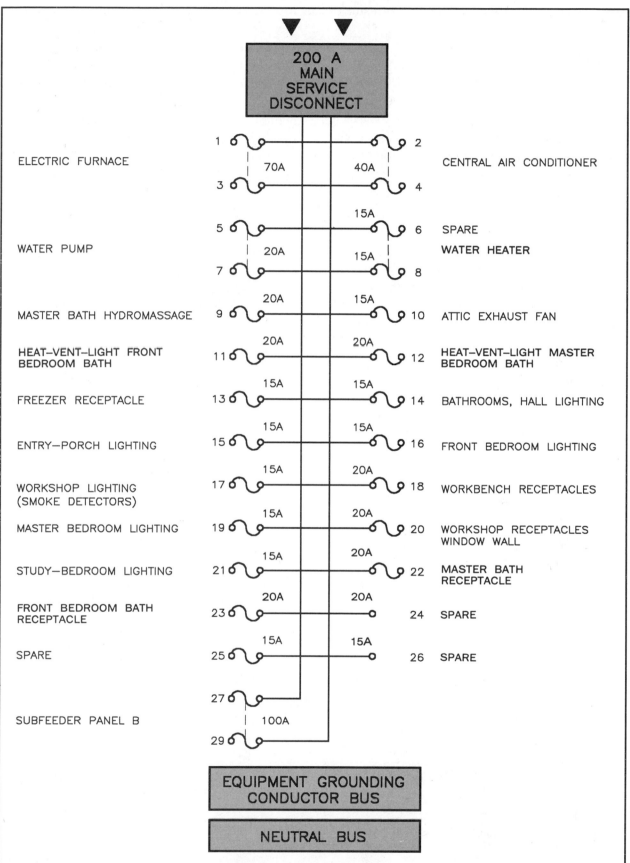

Fig. 28-9   Circuit schedule of main service panel A.

manufacturer too small, and will type a directory on a larger size card, and will attach this to the inside of the panel cover.

Disconnect Means (Panel A)

The requirements for disconnecting electrical services are covered in *Sections 230-70* through *230-83*. *Section 230-71(a)* requires that the service disconnecting means consist of not more than six switches or six circuit breakers mounted in a single enclosure, in a group of separate enclosures in the same location, or in or on a switchboard. This permits the disconnection of all electrical equipment in the house with not more than six hand operations. Some local codes take exception to the six disconnect rule, and require a single main disconnect. Check this out with the local authority that has jurisdiction.

In damp and wet locations, to prevent the accumulation of moisture that could lead to rusting of the enclosure and damage to the equipment inside the enclosure, *Section 373-2(a)* specifies that there be at least a ¼-inch (6.35 mm) air space between the wall and a surface-mounted enclosure. Most disconnect switches, panelboards, meter sockets, and similar equipment have raised mounting holes that provide the necessary clearance. This is clearly visible in the open-meter socket in figure 28-14. The *National Electrical Code®* (in Definitions) considers masonry in direct contact with the earth to be a wet location. ▶ The *Exception* to

Section 373-2(a) allows nonmetallic enclosures to be mounted without an air space on concrete, masonry, tile, and similar surfaces. ◀

Section 384-14 defines a lighting and appliance branch-circuit panelboard as one that has not more than 10% of the overcurrent devices rated less than 30 amperes with provisions for the connection of neutral conductors. *Section 384-16(a)* states that a lighting and appliance branch-circuit panelboard *shall not* have more than two main circuit breakers or two sets of fuses for the panelboards overcurrent protection and disconnecting means.

A lighting and appliance branch-circuit panelboard *shall not* have more than 42 overcurrent devices in it, *Section 384-15*. A two-pole breaker is counted as two overcurrent devices.

In *Section 230-42(b)*, we find that the *minimum* service-equipment rating for a single-family residential service is 100 amperes, three-wire when:

- there are six or more 2-wire branch-circuits.

- the initial computed load is 10 kVA or more.

Service equipment must be marked that it is suitable for use as service equipment, *Section 230-66*.

When nonmetallic-sheathed cable is used, a panelboard must have a terminal bar for attaching the equipment grounding conductors, *Section 384-20*. See figure 28-10.

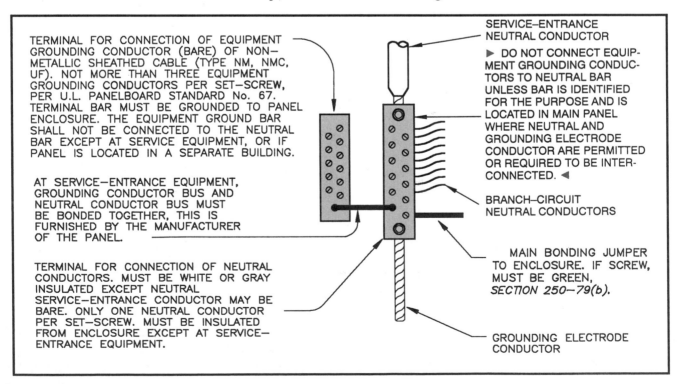

Fig. 28-10   Connections of service neutral, branch-circuit neutral, and equipment grounding conductors at panelboards, *Section 384-20.*

For this residence, Panel A provides a 200-ampere, fusible pullout for the main disconnecting means and circuit breakers for the branch-circuits. This main could have been a 200-ampere circuit breaker. Figure 28-8 is a photo of Panel A. Figure 28-9 is the schedule for Panel A.

Panel A is located in the workshop.

▶ *Working Space.*   To provide for safe working conditions around electrical equipment, *Section 110-16* of the *National Electrical Code®* contains a number of rules that must be followed. For residential installations, here are some issues that address working space.

- A working space of not less than 30 inches wide and 3 feet deep must be provided in front of electrical equipment, such as Panels A and B.

- This working space must be kept clear and not used for storage.

- Do not install electrical panels inside of cabinets, nor above shelving, washers, dryers, freezers, work benches, etc.

- Do not install in bathrooms.

- Do not install electrical panels above or close to sump pump holes.

- The hinged cover (door) on the panel shall be able to be opened to a full 90 degrees.

- A "zone" equal to the width and depth of the electrical equipment, from the floor to a height of 25 feet *or* to the ceiling structural members, must be kept clear of all nonelectrical items (pipes, ducts, brackets, shelves, etc.). This "zone" is dedicated space intended for electrical items only! See *Section 384-4.* ◀

Mounting Height.   Disconnects and panelboards must be mounted so the center of the disconnect handle in its highest position is not higher than 6'7" (2.0 m). See *Section 380-8.*

Headroom.   To provide for safe adequate working space, and easy access to main service equipment and electrical panels, the *NEC®* requires that there be at least 6½ feet of headroom around this equipment. In the past, many service panels were installed in crawl spaces where the home had no basement. This is no longer permitted. This Code rule is found in *Section 110-16(e).*

Lighting Required for Service Equipment and Panels.   Working on electrical equipment with inadequate lighting can result in injury or death. In *Section 110-16(d)*, there is a requirement that illumination be provided for service equipment and panelboards. But the Code does not spell out how much illumination is required. It becomes a judgment call on the part of the electrician and/or electrical inspector. Note that the electrical plans for the workshop show a porcelain pull-chain lampholder to the front of and off to one side of the main service Panel A. In many cases, adjacent lighting, such as the fluorescent ceiling fixtures in the recreation room, is considered to be the required lighting for equipment such as Panel B.

Load Center (Panel B)

When a second load center is installed, it is oftentimes referred to as a subpanel.

For installation where the main service panel is a long distance from areas that have many circuits and/or heavy load concentration, as in the case of the kitchen and laundry of this residence, it is recommended that an additional panel(s) be installed near these loads. This results in the branch-circuit wiring being short. Line losses (voltage and wattage) are considerably less than if the branch-circuits had been run all the way back to the main panel.

The cost of material and labor to install an extra panel should be compared to the cost of running all of the branch-circuit "home runs" back to the main panel.

Panel B is located in the recreation room. It is fed by three No. 3 THHN or THWN conductors run in a 1-inch conduit originating at Panel A. This feeder is protected by a 100-ampere, 2-pole circuit breaker located in Panel A.

▶ DO NOT install panelboards or load centers in bathrooms, *Section 240-24(e).* ◀ Damaging moisture problems may result from the presence of water, and shock hazard may result from both the presence of water and the close proximity of metal faucets, etc.

DO NOT install panelboards or load centers in closets, *Section 240-24(d).* There is a fire hazard because of the presence of ignitable materials.

Panel B is shown in figure 28-11. The circuit schedule for Panel B is found in figure 28-12.

SERVICE-ENTRANCE RACEWAY SIZING

This residence is served with an underground service. The conduit running from the meter to Main Panel A must be sized correctly for the conductors it will contain. In the case of a through-the-roof mast

Fig. 28-11   Typical 240/120-volt, single-phase, 3-wire panelboard, sometimes referred to as a load center. This panel does not have a main disconnect, and is referred to as a *main lug only (MLO)* panel. The branch circuit-breakers are not shown. This is the type of panel that could be used as Panel B in the residence discussed in this text. Many panels have a separate terminal bus on which to terminate the equipment grounding conductors of nonmetallic-sheathed cable. See figure 28-10. *Courtesy* Square D Company, *Group Schneider*.

service, the conduit might be sized for mechanical strength reasons rather than conductor fill. All utilities have requirements for size and guying of mast services. Figure 28-13 shows the calculation for sizing the service-entrance conduit. This calculation is similar to other examples in this text. The conductor dimensions are found in *Chapter 9* of the *National Electrical Code®*

METER/METER BASE

The electric utility must be contacted before the installation of the service equipment begins. They will determine when, where, and how they intend to hook-up to the homeowner's service. Because of the ever-increasing number of wood decks and concrete patios being added to homes, many utilities prefer that the meter be mounted on the side of the house rather than in the rear. If there is a raised deck, the meter could be only a short distance above the deck, making it both subject to physical damage and hard

to read. If concrete patios are poured over underground utility cables, it is very difficult to repair or replace these cables should there be problems. Utilities are happy to furnish manuals and brochures detailing their requirements for services that are not found in the *National Electrical Code®*

The electric utility furnishes the watt-hour meter. The meter base, also referred to as a meter socket, is usually furnished and installed by the electrical contractor, although some utilities furnish the meter base to the electrical contractor for installation.

Figure 28-6 shows a service supplied from an underground service. Figure 28-7 discusses the requirement that the main disconnect must be as close as possible to the point of entrance of the service entrance conductors. Figure 28-14 shows a typical meter socket for an overhead service, usually mounted eye level on the outside wall of the house. The service raceway or service-entrance cable is connected to the threaded hub (boss) on top of the meter socket. Figure 28-15 shows how to seal the raceway where it enters the building to keep out moisture.

Probably more common for residential services is the "pedestal," as shown in figure 28-16. Usually mounted on the outside wall of the house, they could also be installed on the lot line between residential properties. Contact the electric utility on this issue.

For underground services, the utility generally runs the underground lateral service-entrance conductors in a trench from a pad-mount transformer to the line-side terminals of the meter base. The electrician then takes over to complete the service installation by running service-entrance conductors from the load side of the meter to the line side of the main service disconnect.

Quite often, the electric utility and the telephone company have agreements whereby the electric utility will "lay" both power cables and telephone cables in the same trench. The electric utility carries both types of cables on their underground installation vehicles. This is a much more cost effective way to do this, rather than having each utility dig and then backfill their own trenches.

Often overlooked is a requirement in NFPA 54 (Fuel Gas Code), *Section 2.7.2(c)* requiring that gas meters be located at least 3 feet from sources of ignition. Examples of sources of ignition might be an air-conditioner disconnect where an arc can be pro-

PANEL "B"

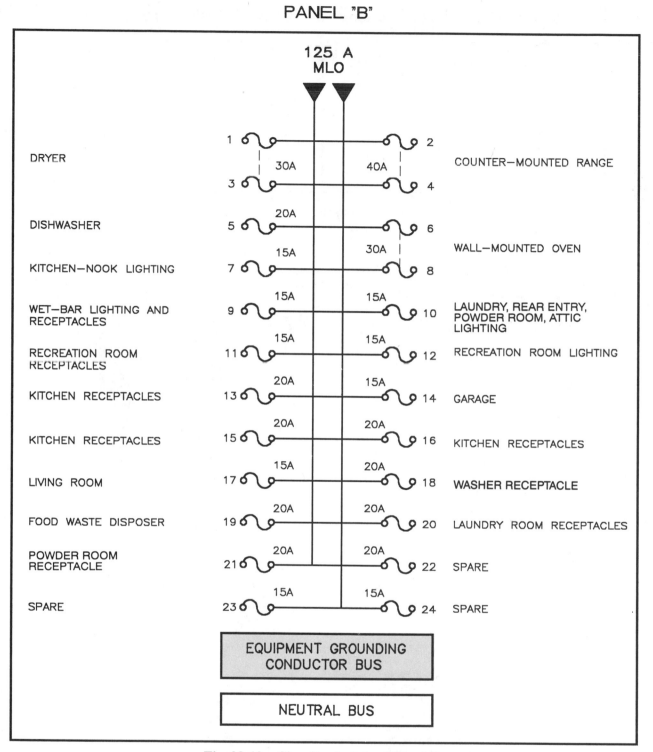

Fig. 28-12   Circuit schedule of Panel B.

duced when the disconnect is opened under load. Furthermore, some electric utilities require a minimum of 3 feet clearance between electric metering equipment and gas meters and gas regulating equipment. It is better to be safe than sorry: Check this issue out with the local electrical inspector if there might be a problem with the installation you are working on.

COST OF USING ELECTRICAL ENERGY

A watt-hour meter is always connected into some part of the service-entrance equipment to record the amount of energy used. Billing by the utility is generally done on a monthly basis. The meter might be mounted on the side of the house, or on a pedestal somewhere on the premise on the lot line. The utility makes this decision. The residence discussed in this

CONDUCTOR SIZE BASED UPON AMPACITY PER *TABLE 310-16, NOTE 3.*		CONDUCTOR SIZE BASED UPON AMPACITY PER *TABLE 310-16.*	
TWO NO. 2/0 THHN COPPER	0.2223 SQ. IN.	TWO NO. 3/0 THHN COPPER	0.2679 SQ. IN.
	0.2223 SQ. IN.		0.2679 SQ. IN.
ONE NO. 1 BARE COPPER	0.0870 SQ. IN.	ONE NO. 1 BARE COPPER	0.0870 SQ. IN.
TOTAL	0.5316 SQ. IN.	TOTAL	0.6228 SQ. IN.
CONDUIT SIZE	1 1/4 INCH	CONDUIT SIZE	1 1/2 INCH

▶ Fig. 28-13   Calculations for determining the size of the electrical metallic tubing service-entrance raceway between Main Panel A and the meter base. Dimensional data found in *Chapter 9, Tables 1, 4, 5,* and *8* of the *NEC.* ◀

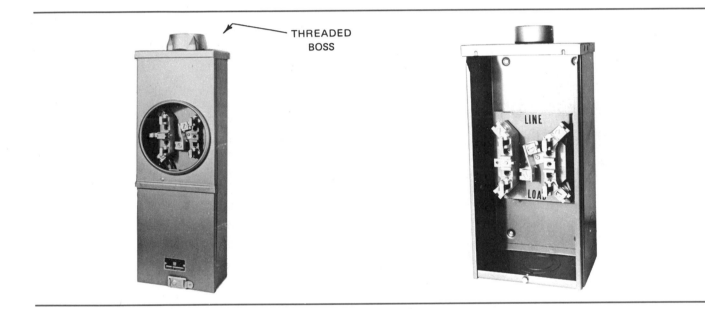

THREADED BOSS

LINE

LOAD

Fig. 28-14   Typical meter sockets.

SECTION 300-5(g) AND SECTION 300-7 STATE THAT WHEN RACEWAYS PASS THROUGH AREAS HAVING GREAT TEMPERATURE DIFFERENCES, SOME MEANS MUST BE PROVIDED TO PREVENT PASSAGE OF AIR BACK AND FORTH THROUGH THE RACEWAY. NOTE THAT OUTSIDE AIR IS DRAWN IN THROUGH THE CONDUIT WHENEVER A DOOR OPENS. COLD OUTSIDE AIR MEETING WARM INSIDE AIR CAUSES THE CONDENSATION OF MOISTURE. THIS CAN RESULT IN RUSTING AND CORROSION OF VITAL ELECTRICAL COMPONENTS. EQUIPMENT HAVING MOVING PARTS, SUCH AS CIRCUIT BREAKERS, SWITCHES, AND CONTROLLERS, IS ESPECIALLY AFFECTED BY MOISTURE. THE SLUGGISH ACTION OF THE MOVING PARTS IN THIS EQUIPMENT IS UNDESIRABLE.

INSULATION OR OTHER TYPE OF SEALING COMPOUND CAN BE INSERTED AS SHOWN TO PREVENT THE PASSAGE OF AIR.

SECTION 230-8 REQUIRES SEALS WHERE SERVICE RACEWAYS ENTER FROM AN UNDERGROUND DISTRIBUTION SYSTEM.

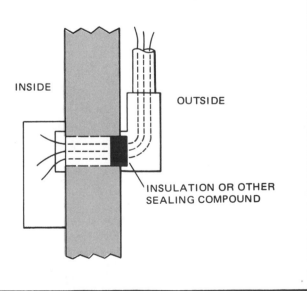

INSIDE

OUTSIDE

INSULATION OR OTHER SEALING COMPOUND

Fig. 28-15   Installation of conduit through a basement wall.

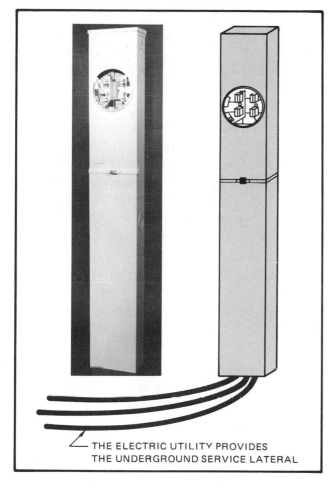

THE ELECTRIC UTILITY PROVIDES
THE UNDERGROUND SERVICE LATERAL

Fig. 28-16   Installing a metal "pedestal" allows for ease of installation of the underground service-entrance conductors by the utility. *Courtesy* **of Milbank Manufacturing Co.**

text has one meter, mounted on the back of the house near the sliding doors of the master bedroom.

The kilowatt (kW) is a convenient unit of electrical power. One thousand watts (W) is equal to one kilowatt. The watt-hour meter measures and records both wattage and time.

For residential metering, most utilities have rates based upon "cents per kilowatt-hour." Usually, the more kilowatt-hours used, the lower the cost per kilowatt-hour.

Some utilities increase their rates during the hot summer months when the air-conditioning load is high. Heavy air-conditioning loads "tax" the utilities' generating capabilities during the hot summer months. The higher rates give homeowners an incentive for not setting their thermostats too low, thus saving energy.

For example:

First 500 kWh @ 6.12 cents per kWh

Over 500 kWh @ 10.64 cents per kWh (summer months)

Over 500 kWh @ 8.64 cents per kWh (winter months)

Some utilities provide a second meter for separate metering of large loads such as electric water heaters or electric heating. Other utilities use electronic meters that have the capability of registering time-of-use power consumption. This was discussed in unit 19.

Burning a 100-watt light bulb for 10 hours is the same as using a 1,000-watt electric heater for 1 hour. Therefore, if the electric rate is 8 cents per kilowatt-hour, the cost to operate the 100-watt bulb for 10 hours would be 8 cents. Likewise, cost of operating the electric heater for 1 hour would also be 8 cents.

EXAMPLE:

$$kWh = \frac{watts \times hours}{1,000} = \frac{100 \times 10}{1,000} = 1 \ kWh$$

EXAMPLE:

$$kWh = \frac{watts \times hours}{1,000} = \frac{1,000 \times 1}{1,000} = 1 \ kWh$$

The cost of the energy used can be calculated as follows:

$$Cost = \frac{watts \times hours \ used \times cost \ per \ kWh}{1,000}$$

EXAMPLE: Find the cost of operating a color television for 8 hours. The television is marked 175 watts. The electric rate is 9.6 cents per kilowatt-hour.

$$Cost = \frac{175 \times 8 \times 0.096}{1,000} = 0.1344 \ (13.4 \ cents)$$

A Typical Electric Bill

Here is what a typical electric bill might look like.

GENERIC ELECTRIC COMPANY			
Anyplace, USA			
Days of	from	to	Due Date
Service	01-28-96	03-01-96	03-25-96
33			
Present reading			23,232
Last reading			21,987
Kilowatt-hours used			1,245
Rate/kWh . . . . 1st 400 kWh @ $0.06168			$24.67
	remaining 845 kWh @ $0.03178		$26.85
Customer charge (single-dwelling)			$ 9.07
Total energy charge			$60.59
State tax			$ 2.88
Total current bill now due			$63.47

Some utilities apply a fuel adjustment charge that may increase or decrease the electric bill. These charges are on a "per kWh" basis. This fuel adjustment charge enables the utility to recover from the consumer extra expenses it might incur for fuel costs used in generating electricity. Fuel charges can vary each time the utility prepares the bill, without having to apply to the regulatory agency for a rate change.

GROUNDING—WHY GROUND?

To appreciate the concepts of proper grounding, let us review some important Code sections.

The two *Fine Print Notes* to *Section 250-1* are extremely important. These notes emphasize the *what* and *why* of grounding and bonding.

FPN No. 1: Systems and circuit conductors are grounded to limit voltages due to lightning, line surges, or unintentional contact with higher voltage lines, and to stabilize the voltage to ground during normal operation. Equipment grounding conductors are bonded to the system grounded conductor to provide a low impedance path for fault current that will facilitate the operation of overcurrent devices under fault conditions.

FPN No. 2: Conductive materials enclosing electrical conductors or equipment, or forming part of such equipment, are grounded to limit the voltage to ground on these materials, and bonded to facilitate overcurrent device operation in case of ground faults. *See Section 110-10.*

Proper grounding means that overcurrent protective devices will operate fast when responding to ground faults. Effective grounding occurs when a low-impedance (ac resistance) ground path is provided. Ohm's law verifies that, in a given circuit, "the lower the impedance, the higher the value of current." As the value of ground-fault current increases, there is an increase in the speed with which a fuse will open or a circuit breaker will trip. This is called an *inverse time* relationship.

Clearing a ground fault or short-circuit is important because arcing damage to electrical equipment as well as conductor insulation damage is closely related to a value called *ampere-squared-seconds* (I^2t), where

I = the current in amperes flowing phase to ground, or phase to phase.

t = the time in seconds that the current is flowing.

Thus, there will be less equipment and/or conductor damage when the fault current is kept to a low value, and when the time that the fault current is allowed to flow is kept to a minimum. The impedance of the circuit determines the *amount* of fault current that will flow. The speed of operation of a fuse or circuit breaker determines the amount of *time* the fault current will flow.

Grounding electrode conductors and equipment grounding conductors carry an insignificant amount of current under normal conditions. However, when a ground fault occurs, these grounding conductors as well as the actual circuit conductors must be capable of carrying whatever value of fault current might flow (how much?) for the time (how long?) it takes for the overcurrent protective device to clear the fault, *Section 250-51.*

▶ This potential hazard is recognized in the *Note* to *Table 250-95*, which states that "Equipment grounding conductors may need to be sized larger than specified in this table in order to comply with *Section 250-51.*" Although this statement is vague, it does call attention to the fact that equipment grounding conductors may need to be sized larger than indicated in the table when high-level ground-fault currents are possible. ◀

What is a high-level ground-fault? To determine the available short-circuit and ground-fault current, a short-circuit calculation is necessary. It is also necessary to know the time/current characteristic of the overcurrent protective device. This is discussed in the *Electrical Wiring, Commercial* text, © Delmar Publishers.

This is referred to as the conductor's "withstand rating."

Tables 250-94 and *250-95* are based upon the fact that copper conductors and their bolted connections can withstand:

- **one ampere**
- **for 5 seconds**
- **for every 30 circular mils**

Thermoplastic insulation on a copper conductor rated 75°C can safely withstand:

- **one ampere**
- **for 5 seconds**
- **for every 42.25 circular mils**

To exceed the previous values will result in damage to the conductors, with possible burn-off and loss of the grounding or bonding path.

When bare equipment grounding conductors are in the same raceway or cable as insulated conductors, always apply the insulated conductor withstand rating limitation.

Properly selected fuses and circuit breakers for normal residential installations generally will protect conductors and other electrical equipment against the types of ground faults and short-circuits to be expected in homes. Fault currents can be quite high in multifamily dwellings (apartments, condos, town houses) making it necessary to take equipment and conductor withstand rating into consideration.

The subject of conductor and equipment withstand rating is covered in greater depth in the *Commercial Wiring* text.

Grounding Electrode System

Article 250, Part H of the Code covers the requirements that establish a *grounding electrode system*. Simply stated, this means that everything is "tied" together. The service-entrance neutral conductor is grounded, this neutral is bonded to the main panel, a grounding electrode conductor is connected to a grounding electrode, and all metal water pipes and gas pipes are bonded together. Should any one of the previous paths become disconnected, the integrity of the grounding system is maintained through the other paths.

Figure 28-5 shows proper grounding and bonding in accordance with *Sections 250-80* and *250-81* of the Code.

By having all metal parts bonded together for a grounding electrode system, potential voltage differences between noncurrent-carrying metal parts is virtually eliminated. In addition, the grounding electrode system serves as a means to bleed off lightning, stabilize voltage of the system, and assures that the overcurrent protective devices will operate.

Figure 28-17 and the following steps illustrate what can happen if the electrical system is not properly grounded and bonded:

1. A "live" wire contacts the gas pipe. The bonding jumper "A" has not been installed.

2. The gas pipe now has 120 volts on it. The pipe is energized. It is "hot."

3. The insulating joint in the gas pipe results in a poor path to ground. Assume a resistance (impedance) of 8 ohms.

4. The 20-ampere overcurrent device does not open:

$$I = \frac{E}{R} = \frac{120}{8} = 15 \text{ amperes}$$

5. If a person touches the "live" gas pipe and the water pipe at the same time, current flows through the person's body. If the body resistance is 12,000 ohms, the current is:

$$I = \frac{E}{R} = \frac{120}{12,000} = 0.01 \text{ ampere}$$

This amount of current passing through a human body can cause death. See unit 6 for a discussion regarding electric shock.

6. The overcurrent device is now "seeing" 15 + 0.01 = 15.01 amperes and does not open.

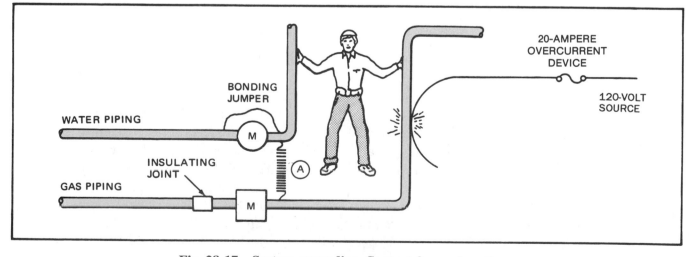

Fig. 28-17 System grounding. See text for explanation.

7. If the grounding electrode system concept had been used, bonding jumper "A" would have kept the voltage difference between the water pipe and the gas pipe at zero. The overcurrent device would have opened. If the bonding jumper was 10 feet (3.05 m) of No. 4 AWG copper wire, the resistance of the jumper is 0.00308 ohm, calculated using the values found in *Table 8, Chapter 9, NEC.* The current would be:

$$I = \frac{E}{R} = \frac{120}{0.00308} = 38,961 \text{ amperes}$$

In an actual installation, the total impedance of *all* parts of the circuit would be much higher than these simple calculations. A much lower current would result. The value of current would be enough to cause the overcurrent device to open.

Section 250-51 state that to be effective, the grounding path must:

1. be permanent and continuous,

2. have the capacity to conduct safely *any* fault current likely to be imposed on it, and

3. have sufficiently low impedance to limit the voltage to ground and to facilitate the operation of the circuit protective device in the circuit.

Section 250-70 states that "bonding shall be provided where necessary to assure electrical continuity and the capacity to conduct safely *any* fault current likely to be imposed."

Section 250-75 states in part that the bonding of metal raceways, enclosures, fittings, etc. that serve as the grounding path "shall be effectively bonded where necessary to assure electrical continuity and the capacity to conduct safely *any* fault current likely to be imposed on them."

The Grounding Electrode.   *Section 250-81* covers the rules for establishing a good grounding electrode system.

Section 250-81(a) states that a metal underground water piping system 10 feet (3.05 m) or longer is an acceptable grounding electrode. The water pipe grounding electrode *must* be supplemented by at least one additional grounding electrode consisting of any one of those listed in *Sections 250-81* and *250-83*. In this residence, a driven ground rod is the supple-

mental grounding electrode, as permitted by *Section 250-83(c)*. At least one additional ground rod must be driven if the resistance to ground of one rod is found to exceed 25 ohms, *Section 250-84*.

Section 250-81(a) requires that a grounding electrode conductor that is the sole (only) connection to a driven ground rod need not be larger than No. 6 AWG copper or No. 4 AWG aluminum.

If the driven ground rod is not selected as the required supplemental grounding electrode, then any one of the following is acceptable:

- the metal frame of a building (this residence is constructed of wood), *Section 250-81(b)*.

- at least 20 feet (6.1 m) of steel reinforcing bars, ½ inch (12.7 mm) minimum diameter, or at least 20 feet (6.1 m) of bare copper conductor not smaller than No. 4 AWG. These must be encased in concrete at least 2-inches (50.8 mm) thick that is in direct contact with the earth, such as near the bottom of a foundation or footing. This is a concrete-encased ground and is commonly referred to as a UFER ground, named after the gentleman who developed the concept. See *Section 250-81(c)*.

- at least 20 feet (6.1 m) of bare copper wire, minimum size of No. 2 AWG, buried directly in the earth at least 2½ feet (762 mm) deep. This is called a *ground ring*. See *Section 250-81(d)*.

- metal underground gas piping. This is *not* permitted to be used as the grounding electrode, *Section 250-83(a)*, but it must be bonded.

- ground plates must be at least 2 square feet (0.186 m²), *Section 250-83(d)*. At least one additional plate must be laid if the resistance of one plate is found to exceed 25 ohms.

Ground Rods

Copper ground rods are commonly used as a grounding electrode as well as stainless steel rods. Galvanized iron or steel pipe and conduit, and non-ferrous metal plates are also permitted. Aluminum rods are *not* permitted. *Section 250-83* covers ground rods. The following text discusses the Code requirements for copper ground rods. Copper ground rods shall:

- be at least ½ inch (12.7 mm) in diameter.

- be installed below the permanent moisture level, if practical.

- not be less than 8 feet (2.44 m) in length.

- be driven to a depth so that at least 8 feet (2.44 m) is in contact with the soil.

- if solid rock is encountered, be driven at not greater than a 45 degree angle from the vertical or laid in a trench that is at least 2½ feet (762 mm) deep.

- be driven so that the upper end of the rod is flush with or just below ground level. If the upper end is exposed, the ground rod, the ground clamp, and the grounding electrode conductor must be protected against physical damage per *Section 250-117.*

- have a resistance to ground of not less than 25 ohms. If the one rod has more than 25 ohms to ground, drive a second rod.

- be at least 6 feet (1.83 m) apart if more than one rod is driven. Actually, it is better to space multiple rods not less than the length of the rod. For example, when driving two 8-foot rods, space them 8 feet apart.

Grounding Conductors

A grounding conductor is defined as "a conductor used to connect equipment or the grounded circuit of a wiring system to a grounding electrode or electrodes." There are two types of grounding conductors.

1. An *equipment grounding conductor* is the conductor that is used to connect noncurrent-carrying metal parts of equipment, raceways, and other enclosures to the system grounded conductor, to the grounding electrode conductor, or to both—at the service equipment. *Section 250-91(b)* recognizes many types of equipment grounding conductors. Copper or aluminum conductors, rigid metal conduit, electrical metallic tubing, the metal armor of armored cable, and flexible metal conduit (has limitations found in the exceptions to *Section 250-91(b)*) all qualify as an equipment ground-

ing conductor. Figures 5-20, 5-21, 5-22, 28-1A, 28-10, and 30-3 illustrate equipment grounding conductors.

Equipment grounding is discussed throughout this text in the appropriate unit where a specific piece of electrical equipment or appliance requires grounding. Some examples of this are the range, cook-top, dryer, food waste disposer, dishwasher, and water pump. Equipment grounding conductors are sized according to *Table 250-95.*

2. A *grounding electrode conductor* is the conductor that connects the grounding electrode (i.e., metal underground water pipe supply) to the equipment grounding conductor and to the grounded conductor of the circuit, or to both—at the service equipment. Reference to grounding electrode conductors is made in figures 28-1, 28-1A, 28-5, 28-10, 28-18, 28-19, and 30-3. Grounding electrode conductors are sized according to *Table 250-94.*

In this residence, a No. 4 AWG armored copper grounding electrode conductor runs from Main Panel A, across the workshop ceiling to the water pipe, where it is terminated with a "listed" ground clamp.

Table 250-94.
Grounding Electrode Conductor for AC Systems

Size of Largest Service-Entrance Conductor or Equivalent Area for Parallel Conductors		Size of Grounding Electrode Conductor	
Copper	Aluminum or Copper-Clad Aluminum	Copper	*Aluminum or Copper-Clad Aluminum
2 or smaller	1/0 or smaller	8	6
1 or 1/0	2/0 or 3/0	6	4
2/0 or 3/0	4/0 or 250 kcmil	4	2
Over 3/0 thru 350 kcmil	Over 250 kcmil thru 500 kcmil	2	1/0
Over 350 kcmil thru 600 kcmil	Over 500 kcmil thru 900 kcmil	1/0	3/0
Over 600 kcmil thru 1100 kcmil	Over 900 kcmil thru 1750 kcmil	2/0	4/0
Over 1100 kcmil	Over 1750 kcmil	3/0	250 kcmil

Where multiple sets of service-entrance conductors are used as permitted in Section 230-40, Exception No. 2, the equivalent size of the largest service-entrance conductor shall be determined by the largest sum of the areas of the corresponding conductors of each set.

Where there are no service-entrance conductors, the grounding electrode conductor size shall be determined by the equivalent size of the largest service-entrance conductor required for the load to be served.

* See installation restrictions in Section 250-92(a).

(FPN): See Section 250-23(b).

Ground Clamps

Ground clamps used for bonding and grounding must be "listed" for the purpose. The use of solder to make up bonding and grounding connections is not acceptable because under high levels of fault current, the solder would probably melt, resulting in loss of the integrity of the bonding and/or grounding path.

Various types of ground clamps are shown in figures 28-20A, 28-20B, 28-21A, and 28-21B. These clamps and their attachment to the grounding electrode must conform to *Section 250-115* of the *NEC*.

Connection of Grounding and Grounded Conductors in Main Panel and Subpanel

According to the second paragraph of *Section 384-20*, equipment *grounding* conductors (these are the bare copper equipment grounding conductors found in nonmetallic-sheathed cable) shall

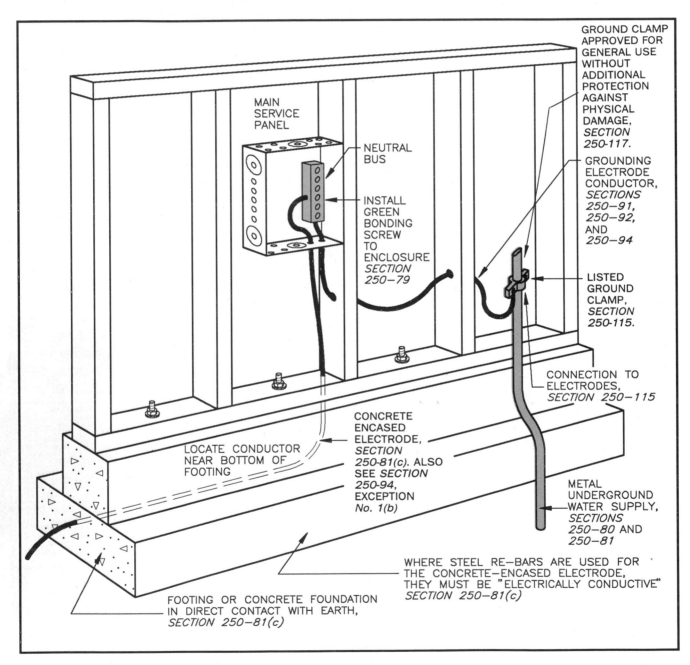

Fig. 28-18 *Section 250-81* of the Code states that if available, an underground metallic water pipe 10 feet or longer in direct contact with the earth shall be used as the grounding electrode. This water pipe ground *must be supplemented* by at least one additional electrode. Here we show a concrete-encased No. 4 AWG bare copper conductor laid near the bottom of the footing, encased by at least 2 inches of concrete. The *minimum* length of the conductor is 20 feet (6.1 m). See *Section 250-81(c)*.

not be connected to the *grounded* conductor (this is the neutral conductor) terminal bar (neutral bar) unless the bar is identified for the purpose, and is located where the *grounded* conductor is connected to the *grounding* electrode conductor. In this residence, this occurs in Main Panel A. The green bonding screw furnished with Panel A is installed in Panel A. This bonds the neutral bar, the ground bus, the grounded neutral conductor, the grounding electrode conductor, and the panel enclosure together. See figure 28-10. *Grounding* conductors and *grounded* conductors are not to be connected together anywhere on the load side of the main service disconnect, *Section 250-61(b)*. The green bonding screw furnished with Panel B will NOT be installed in Panel B. More on this

a little later under BONDING.

Sheet Metal Screws Not Permitted for Connecting Grounding Conductors

▶ Sheet metal screws are *not* permitted to be used as a means of connecting grounding conductors to enclosures, *Section 250-113*. Sheet metal screws do not have the same fine thread that 10/32 machine screws have, which match the tapped 10/32 threaded holes in outlet boxes, device boxes, and other enclosures. Sheet metal screws "force" themselves into a hole instead of nicely threading themselves into a pretapped matching hole. Sheet metal screws have *not* been tested as to their ability to safely carry ground-fault currents as required by *Section 250-51* of the Code. ◀

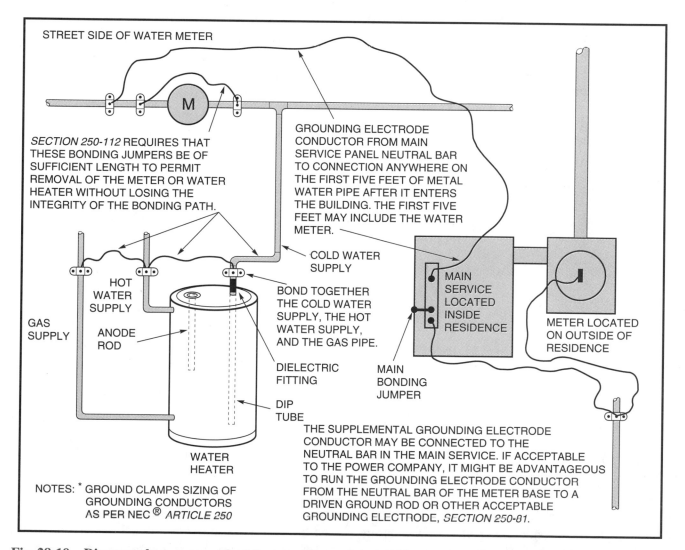

Fig. 28-19 Diagram shows one method that may be used to provide proper grounding and bonding of service-entrance equipment for a typical one-family residence.

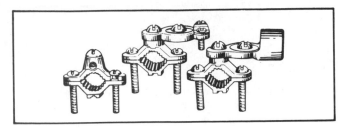

Fig. 28-20A   Typical ground clamps used in residential systems.

Fig. 28-20B   Armored grounding conductor connected with ground clamp to water pipe. *Courtesy* of Thomas & Betts.

Fig. 28-21A   Ground clamp of the type used to bond (jumper) around water meter. *Courtesy* of Thomas & Betts.

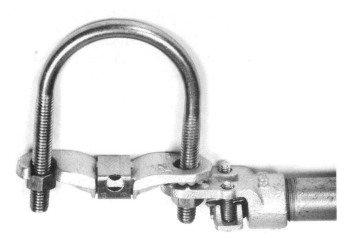

Fig. 28-21B   Ground clamp of the type used to attach ground wire to well casings. *Courtesy* of Thomas & Betts.

1. Equipment bonding "ties" equipment together to keep voltage buildup on the equipment the same, so there will be no difference in potential between the various pieces of equipment. For example, to bond two metal raceways together, an equipment bonding jumper is installed where there is a break in the metal conduit run. See figure 28-22A. Bonding equipment together does NOT necessarily mean that the equipment is properly grounded.

2. A circuit bonding jumper connects conductors of a circuit together. This jumper must have the same or greater current-carrying ability than the circuit conductors. See figure 28-22B.

3. A main bonding jumper is defined as "the connection between the grounded circuit conductor and the equipment grounding conductor at the service." Main bonding jumpers are clearly illustrated in figures 28-1A, 28-5, 28-10, 28-19, and 30-3.

4. Bonding for swimming pools is shown in figure 30-4.

At the main service-entrance equipment, the grounded neutral conductor must be bonded to the metal enclosure. For most residential panels, this main bonding jumper is a bonding screw that is furnished with the panel. This bonding screw is inserted through the neutral bar into a threaded hole in the back of the panel. This bonding screw must be green in color and must be clearly visible after it is in place, *Section 250-79(b)*.

BONDING

Bonding serves a different purpose than grounding. It is "the permanent joining of metallic parts to form an electrically conductive path that will assure electrical continuity and the capacity to conduct safely any current likely to be imposed." Bonding jumpers accomplish this in a number of ways:

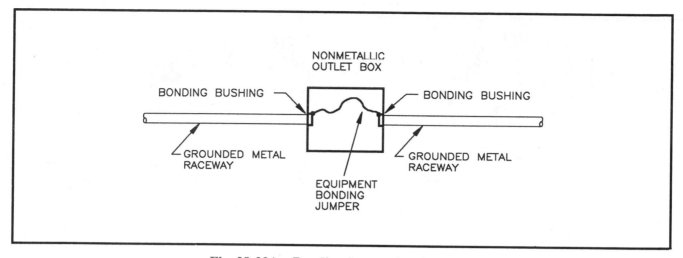

Fig. 28-22A   Bonding jumper (equipment).

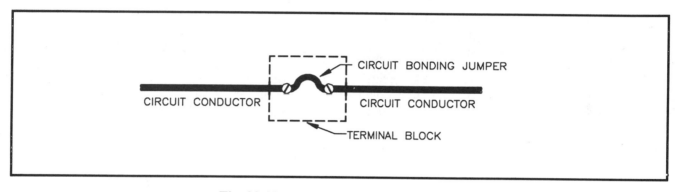

Fig. 28-22B   Bonding jumper (circuit).

The required bonding jumpers for services must not be smaller than the grounding electrode conductor, *Section 250-79*. The lugs on bonding bushings are based on the ampacity of the conductors that would normally be installed in that particular size of raceway. The lugs become larger as the trade size of the bushing increases.

Bonding at service-entrance equipment is very important because service-entrance conductors do not have overcurrent protection at their line side, other than the electric utility company's primary transformer fuses. Overload protection for service-entrance conductors is at their load end. High available fault currents can result in severe arcing, which is a fire hazard. For all practical purposes, available short-circuit current is limited only by (1) the kVA rating and impedance of the transformer that supplies the service equipment, and (2) the size, type, and length of the service conductors between the transformer and service equipment. Fault currents can easily reach 20,000 amperes or more at the main

service in residential installations. Much higher fault current is available in multifamily dwellings such as apartments and condominiums.

Figures 28-5 and 28-19 show the hot and cold metal water pipes bonded together. Some electrical inspectors will consider the hot/cold mixing faucets and the water heater as an acceptable means of bonding the hot and cold water metal pipes together. Most, however, will require a bonding jumper as illustrated in the figures, consistent with the thinking that electrical grounding and bonding shall not be dependent upon the plumbing trade.

Fault current calculations are presented later in this unit. The text *Electrical Wiring—Commercial* covers fault current calculations in greater depth.

Section 250-71(a) lists the parts of the service-entrance equipment that must be bonded. These include the meter base, service raceways, service cable armor, main disconnect. *Section 250-72* lists the methods acceptable for bonding all of the previous equipment. Bonding bushings, figure 28-23,

Fig. 28-23   Insulated bonding bushing with grounding lug.

Fig. 28-24   Insulated bushings.

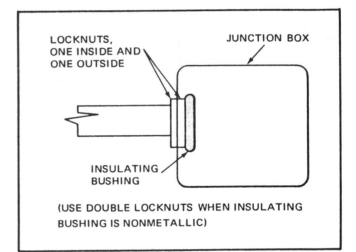

Fig. 28-25   This illustration shows the use of double locknuts plus an insulating bushing. See *Sections 300-4(f)* and *373-6(c)*.

and bonding jumpers are installed to insure a low-impedance path to ground if a fault occurs. Bonding jumpers are required where there are concentric or eccentric knockouts, or where reducing washers are installed. ▶ *Sections 300-4(f)* ◀ and *373-6(c)* state that if No. 4 AWG or larger conductors are installed in a raceway, an insulating bushing or equivalent must be used, figure 28-24. Insulating bushings protect the conductors from abrasion where they pass through the bushing. Combination metal/insulating bushings can be used. Some electrical metallic tubing connectors have an insulated throat.

If the bushing is made of insulating material only, as in figure 28-24, then two locknuts must be used, figure 28-25.

Installing a Grounding Electrode Conductor from Main Service Disconnect to Metal Underground Water Pipe that Will Serve as the Grounding Electrode

Section 250-92(a) states that a No. 4 or larger copper or aluminum grounding electrode conductor may be securely fastened directly to the surface on which it is run, and it must be protected from *severe* physical damage. A No. 6 grounding electrode conductor may be securely fastened directly to the surface on which it is run without physical protection if not subject to physical damage. Grounding conductors smaller than No. 6 must have physical protection. Physical protection can be provided by installing the grounding electrode conductor in rigid metal conduit, intermediate metal conduit, rigid nonmetallic conduit, electrical metallic tubing, or cable armor.

In order to carry high values of ground-fault current, the ground path must have as low an impedance as possible. *Section 250-51* of the Code requires that the ground path be *effective*. This means that the path

must be permanent and continuous, must have the capability of carrying any value of fault current that it might be called upon to carry, and shall have an impedance low enough to limit the voltage to ground and to assure that the protective overcurrent device will operate. *Sections 250-70* and *250-75* have similar requirements. For example, the metal raceway enclosing a grounding electrode conductor must be continuous from the main disconnect to the ground clamp, or be made continuous by proper bonding.

Section 250-71(a)(3) of the Code requires that the metal raceway be bonded at both ends. At first thought, it might appear that simply installing a grounding electrode conductor in a metal raceway makes for a neat and workmanlike installation. However, unless the enclosing metal raceway is properly bonded on both ends, the ground path will be a high impedance path. Bonding the enclosing metal raceway at one end only will result in a ground path impedance approximately twice that of bonding the metal raceway on both ends. Remember—the higher the impedance, the lower the current flow and the longer it takes the overcurrent device to clear the fault.

Past testing (Soares, IAEI, and IEEE data) has shown that for a 300-ampere ground fault, approximately 295 amperes flowed in the metal raceway, and 5 amperes flowed in the enclosed grounding electrode conductor. Another test revealed that for a 590-ampere ground fault, 585 amperes flowed in the metal raceway, and 5 amperes flowed in the enclosed grounding electrode conductor. The tendency for current to flow along the surface of a conductor is called "skin effect." Proper bonding at both ends of a metal raceway is necessary to assure a low impedance path that will safely transfer the high magnitudes of ground-fault current flowing in the metal conduit to the ground bus in the main panel, and/or to the ground clamp.

Figures 28-25A, 28-25B, 28-25C, 28-25D, and 28-25E illustrate accepted methods of installing a

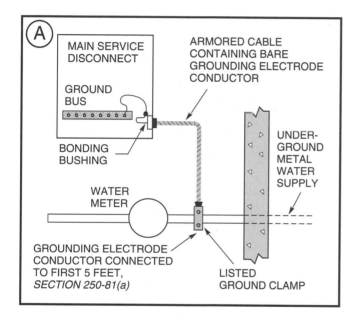

Fig. 28-25A   This installation "Meets Code."

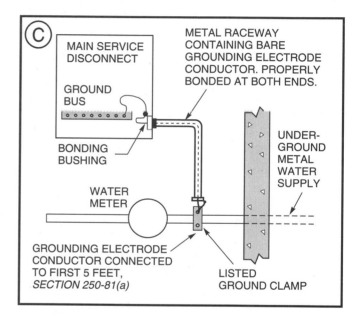

Fig. 28-25C   This installation "Meets Code."

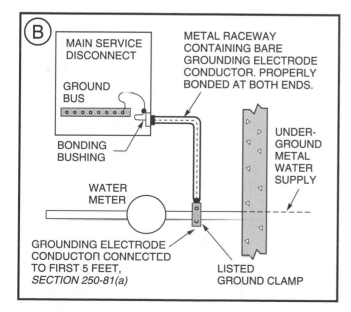

Fig. 28-25B   This installation "Meets Code."

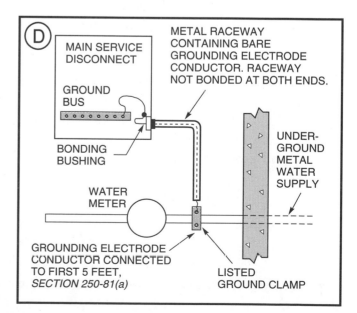

Fig. 28-25D   This installation DOES NOT "Meet Code."

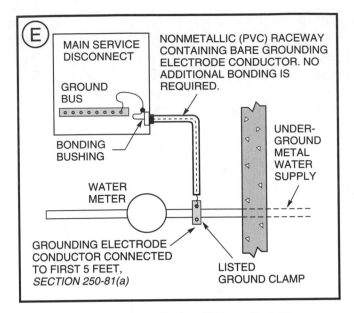

Fig. 28-25E   This installation "Meets Code."

grounding electrode conductor between the main panel and the ground clamp. Additional information concerning grounding electrode conductors can be found in unit 4.

Grounding the Electrical Service When Nonmetallic Water Pipe Is Used

Today it is quite common to find the main water supply to a residence and the interior water piping installed with plastic (PVC) piping.

As previously mentioned, one of the methods used to provide for adequate grounding of services where nonmetallic water supplies are encountered, or where the soil is so dry (sand) that driven ground rods cannot attain a 25-ohm or less resistance ground, *Section 250-84*, or where the metallic underground water piping in direct contact with the earth is less than 10 feet, is to install a concrete-encased electrode, commonly referred to as a UFER ground, figure 28-18.

A UFER ground consists of a bare copper conductor:

- not smaller than No. 4 AWG
- not less than 20 feet (6.1 m) long encased by at least 2 inches (50.8 mm) of concrete in the footing or foundation
- in concrete that is in direct contact with the earth

- lying within and near the bottom of the concrete
- solid or stranded

Some soil conditions are so bad relative to relying on the soil for grounding purposes, that the UFER ground becomes the primary grounding electrode, as opposed to merely being a supplemental ground.

SUMMARY—SERVICE-ENTRANCE EQUIPMENT GROUNDING

Refer to figures 28-18 and 28-19.

- The system must be grounded so the maximum voltage to ground on the ungrounded conductors does not exceed 150 volts, *Section 250-5(b)(1)*.
- All grounding schemes shall be installed so that no objectionable currents will flow over the grounding conductors and other grounding paths, *Section 250-21(a)*.
- The grounding electrode conductor must be connected to the supply side of the service disconnecting means. It must not be connected to any grounded circuit conductor on the load side of the service disconnect, *Section 250-23(a)*.
- The neutral conductor must be grounded, *Section 250-25(a)*.

- Tie (bond) everything together, *Sections 250-80(a)* and *(b)*, *250-71*, *250-81*, and *250-83*.

- The grounding electrode conductor used to connect the grounded neutral conductor to the grounding electrode must not be spliced, *Sections 250-53* and *250-91(a)*.

- Size the grounding electrode conductor according to *Table 250-94*.

- Bond all metal piping together, *Section 250-80(a)*.

- The grounding electrode conductor must be connected to the metal underground water pipe when 10 feet (3.05 m) long or more, including the well casing, *Section 250-81(a)*.

- Connect the grounding electrode conductor to the first 5 feet of metal underground water piping after it enters the building. This will generally be ahead of the water meter, unless the water meter is located in the street outside of the dwelling. The *NEC®* does not permit interior metal water piping to be used to interconnect electrodes and the grounding electrode conductor, *Section 250-81*.

- A metal water pipe ground must be supplemented by another electrode, such as a bare conductor in the footing, *Section 250-81(c)*; a grounding ring, *Section 250-81(d)*; the metal frame of the building IF the metal frame is effectively grounded, *Section 250-81(b)*; a rod or pipe electrodes, *Section 250-83(c)*; or plate electrodes, *Section 250-83(d)*.

- The grounding electrode conductor shall be copper, aluminum, or copper-clad aluminum, *Section 250-91*.

- The grounding electrode conductor may be solid or stranded, insulated or uninsulated, covered or bare, and must not be spliced, *Section 250-91*.

- Bonding shall be provided around all insulating joints or sections of the metal piping system that may be disconnected, *Section 250-112*.

- The connection of the grounding electrode conductor to the grounding electrode shall be accessible, *Section 250-112*.

- It is OK to bury the connection in the ground

using clamps listed for direct soil burial, *Section 250-115*.

- The grounding electrode conductor and bonding jumpers shall be connected tightly using the proper lugs, connectors, clamps, exothermic welding, or other "listed" means. Solder connections are not permitted, *Section 250-115*.

MAIN, FEEDER, AND BRANCH-CIRCUIT OVERCURRENT PROTECTION

The *National Electrical Code®* covers overcurrent protection in *Article 240*. Overcurrent protection for residential main services, feeders, and branch-circuits comes in the form of fuses and circuit breakers, often referred to as the "safety valves" of an electrical circuit.

Fuses and circuit breakers are sized by matching their ampere ratings to conductor ampacities and connected load currents. They sense overload, short-circuit, and ground-fault conditions, and protect the wiring and equipment from reaching dangerous temperatures.

A fuse will function "one-time." It does its job of protecting the circuit.

When a fuse opens, find out what caused it to blow, fix the problem, and replace it with the size and type that will provide proper overcurrent protection. Do not keep replacing a fuse without finding out why the fuse blew in the first place.

When a circuit breaker trips, find out what caused it to trip, fix the problem, then reset it. Do not repeatedly reset the breaker again and again into a fault, as further damage can result.

Overcurrent protective devices are connected in a circuit where the conductors receive their supply, *Section 240-21*. There are some exceptions to this basic rule, also found in *Section 240-21*. A typical example for house wiring is for "taps" that supply built-in ovens and cook-tops. This is covered in unit 20.

Overcurrent devices are connected "in series" with the ungrounded (hot) conductors of a circuit, *Section 240-20*.

The *NEC®* does not allow overcurrent devices to be connected in the grounded conductor, *Section 240-22*. The two exceptions are: (1) where both grounded and ungrounded conductors open simultaneously and (2) where required in *Sections 430-36* and *430-37* for motor overload protection.

Normal loading of a circuit is when the current flowing is within the capability of the circuit and/or the connected equipment. In figure 28-26 we have a 15-ampere circuit carrying 10 amperes.

An overload is a condition where the current flowing is more than the circuit and/or connected equipment is designed to safely carry. Figure 28-27 shows a 15-ampere circuit carrying 20 amperes. A momentary overload is harmless, but a continuous overload condition will cause the conductors and/or equipment to overheat—a potential cause for fire. The current flows through the "intended path"—the conductors and/or equipment.

A short-circuit is a condition when two or more conductors come in contact with one another, resulting in a current flow that bypasses the connected load. A short-circuit might be two "hot" conductors coming together, or it might be a "hot" conductor and a "grounded" conductor coming together. In either case, the current flows outside of the "intended path." The only resistance (impedance) is that of the conductors, the source, and the arc. This low resistance results in high levels of short-circuit current. The heat generated at the point of the arc can result in a fire, figure 28-28.

A ground fault is a condition when a "hot" conductor comes in contact with a grounded surface, such as a grounded metal raceway, metal water pipe, sheet metal, etc. See figure 28-29. The current flows outside of the "intended path." A ground fault can result in a flow of current greater than normal, in which case the overcurrent device will open. A ground fault can also result in a current flow less than normal, in which case the overcurrent device may not open.

Ground-fault circuit interrupters (GFCIs) are an example of a device that protects against very low levels of ground fault passing through the human body when a person touches a "live" wire and a grounded surface. See unit 6.

Large services and feeders require ground-fault protection for certain ampere ratings, voltage ratings, and type of connection. This is discussed in *Electrical Wiring, Commercial*, 9E © Delmar Publishers.

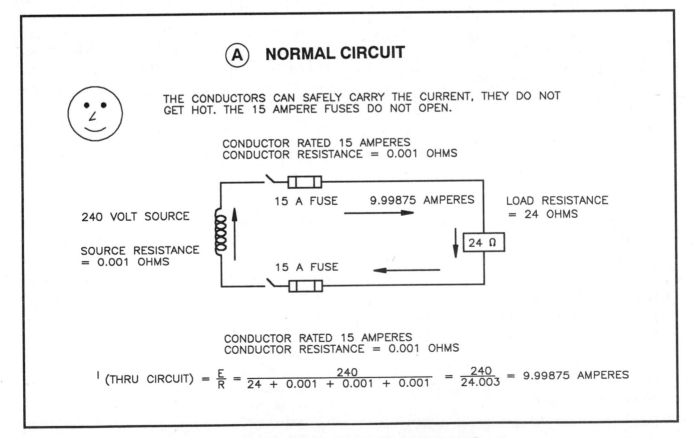

Fig. 28-26   **This is a normally loaded circuit.**

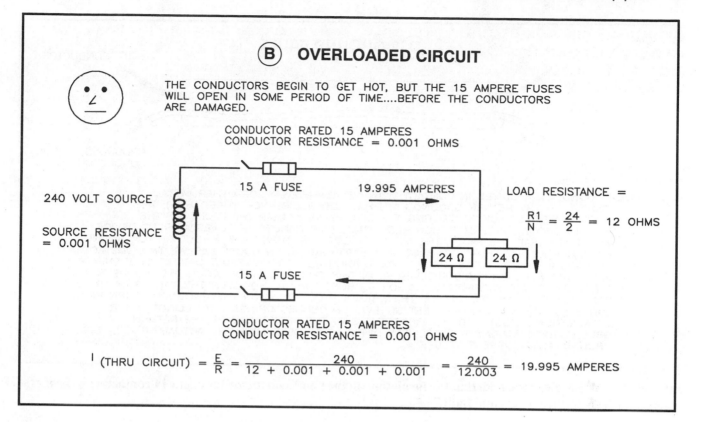

(B)   OVERLOADED CIRCUIT

THE CONDUCTORS BEGIN TO GET HOT, BUT THE 15 AMPERE FUSES WILL OPEN IN SOME PERIOD OF TIME....BEFORE THE CONDUCTORS ARE DAMAGED.

CONDUCTOR RATED 15 AMPERES
CONDUCTOR RESISTANCE = 0.001 OHMS

15 A FUSE   19.995 AMPERES   LOAD RESISTANCE =

240 VOLT SOURCE

$$\frac{R1}{N} = \frac{24}{2} = 12 \text{ OHMS}$$

SOURCE RESISTANCE = 0.001 OHMS

24 Ω   24 Ω

15 A FUSE

CONDUCTOR RATED 15 AMPERES
CONDUCTOR RESISTANCE = 0.001 OHMS

$$I \text{ (THRU CIRCUIT)} = \frac{E}{R} = \frac{240}{12 + 0.001 + 0.001 + 0.001} = \frac{240}{12.003} = 19.995 \text{ AMPERES}$$

Fig. 28-27   This is an overloaded circuit.

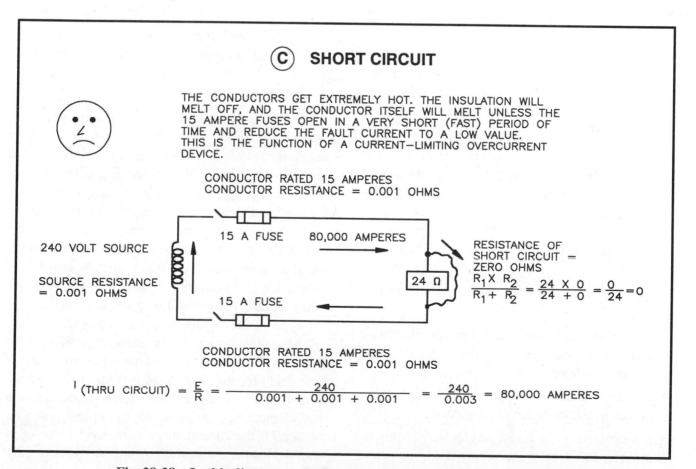

(C)   SHORT CIRCUIT

THE CONDUCTORS GET EXTREMELY HOT. THE INSULATION WILL MELT OFF, AND THE CONDUCTOR ITSELF WILL MELT UNLESS THE 15 AMPERE FUSES OPEN IN A VERY SHORT (FAST) PERIOD OF TIME AND REDUCE THE FAULT CURRENT TO A LOW VALUE. THIS IS THE FUNCTION OF A CURRENT-LIMITING OVERCURRENT DEVICE.

CONDUCTOR RATED 15 AMPERES
CONDUCTOR RESISTANCE = 0.001 OHMS

15 A FUSE   80,000 AMPERES

240 VOLT SOURCE

RESISTANCE OF SHORT CIRCUIT = ZERO OHMS

SOURCE RESISTANCE = 0.001 OHMS

24 Ω

$$\frac{R_1 \times R_2}{R_1 + R_2} = \frac{24 \times 0}{24 + 0} = \frac{0}{24} = 0$$

15 A FUSE

CONDUCTOR RATED 15 AMPERES
CONDUCTOR RESISTANCE = 0.001 OHMS

$$I \text{ (THRU CIRCUIT)} = \frac{E}{R} = \frac{240}{0.001 + 0.001 + 0.001} = \frac{240}{0.003} = 80,000 \text{ AMPERES}$$

Fig. 28-28   In this diagram, note that the connected load is short-circuited.

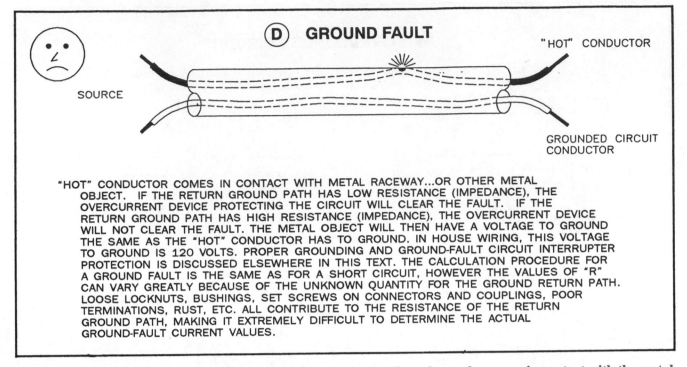

Fig. 28-29   In this diagram, note that the insulation on the "hot" conductor has come in contact with the metal conduit. This is termed a "ground fault."

Plug Fuses, Fuseholders, and Adapters
(Article 240, Part E)

Fuses are a reliable and economical form of overcurrent protection. *Sections 240-50* through *240-54* give the requirements for plug fuses, fuseholders, and adapters. These requirements include the following:

- the protective devices shall not be used in circuits exceeding 125 volts between conductors. An exception to this rule is for a system having a grounded neutral where no conductor is more than 150 volts to ground. This is the case for the 120/240-volt system used in the residence in the plans.

- the fuses shall have ampere ratings of 0 to 30 amperes.

- plug fuses shall have a hexagonal configuration somewhere on the fuse when rated at 15 amperes or less.

- the screw shell of the fuseholder must be connected to the load side of the circuit.

- Edison-base plug fuses may be used only to replace fuses in existing installations where there is no sign of overfusing or tampering.

- all new installations shall be in Type S fuses.

- Type S fuses are classified at 0 to 15 amperes, 16 to 20 amperes, and 21 to 30 amperes. The reason for this classification is given in the following paragraph.

When the electrician installs fusible equipment, the ampere rating of the various circuits must be determined. Based on this rating, an adapter of the proper size is inserted into the Edison-base fuseholder. The proper Type S fuse is then screwed into the adapter. Because of the adapter, the fuseholder is nontamperable and noninterchangeable. For example, assume that a 15-ampere adapter is inserted for the 15-ampere branch-circuits in the residence. It is impossible to substitute a Type S fuse with a larger rating without removing the 15-ampere adapter.

Type S fuses and adapters, figure 28-30, are available with ratings in the range from 3/10 of an ampere to 30 amperes. The Type S fuse may have a time-delay feature. If it does, it is known as a dual-element fuse. Momentary overloads do not cause a dual-element fuse to blow. An example of such an overload is the current surge as an electric motor is started. However, a dual-element fuse opens rapidly if there is a heavy overload or a short-circuit. Figure

28-31 shows a 15-ampere Edison-base plug fuse and a 10-ampere time-delay Edison-base plug fuse.

Dual-Element Fuse (Cartridge and Plug Type)

The most common type of dual-element cartridge fuse is shown in figure 28-32. These fuses are available in 250-volt and 600-volt sizes with ratings from 0 to 600 amperes, and have an interrupting rating of 200,000 amperes.

A dual-element fuse has two fusible elements connected in series. These elements are known as the *overload element* and the *short-circuit element*.

When an excessive current flows, one of the elements opens. The amount of excess current determines which element opens.

The electrical characteristics of these two elements are very different. Thus, a dual-element fuse has a greater range of protection than the single-element fuse.

Overload Element. The overload element opens when excessive heat is developed in the element.

Fig. 28-31   A 10-ampere Type T time-delay Edison base plug fuse and a 15-ampere Type W Edison base plug fuse. *Courtesy* **of Bussmann Division, Cooper Industries.**

TYPE S FUSE
(0-30 AMPERES)

ADAPTER

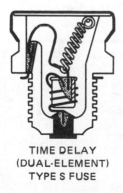

TIME DELAY
(DUAL-ELEMENT)
TYPE S FUSE

Fig. 28-30   Type S fuses and adapter. *Courtesy* **of Bussmann Division, Cooper Industries.**

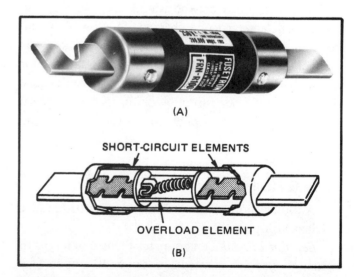

(A)

SHORT-CIRCUIT ELEMENTS

OVERLOAD ELEMENT

(B)

Fig. 28-32   Cartridge-type dual-element fuse (A) is a 250-volt, 100-ampere fuse. The cutaway view in (B) shows the internal parts of the fuse. When this type of fuse has a rejection slot in the blades of 70–600 ampere sizes, or a rejection ring in the end ferrules of the 0–60 ampere sizes, they are UL Class RK1 or RK5 fuses. Photo *courtesy* **of Bussmann Division, Cooper Industries.**

This heat may be the result of a loose connection or poor contact in the fuseholder or excessive current. When the temperature reaches 280°F, a fusible alloy melts and the element opens. However, the mass of the element absorbs a great deal of heat before the cutout opens. A small excess of current will cause this element to open if it continues for a long period of time. This characteristic gives the thermal cutout element a large time-delay on low overloads of about 10 seconds at a current of 500% of the fuse rating. In addition, it provides very accurate protection for prolonged overloads.

Short-Circuit Element. The capacity of the short-circuit element is high enough to prevent it from opening on low overloads. This element is designed to clear the circuit quickly of short-circuits or heavy overloads above 500% of the fuse rating.

Dual-element fuses are used on motor and appliance circuits where the time-lag characteristic of the fuse is required. Single-element fuses do not have this time lag. Such fuses blow as soon as an overcurrent condition occurs. The homeowner may be tempted to install a fuse with a higher ampere rating. Type S fuseholder adapters prevent the use of a higher ampere rating Type S fuse. Thus, the dual-element fuse is recommended for this type of situation.

Sometimes poor connections at the terminals or loose fuse clips can be found because the fuse tubing will appear to be charred. This problem can be dangerous and should be corrected as soon as possible. Quite often, the heat resulting from poor connections in the fuseclips or fuseholder will cause the brass or copper fuseclips or fuseholders to turn from a bright, shiny appearance to that of antique brass or copper.

Class G Fuses

Another style of small dimension cartridge fuse is Class G, figure 28-33.

Section 240-60(a) recognizes the use of a type of cartridge fuse that is rated at not over 60 amperes for circuits of 300 volts or less to ground. The physical size of this fuse is smaller than that of standard cartridge fuses. The time-delay characteristics of this fuse mean that it can handle harmless current surges or momentary overloads without blowing. However, these fuses open very quickly under short-circuit conditions.

These fuses prevent the practice of overfusing because they are size limiting for their ampere ratings. For example, a fuseholder designed to accept a 15-ampere, Class G fuse will not accept a 20-ampere, Class G fuse. In the same manner, a fuseholder designed to accept a 20-ampere, Class G fuse will not accept a 30-ampere, Class G fuse.

Class T Fuses

Another type of physically small fuse is the Class T fuse, figure 28-34. This type of fuse has a high-interrupting rating in sizes from 0 to 600 amperes in both 300-volt and 600-volt ratings. Such fuses are used as the main fuses in a panel having circuit breaker branches, figures 28-8A, 28-37, and 28-38. In this case, the Class T fuses protect the low-interrupting capacity breakers against high-level short-circuit currents.

Most manufacturers list service equipment, metering equipment, and disconnect switches that make use of Class T fuses. This allows the safe installation of this equipment where fault-currents exceed 10,000 amperes, such as on large services, and any service equipment located close to pad-mount transformers, such as are found in apartments, condominiums, shopping centers, and similar locations.

The standard ratings for fuses are given in *Section 240-6*. These ratings range from 1 ampere to 6000 amperes.

Circuit Breakers

Installations in dwellings normally use thermal-magnetic circuit breakers. On a continuous overload, a bimetallic element in such a breaker moves until it unlatches the inner tripping mechanism of

Fig. 28-33   Class G fuses. Photo *courtesy* **of Bussmann Division, Cooper Industries.**

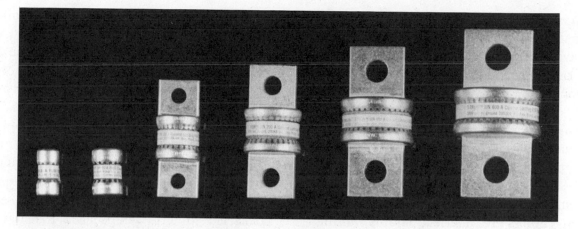

Fig. 28-34   Class T fuses. Illustrated are the 30-, 60-, 100-, 200-, 400-, and 600-ampere, 300-volt ratings. *Courtesy* of Bussmann Division, Cooper Industries.

the breaker. Momentary small overloads do not cause the element to trip the breaker. If the overload is heavy or if there is a short circuit, a magnetic coil in the breaker causes it to interrupt the branch-circuit instantly. Unit 13 covers the effect of tungsten lamp loads on circuit breakers.

Sections 240-80 through *240-83* give the requirements for circuit breakers. The following points are taken from these sections.

- Circuit breakers shall be trip free so that even if the handle is held in the ON position, the internal mechanism will trip to the OFF position.

- Breakers shall indicate clearly whether they are on or off.

- A breaker shall be nontamperable so that it cannot be readjusted (trip point changed) without dismantling the breaker or breaking the seal.

- The rating shall be durably marked on the breaker. For small breakers rated at 100 amperes or less and 600 volts or less, the rating must be molded, stamped, or etched on the handle or on another part of the breaker that will be visible after the cover of the panel is installed.

- Every breaker with an interrupting rating other than 5000 amperes shall have this rating marked on the breaker.

- Circuit breakers rated at 120 volts and 277 volts and used for fluorescent loads, shall not be used as switches unless marked "SWD."

Most circuit breakers are ambient temperature compensated. This means that the tripping point of

the breaker is not affected by an increase in the surrounding temperature. An ambient-compensated breaker has two elements. One element heats up due to the current passing through it and the heat in the surrounding area. The other element heats up because of the surrounding air only. The actions of these elements oppose each other. Thus, as the tripping element tends to lower its tripping point because of external heat, the second element opposes the tripping element and stabilizes the tripping point. As a result, the current through the tripping element is the only factor that causes the element to open the circuit. It is a good practice to turn the breaker on and off periodically to "exercise" its moving parts.

Section 240-6 gives the standard ampere ratings of circuit breakers.

INTERRUPTING RATINGS FOR FUSES AND CIRCUIT BREAKERS

Section 110-9 states that all fuses, circuit breakers, and all other electrical devices "intended to break current shall have an interrupting rating sufficient for the nominal circuit voltage, voltage to ground, and the current which is available at the line terminals of the equipment."

▶ According to *Section 110-10*, all overcurrent devices, the total circuit impedance, and the withstand capability of all circuit components (wires, contactors, and so on), must be selected so there will not be extensive damage in the event of a fault, either line-to-line or line-to-ground. ◀

Section 230-65 of the Code states that "service equipment shall be suitable for the short-circuit current available at its supply terminals."

The overcurrent protective device must be able to interrupt the current that may flow under any condition (overload or short circuit). Such interruption must be made with complete safety to personnel and without damage to the panel or switch in which the overcurrent device is installed.

> **Overcurrent devices with inadequate interrupting ratings are, in effect, bombs waiting for a short-circuit to trigger them into an explosion. Personal injury may result and serious damage will be done to the electrical equipment.**

Class G fuses have an interrupting rating of 100,000 amperes root mean square (rms) symmetrical. Plug fuses can interrupt no more than 10,000 amperes rms symmetrical. Cartridge dual-element fuses (figure 28-32) and Class T fuses (figure 28-34) have interrupting ratings of 200,000 amperes rms symmetrical. These interrupting ratings are listed by Underwriters Laboratories. The interrupting rating of a fuse is marked on its label when the rating is other than 10,000 amperes.

Short-Circuit Currents

This text does not cover in detail the methods of calculating short-circuit currents. (See *Electrical Wiring—Commercial*.) The ratings required to determine the maximum available short-circuit current delivered by a transformer are the kVA and impedance values of the transformer. The size and length of wire installed between the transformer and the overcurrent device must be considered as well.

The transformers used in modern electrical installations are efficient and have very low impedance values. A low-impedance transformer having a given kVA rating delivers more short-circuit current than a transformer with the same kVA rating and a higher impedance. When an electrical service is connected to a low-impedance transformer, the problem of available short-circuit current is very serious. *Section 230-65* gives the interrupting rating requirements for services.

The examples given in figures 28-35, 28-36, 28-37, and 28-38 show that the available short-circuit current at the transformer secondary terminals is 20,800 amperes.

Determining Short-Circuit Current

The local electric utility and the electrical inspector are good sources of information when short-circuit current is to be determined.

A simplified method is given as follows to determine the approximate available short-circuit current at the terminals of a transformer.

1. Determine the normal full-load secondary current delivered by the transformer.
   For single-phase transformers:

$$I = \frac{\text{kVA} \times 1000}{E}$$

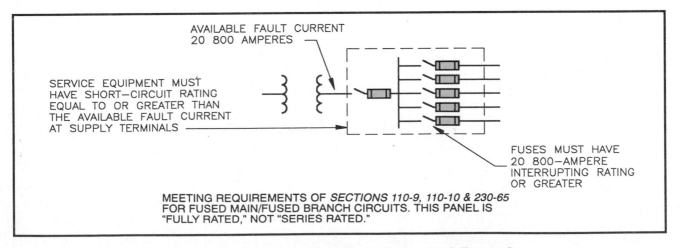

Fig. 28-35   **Fused main/fused branch circuits, fully-rated.**

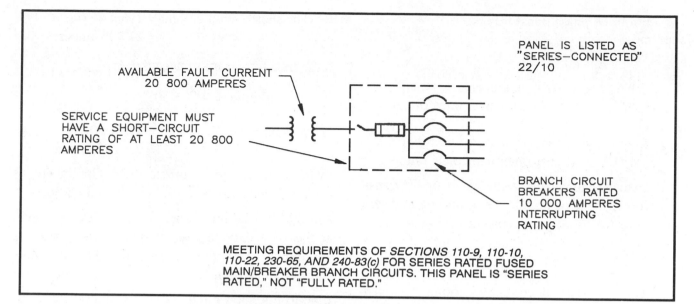

Fig. 28-36   Fused main/breaker branch circuits, series-rated.

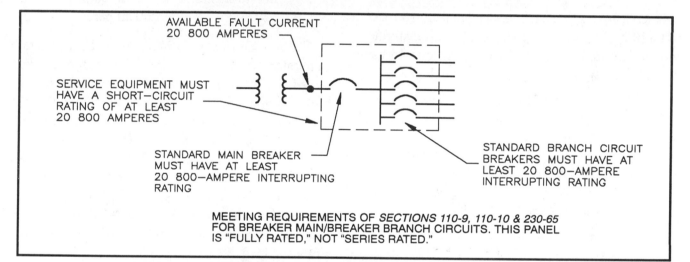

Fig. 28-37   Breaker main/breaker branch circuits, fully-rated.

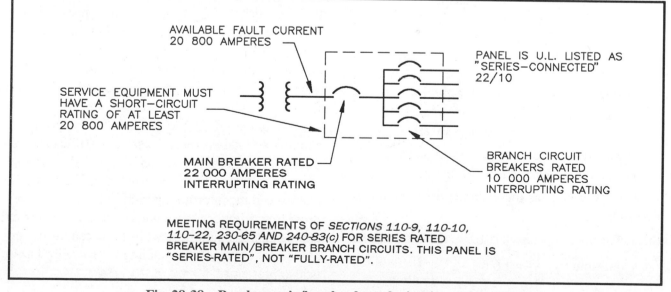

Fig. 28-38   Breaker main/breaker branch circuits, series-rated.

For three-phase transformers:

$$I = \frac{kVA \times 1000}{E \times 1.73}$$

Where I = current, in amperes
kVA = kilovolt-amperes (transformer nameplate rating)
E = secondary line-to-line voltage (transformer nameplate rating)

2. Using the impedance value given on the transformer nameplate, find the multiplier to determine the short-circuit current.

$$\text{Multiplier} = \frac{100}{\text{percent impedance}}$$

3. The short-circuit current = normal full-load secondary current × multiplier.

EXAMPLE: A transformer is rated at 100 kVA and 120/240 volts. It is a single-phase transformer with an impedance of 1% (from the transformer nameplate). Find the short-circuit current.

For a single-phase transformer:

$$I = \frac{kVA \times 1000}{E}$$

$$I = \frac{100 \times 1000}{240}$$

= 416 amperes, full-load current

The multiplier for a transformer impedance of one percent is:

$$\text{multiplier} = \frac{100}{\text{percent impedance}}$$

$$= \frac{100}{1}$$

multiplier = 100

The short-circuit current = $I \times$ multiplier
= 416 amperes × 100
= 41 600 amperes

Thus, the available short-circuit current at the terminals of the transformer is 41 600 amperes.

This value decreases as the distance from the transformer increases.

If the transformer impedance is 1.5%, the multiplier is:

$$\text{multiplier} = \frac{100}{1.5} = 66.6$$

The short-circuit current = 416 x 66.6
= 27 706 amperes

If the transformer impedance is 2%:

$$\text{multiplier} = \frac{100}{2} = 50$$

The short-circuit current = 416 x 50
= 20 800 amperes

(*Note:* A short-circuit current of 20 800 amperes is used in figures 28-35, 28-36, 28-37, and 28-38.)

CAUTION: The line-to-neutral short-circuit current at the secondary of a single-phase transformer is approximately 1½ times greater than the line-to-line short-circuit current. For the previous example,

Line-to-line short-circuit current = 20 800 amperes

Line-to-neutral short-circuit current
= 20 800 × 1.5 = 31 200 amperes

It is significant to note that Underwriters Laboratories Standard 1561 allows the marked impedance on the nameplate of a transformer to vary plus or minus (±) 10% from the actual impedance value of the transformer. Thus, for a transformer marked "2%Z," the actual impedance could be as low as 1.8%Z to as much as 2.2%Z. Therefore, the available short-circuit current at the secondary of a transformer as calculated previously should be increased by 10% if one wishes to be on the safe side when considering the interrupting rating of the fuses and breakers to be installed.

An excellent point-to-point method for calculating fault currents at various distances from the secondary of a transformer that uses different size conductors is covered in detail in *Electrical Wiring—Commercial*, 9th Edition. It is simple and accurate. The point-to-point method can be used to compute both single-phase and three-phase faults.

Calculating fault-currents is just as important as calculating load-currents. Overloads will cause conductors and equipment to run hot, will shorten their life, and will eventually destroy them. Fault currents that exceed the interrupting rating of the overcurrent protective devices can cause violent electrical explosions of equipment with the potential of serious injury to personnel standing near it. Fire hazard is also present.

PANELS AND LOAD CENTERS
Fused Main/Fused Branches (figure 28-35)
(Fully-Rated)

- must be listed and marked as suitable for use as service equipment when used as service equipment. Does not need this service equipment marking when used as a subpanel that has over-current protection ahead of the panel.

- must be sized to satisfy the required ampacity as determined by service-entrance calculations and/or local codes.

- main and branch fuses must have interrupting rating adequate for the available fault current.

- panel must have short-circuit rating adequate for the available fault current.

- for overload conditions on a branch-circuit, only the branch-circuit fuse(s) will open. All other circuits remain energized. System is selective.[1]

- for short-circuits (line-to-line or line-to-neutral) and for ground faults (line-to-ground) only the fuse protecting the faulted circuit opens. All other circuits remain energized. System is selective.[1]

Fused Main/Breaker Branches (figure 28-36)
(Series-Rated)

- must be listed and marked as suitable for use as service equipment when used as service equipment. Does not need this service equipment marking when used as a subpanel that has over-current protection ahead of the panel.

- must be marked and listed "Series-Rated."

- must be sized to satisfy the required ampacity as determined by service-entrance calculations and/or local codes.

- main fuses must have interrupting rating adequate for the available fault circuit.

- panel must have short-circuit rating adequate for the available fault current.

- main fuses must be current-limiting, such as Class T fuses, so that if a fault exceeding the branch breaker's interrupting rating occurs,

the fuse will limit the let-through fault current to a value less than the maximum interrupting rating of the branch breaker. Branch-circuit breakers generally have 10,000 amperes interrupting rating.

- for branch-circuit faults (L-N or L-G) exceeding the branch breaker's interrupting rating, one fuse opens, deenergizing one-half of the panel. System is nonselective.[1]

- for branch-circuit faults (L-L) exceeding the branch breaker's interrupting rating, both main fuses open, deenergizing the entire panel. System is nonselective.[1]

- check time-current curves of fuses and breakers and check the unlatching time of breakers to determine at what point in "time" and "current" the breakers are slower than the current-limiting fuses.

- for normal overloads and low-level faults, this system is selective.[1]

Breaker Main/Breaker Branches (figure 28-37)
(Fully-Rated)

- must be listed and marked as suitable for use as service equipment when used as service equipment. Does not need this service equipment marking when used as a subpanel that has over-current protection ahead of the panel.

- must be sized to satisfy the required ampacity as determined by service-entrance calculations and/or local codes.

- main and branch breakers must have interrupting rating adequate for the available fault current.

- panel must have short-circuit rating adequate for the available fault current.

[1] The terms and theory behind selective and nonselective systems are covered in detail in *Electrical Wiring Commercial*, 9th Edition. In all instances, it is advisable to check the time-current curves of fuses and circuit breakers to be installed. Also check the circuit breaker manufacturer's "unlatching time" data, as this will help predetermine at what values of current the system will be selective or nonselective.

- for overload conditions and low-level faults on a branch-circuit, only the branch breaker will trip off. All other circuits remain energized. System is selective.[1]

- for branch-circuit faults (short-circuits or ground faults) that exceed the instant trip setting of the main breaker, both the branch breaker *and* the main breaker trip OFF. Entire panel is deenergized. System is nonselective.[1]

- check manufacturer's data for time-current characteristic curves and unlatching times of the breakers.

Breaker Main/Breaker Branches (figure 28-38) (Series-Rated)

- must be listed and marked as suitable for use as service equipment when used as service equipment. Does not need this service equipment marking when used as a subpanel that has overcurrent protection ahead of the panel.

- must be sized to satisfy the required ampacity as determined by service-entrance calculations and/or local codes.

- main breaker will have interrupting rating higher than the branch breaker. Example: main C/B has 22,000-ampere interrupting rating, whereas branch C/B has 10,000-ampere interrupting rating.

- must be marked and listed "Series-Rated."

- panel must be listed and marked with its maximum fault current rating.

- for branch-circuit faults (short-circuits or ground faults) that exceed the instant trip setting of the main breaker, both the branch breaker *and* main breaker trip OFF. Entire panel is deenergized. System is nonselective.[1]

- for overload conditions and low-level faults on a branch, only the branch breaker will trip. System is selective.[1]

- obtain time-current curves and unlatching time data from the manufacturer.

- the panel's series rating is attained because when a heavy fault occurs, both main and branch breakers open. The two arcs in series add impedance; thus the fault current is reduced to a level less than the 10,000 A.I.C. branch breaker.

- when panels containing low interrupting rating circuit breakers are separated from the equipment in which the higher interrupting rating circuit breakers protecting the panel are installed, the system becomes a field-installed "Series-rated System." This panel must be field marked by the installer electrician, and this marking must be readily visible, *Section 110-22*. This marking calls attention to the fact that the actual available fault current exceeds the interrupting rating of the breakers in the panel, and that the panel is protected by properly selected circuit breakers back at the main equipment. Also see *Section 240-83(c)*.

CAUTION—SERIES-RATED SYSTEM

_____ AMPERES AVAILABLE

IDENTIFIED REPLACEMENT COMPONENT REQUIRED

- CAUTION: The above field labeling is required for any series rated systems—fuse/breaker, or breaker/breaker. Closely check the manufacturer's catalog numbers to be sure that the circuit breaker combination you are intending to install "in series" is listed by Underwriters Laboratories for that purpose. This is extremely important because there are many circuit breakers that have the same physical size and dimensions, yet have different interrupting and voltage ratings, and have not been tested as "series-rated" devices. To install these as a series combination constitutes a violation of *Section 110-3(b)* and could be dangerous.

REVIEW

Note: Refer to the Code or the plans where necessary.

1. Where does an overhead service start and end? _____

   _____

   _____

   _____

2. What are service-drop conductors? _____

   _____

   _____

   _____

3. Who is responsible for determining the service location? _____

4. a. The service head must be located (above) (below) the point where the service-drop conductors are spliced to the service-entrance conductors. Circle one.

   b. What Code section provides the answer to part (a)? _____

5. What is a mast-type service entrance? _____

   _____

   _____

   _____

   _____

   _____

   _____

6. a. What size and type of conductors are installed for this service? _____

   _____

   b. What size conduit is installed? _____

   _____

   c. What size grounding electrode conductor is installed? (not neutral) _____

   d. Is the grounding electrode conductor insulated, armored, or bare? _____

7. How and where is the grounding electrode conductor attached to the water pipe?

   _____

   _____

8. When a conduit is extended through a roof, must it be guyed? _____

   _____

   _____

9. What are the minimum distances or clearances for the following?

   a. Service-drop clearance over private driveway _____

   b. Service-drop clearance over private sidewalks _____

   c. Service-drop clearance over alleys _____

   d. Service-drop clearance over a roof having a roof pitch of not less than 4/12. (Voltage between conductors does not exceed 300 volts.) _____

      _____

   e. Service-drop horizontal clearance from a porch _____

   f. Service-drop clearance from a fence that can be climbed _____

10. What size ungrounded conductors are installed for each of the following residential services? (Use Type THWN copper conductors.) See *Note 3* to *Table 310-16*.

    a. 60-ampere service   No. _____ THWN copper

    b. 100-ampere service   No. _____ THWN copper

    c. 200-ampere service   No. _____ THWN copper

11. What size grounding electrode conductors are installed for the services listed in question 10? *(See Table 250-94.)*

    a. 60-ampere service   No. _____ AWG grounding conductor

    b. 100-ampere service   No. _____ AWG grounding conductor

    c. 200-ampere service   No. _____ AWG grounding conductor

12. What is the recommended height of a meter socket from the ground? _____

    _____

    _____

13. a. May the bare grounded neutral conductor of a service be buried directly in the ground? _____

    b. What section of the Code covers this? _____

14. What exceptions are made regarding the use of bare neutral conductors installed underground? _____

    _____

    _____

    _____

    _____

15. How far must mechanical protection be provided when underground service conductors are carried up a pole? _____

16. a. A method of service disconnect may consist of how many switches or circuit breakers? _____

    b. Must these devices be in one enclosure? _____

    c. What type of main disconnect is provided in this residence? _____

    _____

17. Complete the following table by filling in the columns with the appropriate information.

	CIRCUIT NUMBER	AMPERE RATING	POLES	VOLTS	WIRE SIZE
A. LIVING ROOM RECEPTACLE OUTLETS					
B. WORKBENCH RECEPTACLE OUTLETS					
C. WATER PUMP					
D. ATTIC EXHAUST FAN					
E. KITCHEN LIGHTING					
F. HYDROMASSAGE TUB					
G. ATTIC LIGHTING					
H. COUNTER—MOUNTED COOKING UNIT					
I. ELECTRIC FURNACE					

18. a. What size conductors supply Panel B? _____

    b. What size conduit? _____

    c. Is this conduit run in the form of electrical metallic tubing or rigid conduit?

    _____

    d. What size overcurrent device protects the feeders to Panel B? _____

19. How many electric meters are provided for this residence? _____

20. a. According to the Code, is it permissible to ground rural service-entrance systems and equipment to driven ground rods only when a metallic water system is available? _____

    b. What section of the Code applies? _____

21. What table of the Code lists the sizes of grounding electrode conductors to be used for service entrances of various sizes similar to the type found in this residence?

_____

22. *Section 250-81* requires that a supplemental ground be provided if the available grounding electrode is: (Circle correct answer.)

    a. water pipe

    b. building steel

    c. concrete encases ground

23. Do the following conductors require mechanical protection?

    a. No. 8 grounding electrode conductor _____

    b. No. 6 grounding electrode conductor _____

    c. No. 4 grounding electrode conductor _____

24. Why is bonding of service-entrance equipment necessary? _____

_____

_____

_____

_____

_____

_____

_____

_____

25. What special types of bushings are required on service entrances? _____

_____

26. When No. 4 AWG conductors or larger are installed in conduit, what additional provision is required on the conduit ends? _____

_____

_____

_____

_____

27. What minimum size copper bonding jumpers must be installed to bond properly the
electrical service for the residence discussed in this text? _____

_____

_____

_____

_____

28. a. What is a Type S fuse? _____

        _____

    b. Where must Type S fuses be installed? _____

        _____

29. a. What is the maximum voltage permitted between conductors when using plug fuses?

        _____

    b. May plug fuses (Type S) be installed in a switch that disconnects a 120/240-volt
clothes dryer? _____

    c. Give a reason for the answer to (b). _____

        _____

30. Will a 20-ampere, Type S fuse fit properly into a 15-ampere adapter? _____

31. What part of a circuit breaker causes the breaker to trip

    a. on an overload? _____   b. on a short circuit? _____

32. What is meant by an ambient-compensated circuit breaker? _____

_____

_____

_____

_____

_____

_____

_____

_____

33. List the standard sizes of fuses and circuit breakers up to and including 100 amperes.

_____

34. Using the method shown in this unit, what is the approximate short-circuit current available at the terminals of a 50-kVA single-phase transformer rated 120/240 volts? The transformer impedance is 1%.

    a. line-to-line?

    b. line-to-neutral?

35. Where is the main panel for this residence located? _____

36. a. On what type of wall is Panel A fastened? _____

    b. On what type of wall is Panel B fastened? _____

37. State four possible combinations of service equipment which meet the requirements of *Section 230-65* of the Code.

    a. _____

_____

_____

    b. _____

_____

_____

    c. _____

_____

_____

    d. _____

_____

_____

38. When conduits pass through the wall from outside to inside, the conduit must be _____ to prevent air circulation through the conduit.

39. Briefly explain why electrical systems and equipment are grounded. _____

_____

_____

_____

_____

_____

40. What Code section states that all overcurrent devices must have adequate interrupting ratings for the current to be interrupted? _____

_____

41. All electrical components have some sort of "withstand rating." This rating indicates the ability of the component to withstand fault currents for the time required by the overcurrent device to open the circuit. What Code section refers to withstand ratings with reference to overcurrent protection? _____

42. Arcing fault damage is closely related to the value of _____

43. In general, systems are grounded so that the maximum voltage to ground does not exceed (Circle one)
   a. 120 volts.   b. 150 volts.   c. 300 volts.

44. To insure a complete grounding electrode system (Circle one)
   a. everything must be bonded together.
   b. all metal pipes and conduits must be isolated from one another.
   c. the service neutral is grounded to the water pipe only.

45. An electric clothes dryer is rated at 5700 watts. The electric rate is 10.091 cents per kilowatt-hour. The dryer is used continuously for 3 hours. Find the cost of operation, assuming the heating element is on continuously.

46. A heating cable rated at 750 watts is used continuously for 72 hours to prevent snow from freezing in the gutters of the house. The electric rate is 8.907 cents per kilowatt-hour. Find the cost of operation.

47. *Section 230-65* of the Code requires that the service equipment (breakers, fuses, and the panel itself) be rated equal to or greater than _____

_____

48. The utility company has provided a letter to the contractor stating that the available fault current at the line side of the main service-entrance equipment in a residence is 17,000 amperes RMS symmetrical, line-to-line. In the space following each statement, write in "Meets Code" or "Violation" of *Section 230-65* of the Code.

   a. Main breaker 10,000-ampere interrupting rating; branch breakers 10,000-ampere interrupting rating. _____

   b. Main current-limiting fuse having a 200,000-ampere interrupting rating; branch breakers having 10,000-ampere interrupting rating. The panel is "Series-Rated."

   _____

   c. Main breaker has 22,000-ampere interrupting rating; branch breakers have 10,000-ampere interrupting rating. The panel is marked "Series-Connected."

   _____

49. While working on the main panel with the panel energized, the electrician inadvertently causes a direct short circuit (line-to-ground) on one of the branch-circuit breakers. The available fault current at the main service equipment is rather high. The panel is labeled "Series-Connected." The main breaker is rated 100 amperes. What will happen?

	TRUE	FALSE
a. Only the branch breaker will trip off.	_____	_____
b. The branch breaker will not trip off.	_____	_____
c. The main breaker and the branch breaker will probably both trip off, resulting in a total power outage.	_____	_____
d. All of the 120-volt connected loads in the panel will lose power.	_____	_____
e. All of the 240-volt connected loads and one-half of the 120-volt connected loads will lose power. There will not be a total power outage.	_____	_____
f. One fuse will open; therefore, all 240-volt connected loads and one-half of the 120-volt connected loads will lose power. There will not be a total power outage.	_____	_____

50. Repeat question 49 on a main service panel that consists of 100-ampere main current-limiting fuses and breakers for the branch-circuits. (Mark TRUE or FALSE.)

   a. _____

   b. _____

   c. _____

   d. _____

   e. _____

   f. _____

51. Here are five commonly used terms in the electrical industry. Enter the letter of the term before its definition.

    a. Grounding electrode conductor

    b. Main bonding jumper

    c. Grounded circuit conductor

    d. Equipment grounding conductor

    e. Lateral service conductors

    _____ This is the neutral conductor.

    _____ This is the term used to define underground service-entrance conductors that run between the meter and the utilities connection.

    _____ This is the conductor (sometimes a large threaded screw) that connects the neutral bar in the service equipment to the service-entrance enclosure.

    _____ This is the conductor that runs between the neutral bar in the main service equipment to the grounding electrode (water pipe, ground rod, etc.).

    _____ This is the bare copper conductor found in nonmetallic-sheathed cable.

52. The electrician mounted the disconnect switch for a central air-conditioning unit 8 feet above the ground. This was easy to do because all he had to do was run the conduit across the ceiling joists in the basement, through the outside wall, and directly into the back of the disconnect switch. The electrical inspector turned the job down, citing *Section* _____ of the *NEC*® that requires that the disconnecting operating handle must not be higher than _____ feet above the ground.

53. *Section* _____ of the *NEC*® prohibits the connecting of fuses and/or circuit breakers into the grounded circuit conductor.

54. In general, overcurrent protective devices are inserted where the conductor receives its

    _____ .

55. According to the *NEC*® definition of a wet location, a basement cement wall that is in direct contact with the earth is considered to be a wet location. A panelboard mounted on this wall must have at least (¼ inch) (½ inch) (1 inch) space between the wall and the panel. Underline the correct answer.

56. When using re-bars as the concrete encased electrode, the re-bars must be (a) insulated with plastic material so they will not rust, or (b) electrically conductive so as to provide excellent conductivity. Underline the correct answer.

57. The circuit directory in a panelboard (may) (shall) be filled out according to *Section* _____ of the *National Electrical Code*® Circle the correct answer, and enter the correct Section number.

58. A main service panel is located in a dark corner of a basement, far from the basement light. In your opinion, does this installation meet the requirements of *Section 110-16(e)*?

    _____ .

59. a. Does the *NEC*® permit panelboards to be installed in clothes closets? _____

    b. Does the *NEC*® permit panelboards to be installed in bathrooms? _____

60. Does the electric utility in your area allow the location of the meter to be on the back side of a residence? _____

61. The term used when the utility charges different rates during different periods during the day is _____

62. A ground rod is driven below the meter outside of the house. An armored grounding electrode conductor connects between the meter base and the ground rod. The ground clamp is buried under the surface of the soil. This is permitted if the ground clamp is _____ for direct burial according to *Section* _____ .

63. Copper ground rods (grounding electrode) are the most commonly used. These rods must:

    a. be at least _____ inch in diameter.

    b. be driven to a depth of at least _____ feet unless solid rock is encountered, in which case the rod may be driven at a _____ degree angle, or it may be laid in a trench that is at least _____ feet deep.

    c. be separated by at least _____ feet when more than one rod is driven.

    d. have a ground resistance of not over _____ ohms for one rod.

64. What section of the Code prohibits the use of sheet metal screws as a means of attaching grounding conductors to enclosures? *Section* _____

UNIT 29

Service-Entrance Calculations

OBJECTIVES

After studying this unit, the student will be able to

- determine the total calculated load of the residence.
- calculate the size of the service entrance, including the size of the neutral conductors.
- understand the Code requirements for services, *Article 230*.
- understand how to read a watt-hour meter.
- fully understand special Code rules for single-family dwelling service-entrance conductor sizing.
- derate service-entrance conductor if installation is located in extremely hot climate.
- do an optional calculation for computing the required size of service-entrance conductors for residence.

The various load values determined in earlier units of this text are now used to illustrate the proper method of determining the size of the service-entrance conductors for the residence. The calculations are based on *National Electrical Code®* requirements. The student must check local and state electrical codes for any variations in requirements that may take precedence over the *National Electrical Code®*

SIZE OF SERVICE-ENTRANCE CONDUCTORS AND SERVICE DISCONNECTING MEANS

The size and rating of service-entrance conductors are covered in *Section 230-42. Section 230-42(a)* states that all computations shall be done according to *Article 220. Section 230-42(b)(1)* and *(2)* state that the service-entrance conductors shall not be smaller than:

- 100 ampere, three-wire, for a single-family dwelling with six or more two-wire circuits.

- 100 ampere, three-wire, for a single-family dwelling with an initial computed load of 10 kVA or more.

The sizing of the service disconnecting means is covered in *Section 230-79(c)*, which repeats the previous requirements.

An electric range has a rating of at least 8 kW, *Table 220-19*. Add to this two small appliance circuits, a laundry circuit, a clothes dryer, and a number of lighting circuits. It is obvious that there are few, if any, single-family dwellings for which a service smaller than 100 amperes can be installed.

Two methods for calculating services on homes are permitted by the *National Electrical Code®* Method 1 is outlined in *Article 220, Parts A* and *B*. Method 2 is given in *Article 220, Part C*.

Method 1—*Article 220, Parts A and B*

All of the volt-ampere values in the **Single-family Dwelling Service-Entrance Calculations** were taken from previous units of this text.

Single-Family Dwelling Service-Entrance Calculations

1. General Lighting Load *(Section 220-3(b))*

3,232 sq. ft. @ 3 VA per sq. ft. = 9,696 VA

2. Minimum Number of Lighting Branch-Circuits

9,696 VA ÷ 120 volts = 80.8 amperes

then, $\dfrac{amperes}{15} = \dfrac{80.8}{15} = 5.387$ (round up to 6) 15-ampere branch-circuits

3. Small Appliance Load *(Sections 220-4(b) & 220-16(a))*
(Minimum of two 20-ampere branch-circuits)

Kitchen	3
Clothes Washer	1*
Workshop	2**

6   branch-circuits @ 1,500 VA each = 9,000 VA

4. Laundry Branch-Circuit *(Sections 220-4(c) & 220-16(b))*
(Minimum of one 20-ampere branch-circuit)

1   branch-circuit(s) @ 1,500 VA = 1,500 VA

5. Total General Lighting, Small Appliance, and Laundry Load

Lines 1 + 3 + 4 = 20,196 VA

6. Net computed General Lighting, Small Appliance, and Laundry Loads (less ranges, ovens, and "fastened-in-place" appliances). Apply Demand Factors from *Table 220-11.*

a. First 3,000 VA @ 100%   = 3,000 VA

b. Line 5 20,196 – 3,000 = 17,196 @ 35% = <u>6,019 VA</u>

Total a + b = 9,019 VA

7. Electric Range, Wall-Mounted Ovens, Counter-Mounted Cooking Units *(Table 220-19)*

Wall-mounted oven	7,450 VA
Counter-mounted cooking unit	<u>6,600 VA</u>
Total	14,050 VA (14 kW)

14 kW exceeds 12 kW by 2 kW

2 kW × 5% = 10% increase, therefore: 8 kW + 0.8 kW = 8.8 kW = 8,800 VA

---

* This is in addition to the required minimum of one laundry branch-circuit. Thus, the laundry is served by a 20-ampere branch-circuit B18 for the automatic washer and another 20-ampere branch-circuit B20 supplying two receptacles for plugging in an electric iron, steamer, or clothes press.

** Although not true small appliance circuits as defined by the *NEC*, it is better to add these circuits in as 1,500 VA small appliance circuits rather than including them in the general lighting load calculations. This results in a more adequate load calculation.

8. **Electric Dryer** *(Table 220-18)* = 5,700 VA

9. **Electric Furnace** *(Section 220-15),*
   Air Conditioner, Heat Pump *(Article 440)*
      Air conditioner: $30 \times 240 = 7{,}200$ VA
      Electric Furnace: 13,000 VA
      (Enter largest value, *Section 220-21*) = 13,000 VA

10. **Net Computed General Lighting, Small Appliance, Laundry,**
    Ranges, Ovens, Cook Tops Units, HVAC.
       Lines 6 + 7 + 8 + 9 = 36,519 VA

11. **List "Fastened-in-Place" Appliances** *in addition* **to**
    Electric Ranges, Air Conditioners, Clothes Dryers, Space Heaters.

Appliance	VA Load
Water Heater	3,000 VA
Dishwasher*	1,000 VA

   * Motor: $7.2 \times 120 = 864$ VA
     Heater: 1,000 VA (watts)
     Enter largest value, *Section 220-221.*

Garage Door Opener $5.8 \times 120 =$	696 VA
Food Waste Disposer $7.2 \times 120 =$	864 VA
Water Pump 8×240 =	1,920 VA
Hydromassage Tub 10×120 =	1,200 VA
Attic exhaust fan 5.8×120 =	696 VA
Heat/Vent/Lights $1,500 \times 2$ =	3,000 VA
Freezer 5.8×120 =	696 VA
Total	13,072 VA

12. **Apply 75% Demand Factor** *(Section 220-17)* **if Four or More "Fastened-in-Place" Appliances.**
    If Less than Four, Figure at 100%. ▶ Do not include electric ranges, clothes dryers, space heat-
    ing, or air conditioning equipment. ◀
       Line 11 $13{,}072 \times 0.75 = 9{,}804$ VA

13. **Total Computed Load (Lighting, Small Appliance, Ranges, Dryers, HVAC, "Fastened-in-**
    Place" Appliances)
       Line 10 36,519 + Line 12 9,804 = 46,323 VA

14. **Add 25% of Largest Motor** *(Section 220-14)*
       This is the water pump motor:
          $8 \times 240 \times 0.25 = 480$ VA

15. **Grand Total Line 13 + Line 14** = 46,803 VA

16. Minimum Ampacity for Ungrounded Service-Entrance Conductors

$$\text{Amperes} = \frac{\text{Line } 15}{240} = \frac{46,803}{240} = 195 \text{ amperes}$$

17. Minimum Ampacity for Neutral Service-Entrance Conductor, *Section 220-22, Note 3* **to** *Table 310-16.* **Do Not Include Straight 240-Volt Loads.**

    a. Line 6   9,019 VA

    b. Line 7 $8,800 \times 0.70$ = 6,160 VA

    c. Line 8 $5,700 \times 0.70$ = 3,990 VA

    d. Line 11 (include only 120-volt loads)

Freezer	696 VA	
Food Waste Disposer	864 VA	
Garage Door Opener	696 VA	
Heat/Vent/Light	3,000 VA	
Attic Fan	696 VA	
Hydromassage tub	1,200 VA	
Dishwasher heater	1,000 VA	
Total	8,152 VA	

    e. Line d total @ 75% Demand Factor: $8,152 \times 0.75 =$ 6,114 VA

    f. Add 25% of largest 120-volt motor
       This is food waste disposer:
       $7.2 \times 120 \times 0.25$ =   216 VA

    g. Total a + b + c + e + f   = 25,499 VA

$$\text{Amperes} = \frac{\text{volt} - \text{amperes}}{\text{volts}} = \frac{25,499}{240} = 106.245 \text{ amperes}$$

18. Ungrounded Conductor Size (copper) *(Note 3, Table 310-16)*   2/0THW

Note: This could be a No. 3/0 THW, THHW, THWN, XHHW, or THHN per *Table 310-16*, or a No. 2/0 (same types) using *Note 3* to *Table 310-16*. ► *Note 3*, following *Table 310-16* may only be used for 120/240-volt, 3-wire, single-phase service-entrance conductors, service lateral conductors, and feeder conductors that serve as the main power feeder to a dwelling unit. ◄ This table *cannot* be applied to the feeder to Panel B because that feeder carries only part of the load in the residence.

19. Neutral Conductor Size (copper) *(Section 220-22)*   1

Note: The previous calculations indicate that a No. 2 AWG copper neutral conductor would be adequate based upon the unbalanced load calculations. ► However, *Note 3* of *Table 310-16* states that the neutral conductor is permitted to be smaller than the ungrounded "hot" conductors if the requirements of *Sections 215-2, 220-22,* and *230-42* are met. ◄ The Specifications for this residence specify that a No. 1 neutral conductor be installed.

20. Grounding Electrode Conductor Size (copper) *(Table 250-94)*   4

21. Conduit Size (See fig. 28-13 for calculations)   1¼ inch

Table 29-1 replicates *Note 3* to *Table 310-16* in the *NEC*.® The current ratings for a given size wire are higher than the allowable ampacities listed in *Table 310-16* because of the tremendous diversity of electricity use in homes.

Be careful. Locations exposed to the weather are considered to be wet locations, per the definition in the *NEC*.® Therefore, most electrical inspectors will require that service-entrance conductors, from the line side terminals of the service disconnect to the service head, be suitable for wet locations. This would mean that the conductors must have a "W" designation in their type lettering.

High Temperatures

In certain parts of the country, such as the southwestern desert climates where extremely hot temperatures are common, the "authority having jurisdiction" may require that service-entrance conductors and any other conductors exposed to direct sunlight be corrected (derated) according to the "Ampacity Correction Factors" found below *Table 310-16*. See figure 29-1.

EXAMPLE: If the U.S. weather bureau lists the average summer temperature as 113°F (45°C), then a correction factor of 0.82 must be applied. For instance, a 3/0 XHHW copper conductor per *Table 310-16* is 200 amperes at 86°F (30°C). At 113°F (45°C) the conductor's ampacity is

$$200 \times 0.82 = 164 \text{ amperes}$$

A properly sized conductor capable of carrying 200 amperes safely requires

$$\frac{200}{0.82} = 243.9 \text{ amperes}$$

Therefore, according to *Table 310-16* and the applied correction factor a 250 kcmil Type XHHW copper conductor is required.

Check this out with the electrical inspection department where this situation might be encountered.

Neutral Conductor

Why is the neutral conductor permitted to be smaller than the "hot" ungrounded conductors?

The neutral conductor must be able to carry the computed maximum neutral current, *Section 220-22*. Straight 240-volt loads are not included in

▶ SPECIAL AMPACITY RATINGS FOR RESIDENTIAL 120/240-VOLT, 3-WIRE, SINGLE-PHASE SERVICE-ENTRANCE CONDUCTORS, SERVICE LATERAL CONDUCTORS, AND FEEDER CONDUCTORS THAT SERVE AS THE MAIN POWER FEEDER TO A DWELLING UNIT. ◀

COPPER CONDUCTOR (AWG) FOR INSULATION FOR RH-RHH-RHW-THW-THWN-THHW-THHN-XHHW USE	ALUMINUM OR COPPER-CLAD ALUMINUM CONDUCTORS (AWG)	ALLOWABLE SPECIAL AMPACITY
4	2	100
3	1	110
2	1/0	125
1	2/0	150
1/0	3/0	175
2/0	4/0	200
3/0	250 kcmil	225
4/0	300 kcmil	250
250 kcmil	350 kcmil	300
350 kcmil	500 kcmil	350
400 kcmil	600 kcmil	400

▶ **Table 29-1   This table is similar to the table in *Note 3* to *Table 310-16*, and is *only* for residential 120/240-volt, 3-wire, single-phase service-entrance conductors, service lateral conductors, and residential feeder conductors that carry the main power to the dwelling. This table is based on the great diversity of electrical use in homes. In the residence discussed in this text, this table cannot be used to determine the feeder size to Panel B because the feeder to Panel B does *not* carry the main power to the home. The neutral conductor may be smaller than the phase conductors when maximum unbalanced load calculations prove that the requirements of *Section 215-1, 220-22*, and *240-42* have been met. Reduced neutrals require 240-volt loads, such as electric ranges, dryers, water heaters, central air-conditioners, or heat pumps, where none of or only some of the load current passes through the neutral conductor. ◀**

the calculation of neutral conductors. The previous calculations resulted in a computed load of 195 amperes for the ungrounded conductors, and 106 amperes for the neutral conductor. Referring to *Table 310-16*, this could be two No. 3/0 THW, THHW, THWN, XHHW, or THHN copper conductors and one No. 2 neutral conductor. The neutral conductor may be insulated or bare per *Sections 230-30* and *230-41*.

▶ *Note 3* following *Table 310-16* provides special consideration for 120/240-volt, 3-wire, single-phase service-entrance conductors, service lateral conductors, and feeder conductors that serve as the main

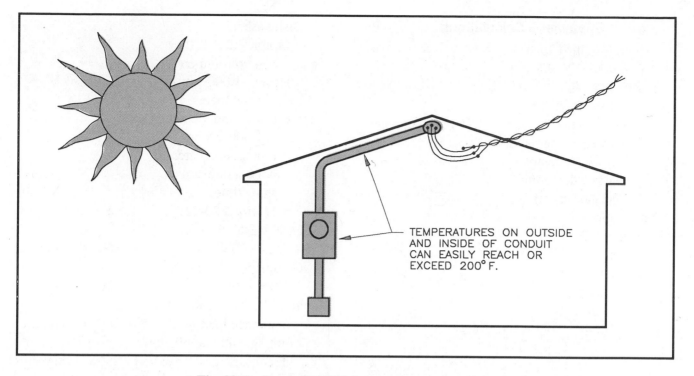

Fig. 29-1   Example of high-temperature location.

power feeder to a dwelling unit. For a 200-ampere residential service, No. 2/0 THW, THHW, THWN, XHHW, or THHN copper conductors are adequate. See Table 29-2.

Note 3 also permits the neutral conductor to be smaller than the "hot" ungrounded conductors, but only if it can be verified by calculations that the requirements of *Sections 215-2, 220-22,* and *230-42* are met. ◀ We verified this in our previous calculations for the residence discussed in this text.

The Specifications for this residence call for two No. 2/0 ungrounded "hot" conductors and one No. 1 neutral conductor.

The service neutral conductor serves another purpose. In addition to carrying the unbalanced current, the neutral conductor must also be capable of safely carrying any fault current that it might be called upon to carry, such as a line-to-ground fault in the service equipment. Sizing the neutral conductor only for the normal neutral unbalance current could result in a neutral conductor too small to safely carry fault currents. The neutral conductor might burn off, causing serious voltage problems, damage to equipment, and the creation of fire and shock hazards.

In sizing neutral conductors for dwellings, this generally is not a problem. But installing reduced size neutrals on commercial and industrial services can present a major problem. Refer to *Sections 230-23, 230-31, 230-42(c), 250-1, 250-23(b), 250-51, 250-70,* and *250-75.* This topic is covered in *Electrical Wiring, Commercial,* 9E © Delmar Publishers.

SERVICE-ENTRANCE CONDUCTOR SIZE						
	COPPER			ALUMINUM		
SIZE (AMPERES)	PHASE (HOT) CONDUCTORS	NEUTRAL CONDUCTOR	CONDUIT SIZE (INCHES)	PHASE (HOT) CONDUCTORS	NEUTRAL CONDUCTOR	CONDUIT SIZE (INCHES)
100	#2 AWG	#4 AWG	1 1/2	#1 AWG	#2 AWG	1 1/2
125	#1 AWG	#4 AWG	1 1/2	00 AWG	#1 AWG	1 1/2
150	0 AWG	#4 AWG	1 1/2	000 AWG	0 AWG	2
175	00 AWG	#2 AWG	2	0000 AWG	00 AWG	2
200	000 AWG	#2 AWG	2	250 kcmil	000 AWG	2

Table 29-2   Table of service-entrance conductor sizing used in some cities so that service-entrance calculations do not have to be made for each service-entrance installation.

Subpanel B Calculations

General lighting load
(Sections 220-3b,
220-11)
1,420 sq. ft. @ 3 volt-
amperes per sq. ft. =   4,260 volt-amperes
(kitchen, living room,
laundry, rear entry
hall, powder room,
recreation room)
Small appliance load
(Section 220-16)
Kitchen   3
Laundry   1
Automatic
washer   1
Total   5 @ 1,500 VA
per
circuit =   7,500 volt-amperes
Total general lighting and
small appliance load = 11,760 volt-amperes
Application of demand
factors (Table 220-11)
3,000 volt-amperes @
100% =   3,000 volt-amperes
11,760 – 3,000 = 8,760
@ 35% =   3,066 volt-amperes

A Net computed load (less
range and "fastened-in-
place" appliances)   6,066 volt-amperes

Wall-mounted oven and
counter-mounted range:
Table 220-19, Note 4:
Wall-mounted oven   7,450 volt-amperes
Counter-mounted range   6,600 volt-amperes
Total   14,050 volt-amperes

14 kW exceeds 12 kW by 2 kW:
2 kW × 5% = 10%
increase
8 kW × 0.10 =
0.8 kW
(Column A, Table 220-19)

B 8 + 0.8 = 8.8 kW =   8,800 volt-amperes

C Net computed load (with
range) = A + B = C   14,866 volt-amperes

Dryer (Section 220-18)   5,700 volt-amperes

Dishwasher
Motor: 7.2 × 120 =
864 volt-amperes
Heater: 1,000 watts
(volt-amperes)
(maximum demand
is 1,000 VA because
motor and heater do
not operate at the
same time,
Section 220-21)   1,000 volt-amperes
Food waste disposer
7.2 × 120 =   864 volt-amperes
Garage door opener
5.8 × 120 =   696 volt-amperes
Total "fastened-in-place"
appliance load   8,260 volt-amperes
Lighting, small appliance
circuits, ranges, dryer = 14,866 volt-amperes

D Add 25% of largest motor.
7.2 × 120 × 0.25 =   216 volt-amperes

E Net computed load
(lighting, small appli-
ance circuits, ranges,
dryer, dishwasher, food
waste disposer, garage
door opener)   23,342 volt-amperes

$$\text{Amperes} = \frac{\text{volt-amperes}}{\text{volts}} = \frac{23,342}{240} = 97.3 \text{ amperes}$$

The feeder conductors supplying Panel B could be No. 3 THHW, THW, THHN, or XHHW per *Table 310-16*. We have not shown the calculations for the neutral to Panel B. The specifications and also figure 28-1A require that three No. 3 THHN feeder conductors supply Panel B.

Method 2 (Optional Calculations — Dwellings)—*Article 220, Part C*

A second method for determining the load for a one-family dwelling is given in *Section 220-30*. This method simplifies the calculations, but may be used only when the service-entrance conductors have an ampacity of 100 amperes or more. In most cases, the service-entrance conductors are smaller than those permitted by *Article 220, Parts A* and *B*.

Let's take a look at *Part C* of *Article 220*. This is an alternative method of computing service loads and feeder loads. It is referred to as the *optional method*.

Section 220-30(a) tells us that we are permitted to calculate service-entrance conductors and feeder conductors (both phase conductors and the neutral conductor) using *Table 220-30*. *Section 220-30* addresses single-family dwellings. *Section 220-32* addresses multifamily dwellings.

Section 220-30(b) tells us to

1. include 1500 volt-amperes for each 20-ampere small appliance circuit.

2. include 3 volt-amperes per square foot for lighting and general-use receptacles.

3. include the nameplate rating of appliances:

   - that are fastened in place
   - that are permanently connected
   - that are located to be connected to a specific circuit

   Appliances listed in *Section 220-30(b)(3)* are ranges, wall-mounted ovens, counter-mounted cooking units, clothes dryers, and water heaters.

4. include nameplate ampere rating or kVA for motors and all low-power-factor loads. The intent of the reference to low power factor is to address such loads as low-cost, low-power-factor fluorescent ballasts of the type that might be used in the recessed lay-in fluorescent fixtures in the recreation room. It is always recommended that high-power-factor ballasts be installed.

The first sentence of *Table 220-30* tells us to select the largest load from a list of five types of load possibilities. Note that the list of five possibilities references air conditioners, heat pumps, central electric heat (furnaces), and separately controlled electric space heaters, such as electric baseboard heating units.

So let's begin our optional calculation for the residence discussed in this text. The residence has an air conditioner and an electric furnace.

Air conditioner (refer to *Table 220-30, item 1*)

$$30 \times 240 = 7,200 \text{ volt-amperes}$$

Electric furnace (refer to *Table 220-30, item 2*)

$$13,000 \times 0.65 = 8,450 \text{ volt-amperes}$$

Therefore we will select the electric furnace load for our calculations because it is the largest load. It is

also a noncoincidental load, as defined in *Section 220-21*. We can omit the air-conditioner load from our calculations from here on.

We now add up all of the other loads:

General lighting load 3,232 sq. ft. @ 3 volt-amperes per sq. ft. =	9,696 volt-amperes
Small appliance circuits (7) @ 1,500 volt-amperes each	10,500 volt-amperes
Wall-mounted oven (nameplate rating)	6,600 volt-amperes
Counter-mounted cooking unit (nameplate rating)	7,450 volt-amperes
Water heater (nameplate rating)	3,000 volt-amperes
Clothes dryer (nameplate rating)	5,700 volt-amperes
Dishwasher (maximum demand, heater only)	1,000 volt-amperes
Food waste disposer 7.2 × 120	864 volt-amperes
Water pump 8 × 240	1,920 volt-amperes
Garage door opener 5.8 × 120	696 volt-amperes
Heat-vent-lights (2) 1475 × 2	2,950 volt-amperes
Attic exhaust fan 5.8 ×120	696 volt-amperes
Hydromassage tub 10 × 120	1,200 volt-amperes
Total other loads	52,272 volt-amperes

We can now complete our optional calculation:

Enter electric furnace load 13,000 × 0.65 =	8,450 volt-amperes
Plus first 10 kVA of all other loads at 100%	10,000 volt-amperes
Plus remainder all other loads at 40%: 52,272 − 10,000 = 42,272 42,272 × 0.40	16,909 volt-amperes
Total	35,359 volt-amperes

$$\text{Amperes} = \frac{\text{volt-amperes}}{\text{volts}} = \frac{35,359}{240} = 147.3 \text{ amperes}$$

No. 1 THW or THWN conductors or equivalent could be installed for this service, *Table 310-16, Note 3.* However, the residence in the plans is to have a full 200-ampere, 120/240-volt service consisting of two No. 2/0 THW, THHN, THWN, or XHHW phase conductors and one No. 1 bare neutral conductor.

Service-Entrance Conductor Size Table

Table 29-2 has been taken from one major city that prefers to show minimum service-entrance conductor size requirements rather than having electrical contractors make calculations each and every time they install a service.

Note that the conductor sizes in Table 29-2 are larger than conductor sizes permitted by *Note 3* to *Table 310-16* of the *National Electrical Code.®* But because the Code is considered to be a minimum standard, it is within the realm of local authorities to publish code requirements specifically for their communities.

Main Disconnect

The main service disconnecting means in this residence is rated 200 amperes. This was discussed in unit 28.

Grounding Electrode Conductor

Grounding electrode conductors are sized ac-

cording to *Table 250-94* of the Code.

As an example, this residence is supplied by No. 2/0 AWG copper service-entrance conductors. Checking *Table 250-94*, we find that the minimum grounding electrode conductor must be a No. 4 AWG copper conductor.

The grounding electrode conductor for the service in this residence is a No. 4 AWG copper armored ground cable. This was discussed in unit 28.

READING THE METER

Figure 29-2 shows a typical single-phase watt-hour meters with five dials. From right to left, the dials represent single units, tens, hundreds, thousands, and tens of thousands in kilowatt-hours.

Starting with the first dial, record the last number the pointer has passed. Continue doing this with each dial until the full reading is obtained. The reading on the five-dial meter in figure 29-3 is 18672 kilowatt-hours.

If the meter reads 18975 one month later, figure 29-4, by subtracting the previous reading of 18672 it is found that 303 kilowatt-hours were used during the month.

The number is multiplied by the rate per kilowatt-hour, and the power company bills the con-

Table 250-94.
Grounding Electrode Conductor for AC Systems

Size of Largest Service-Entrance Conductor or Equivalent Area for Parallel Conductors		Size of Grounding Electrode Conductor	
Copper	Aluminum or Copper-Clad Aluminum	Copper	*Aluminum or Copper-Clad Aluminum
2 or smaller	1/0 or smaller	8	6
1 or 1/0	2/0 or 3/0	6	4
2/0 or 3/0	4/0 or 250 kcmil	4	2
Over 3/0 thru 350 kcmil	Over 250 kcmil thru 500 kcmil	2	1/0
Over 350 kcmil thru 600 kcmil	Over 500 kcmil thru 900 kcmil	1/0	3/0
Over 600 kcmil thru 1100 kcmil	Over 900 kcmil thru 1750 kcmil	2/0	4/0
Over 1100 kcmil	Over 1750 kcmil	3/0	250 kcmil

Where multiple sets of service-entrance conductors are used as permitted in Section 230-40, Exception No. 2, the equivalent size of the largest service-entrance conductor shall be determined by the largest sum of the areas of the corresponding conductors of each set.

Where there are no service-entrance conductors, the grounding electrode conductor size shall be determined by the equivalent size of the largest service-entrance conductor required for the load to be served.

* See installation restrictions in Section 250-92(a).

(FPN): See Section 250-23(b).

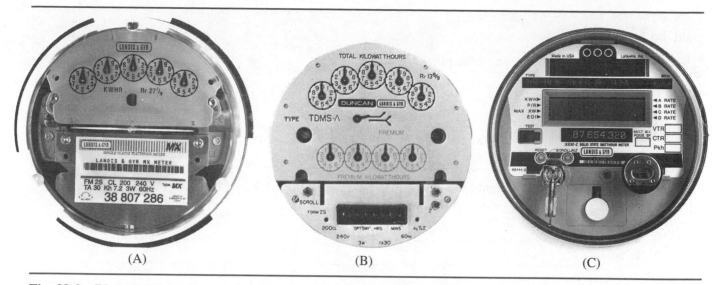

(A)   (B)   (C)

Fig. 29-2   Photo "A" shows a typical single-phase watt-hour meter. Photo "B" shows a single-phase watt-hour meter that has two sets of dials. One registers the total kilowatt-hours. The second registers premium time kilowatt-hours. Photo "C" shows a programmable electronic digital display watt-hour meter that is capable of registering four different "Time-Of-Use" rates. *Courtesy* **of Landis & Gyr Energy Management, Inc.**

Fig. 29-3   The reading of this five-dial meter is 18672 kilowatt-hours.

Fig. 29-4   One month later, the meter reads 18975 kilowatt-hours, indicating that 303 kilowatt-hours were used during the month.

sumer for the energy used. The utility may also add a fuel adjustment charge.

TYPES OF WATT-HOUR METERS

Figure 29-2 shows three different types of watt-hour meters. Wiring diagrams for the metering and control of electric water heaters, air conditioners, and heat pumps were discussed in unit 19. Electric utilities across the country have different requirements for hooking up these loads. Consult your utility for their latest information.

REVIEW

Note: Refer to the Code or the plans where necessary.

1. When a service-entrance calculation results in a value of 10 kW or more, what is the minimum size service required by the Code? _____

2. a. What is the unit load per square foot for the general lighting load of a residence?

   _____

   b. What are the demand factors for the general lighting load in dwellings? _____

   _____

   _____

   _____

   _____

   _____

3. a. What is the ampere rating of the circuits that are provided for the small appliance loads? _____

   b. What is the minimum number of small appliance circuits permitted by the Code?

   _____

   c. How many small appliance circuits are included in this residence? _____

4. Why is the air-conditioning load for this residence omitted in the service calculations?

   _____

   _____

   _____

   _____

   _____

   _____

5. What demand factor may be applied when four or more fixed appliances are connected to a service, in addition to an electric range, air conditioner, clothes dryer, or space heating equipment? *(Section 220-17)* _____

   _____

6. What load may be used for an electric range rated at not over 12 kW? *(Table 220-19)*

   _____

7. What is the computed load for an electric range rated at 16 kW? *(Table 220-19)* Show calculations.

8. What is the computed load when fixed electric heating is used in a residence? *(Section 220-15)* _____

9. On what basis is the neutral conductor of a service entrance determined? _____

_____

_____

_____

_____

10. Why is it permissible to omit an electric space heater, water heater, and certain other 240-volt equipment when calculating the neutral service-entrance conductor for a residence? _____

_____

_____

_____

11. a. What section of the Code contains an optional method for determining residential service-entrance loads? _____

b. Is this section applicable to a two-family residence? _____

12. Calculate the minimum size of service-entrance conductors required for a residence containing the following: floor area 24' × 38' (7.3 m x 11.6 m); 12-kW electric range; 5-kW dryer consisting of a 4-kW, 240-V heating element, a 120-V motor, a 120-V light (combined motor and light is 1 kW); 2200-W, 120-V sauna heater; 12-kW, 240-V electric heat (six units); 12-A, 240-V air conditioner; 3-kW, 240-V water heater. Determine the sizes of the ungrounded conductors and the neutral conductor. Use Type THWN conductors. Be sure to include the small appliance circuits and the laundry circuit. Use Method 1.

Two No. _____ THWN ungrounded conductors

One No. _____ THWN neutral (or bare neutral if permitted)

One No. _____ AWG grounding electrode conductor to water meter

STUDENT CALCULATIONS

13. Read the meter shown. Last month's reading was 22796. How many kilowatt-hours of electricity were used for the current month?

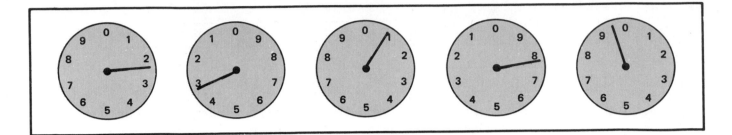

14. After studying this unit, we realize that electric utilities across the country have many ways to meter residential customers. They may have lower rates during certain hours of the day. A common term used to describe these programs is _____

_____

UNIT 30

Swimming Pools, Spas, Hot Tubs

OBJECTIVES

After studying this unit, the student will be able to

* recognize the importance of proper swimming pool wiring with regard to human safety.
* discuss the hazards of electrical shock associated with faulty wiring in, on, or near pools.
* describe the differences between permanently installed pools and pools that are portable or storable.
* understand and apply the basic Code requirements for the wiring of swimming pools, spas, hot tubs, and hydromassage bathtubs.

POOL WIRING *(ARTICLE 680 NEC®)*

Because of the complexity of swimming pool wiring, a detailed drawing of the *National Electrical Code®* requirements for swimming pools appears on Sheet 10 of 10 in the Plans found in the back of this text. You will want to refer to this drawing continually as you study this unit.

To protect people against the hazards of electric shock associated with swimming pools (water and electricity are not a good combination), extreme care is required when wiring the equipment in and around pools. The basic code requirements found in Chapters 1 through 4 of the *NEC®* are to be followed, but, in addition, *Article 680* amends the requirements of Chapters 1 through 4 with many special rules. These rules are unique to swimming pools, wading pools, hydromassage bathtubs, therapeutic pools, decorative pools, hot tubs, and spas.

ELECTRICAL HAZARDS

A person can suffer an immobilizing or lethal shock in a residential-type pool in either of two ways.

1. An electrical shock can be transmitted to someone in a pool who touches a "live" wire, or the "live" casing or enclosure of an appliance, such as a hair dryer, radio, or extension cord, among others, figure 30-1.

2. In the event that an appliance falls into the pool, an electrical shock can be transmitted to a person in the water, by means of voltage gradients in the water. Refer to figure 30-2 for an illustration of this life-threatening hazard.

As shown in the figure, "rings" of voltage radiate outwardly from the radio to the pool walls. These rings can be likened to the rings that form when a rock is thrown into the water. The voltage rings or gradients range from 120 volts at the radio to zero volts at the pool walls. The pool walls are assumed to be at ground or zero potential. The gradients, in varying degrees, are found in the entire body of water.

Figure 30-2 shows voltage gradients in the pool of 90 volts and 60 volts. This figure is a simplification of the actual situation in which there are many

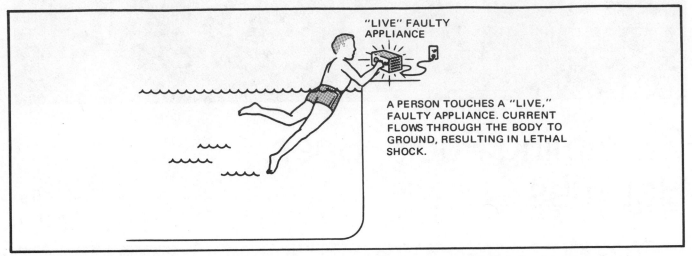

Fig. 30-1   Touching a "live," faulty appliance can cause lethal shock.

voltage gradients. In this case, the voltage differential, 30 volts, is an extremely hazardous value. **The person in the pool, who is surrounded by these voltage gradients, is subject to severe shock, immobilization (which can result in drowning), or actual electrocution.** Tests conducted over the years have shown that a voltage gradient of 1½ volts per foot can cause paralysis.

Study the Code Requirements for Swimming Pool Wiring appearing in the plans. You will note that, in general, underwater swimming-pool lighting fixtures must be positioned at least 18 inches below the normal water level. The reasoning behind this rule is that when a person is in the water close to an underwater lighting fixture, the relative position of the fixture is well below the person's heart. Fixtures that are permitted to be installed no less than 4 inches below the normal water level have undergone tougher, abnormal impact tests, such as those tests given to cleaning tools or other mechanical objects. The lenses on these fixtures can withstand these abnormal impact tests, whereas the standard underwater fixtures that are required to be 18 inches below the water level are subjected to normal impact tests that duplicate the impact of a person kicking the lens.

The shock hazard to a person in a pool is quite different from that of the normal "touch" shock

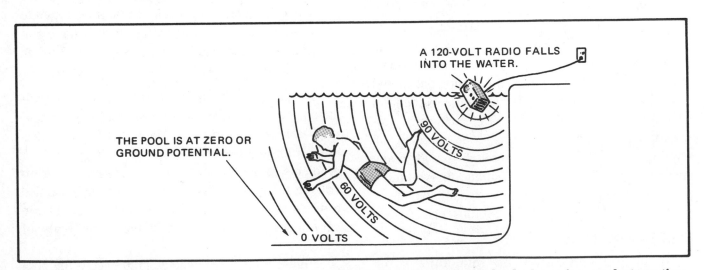

Fig. 30-2   Voltage gradients surrounding a person in the pool can cause severe shock, drowning, or electrocution.

hazard to a person not submersed in water. The water makes contact with the entire skin surface of the body rather than just at one "touch" point. Also, skin wounds, such as cuts and scratches, reduce the body's resistance to shock to a much lower value than that of the skin alone. Body openings such as ears, nose, and mouth further reduce the body resistance. As Ohm's law states, for a given voltage, the lower the resistance, the higher the current.

Another hazard associated with spas and hot tubs is prolonged immersion in hot water. If the water is too hot and/or the immersion too long, lethargy (drowsiness, hyperthermia) can set in. This can increase the risk of drowning. The Underwriters Laboratories Standard #1563 establishes maximum water temperature at 104°F. The suggested maximum time of immersion is generally 15 minutes.

Instructions furnished with spas and hot tubs specify the maximum temperature and time permitted for using the spa or hot tub.

Hot water, in combination with drugs and alcohol, presents a real hazard to life. Caution must be observed at all times.

The figures that follow show the *National Electrical Code®* rules that relate to safety procedures for wiring pools.

Wiring Methods

Throughout *Article 680* of the Code, you will find that the wiring must be installed in rigid metal conduit, intermediate metal conduit, rigid nonmetallic conduit, Type MC cable, or, in some cases, electrical metallic tubing. For wet-niche fixtures, brass or other corrosion-resistant metal must be used.

Of the many Code rules covering pools, note in *Section 680-25(c)* that the circuit conductors and the equipment grounding conductor for pool-associated motors must be installed in rigid metal conduit, rigid nonmetallic conduit, or Type MC cable. Electrical metallic tubing (EMT) is permitted where installed on or within buildings.

Section 680-25(c), Exception No. 3 allows any type of wiring method recognized in *Chapter 3* of the Code for that portion of the interior wiring of one-family dwelling installations that supplies pool-associated pump motors, ▶ provided the wiring method has an insulated or covered equipment grounding conductor not smaller than No. 12

AWG. The outer jacket of any wiring method recognized in *Chapter 2* of the *NEC®* such as non-metallic-sheathed cable, is the covering referred to above. ◀

Ground-Fault Circuit Interrupters

Many lives have been saved because circuits have been protected against ground faults with ground-fault circuit interrupters, commonly referred to as GFCIs. Throughout *Article 680* you will find reference to the need for GFCIs, either the receptacle type or circuit breaker type. Generally, once you begin to run wires beyond (on the load side) a GFCI device, you are not permitted to have other conductors in the same raceway, box, or enclosure. The exception to this is in the case when the GFCI is installed in a panelboard, when obviously there will be a "mix" of other circuit conductors that are not GFCI protected.

CODE-DEFINED POOLS

▶ The Code in *Section 680-4* describes a *permanently installed swimming, wading, or therapeutic pool* as one that is constructed in the ground or partially in the ground, other pools that have a water depth of greater than 42 inches (1.07 m), and all pools installed outside of a building regardless of water depth and whether or not it is served by electrical circuits of any nature.

The Code in *Section 680-4* describes a *storable swimming pool or wading pool* as one that is constructed above the ground, is capable of holding a water depth of not over 42 inches (1.07 m), or a pool with nonmetallic, molded polymeric walls, or inflatable fabric walls regardless of dimension. ◀ The construction of this type of pool is such that it can be readily disassembled for storage, and reassembled to its original form.

Cord-connected filter pumps for storable pools are generally "double-insulated." The cord itself has an equipment grounding conductor in it. The attachment plug cap on the cord is the 3-prong type. All of the electrical equipment associated with a storable pool must be GFCI protected.

Storable pools are covered in *Article 680, Part C* of the *NEC®*

GROUNDING AND BONDING OF SWIMMING POOLS

Grounding (Section 680-24)

Section 680-24 of the Code requires that all of the following items *must* be grounded, as illustrated in figure 30-3:

- wet-niche and dry-niche lighting fixtures.

- dry-niche underwater lighting fixtures.

- all electrical equipment within 5 feet (1.52 m) of the inside wall of the pool.

- all electrical equipment associated with the recirculating system of the pool.

- junction boxes.

- transformer enclosures.

- ground-fault circuit interrupters.

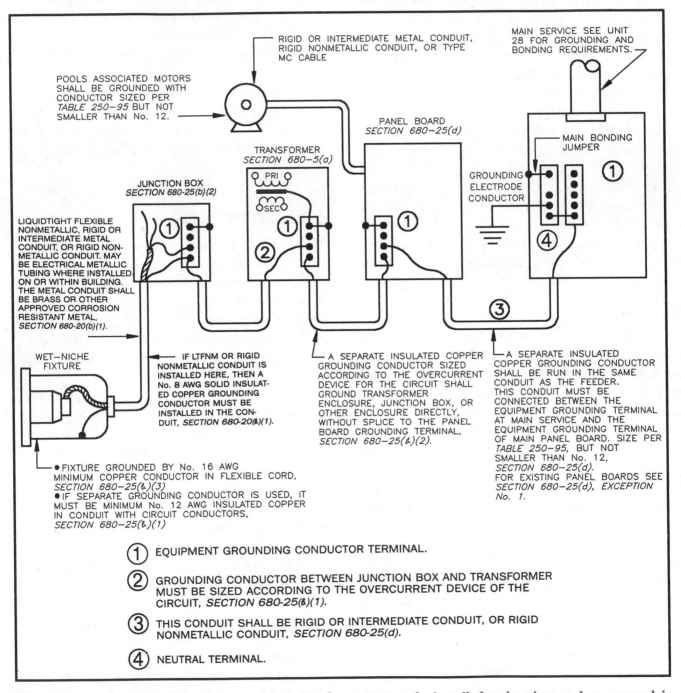

Fig. 30-3 Grounding of important metal parts of a permanently installed swimming pool as covered in *Sections 680-24* and *680-25*. Also refer to Plan 10/10.

- panelboards not part of the service equipment that supply the electrical equipment associated with the pool.

Proper grounding and bonding ensures that all of the metal parts in and around the pool area are at the same ground potential, thus reducing the shock hazard. Proper grounding and bonding practices also facilitate the opening of the overcurrent protective device (fuse or circuit breaker) in the event of a fault in the circuit. Grounding is covered in other units of this text.

Grounding Conductors

Grounding conductors *must*:

- be run in the same conduit with the circuit conductors, or be part of an approved flexible cord assembly, as used for the connection of wet-niche underwater lighting fixtures.

- terminate on equipment grounding terminals provided in the junction box, transformer, ground-fault circuit interrupter, subpanel, or other specific equipment.

It is important to note that metal conduit by itself is NOT considered to be an adequate grounding means for equipment grounding in and around pools.

Bonding *(Section 680-22)*

Section 680-22 of the Code requires that all metal parts of a pool installation *must* be bonded together by connecting the parts to a common bonding grid, figure 30-4. The No. 8 or larger solid copper bonding wire need not be connected to any service equipment, a remote panelboard, or any grounding electrode. It need only "tie everything together." This helps keep all metal parts in and around the pool at the same voltage potential, reducing the shock hazard brought about by stray voltages and voltage gradients.

The bonding grid, *Section 680-22(b)*, may be:

- structural steel reinforcing rods in the concrete where the rods are bonded together with tightly twisted steel tie wires, or

- the wall of a bolted or welded metal pool, or

- a solid copper wire no smaller than No. 8.

▶ Double-insulated pumps that are used with permanently installed pools are permitted to be plug-and cord-connected with a three-wire cord that contains the circuit conductors plus the equipment grounding conductor where the equipment grounding conductor serves only to ground the internal, nonaccessible, noncurrent-carrying metal parts of the pump. These pumps are "listed" and clearly marked as being "double-insulated." These pumps *do not* have to be bonded to the common bonding grid as illustrated in figure 30-4. See *Section 680-28*. ◀

Bonding Conductors: Bonding conductors:

- need *not* be installed in conduit.

- may be connected directly to the equipment that requires bonding, by means of brass, copper, or copper-alloy clamps.

LIGHTING FIXTURES UNDERWATER

Article 680, Part B covers underwater lighting fixtures for permanently installed pools.

There are three types of underwater lighting fixtures defined in *Section 680-4* of the Code.

1. **Dry-niche:** Dry-niche lighting fixtures are mounted in the side walls of the pool with the top of the lens not less than 18 inches below the normal water line, unless otherwise marked. They are designed to be relamped from the rear. These fixtures are waterproof. They are designed so water does not enter the niche. They have a means for connecting equipment grounding conductor(s) and have a provision to drain out any water that might accumulate in the fixture.

2. **Wet-niche:** Wet-niche lighting fixtures are mounted in the side walls of the pool with the top of the lens not less than 18 inches below the normal water line, unless otherwise marked. They are designed to be relamped from the front. The supply cord is stored inside the fixture. The cord should be long enough to reach the top of the pool deck for lamp replacement. The forming shell to which the fixture attaches is intended to fill up with pool water. The fixture is completely surrounded by water.

3. **No-niche:** No-niche lighting fixtures are mounted on a bracket on the inside wall of the pool with the top of the lens not less than 18 inches below the normal water line, unless otherwise marked. They are designed to be relamped from the front. The supply conduit terminates on this bracket. The supply cord runs through this conduit to a deck box for connection to the circuit wiring. The extra supply cord is stored in the space behind the no-niche fixture. The cord should be long enough to reach the top of the pool deck

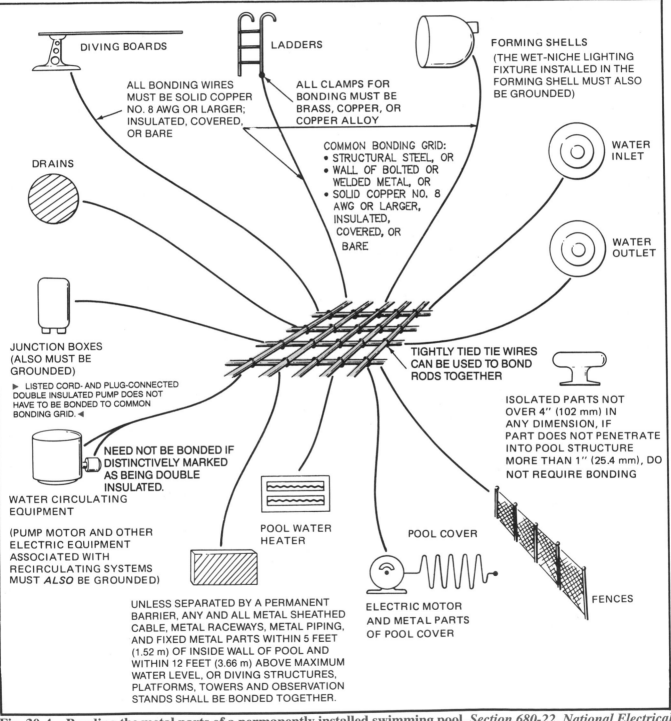

Fig. 30-4   Bonding the metal parts of a permanently installed swimming pool, *Section 680-22, National Electrical Code.®* **In addition to bonding, some of the above items must also be grounded. Refer to** *Sections 680-24* **and** *25* **for specifics. Also see Plan 10/10.**

for lamp replacement. This type of fixture is mainly installed in retrofit installations and for aboveground pools.

ELECTRIC HEATING OF SWIMMING POOL DECKS
(Section 680-27)

The Code requirements for the electric heating units are illustrated in figure 30-5.

Permanently wired radiant heaters:

- shall be suitably guarded.
- shall be securely fastened.
- shall not be located over the pool.
- shall be at least 5 feet (1.52 m) back from the inside walls of the pool.
- shall be mounted at least 12 feet (3.66 m) vertically above pool deck unless approved otherwise.

Unit heaters:

- shall be rigidly mounted on the structure.
- shall be totally enclosed or guarded.

- shall not be located over the pool.
- shall be at least 5 feet (1.52 m) back from the inside walls of the pool.

Radiant heat cables:

- are not permitted to be embedded in or below the pool.

Circuit sizing for electric heaters:

The branch-circuit conductors and the branch-circuit overcurrent protective devices shall be rated not less than 125% of the heater's nameplate rating according to *Section 680-9* of the Code.

SPAS AND HOT TUBS *(Article 680, Part D)*

▶ There are basically two types of spas and hot tubs.

1. **Self-contained spa or hot tub:** A factory fabricated unit consisting of a spa or hot tub vessel with all water circulating, heating, and control equipment integral to the unit. Equipment may include pumps, air blowers, heaters, lights, controls, sanitizers, or generators.

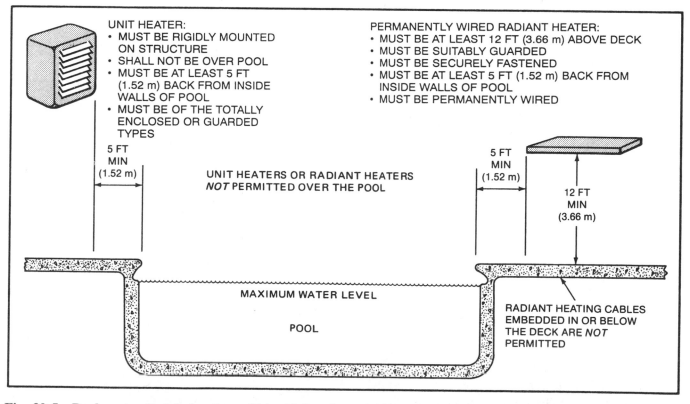

Fig. 30-5 Deck-area electric heating within 20 feet from inside edge of swimming pools, *Sections 680-9* **and** *680-27, National Electrical Code.*

2. **Packaged spa or hot tub assembly:** A factory fabricated unit consisting of water circulating, heating, and control equipment mounted on a common base, intended to operate a spa or hot tub. Equipment may include pumps, air blowers, heaters, lights, controls, sanitizers, or generators. ◄

The basic difference between a spa and a hot tub is that a hot tub is constructed of wood such as redwood, teak, cypress, or oak, whereas a spa is made of plastics, fiberglass, concrete, tile, or other man-made products.

Spas and hot tubs contain electrical equipment for heating and recirculating water.

Spas and hot tubs are intended to be filled and used. They are not drained after each use, as in the case of a regular bathtub.

Some spas and hot tubs are furnished with a cord-and-plug set when shipped by the manufacturer. Where these can be converted from cord-and-plug connection to "hard wiring" they are listed and identified as "convertible" spas and hot tubs. The manufacturer must furnish clear instructions as to how the conversion from cord-and-plug set to "hard wiring" is to be done.

Outdoor Installations, *Section 680-40*

Spas and hot tubs installed outdoors *must* conform to the installation requirements discussed previously for regular swimming pools, *Article 680, Parts A and B.*

Exception No. 1 to *Section 680-40* states that the metal bands and hoops that secure the wood staves of a spa or hot tub do not have to be grounded.

Listed packaged units are permitted to have a cord not longer than 15 feet (4.56 m). Such units must be GFCI protected, *Section 680-40, Exception No. 2.*

Listed packaged units that have a factory-installed remote-control panel may be connected with liquid-tight flexible conduit not over 3 feet (914 mm) long.

Indoor Installation, *Section 680-41*

Spas and hot tubs installed indoors *must* conform to the following requirements:

- must be connected using the wiring methods covered in *Chapter 3* of the Code.
- may be cord- and plug-connected if rated 20 amperes or less.

Receptacles, *Section 680-41(a):*

- at least one receptacle outlet must be installed at least 5 feet (1.52 m), but not more than 10 feet, (3.05 m) from the inside wall of the spa or hot tub.
- any other receptacles installed on the property must be located at least 5 feet (1.52 m) from the inside walls of the spa or hot tub.
- any 125-volt receptacle located within 10 feet (3.05 m) of the inside walls of the spas or hot tub must be GFCI protected.
- any receptacles that supply power to the spa or hot tub must be GFCI protected no matter how far they are from the spas or hot tub.

Lighting Fixtures, Lighting Outlets, and Ceiling Fans, *Section 680-41(b):*

- all lighting outlets, fixtures, and ceiling fans located within 5 feet (1.52 m) from the inside wall of the spa or hot tub or directly over the spa or hot tub must be at least 7½ feet (2.29 m) above the maximum water level and must be GFCI protected.

   There are a few exceptions to this:

   1. GFCI protection is not required if the fixture or ceiling fan is 12 feet (3.66 m) or more above maximum water level.

   2. Recessed lighting fixtures may be installed less than 7½ feet (2.29 m) above the hot tub or spa if the fixture:

      - is GFCI protected, and
      - has a glass or plastic lens, and
      - has a nonmetallic trim suitable for use in damp locations, or
      - has an electrically isolated metal trim suitable for use in damp locations.

   3. Surface-mounted fixtures may be installed less than 7½ feet (2.29 m) above the hot tub or spa if the fixture:

- is GFCI protected, and

- has a plastic or glass globe, and

- has a nonmetallic body, or

- has a metallic body that is isolated from contact.

- is suitable for use in damp locations.

- if any underwater lighting is to be installed, the rules discussed previously for regular swimming pools apply.

Wall Switches, *Section 680-41(c)*:

- wall switches must be located at least 5 feet (1.52 m) from the inside walls of the spa or hot tub.

Grounding, *Section 680-41(f)* and *(g)*:

The requirements for grounding spas and hot tubs are:

- ground all electrical equipment within 5 feet (1.52 m) of the inside wall of the spa or hot tub.

- ground all electrical equipment associated with the circulating system, including the pump motor.

- grounding *must* conform to all of the applicable Code rules of *Article 250*.

- all equipment shall be connected using wiring methods as found in *Chapter 3* of the Code.

- if equipment is connected by means of a flexible cord, the grounding conductor must be part of the flexible cord, and must be fastened to a fixed metal part of the equipment.

Bonding, *Sections 680-41(d)* and *(e)*:

The bonding requirements for spas and hot tubs are similar to those for regular swimming pools. Bond together:

- all metal fittings within or attached to the spa or hot tub.

- all metal parts of electrical equipment associated with the circulating system, including pump motors.

- all metal pipes, conduits, and metal surfaces within 5 feet (1.52 m) of the inside wall of the spa or hot tub. This bonding is not required if the materials are separated from the spa or hot tub by a permanent barrier, such as a wall or building.

- all electrical devices and controls *not* associated with the spa or hot tub located less than 5 feet from the inside wall of the spa or hot tub. If located 5 feet (1.52 m) or more from the hot tub or spa, bonding is not required.

Bonding is to be accomplished by means of threaded metal piping and fittings, by metal-to-metal mounting on a common base or frame, or by means of a No. 8 or larger, solid, insulated, covered or bare bonding jumper.

Electric Water Heaters, Section 680-41(h):

- must be listed for the purpose.

- must have their heating elements divided into loads not over 48 amperes, and protected at not over 60 amperes. This requirement is usually met by the manufacturer of the heater.

- must have branch-circuit conductors rated not less than 125% of the total load as indicated on the nameplate of the heater.

Ground-Fault Protection, Section 680-42:

▶ Any and all 125-volt receptacles that are located within 5 feet (1.52 m) of the inside walls of a hydromassage tub must be GFCI protected, *Section 680-71*. ◀

DISCONNECTING MEANS

▶ To provide for safety of personnel, *Section 680-12* requires that a disconnecting means must be installed (1) within sight, (2) in an accessible location, and (3) at least five feet (1.52 m) horizontally from the inside walls of a pool, spa, or hot tub equipment. ◀

HYDROMASSAGE BATHTUBS
(Article 680, Part G)

▶ An outlet that supplies a self-contained spa or hot tub, or a packaged spa or hot tub assembly must be GFCI protected, *Section 680-42*. The exception to this is for a "listed" packaged assembly or "listed" self-contained unit that has integral built-in GFCI protection already provided for all of the electrical parts within the assembly, and is so marked. ◄

Hydromassage bathtubs, together with their associated electrical components, must be GFCI protected.

Because a hydromassage bathtub does not constitute any more of a shock hazard than a regular bathtub, the *National Electrical Code*® permits all other wiring (fixtures, switches, receptacles, and other equipment) in the same room but not directly associated with the hydromassage bathtub to be installed according to all the normal Code requirements covering installation of that equipment in bathrooms.

This residence has a hydromassage tub in the master bathroom. It is a prewired unit furnished with a 3-foot length of ½-inch watertight conduit that contains a black, a white, and a green equipment grounding conductor.

The hydromassage tub is powered by a ½-hp, 115-volt, 3450-r/min, single-phase motor rated at 10 amperes.

A hydromassage tub and all of its associated equipment generally provided by the manufacturer must be connected to a circuit that has GFCI protection.

The hydromassage tub is connected to a separate 120-volt, 20-ampere Circuit A9. For more details on the circuitry for the hydromassage bathtub see unit 22.

A hydromassage tub is also known as a whirlpool tub or whirlpool bathtub.

Figure 22-12 illustrates one manufacturer's current production-model hydromassage bathtub.

By definition in *Article 680* of the Code, a hydromassage bathtub is intended to be filled (used), then drained after each use, whereas the spa or hot tub is filled with water and not drained after each use.

FOUNTAINS *(Article 680, Part E)*

The home discussed in this text does not have a fountain, however, we will discuss two key issues. It is significant to note that self-contained, portable fountains having no dimension over 5 feet (1.52 m) do not have to conform to *Part E* of *Section 680*. When fountains share water with a regular swimming pool, the fountain wiring must conform to pool wiring requirements.

UNDERWRITERS LABORATORIES STANDARDS

The U.L. standards of interest are:

UL 676	Underwater Lighting Fixtures
UL 943	Ground-Fault Circuit Interrupters
UL 1081	Swimming Pumps, Filters, and Chlorinators
UL 1241	Junction Boxes for Underwater Lighting Fixtures
UL 1261	Electric Water Heaters for Pools and Spas
UL 1563	Electric Hot Tubs, Spas, and Associated Equipment

SUMMARY

The figures in this unit and a plan-sized diagram present a detailed overview of the Code requirements for pool wiring.

REVIEW

Note: Refer to the Code or the plans where necessary.

1. The article of the *National Electrical Code*® that covers most of the requirements for wiring of swimming pools is *Article* _____ .

2. Name the two ways in which a person may sustain an electrical shock when in a pool.

_____

_____

_____

_____

3. The Code in *Article 680* discusses two types of pools. Name and describe each type.

_____

_____

_____

_____

_____

4. Use *Section 680-24* of the Code to determine if the following items must be grounded.

	TRUE	FALSE
a. Wet-and-dry-niche lighting fixtures.	_____	_____
b. Electrical equipment located within 5 feet (1.52 m) of inside walls of pool.	_____	_____
c. Electrical equipment located within 10 feet (3.05 m) of inside wall of pool.	_____	_____
d. Recirculating equipment and pumps.	_____	_____
e. Lighting fixtures and ceiling fans installed more than 15 feet (4.56 m) from inside walls of pool.	_____	_____
f. Junction boxes, transformers, and GFCI enclosures.	_____	_____
g. Panelboards that supply the electrical equipment for the pool.	_____	_____
h. Panelboards 20 feet (6.1 m) from the pool that do not supply the electrical equipment for the pool.	_____	_____

5. Grounding conductors (must) (may) be run in the same conduit as the circuit conductors. Circle correct answer.

6. Grounding conductors (may) (may not) be spliced with wire-nut types of wire connectors. Circle correct answer.

7. The purpose of grounding and bonding is to _____

_____

_____

_____

8. What parts of a permanently installed pool must be bonded together? _____

_____

_____

_____

_____

_____

_____

_____

9. May electrical wires be run above the pool? Explain. _____

_____

_____

10. What is the closest distance that a receptacle may be installed to the inside wall of a pool? _____

_____

_____

_____

_____

_____

11. Receptacles located within 15 feet (4.56 m) from the inside wall of a pool must be protected by a _____

12. Lighting fixtures installed over a pool must be mounted at least (10 feet) (3.05 m), (12 feet) (3.66 m), (15 feet) (4.56 m), above the maximum water level. Circle correct answer.

13. Grounding conductor terminations in wet-niche metal-forming shells, as well as the conduits entering junction boxes or transformer enclosures where the conduit runs directly to the wet-niche lighting fixture, must be _____ with a (an) _____ to prevent corrosion to the terminal and to prevent the passage of air through the conduit, which could result in corrosion.

14. Wet-niche lighting fixtures are accessible (from the inside of the pool) (from a tunnel) (on top of a pole). Circle correct answer.

15. Wet-niche lighting fixtures operating at above 15 volts must be protected by _____

_____

16. Dry-niche lighting fixtures are accessible (from a tunnel, or passageway, or deck) (from the inside of the pool) (on top of a pole). Circle correct answer.

17. In general, it is *not* permitted to install conduits under the pool or within 5 feet (1.52 m) measured horizontally from the inside wall of the pool. True or false? Explain.

_____

_____

_____

_____

_____

_____

_____

18. Junction boxes, wet-niche lighting fixtures, and transformer and GFCI enclosures have one thing in common. They all (are made of bronze) (have threaded hubs) (must be mounted at least 8 inches, or 203 mm, above the deck). Circle correct answer.

19. Lighting fixtures are permitted to be mounted less than 5 feet (1.52 m) measured horizontally from the inside wall of the pool only if they are (made of plastic) (rigidly fastened to an existing structure) (controlled by a wall switch). Circle correct answer.

20. The Code permits radiant heating cable to be buried in or below the deck of a pool. True or false? _____

21. For indoor spas and hot tubs, the following statements are either true or false. Check one.

	TRUE	FALSE
a. Receptacles may be installed within 5 feet (1.52 m) from the edge of the spa or hot tub.	_____	_____
b. All receptacles within 10 feet (3.05 m) of the spa or hot tub must be GFCI protected.	_____	_____
c. Any receptacles that supply power to pool equipment must be GFCI protected.	_____	_____
d. Wall switches must be located at least 10 feet (3.05 m) from the pool.	_____	_____
e. Lighting fixtures above the pool or within 5 feet (1.52 m) from the inside wall of the pool must be GFCI protected.	_____	_____

22. Bonding and grounding of electrical equipment in and around spas and hot tubs (are required by the Code) (are not required by the Code) (are decided by the electrician). Circle correct answer.

23. Where a spa or hot tub is installed in an existing bathroom, an existing receptacle outlet within 5 feet (1.52 m) of the tub is permitted to remain, but only if the circuit feeding the receptacle outlet has _____ protection.

24. a. Hydromassage bathtubs and their associated electric components (shall) (shall not) be protected by a ground-fault circuit interrupter. Circle the correct answer.

    b. All receptacles located within 5 feet (1.52 m) of a hydromassage tub (shall) (shall not) be protected by a ground-fault circuit interrupter. Circle the correct answer.

25. What section of the Code requires that no-niche underwater lighting fixtures be grounded? *Section* _____ .

26. Underwater pool lighting fixtures shall be mounted so that the top of the lens is not less than _____ inches below the normal water line.

27. What section of the Code tells us that all pool associated pump motors shall be grounded? *Section* _____ .

28. What section of the Code tells us that double insulated pumps for permanently installed pools do not have to be bonded to the common bonding grid? *Section* _____ .

UNIT 31

SMART HOUSE® System Wiring

OBJECTIVES

After studying this unit, the student will be able to

- understand the basics of SMART HOUSE® wiring.
- understand how all general Code rules apply to SMART HOUSE® wiring, plus the special rules found in *Article 780, NEC®*
- understand the special terminology used in conjunction with SMART HOUSE® wiring.
- understand the differences between SMART HOUSE® and SMART-REDI® installations.
- identify symbols unique to SMART HOUSE® systems.

INTRODUCTION

This unit is intended as a brief overview of one system of home automation referred to as a SMART HOUSE® The text touches on the highlights of such a system. Complete information is available from the SMART HOUSE® L.P., 401 "J" Prince George Boulevard, Upper Marlboro, MD 20772-8731.

Everyone is familiar with individual remote control of television sets, video cassette and/or compact disc players, overhead door openers, etc. Now, instead of separate, stand-alone controls for each of these items, one integrated system can be installed that will serve all.

The world of micro-electronics enables electrical and electronic equipment such as appliances, low-voltage signaling, video, stereo, cable television, heating and cooling, telephone, security systems, smoke and heat detectors, and similar devices to be interconnected and controlled through a unique wiring system.

One such integrated system, called the SMART HOUSE® System brings together security, energy management, entertainment, communications, and lighting control. This system can be controlled from inside the house, using wall-mounted control panels,

infrared controls, or touch-screen television sets. The system can also be controlled from inside or outside of the house using touchtone telephones.

The SMART HOUSE® concept was developed by the National Association of Home Builders. The products and the design recommendations for SMART HOUSE® wiring change as electricians, installers, builders, and end-users gain experience and offer feedback to the parent organization, the National Association of Home Builders.

The SMART HOUSE® System has been modified and improved since it first entered the marketplace in mid-1991. SMART HOUSE® Consortium Participating Manufacturers are in the process of deploying their second generation of products. There have also been refinements in the design and layout of a SMART HOUSE® System. These product and design changes have been driven by feedback from electricians, installers, builders, and end-users of the SMART HOUSE® System.

The primary changes in the second generation of products have been to streamline installation time and manufacturing costs. This is most noticeable in the change over to RJ-45 and RJ-12 type modular connectors on the cable runs from SMART blocks to switches.

The SMART HOUSE® System is an integrated system consisting of the wiring, outlets, and controls needed to provide remote and programmable control of typical house systems and appliances. The SMART HOUSE® wiring and respective outlets are designed for installation during new home construction, using hybrid cables that contain both No. 12 AWG power conductors and dedicated control conductors. Additional communications cables and applications cables are also installed during the rough-in stage.

As you will learn in this unit, the wiring of a SMART HOUSE® combines conventional house wiring methods with the high tech world of electronics. Where it makes sense, receptacles and lighting can be controlled by SMART HOUSE® components. Conventional receptacles and lighting outlets are installed where there is no real need to provide the features and advantages of SMART HOUSE® control. An example might be in a bedroom, where the lighting and two receptacles are SMART HOUSE® controlled, and the remaining receptacles are wired in the conventional manner. The options are unlimited.

A SMART HOUSE® System is wired totally with SMART HOUSE® components, and makes the house fully operational as an intelligent home. A SMART-REDI® house is a house that has been wired with the proper SMART HOUSE® cables, device boxes, etc., but for the present does not have all of the SMART HOUSE® electronic devices, system controllers, and appliances installed. These can be installed at a later date. SMART-REDI® wiring enables the house to be completely finished as far as general construction is concerned, yet to have the ability and flexibility to convert the electrical/electronic systems to a complete SMART HOUSE® installation at some future date. Such a later-date conversion is easily accomplished because all of the concealed hybrid cables and boxes have been installed during the earlier rough-in period.

This unit will familiarize the student with SMART HOUSE® installations. A separate, fully detailed, comprehensive, and illustrated text titled *SMART HOUSE® Wiring* is available from Delmar Publishers.

Figure 31-1 illustrates a three-dimensional view of a very simple SMART HOUSE® installation, showing the various types of cables and outlets used in SMART HOUSE® wiring.

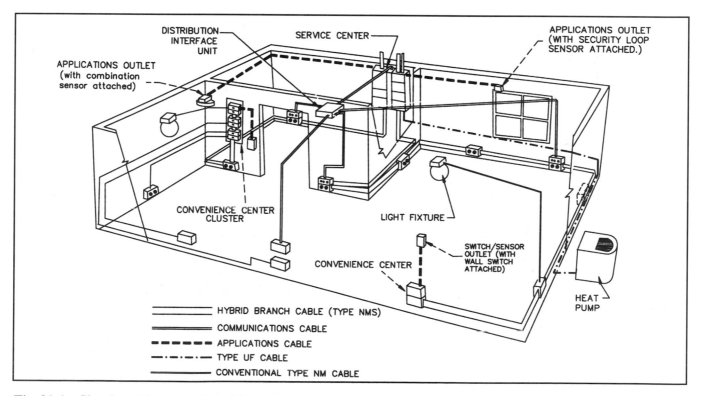

Fig. 31-1   Simple cable and outlet wiring layout for SMART HOUSE® installations. Diagram *courtesy* **of SMART HOUSE® L.P.**

TESTING AND LISTING

Various leading electronic and electrical manufacturers have designed their products to conform to the specifications of the SMART HOUSE® System concepts. All of these products meet specific Underwriters Laboratories requirements.

CODE REQUIREMENTS

All of the *National Electrical Code®* rules discussed throughout this text apply to SMART HOUSE® installations. The rating and number of circuits, location of lighting and receptacle outlets, location of switches, load calculations, ground-fault circuit interrupter (GFCI) requirements, etc., are the same for both regular houses and SMART HOUSES® In addition, *Article 780*, Closed Loop and Programmed Power Distribution, supplements the general code rules.

For example, the electric clothes dryer circuit sizing is determined by the same method discussed in unit 15, the load calculations for electric ranges and counter-mounted cooking appliances are calculated as discussed in unit 20, the lighting branch-circuits and the appliance circuits are calculated as discussed in the units covering the various rooms in the dwelling. Basically, nothing has changed, except that through the use of SMART HOUSE® wiring components, we now have the additional features and benefits of being able to monitor, signal, remote control, and interchange information from one appliance to another, or transmit information from one location to another.

SMART® appliances have integral electronic chips that enable the appliance to communicate with other components in the system. For example, a SMART electric clothes dryer installed in a SMART HOUSE® can send a signal that the dryer has completed its drying cycle to any or all television sets in the house.

SMART HOUSE® COMPONENTS

The basic concept of SMART HOUSE® wiring is to install a family of three multipurpose cables that carry 120-volt ac power, 12-volt dc power, telephone, audio/video, and low-voltage control signals from the service center to outlets and fixed-in-place appliances throughout the house. In addition, larger size conventional wiring is installed from the main load-center to fixed-in-place appliances such as electric furnaces, electric ranges, electric clothes dryers, and similar heavy electrical loads.

Hybrid Branch Cable

▶ This cable is UL listed and is marked Type NMS. Reference to Type NMS cable is found in *Section 336-25* of the *National Electrical Code®* ◀ It contains three No. 14 AWG copper conductors (hot, neutral, equipment ground) for 15-ampere circuits, or three No. 12 AWG copper conductors (hot, neutral, equipment ground) for 20-ampere circuits. See figure 31-2. These power wires are tested to UL Standard 719 (nonmetallic-sheathed cable). Hybrid branch cable also contains six No. 22 AWG copper low-voltage control wires. This hybrid cable carries the 120-volt power and low-voltage control/communication circuitry between the service center and combination wall outlets called convenience centers. It is then run from convenience center to convenience center in the same way that nonmetallic-sheathed cable is run between outlets.

All of these conductors are enclosed in a tough outer jacket, similar in appearance to standard nonmetallic-sheathed cable. Hybrid branch cables are color coded white for the No. 12 AWG size.

Hybrid branch cable is installed according to the requirements of *Article 336* and *Article 780* of the *National Electrical Code®* Generally, 1-inch diameter holes are drilled through wood studs and joists for the hybrid cable. A special tool is used to make all of the connections at the termination points.

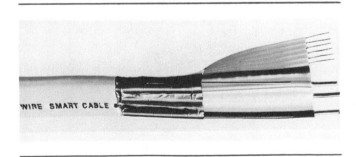
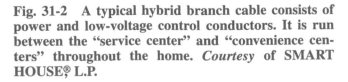

Fig. 31-2 A typical hybrid branch cable consists of power and low-voltage control conductors. It is run between the "service center" and "convenience centers" throughout the home. *Courtesy* of SMART HOUSE® L.P.

To terminate a hybrid branch cable, the white outer jacket is removed, exposing the insulated conductors inside. The folded construction is flattened out and placed between the two halves of a special cable tap device. When clamped with the special tool, prongs pierce the insulation to make contact with the conductors. Proper polarity is assured. It is impossible to reverse the hot and neutral conductors, or mix up power and control wires. Also, the connection for the equipment grounding conductor is automatic. All terminations are completely insulated. No connections are made by stripping off conductor insulation and using wire nuts or screw terminals.

The cable tap device then serves as the backplate inside of a convenience center, where the receptacles and devices to be mounted in the convenience center merely plug into the cable tap device.

Communications Cable

The communications cable, figure 31-3, provides the means of distributing television, audio, telephone, and other data communication circuits throughout the house. Communications cable contains eight (four twisted pairs) telephone wires plus two coaxial cables, all enclosed in a tough plastic outer jacket color-coded brown. Communications cables are run from the service center to distribution interface units (DIUs). In some instances, separate coax and telephone cables are installed instead of the communications cable.

From the DIUs, which are really four-way splitters for coaxial cables and telephone conductors, coaxial cables and telephone conductors are run to selected convenience centers throughout the house. DIUs are generally located in out-of-the-way places such as in garages, basements, attics, closets, or other inconspicuous locations.

The terminations for the coaxial cables are made using standard F-type connectors. The telephone wires are connected using conventional push-down connectors commonly used in telephone installations.

Applications Cable

Applications cable, figure 31-4, provides the means for distributing control signals from the system controller and 12-volt dc power from the continuous power supply to switches, control panels, sensors, and other devices that do not operate at 120 volts.

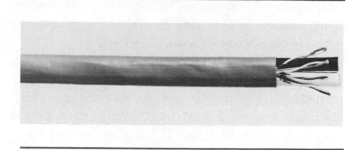

Fig. 31-3   Communications cable. Note that it contains four pairs of telephone conductors plus two coaxial cables. *Courtesy* of SMART HOUSE® L.P.

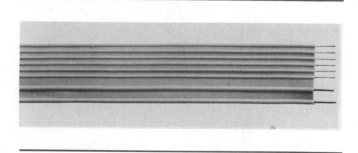

Fig. 31-4   Applications cable. Note that it contains two No. 18 AWG copper conductors for the 12-volt dc circuitry and six No. 24 AWG copper conductors for low-voltage control. *Courtesy* of SMART HOUSE® L.P.

Applications cable contains two No. 18 AWG copper conductors for the 12-volt dc application plus six No. 24 AWG copper conductors for low-voltage control. The outer jacket is color coded orange.

Termination for the applications cable conductors uses the insulation displacement method similar to that of the hybrid branch cable termination method.

In some installations, the applications cable between SMART blocks to switches can be less expensive ribbon cable or twisted pairs of 24-gauge wire. Instead of the insulation piercing method, RJ-45 and RJ-12 connectors are used.

Conventional Wiring

Conventional wiring methods such as nonmetallic sheathed cable (Romex), armored cable (BX), underground feeder cable (UF), or electrical metallic tubing are used to supply appliances such as electric ranges, electric water heaters, electric clothes dryers, electric furnaces, heat pumps, and similar appliances that require a dedicated circuit. A separate applications cable is run with the conventional wiring method to provide a communications link from the SMART appliance or lighting outlet to the system controller.

In a SMART HOUSE® System, conventional wiring methods are a smaller portion of the total installation. Lighting fixtures, for instance, are wired by merely running a No. 14/2 AWG nonmetallic-sheathed cable from the lighting outlet box to a nearby convenience center. An integral part of the convenience center is the switching relay or a dimmer control.

Examples and code installation requirements for nonmetallic-sheathed cable, armored cable, and Type UF cable are found elsewhere in this text.

Service Center

The service center is the brain of the entire installation, figure 31-5. Electric utility power (service-entrance), telephone, and television feed into the service center and are then distributed from the service center to the many locations throughout the house as required. For ease of installation, all of the service-center components are mounted on a ½-inch plywood backboard, located on a dedicated wall space agreed upon by the utilities involved. In almost all cases, the electric utility will determine the point where they will provide service.

A typical service center consists of the following components:

- **Load Center:** The load center is much the same as a regular load center. What differentiates the SMART HOUSE® load center is that in addition to providing the main disconnecting means and branch-circuit overcurrent protective devices, it also provides surge protection and 30 mA ground-fault protection on all 120-volt branch-circuits. The SMART HOUSE® load center also has a large opening on one side that matches a similar opening in another component called a cable manager, sometimes referred to as a header raceway.

  The load calculations for the service equipment as well as for feeders and branch-circuits are the same as in a conventional residence.

  As with all installations, the electrician must provide proper overcurrent protection for the main and branch-circuits. The load-center overcurrent devices must have adequate interrupting ratings in accordance with *Sections 110-9, 110-10,* and *230-65* of the Code. Larger homes

will tend to have larger load centers, requiring the utility to install larger transformers to supply the load. Larger (higher kVA rating) transformers associated with low transformer impedance characteristics result in high levels of available short-circuit current at the main service-entrance equipment. Always take available short-circuit currents into consideration to be sure the installation is in compliance with the Code.

See unit 28 for more information regarding available short-circuit currents.

- **System Controller:** The system controller is the computer of the system, containing both the hardware and the software. The system controller allows homeowners to program system functions and events similar to programming a VCR or microwave oven. All applications cables and the No. 22 AWG control wires of the hybrid branch cables terminate in the system controller.

- **Coaxial Headend:** The coaxial headend is the gateway for cable, satellite, or antenna television signals. Distribution to the many convenience centers throughout the house starts here. The coaxial network also allows signals generated within the house, such as from VCRs or security cameras, to be watched on any television set in the home.

- **12-Volt DC Power Supply:** The 12 VDC power supply (UPS) provides power to the system controller. It also supplies power to the dedicated circuits supplied by applications cable circuits where 120-volt power is not required. An example of this would be security system smoke detectors, heat detectors, motion detectors, and other security window and door sensors.

- **Remote-Control Device (RCD):** The RCD is an optional control unit that can control up to eight circuit breakers in the load center. The signal is sent through the system controller. The RCD allows the homeowner or utility companies to turn major appliances OFF/ON for energy management purposes. This might include water heaters, heat pumps, and similar large loads. The RCD also allows the homeowner to telephone home and command the thermostat to be turned to a higher temperature setting prior to the homeowner's return home. Remote control can be done using telephone lines, either within the home, outside the home, or even long distance.

• **Telephone Gateway:**   The telephone gateway is the communications nerve center for SMART HOUSE® Telephone signals are distributed through the home via the telephone gateway. It can handle up to four outside telephone lines at one time.

The telephone gateway is connected to the system controller, allowing the SMART HOUSE® System to be controlled by telephone from either inside or outside of the home. In effect, any touch-tone telephone becomes a mini control panel.

• **Cable Manager:**   The cable manager is an enclosure that serves as a junction box that allows easy interconnection of the hybrid branch cables, communications cables, load center conductors, systems controller conductors, coaxial

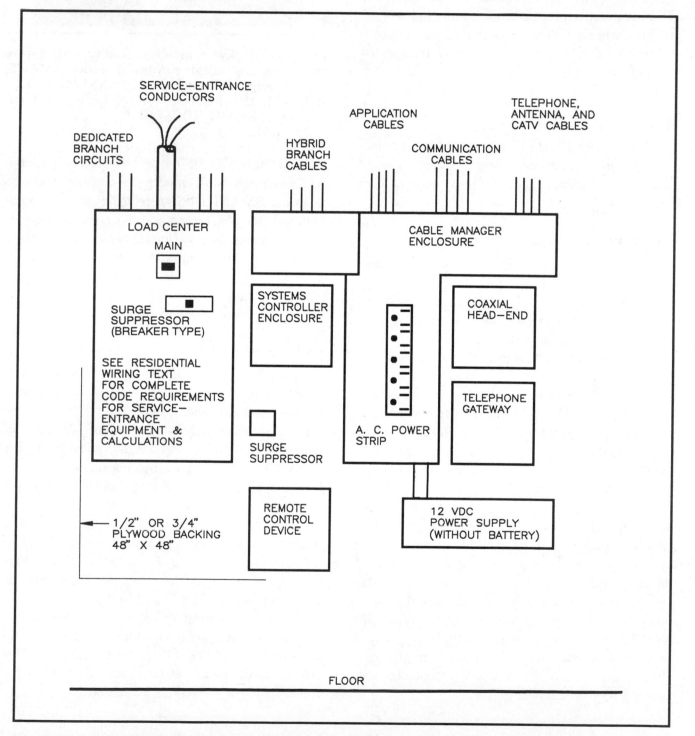

Fig. 31-5   A SMART HOUSE® "service center" schematic layout of the various components required for this type of installation.

conductors, telephone, antenna, and CATV cables, and the 12 VDC power supply.

- **I/O Module (SMART BRIDGE®):** An input/output module is an optional unit for connecting signals from 8 input devices such as switches and sensors, and 8 output devices such as relays and thermostats to the system controller. The system controller can control such things as power usage in the home by changing the thermostat setting and switching lights "on" only when the home is occupied. Remote control can also be done over telephone lines within and outside the home.

SMART HOUSE® OUTLETS

The SMART HOUSE® uses three types of electrical outlets, called convenience centers, switch/sensors, and applications outlets. Let's look at them in more detail.

Convenience Centers

A convenience center is a specially configured two-gang wall outlet that can contain combinations of duplex receptacles, 15-ampere lighting control relays and 5-ampere dimmer control relays. Thus, a convenience center provides 120-volt power, television, telephone, and control communications, all from one location.

The wall boxes that are installed for the convenience centers and the wiring to the convenience centers are the same for a SMART HOUSE® and a SMART-REDI® house. Standard receptacles, both non-GFCI and GFCI, are installed for SMART-REDI® homes. See figure 31-6.

For the SMART HOUSE® System, both non-GFCI- and GFCI-type receptacles having the familiar power prongs and six additional pins provide low-voltage control signals to the new generation of "smart" appliances. A SMART receptacle includes a relay, controlled by an electronic chip, that communicates with the system controller to accept ON/OFF commands to control the outlet.

Ordinary, existing "non-smart" appliances, when plugged into the SMART receptacle, will receive 120-volt ac power only.

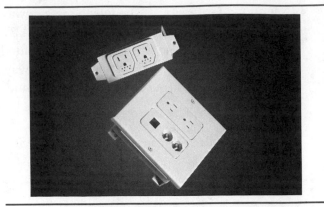

Fig. 31-6 Typical convenience center with duplex receptacle and telephone/coaxial outlet. SMART-REDI® receptacle is shown at top; full SMART receptacle with six additional control pins at bottom. *Courtesy* **of SMART HOUSE; L.P.**

IMPORTANT: In those locations where the *NEC®* requires GFCI-protected receptacles, be sure to install SMART GFCI receptacles at each location. SMART GFCI receptacles are connected in parallel by tapping onto the hybrid branch cable or to a 15-ampere lighting control relay. Because of this, the SMART GFCI receptacles do not provide "feed-thru" protection for other downstream receptacles as with conventional wiring methods.

Switch/Sensor Outlet

This device takes the place of a standard wall switch. It accommodates low-voltage, programmable wall switches and system sensors. These low-voltage switches can control a specific lighting fixture relay, lighting dimmer and relay, or duplex receptacle located in a nearby convenience center.

An applications cable is run between the switch sensor outlet and a convenience center.

The 120-volt wiring runs from a lighting fixture, fan, or other 120-volt equipment to be controlled, to the solid-state switching device or solid-state dimmer/switching device in the nearby convenience center. See figure 31-7(A) and (B).

No special wiring is needed for multilocation switching. The system is simply programmed to function for whatever switch control is desired. This can be reprogrammed at some later date, using the control panel or a touchtone telephone.

(A)

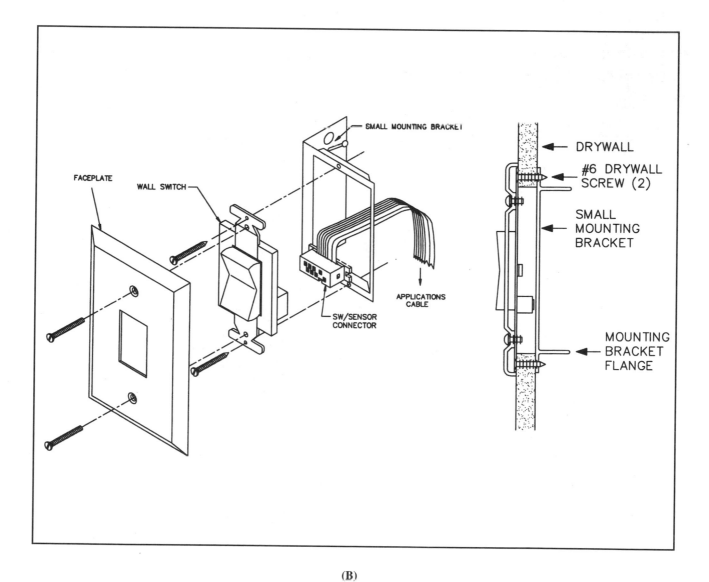

(B)

Fig. 31-7   (A) Photo showing four SMART HOUSE® switch/sensor outlets. (B) The exploded view shows how the connections are made. *Courtesy* of SMART HOUSE® L.P.

Applications Outlet

The applications outlet is similar to a switch/sensor outlet, but is connected to a dedicated circuit called the applications bus, contained in the applications cable. Applications outlets are installed for sensors, switches, smoke and heat detectors, security sensors, and other devices that require 12 VDC continuous supply and/or control signals from the system controller. See figure 31-8.

Applications outlets are installed during the rough-in stages of a SMART-REDI® house and are activated any time in the future when the house is to be converted to full SMART HOUSE® System status.

Modular Components

The modular design of all three SMART HOUSE® outlets makes it easy to change switches and sensors, or to upgrade from a SMART-REDI® standard receptacle to the SMART® receptacle at a later date. Each convenience center, switch/sensor outlet, and applications outlet contains a backplate called a cable tap, which connects to all conductors of the hybrid cable and has plug-in sockets for receptacles, switches, and sensors. The components are modular and can easily be plugged and unplugged from the backplate, without ever touching the actual wiring.

SMART HOUSE® Wiring Symbols

The following legend shows the symbols for devices used in SMART HOUSE® installations that differ from the symbols for standard electrical devices.

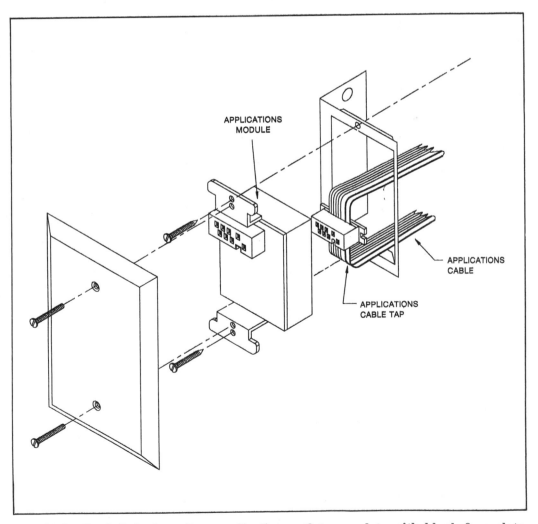

Fig. 31-8 Exploded view of an application outlet, complete with blank face plate, applications module, applications cable tap device, and applications cable. *Courtesy* of **SMART HOUSE, L.P.**

SYMBOL	DESCRIPTION	SYMBOL	DESCRIPTION
◇	CONVENIENCE CENTER: A SPECIAL 2-GANG BOX THAT WILL ACCEPT ANY COMBINATION OF TWO OF THE FOLLOWING COMPONENTS.	Ⓐ	APPLICATIONS OUTLET
◁	• DUPLEX RECEPTACLE	$	SWITCH/SENSOR OUTLET
◁ GFCI	• GFCI DUPLEX RECEPTACLE	$$	DUAL SWITCH/SENSOR OUTLET
◁ S	• SWITCHED DUPLEX RECEPTACLE	$$$$	QUAD SWITCH/SENSOR OUTLET
◀	• TELECOM/COAX OUTLET	◇S	SENSOR OUTLET
◁	• FIXTURE/SWITCH BLOCK	⬠DIU	DISTRIBUTION INTERFACE UNIT
◁ D	• DIMMER/SWITCH BLOCK	⌒	HYBRID BRANCH CABLE
		- - -	COMMUNICATIONS
		- - -·	APPLICATIONS CABLE
◁	• BLANK (NO BLOCK INSTALLED)	·········	CLASS 2 CABLE
		— - —	TYPE NM-B CABLE

REVIEW

Note: Refer to the Code or the plans where necessary.

1. SMART HOUSE® wiring consists of several types of cables and devices not used in conventional house wiring. SMART HOUSE® wiring (must) (need not) comply with the requirements of the *National Electrical Code.*® Circle the correct answer.

2. What is the basic difference between a SMART HOUSE® and a SMART-REDI® house?

_____

_____

_____

_____

_____

_____

3. When installing appliances (such as electric ranges, electric clothes dryers, and similar appliances), the actual load calculations and installation of the branch-circuit wiring must be determined using the *National Electrical Code®* as the basis—just as one would make the calculations for a conventional dwelling. This statement is TRUE _____ FALSE _____ . Mark an "X" in the space indicating the correct answer.

4. Name the three types of special cables used in wiring a SMART HOUSE® System.

   a. _____

   b. _____

   c. _____

5. Name the components that make up a service center for a full capability SMART HOUSE® System.

   a. _____

   b. _____

   c. _____

   d. _____

   e. _____

   f. _____

   g. _____

   h. _____

   i. _____

6. SMART HOUSE® wall outlets are called _____

   _____

7. Because of the many features and benefits possible, a home with a SMART HOUSE® System lends itself to being furnished with more electrical appliances and other electrical items than a conventional home. Service-entrance equipment can be 400 amperes or larger. The power company might install a larger than normal transformer to adequately serve the connected load. This can result in available fault-currents at the main panel to be considerably higher than that found at the line-side terminals of the main panel of a conventional home. To assure that the installation is safe and in compliance with the *National Electrical Code,®* the electrician must be sure that the installation of the service-equipment conforms to the requirements of Sections _____ of the Code.

8. Six unique symbols are used on a SMART HOUSE® electrical floor plan that are not found on conventional electrical plans. Match the symbols on the following page with the correct matching statement by inserting the letter of the symbol in the space provided in front of the matching statements.

   a. _____   _____ Distribution Interface Unit

   b. _____   _____ Convenience Center with Terminator

   c. _____   _____ Switch/Sensor Outlet

   d. _____   _____ Applications Outlet

   e. _____   _____ Convenience Center with Coaxial tap

   f. _____   _____ Convenience Center

SYMBOL	DESCRIPTION
A	___DISTRIBUTION INTERFACE UNIT
B	___CONVENIENCE CENTER WITH TERMINATOR
C	___SWITCH/SENSOR OUTLET
D	___APPLICATIONS OUTLET
E	___CONVENIENCE CENTER WITH COAXIAL TAP
F	___CONVENIENCE CENTER

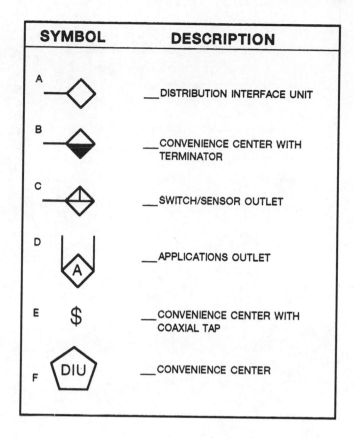

9. Circle the correct dc voltage used in the SMART HOUSE® System.

   a. 12 volts

   b. 24 volts

   c. 65 volts

Specifications for Electrical Work Single-Family Dwelling

1. **GENERAL:** The "General Clause and Conditions" shall be and are hereby made a part of this division.

2. **SCOPE:** The electrical contractor shall furnish and install a complete electrical system as shown on the drawings and/or in the specifications. Where there is no mention of the responsible party to furnish, install, or wire for a specific item on the electrical drawings, the electrical contractor will be responsible completely for all purchases and labor for a complete operating system for this item.

3. **WORKMANSHIP:** All work shall be executed in a neat and workmanlike manner. All exposed conduits shall be routed parallel or perpendicular to walls and structural members. Junction boxes shall be securely fastened, set true and plumb, and flush with finished surface when wiring method is concealed.

4. **LOCATION OF OUTLETS:** The electrical contractor shall verify location, heights, outlet and switch arrangements, and equipment prior to rough-in. No additions to the contract sum will be permitted for outlets in wrong locations, in conflict with other work, and so on. The owner reserves the right to relocate any device up to 10 feet prior to rough-in, without any charge by the electrical contractor.

5. **CODES:** The electrical installation is to be in accordance with the latest edition of the *National Electrical Code*,® all local electrical codes, and the utility company's requirements.

6. **MATERIALS:** All materials shall be new and shall be listed and bear the appropriate label of Underwriters Laboratories, Inc., or another nationally recognized testing laboratory for the specific purpose. The material shall be of the size and type specified on the drawings and/or in the specifications.

7. **WIRING METHOD:** Wiring, unless otherwise specified, shall be nonmetallic-sheathed cable, armored cable, or electrical metallic tubing, adequately sized and installed according to the latest edition of the *National Electrical Code*® and local ordinances.

8. **PERMITS AND INSPECTION FEES:** The electrical contractor shall pay for all permit fees, plan review fees, license fees, inspection fees, and taxes applicable to the electrical installation and shall be included in the base bid as part of this contract.

9. **TEMPORARY WIRING:** The electrical contractor shall furnish and install all temporary wiring for hand-held tools and construction lighting per latest OSHA standards and include all cost in base bid.

10. **WORKSHOP:** Workshop wiring is to be installed in electrical metallic tubing using steel compression gland fittings.

11. **NUMBER OF OUTLETS PER CIRCUIT:** In general, not more than (10) lighting and/or receptacle outlets shall be connected to any one lighting branch-circuit. Exceptions may be made in the case of low-current-consuming outlets.

12. **CONDUCTOR SIZE:** General lighting branch-circuits shall be No. 14 AWG copper protected by 15-ampere overcurrent devices.

    Small appliance circuits shall be No. 12 AWG copper protected by 20-ampere overcurrent devices. All other circuits: wire and overcurrent device as required by the Code.

13. **LOAD BALANCING:** The electrical contractor shall connect all loads, branch-circuits, and feeders per Panel Schedule, but shall verify and modify these connections as required to balance connected and computed loads to within 10% variation.

14. **SPARE CONDUITS:** Furnish and install two empty ½-inch thinwall (EMT) conduits between workshop and attic for future use.

15. **GUARANTEE OF INSTALLATION:** The electrical contractor shall guarantee all work and materials for a period of one full year after final acceptance by the architect/engineer, electrical inspector, and owner.

16. **APPLIANCE CONNECTIONS:** The electrical contractor shall furnish all wiring materials and make all final electrical connections for all permanently installed appliances such as, but not limited to, furnace, water heater, water pump, built-in ovens and ranges, food waste disposer, dishwasher, and clothes dryer.

    These appliances are to be furnished by owner.

17. **CHIMES:** Furnish and install two (2) two-tone door chimes where indicated on the plans, complete with two (2) push buttons and suitable chime transformer. Allow $150.00 for above items. Chimes and buttons to be selected by owner.

18. **DIMMERS:** Furnish and install dimmer switches where indicated.

19. **EXHAUST FANS:** Furnish, install, and provide connections for all exhaust fans indicated on the plans, including, but not limited to, ducts, louvers, trims, speed controls, and lamps. Included are recreation room, laundry, rear entry powder room, range hood, and bedroom hall ceiling fan. Allow a sum of $700.00 in base bid for this. This allowance does not include the two bathroom heat/vent/light fixtures.

20. **FIXTURES:** A fixture allowance of $1,500.00 shall be included in the electrical contractor's bid for all surface-mounted fixtures and post lamp, not including the ceiling paddle fans. Fixtures to be selected by owner. This allowance includes the three (3) medicine cabinets.

    The ceiling paddle fans, speed controls, and lamps for same shall be furnished by the owner and installed by the contractor.

    The electrical contractor shall:

    a. furnish and install all incandescent and fluorescent recessed fixtures.

    b. install all surface, recessed, track, strip, pendant, hanging fixtures, and post lamp.

    c. furnish and install two 2-lamp fluorescent fixtures above the workbench.

    d. furnish and install all lamps except as noted. Fluorescent lamps shall be Watt-Mizer F40 SP30/RS/WM or equivalent. Incandescent lamps shall be energy-efficient Watt-Mizer lamps or equivalent. Lamps for ceiling paddle fans to be furnished by owner.

    e. furnish and install all porcelain pull-chain and keyless lampholders.

    This fixture allowance does not include the two bathroom ceiling heat/vent/light fixtures.

    See Clause 21.

21. **HEAT/VENT/LIGHT CEILING FIXTURES:**   Furnish and install two heat/vent/light units where indicated on the plans complete with switch assembly, ducts, louvers, required to perform the heating, venting, and lighting operations as recommended by the manufacturer.

22. **PLUG-IN STRIP:**   Where noted in the workshop, furnish and install multioutlet assembly with six outlets.

23. **SWITCHES, RECEPTACLES, AND FACEPLATES:**   All flush switches shall be of the quiet ac-rated toggle type. They shall be mounted 46 inches to center above the finished floor unless otherwise noted.

    Receptacle outlets shall be mounted 12 inches to center above the finished floor unless otherwise noted. All convenience receptacles shall be of the grounding type. Furnish and install where indicated, ground-fault circuit interrupter receptacles to provide ground-fault circuit protection as required by the *National Electrical Code.*® All wiring devices are to be provided with ivory handles or faces and shall be trimmed with ivory faceplates except in the kitchen, where chrome-plated steel faceplates shall be used.

    Receptacle outlets, where indicated, shall be of the split-circuit design.

24. **TELEVISION OUTLETS:**   Furnish and install 4-inch square, 1½-inch-deep outlet boxes with single-gang raised plaster covers at each television outlet where noted on the plans. Mount at the same height as receptacle outlets. Furnish and install 75-ohm coaxial cable to each television outlet from a point in the workshop near the main service-entrance switch. Allow 6 feet (1.83 m) of cable in workshop. Furnish and install television plug-in jacks at each location. Faceplates are to match other faceplates in home. All remaining work done by others.

25. **TELEPHONES:**   Furnish and install a 3-inch-deep device box or 4-inch square box, 1½ inches deep, with suitable single-gang raised plaster cover at each telephone location, as indicated on the plans.

    Furnish and install four-conductor No. 18 AWG copper telephone cables from the telephone company's point of demarcation near service-entrance panel to each designated telephone location. Terminate in proper modular jack, complete with faceplates that match the electrical device faceplates. Allow 6 feet of cable to hang below ceiling joists. Telephone company to furnish, install, and connect any and all equipment (including grounding connection) up to and including their Standard Network Interface device.

    Installation shall be in accordance with any and all applicable *National Electrical Code*® and local code regulations.

26. **SERVICE ENTRANCE:**   Furnish and install in the workshop where indicated on the plans, one (1) 200-ampere, 120/240-volt, single-phase main pull-out type panel, complete with 200-ampere Class T fuses.

    Furnish and install all active and spare branch-circuit breakers as indicated in panel schedule. Panel to be series-rated for 100 000 amperes interrupting rating rms symmetrical.

27. Service-entrance underground lateral conductors to be furnished and installed by utility. Meter equipment (pedestal type) to be furnished by utility and installed by electrical contractor where indicated on plans. Electrical contractor to furnish and install all panels, conduits, fittings, conductors, and other materials required to complete the service-entrance installation from the demarcation point of the utility's equipment to and including the Main Panel.

28. Service-entrance conductors supplied by the electrical contractor shall be two (2) No. 2/0 THW, THHW, THWN, or XHHW phase conductors and one (1) No. 1 bare neutral conductor. Install 1½-inch electrical metallic tubing from Main Panel A to meter pedestal.

29. Bond and ground service-entrance equipment in accordance with latest edition of the *National Electrical Code,*® local, and utility code requirements. Install No. 4 AWG copper grounding electrode conductor.

30. **SUBPANEL:**   Furnish and install in recreation room where indicated on the plans, one (1) 20-circuit, 120/240-volt, single-phase, three-wire circuit breaker MLO load center complete with active and spare branch-circuit breakers as indicated in panel schedule. Load center to be rated 125 amperes, series-rated for 22 000 amperes interrupting rating rms symmetrical.

    Feed panel with three (3) No. 3 THHN or THWN conductors from 100-ampere, 2-pole breaker in main panel A. Install conductors in 1-inch electrical metallic tubing.

31. **CIRCUIT IDENTIFICATION:**   All panelboards shall be furnished with typed-card directories with proper designation of the branch-circuit loads, feeder loads and equipment served. The directories shall be located in the panel in a holder for clear viewing.

32. The electrical contractor shall seal and weatherproof all penetrations through foundations, exterior walls, and roofs.

33. Upon completion of the installation, the electrical contractor shall review and check the entire installation, clean equipment and devices, and remove surplus materials and rubbish from the owner's property, leaving the work in neat and clean order and in complete working condition. The electrical contractor shall be responsible for the removal of any cartons, debris, and rubbish for equipment installed by the electrical contractor, including equipment furnished by the owner or others and removed from the carton by the electrical contractor.

34. **SMOKE DETECTORS:**   Furnish and install all smoke detectors and associated wiring per manufacturer's instructions and all codes. See Clause 5. Detectors to be of the AC/DC type. Interconnect to provide simultaneous signaling.

35. **ALTERNATIVE BID   Low-Voltage, Remote-Control System:**   The electrical contractor shall submit an alternative bid on the following:

    Furnish and install a complete low-voltage remote-control system to accomplish the same results as would be obtained with the conventional switching arrangement as indicated on the electrical plans.

    In addition, furnish and install two 8-position master selector switches, one in the master bedroom and one in the front hall or as directed by the architect or owner. Outlets to be controlled by this switch to be selected by the owner.

    All low-voltage wiring to conform to the *National Electrical Code.*®

36. **SPECIAL-PURPOSE OUTLETS:**   Install, provide, and connect all wiring for all special-purpose outlets. Upon completion of the job, all fixtures and appliances shall be operating properly. See plans and other sections of the specifications for information as to who is to furnish the fixtures and appliances.

Schedule of Special-Purpose Outlets

SYMBOL	DESCRIPTION	VOLTS	HORSE-POWER	APPLIANCE AMPERE RATING	TOTAL APPLIANCE WATTAGE RATING (OR VA)	CIRCUIT AMPERE RATING	POLES	WIRE SIZE THHN	CIRCUIT NUMBER	COMMENTS
(▲) A	Hydromassage tub, master bedroom	120	1/2	10	1200	20	1	12	A9	Connect to Class "A" GFCI. Separate circuit.
(▲) B	Water pump	240	1	8	1920	20	2	12	A(5–7)	Run circuit to disconnect switch on wall adjacent to pump; protect with Fusetron dual-element time-delay fuses sized at 125% of motor's F.L.A.
(▲) C	Water heater: top element 2000W. Bottom element 3000W.	240	–0–	8.33 12.50 20.83	2000 3000 5000	20	2	12	A(6–8)	Connected for limited demand.
(▲) D	Dryer	120/240	120V 1/6 Motor Only	23.75	5700 Total	30	2	10	B(1–3)	Provide flush mounted 30-A dryer receptacle.
(▲) E	Overhead garage door opener	120	1/4	5.8	696	15	1	14	B14	Unit comes with 3W cord. Provide box-cover unit (fuse/switch). Install Fustat Type S fuse, 8 amperes. Unit has integral protection. Fustat fuses are additional "back-up" protection. Connect to garage lighting circuit.
(▲) F	Wall-mounted oven	120/240	–0–	27.5	6600	30	2	10	B(6–8)	
(▲) G	Countertop range	120/240	–0–	31	7450	40	2	8	B(2–4)	
(▲) H	Food waste disposer	120	1/3	7.2	864	20	1	12	B19	Controlled by S.P. switch on wall to the right of sink.
(▲) I	Dishwasher	120	1/3 Motor Only	Motor 7.20 Htr. 8.33 Total 15.53	1000W 864 1864	20	1	12	B5	
(▲) J	Heat/vent/light master B.R. bath	120	–0–	12.5	1500	20	1	12	A12	
(▲) K	Heat/vent/light front B.R. bath	120	–0–	12.5	1500	20	1	12	A11	
(▲) L	Attic exhaust fan	120	1/4	5.8	696	15	1	14	A10	Run circuit to 4" square box. Locate near fan in attic. Provide box-cover unit. Install Fustat Type S fuse, 8 amperes. Unit has integral protection. Fustat fuses provide additional back-up protection.
(▲) M	Electric furnace	240	1/3 Motor	Motor 3.5 Htr. 50.7 Total 54.2	13000	70	2	4	A(1–3)	The overcurrent device and branch-circuit conductors shall not be less than 125% of the total load of the heaters and motor. 54.2 × 1.25 = 67.75 (Section 424-3(b)).
(▲) N	Air conditioner	240	–0–	30	7200	40	2	8	A(2–4)	Compressor rated load amperes 27.8 Compressor locked rotor amperes 135.0 Condenser fan full load amperes 2.2 Condenser locked rotor amperes 4.5 Branch-circuit dual element fuse 40.0 Minimum circuit ampacity 37.5
(▲) O	Freezer	120	1/4	5.8	696	15	1	14	A13	Install single receptacle outlet. Do not provide GFCI protection.

APPENDIX

USEFUL FORMULAS

TO FIND	SINGLE PHASE	THREE PHASE	DIRECT CURRENT
AMPERES when kVA is known	$\dfrac{kVA \times 1000}{E}$	$\dfrac{kVA \times 1000}{E \times 1.73}$	not applicable
AMPERES when horsepower is known	$\dfrac{hp \times 746}{E \times \% \, eff. \times pf}$	$\dfrac{hp \times 746}{E \times 1.73 \times \% \, eff. \times pf}$	$\dfrac{hp \times 746}{E \times \% \, eff.}$
AMPERES when kilowatts are known	$\dfrac{kW \times 1000}{E \times pf}$	$\dfrac{kW \times 1000}{E \times 1.73 \times pf}$	$\dfrac{kW \times 1000}{E}$
KILOWATTS	$\dfrac{I \times E \times pf}{1000}$	$\dfrac{I \times E \times 1.73 \times pf}{1000}$	$\dfrac{I \times E}{1000}$
KILOVOLT AMPERES	$\dfrac{I \times E}{1000}$	$\dfrac{I \times E \times 1.73}{1000}$	not applicable
HORSEPOWER	$\dfrac{I \times E \times \% \, eff. \times pf}{746}$	$\dfrac{I \times E \times 1.73 \times \% \, eff. \times pf}{746}$	$\dfrac{I \times E \times \% \, eff.}{746}$
WATTS	$E \times I \times pf$	$E \times I \times 1.73 \times pf$	$E \times I$

I = amperes	E = volts	kW = kilowatts	kVA = kilovolt-amperes
hp = horsepower		% eff. = percent efficiency e.g., 90% eff. is 0.90	pf = power factor e.g., 95% pf is 0.95

EQUATIONS BASED ON OHM'S LAW:

P = POWER, IN WATTS
I = CURRENT, IN AMPERES
R = RESISTANCE, IN OHMS
E = ELECTROMOTIVE FORCE, IN VOLTS

METRIC SYSTEM OF MEASUREMENT

The following table provides useful conversions of English Customary terms to SI terms, and SI terms to English Customary terms. The metric system is known as the International System of Units (SI), taken from the French *Le Système International d'Unites*. Whenever the slant line / is found, say it as "per." Be practical when using the values in the following table. Present-day use of calculators and computers provide many "places" beyond the decimal point. You must decide how accurate your results must be when performing a calculation. When converting centimeters to feet, for example, one could be reasonably accurate by rounding the value of 0.03281 to 0.033. When a manufacturer describes a product's measurement using metrics, but does not physically change the product, it is referred to as a "soft conversion." When a manufacturer actually changes the physical measurements of the product to standard metric size or a rational whole number of metric units, it is referred to as a "hard conversion." When rounding off numbers, be sure to round off in such a manner that the final result does not violate a "maximum" or "minimum" value, such as might be required by the *National Electrical Code.*® The rule for discarding digits is that if the digits to be discarded begin with a 5 or more, increase the last digit retained by one. For example: If rounded to three digits, 6.3745 would become 6.37. If rounded to four digits, it would become 6.375. The following information has been developed from the latest U.S. Department of Commerce, National Institute of Standards and Technology publications covering the metric system.

Multiply This Unit(s)	By This Factor	To Obtain This Unit(s)
acre	4 046.9	square meters (m²)
acre	43 560	square feet (ft²)
ampere hour	3 600	coulombs (C)
angstrom	0.1	nanometer (nm)
atmosphere	101.325	kilopascals (kPa)
atmosphere	33.9	feet of water (at 4°C)
atmosphere	29.92	inches of mercury (at 0°C)
atmosphere	0.76	meter of mercury (at 0°C)
atmosphere	0.007 348	ton per square inch
atmosphere	1.058	tons per square foot
atmosphere	1.0333	kilograms per square centimeter
atmosphere	10 333	kilograms per square meter
atmosphere	14.7	pounds per square inch
atmosphere	100	kilopascals (kPa)
bar	0.158 987 3	cubic meter (m³)
barrel (oil, 42 U.S. gallons)	158.987 3	liters (L)
barrel (oil, 42 U.S. gallons)	0.002 359 737	cubic meter (m³)
board foot	0.035 239 07	cubic meter (m³)
bushel	778.16	foot-pounds
Btu	252	gram-calories
Btu	0.000 393 1	horsepower-hour
Btu	1 054.8	joules (J)
Btu	1.055 056	kilojoules (kJ)
Btu	0.000 293 1	kilowatt-hour (kWh)
Btu	0.000 393 1	horsepower (hp)
Btu per hour	0.293 071 1	watt (W)
Btu per hour	1.899 108	Kilojoules per kelvin (kJ/K)
Btu per degree Fahrenheit	2.326	Kilojoules per kilogram (kJ/kg)
Btu per pound	1.055 056	kilowatts (kW)
Btu per second	4.184	joules (J)
calorie	0.003 968 3	Btu
calorie, gram	10.763 9	candelas per meter squared (cd/m²)
candela per feet squared (cd/ft²)		

Note: The former term *candlepower* has been replaced with the term candela.

candela per meter squared (cd/m²)	0.092 903	candela per feet squared (cd/ft²)
candela per meter squared (cd/m²)	0.291 864	footlambert
candela per square inch (cd/in²)	1 550.003	candelas per square meter (cd/m²)

Celsius = (Fahrenheit − 32) × 5/9
Celsius = (Fahrenheit − 32) × 0.555555
Celsius = (0.556 × Fahrenheit) − 17.8

Note: The term *centigrade* was officially discontinued in 1948, and was replaced by the term *Celsius.* The term *centigrade* may still be found in some publications.

Multiply This Unit(s)	By This Factor	To Obtain This Unit(s)
centimeter (cm)	0.032 81	foot (ft)
centimeter (cm)	0.393 7	inch (in)
centimeter (cm)	0.01	meter (m)
centimeter (cm)	10	millimeters (mm)
centimeter (cm)	393.7	mils
centimeter (cm)	0.010 94	yard
circular mil	0.000 005 067	square centimeter (cm²)
circular mil	0.785 4	square mil
circular mil	0.000 000 785 4	square inch (in²)
cubic centimeter (cm³)	0.061 02	cubic inch (in³)
cubic foot per second	0.028 316 85	cubic meter per second (m³/s)
cubic foot per minute	0.000 471 947	cubic meter per second (m³/s)
cubic foot per minute	0.471 947	liter per second (L/s)
cubic inch (in³)	16.39	cubic centimeters (cm³)
cubic inch (in³)	0.000 578 7	cubic foot (ft³)
cubic meter (m³)	35.31	cubic feet (ft³)
cubic yard per minute	12.742 58	liters per second (L/s)
cup (c)	0.236 56	liter (L)
decimeter	0.1	meter (m)
dekameter	10	meters (m)
Fahrenheit = (9/5 Celsius) + 32		
Fahrenheit = (Celsius × 1.8) + 32		
fathom	1.828 804	meters (m)
fathom	6.0	feet
foot	30.48	centimeters (cm)
foot	12	inches
foot	0.000 304 8	kilometer (km)
foot	0.304 8	meter (m)
foot	0.000 189 4	mile (statute)
foot	304.8	millimeters (mm)
foot	12 000	mils
foot	0.333 33	yard
foot, cubic (ft³)	0.028 316 85	cubic meter (m³)
foot, cubic (ft³)	28.316 85	liters (L)
foot, board	0.002 359 737	cubic meter (m³)
cubic feet per second (ft³/s)	0.028 316 85	cubic meter per second (m³/s)
cubic feet per minute (ft³/min)	0.000 471 947	cubic meter per second (m³/s)
cubic feet per minute (ft³/min)	0.471 947	liter per second (L/s)
foot, square (ft²)	0.092 903	square meter (m²)
footcandle	10.763 91	lux (lx)
footlambert	3.426 259	candelas per square meter (cd/m²)
foot of water	2.988 98	kilopascals (kPa)
foot pound	0.001 286	Btu
foot pound-force	1 055.06	joules (J)
foot pound-force per second	1.355 818	joules (J)
foot pound-force per second	1.355 818	watts (W)
foot per second	0.304 8	meter per second (m/s)
foot per second squared	0.304 8	meter per second squared (m/s²)
gallon (U.S. liquid)	3.785 412	liters (L)
gallons per day	3.785 412	liters per day (L/d)
gallons per hour	1.051 50	milliliters per second (mL/s)
gallons per minute	0.063 090 2	liter per second (L/s)
gauss	6.452	lines per square inch
gauss	0.1	millitesla (mT)
gauss	0.000 000 064 52	weber per square inch
grain	64.798 91	milligrams (mg)
gram (g) (a little more than the weight of a paper clip)	0.035 274	ounce (avoirdupois)
gram (g)	0.002 204 6	pound (avoirdupois)
gram per meter (g/m)	3.547 99	pounds per mile (lb/mile)
grams per square meter (g/m²)	0.003 277 06	ounces per square foot (oz/ft²)
grams per square meter (g/m²)	0.029 494	ounces per square yard (oz/yd²)
gravity (standard acceleration)	9.806 65	meters per second squared (m/s²)
quart (U.S. liquid)	0.946 352 9	liter (L)
horsepower (550 ft•lbf/s)	0.745 7	kilowatt (kW)
horsepower	745.7	watts (W)
horsepower hours	2.684 520	megajoules (MJ)
inch per second squared (in/s²)	0.025 4	meter per second squared (m/s²)

Multiply This Unit(s)	By This Factor	To Obtain This Unit(s)
inch	2.54	centimeters (cm)
inch	0.254	decimeter (dm)
inch	0.025 4	meter (m)
inch	25.4	millimeters (mm)
inch	1 000	mils
inch	0.027 78	yard
inch, cubic (in³)	16 387.1	cubic millimeters (mm³)
inch, cubic (in³)	16.387 06	cubic centimeters (cm³)
inch, cubic (in³)	645.16	square millimeters (mm²)
inches of mercury	3.386 38	kilopascals (kPa)
inches of mercury	0.033 42	atmosphere
inches of mercury	1.133	feet of water
inches of water	0.248 84	kilopascal (kPa)
inches of water	0.073 55	inch of mercury
joule (J)	0.737 562	foot pound-force (ft•lbf)
kilocandela per meter squared (kcd/m²)	0.314 159	lambert˙
kilogram (kg)	2.204 62	pounds (avoirdupois)
kilogram (kg)	35.274	ounces (avoirdupois)
kilogram per meter (kg/m)	0.671 969	pound per foot (lb/ft)
kilogram per square meter (kg/m²)	0.204 816	pound per square foot (lb/ft²)
kilogram meter squared (kg•m²)	23.730 4	pounds foot squared (lb•ft²)
kilogram meter squared (kg•m²)	3 417.17	pounds inch squared (lb•in²)
kilogram per cubic meter (kg/m³)	0.062 428	pound per cubic foot (lb/ft³)
kilogram per cubic meter (kg/m³)	1.685 56	pound per cubic yard (lb/yd³)
kilogram per second (kg/s)	2.204 62	pounds per second (lb/s)
kilojoule (kJ)	0.947 817	Btu
kilometer (km)	1 000	meters (m)
kilometer (km)	0.621 371	mile (statute)
kilometer (km)	1 000 000	millimeters (mm)
kilometer (km)	1 093.6	yards
kilometer per hour (km/h)	0.621 371	mile per hour (mph)
kilometer squared (km²)	0.386 101	square mile (mile²)
kilopound-force per square inch	6.894 757	megapascals (MPa)
kilowatts (kW)	56.921	Btus per minute
kilowatts (kW)	1.341 02	horsepower (hp)
kilowatts (kW)	1 000	watts (W)
kilowatt-hour (kWh)	3 413	Btus
kilowatt-hour (kWh)	3.6	megajoules (MJ)
knots	1.852	kilometers per hour (km/h)
lamberts˙	3 183.099	candelas per square meter (cd/m²)
lamberts˙	3.183 01	kilocandelas per square meter (kcd/m²)
liter (L)	0.035 314 7	cubic foot (ft³)
liter (L)	0.264 172	gallon (U.S. liquid)
liter (L)	2.113	pints (U.S. liquid)
liter (L)	1.056 69	quarts (U.S. liquid)
liter per second (L/s)	2.118 88	cubic feet per minute (ft³/min)
liter per second (L/s)	15.850 3	gallons per minute (gal/min)
liter per second (L/s)	951.022	gallons per hour (gal/hr)
lumen per square foot (lm/ft²)	10.763 9	lux (lx)
lumen per square foot (lm/ft²)	1.0	footcandles
lumen per square foot (lm/ft²)	10.763	lumens per square meter (lm/m²)
lumen per square meter (lm/m²)	1.0	lux (lx)
lux (lx)	0.092 903	lumen per square foot (foot-candle)
maxwell	10	nanowebers (nWb)
megajoule (MJ)	0.277 778	kilowatt hours (kWh)
meter (m)	100	centimeters (cm)
meter (m)	0.546 81	fathom
meter (m)	3.2809	feet
meter (m)	39.37	inches
meter (m)	0.001	kilometer (km)
meter (m)	0.000 621 4	mile (statute)
meter (m)	1 000	millimeters (mm)
meter (m)	1.093 61	yards
meter, cubic (m³)	1.307 95	cubic yards (yd³)
meter, cubic (m³)	35.314 7	cubic feet (ft³)
meter, cubic (m³)	423.776	board feet

Multiply This Unit(s)	By This Factor	To Obtain This Unit(s)
meter per second (m/s)	3.280 84	feet per second (ft/s)
meter per second (m/s)	2.236 94	miles per hour (mph)
meter, squared (m²)	1.195 99	square yards (yd²)
meter, squared (m²)	10.763 9	square feet (ft²)
mho per centimeter (mho/cm)	100	siemens per meter (S/m)

Note: The older term *mho* has been replaced with *siemens*. The term *mho* may still be found in some publications.

micro inch	0.025 4	micrometer (μm)
mil	25.4	micrometers (μm)
mil	0.025 4	millimeter (mm)
miles per hour	1.609 344	kilometers per hour (km/h)
miles per hour	0.447 04	meter per second (m/s)
miles per gallon	0.425 143 7	kilometer per liter (km/L)
miles	1.609 344	kilometers (km)
miles	5 280	feet
miles	1 609	meters (m)
miles	1 760	yards
miles (nautical)	1.852	kilometers (km)
miles squared	2.590 000	kilometers squared (km²)
millibar	0.1	kilopascal (kPa)
milliliter (mL)	0.061 023 7	cubic inch (in³)
milliliter (mL)	0.033 814	fluid ounce (U.S.)
millimeter (about the thickness of a dime)	0.1	centimeter (cm)
millimeter (mm)	0.003 280 8	foot
millimeter (mm)	0.039 370 1	inch
millimeter (mm)	0.001	meter (m)
millimeter (mm)	39.37	mils
millimeter (mm)	0.001 094	yard
millimeter squared (mm²)	0.001 550	square inch (in²)
millimeter cubed (mm³)	0.000 061 023 7	cubic inches (in³)
millimeter of mercury	0.133 322 4	kilopascal (kPa)
ohm	0.000 001	megohm
ohm	1 000 000	micro ohms
ohm circular mil per foot	1.662 426	nano ohms meter (nΩ•m)
oersted	79.577 47	amperes per meter (A/m)
ounce (avoirdupois)	28.349 52	grams (g)
ounce (avoirdupois)	0.062 5	pound (avoirdupois)
ounce, fluid	29.573 53	milliliters (mL)
ounce (troy)	31.103 48	grams (g)
ounce per foot squared (oz/ft²)	305.152	grams per meter squared (g/m²)
ounces per gallon (U.S. liquid)	7.489 152	grams per liter (g/L)
ounce per yard squared (oz/yd²)	33.905 7	grams per meter squared (g/m²)
pica	4.217 5	millimeters (mm)
pint (U.S. liquid)	0.473 176 5	liter (L)
pint (U.S. liquid)	473.177	milliliters (mL)
pound (avoirdupois)	453.592	grams (g)
pound (avoirdupois)	0.453 592	kilograms (kg)
pound (avoirdupois)	16	ounces (avoirdupois)
poundal	0.138 255	newton (N)
pound foot (lb•ft)	0.138 255	kilogram meter (kg•m)
pound foot per second	0.138 255	kilogram meter per second (kg•m/s)
pound foot squared (lb•ft²)	0.042 140 1	kilogram meter squared (kg•m²)
pound-force	4.448 222	newtons (N)
pound-force foot	1.355 818	newton meters (N•m)
pound-force inch	0.112 984 8	newton meter (N•m)
pound-force per square inch	6.894 757	kilopascals (kPA)
pound-force per square foot	0.047 880 26	kilopascal (kPa)
pound per cubic foot (lb/ft³)	16.018 46	kilograms per cubic meter (kg/m³)
pound per foot (lb/ft)	1.488 16	kilograms per meter (kg/m)
pound per foot squared (lb/ft²)	4.882 43	kilograms per meter squared (kg/m²)
pound per foot cubed (lb/ft³)	16.018 5	kilograms per meter cubed (kg/m³)
pound per gallon (U.S. liquid)	119.826 4	grams per liter (g/L)
pound per second (lb/s)	0.453 592	kilogram per second (kg/s)
pound inch squared (lb•in²)	292.640	kilograms millimeter squared (kg•mm²)
pound per mile	0.281 849	gram per meter (g/m)
pound square foot (lb•ft²)	0.042 140 11	kilogram square meter (kg/m²)

Multiply This Unit(s)	By This Factor	To Obtain This Unit(s)
pound per cubic yard (lb/yd³)	0.593 276	kilogram per cubic meter (kg/m³)
quart (U.S. liquid)	946.353	milliliters (mL)
square centimeter (cm²)	197 300	circular mils
square centimeter (cm²)	0.001 076	square foot (ft²)
square centimeter (cm²)	0.155	square inch (in²)
square centimeter (cm²)	0.000 1	square meter (m²)
square centimeter (cm²)	0.000 119 6	square yard (yd²)
square foot (ft²)	144	square inches (in²)
square foot (ft²)	0.092 903 04	square meter (m²)
square foot (ft²)	0.111 1	square yard (yd²)
square inch (in²)	1 273 000	circular mils
square inch (in²)	6.4516	square centimeters (cm²)
square inch (in²)	0.006 944	square foot (ft²)
square inch (in²)	645.16	square millimeters (mm²)
square inch (in²)	1 000 000	square mils (m²)
square meter (m²)	10.764	square feet (ft²)
square meter (m²)	1 550	square inches (in²)
square meter (m²)	0.000 000 386 1	square mile
square meter (m²)	1.196	square yards (yd²)
square mil	1.273	circular mils
square mil	0.000 001	square inch (in²)
square mile	2.589 988	square kilometers (km²)
square millimeter (mm²)	1 973	circular mils
square yard (yd²)	0.836 127 4	square meter (m²)
tablespoon (tbsp)	14.786 75	milliliters (mL)
teaspoon (tsp)	4.928 916 7	milliliters (mL)
therm	105.480 4	megajoules (MJ)
ton (long) (2 240 lb)	1 016.047	kilograms (kg)
ton (long) (2 240 lb)	1.016 047	metric tons (t)
ton, metric	2 204.62	pounds (avoirdupois)
ton, metric	1.102 31	tons, short (2 000 lb)
ton, refrigeration	12 000	Btus per hour
ton, refrigeration	4.716 095 9	horsepower-hours
ton, refrigeration	3.516 85	kilowatts (kW)
ton (short) (2 000 lb)	907.185	kilograms (kg)
ton (short) (2 000 lb)	0.907 185	metric ton (t)
ton per cubic meter (t/m³)	0.842 778	ton per cubic yard (ton/yd³)
ton per cubic yard (ton/yd³)	1.186 55	tons per cubic meter (ton/m³)
torr	133.322 4	pascals (Pa)
watt (W)	3.412 14	Btus per hour (Btu/hr)
watt (W)	0.001 341	horsepower
watt (W)	0.001	kilowatt (kW)
watt-hour (Wh)	3.413	Btus
watt-hour (Wh)	0.001 341	horsepower-hour
watt-hour (Wh)	0.001	kilowatt-hour (kW/h)
yard	91.44	centimeters (cm)
yard	3	feet
yard	36	inches
yard	0.000 914 4	kilometer (km)
yard	0.914 4	meter (m)
yard	914.4	millimeters
yard, cubic (yd³)	0.764 55	cubic meter (m³)
yard, squared (yd²)	0.836 127	meter squared (m²)

˙ These terms are no longer used, but may still be found in some publications.

SINGLE-FAMILY DWELLING SERVICE-ENTRANCE CALCULATIONS

1. General Lighting Load *(Section 220-3(b))*

_____ sq. ft. @ 3 VA per sq. ft.   = _____ VA

2. Minimum Number of Lighting Branch-Circuits

$$\frac{\text{Line 1 VA}}{120} = \frac{\qquad}{120} \qquad\qquad = \underline{\qquad}\ \text{amperes}$$

then, $\dfrac{\text{amperes}}{15} = \dfrac{\qquad}{15} = \underline{\qquad}$ 15-ampere branch-circuits

3. Small Appliance Load *(Sections 220-4(b) and 220-16(a))*
(Minimum of two 20-ampere branch-circuits)

_____ branch-circuits @ 1,500 VA each   = _____ VA

4. Laundry Branch-Circuit *(Sections 220-4(c) and 220-16(b))*
(Minimum of one 20-ampere branch-circuit)

_____ branch-circuits @ 1,500 VA each   = _____ VA

5. Total General Lighting, Small Appliance, and Laundry Load
Lines 1 + 3 + 4   = _____ VA

6. Net Computed General Lighting, Small Appliance, and Laundry Loads. Apply Demand Factors from *Table 220-11.* **Do not include ranges, ovens, and "fastened-in-place" appliances.**

a. First 3,000 VA @ 100%   = _3,000_ VA

b. Line 5 _____ − 3,000 = _____ @ 35% = _____ VA

Total a + b = _____ VA

7. Electric Range, Wall-Mounted Ovens, Counter-Mounted Cooking Units *(Table 220-19)*   = _____ VA

8. Electric Dryer *(Table 220-18)*   = _____ VA

9. Electric Furnace *(Section 220-15)*
Air Conditioner, Heat Pump *(Article 440)*
(Enter largest value, *Section 220-21*)   = _____ VA

10. Net Computed General Lighting, Small Appliance, Laundry, Ranges, Ovens, Cook Tops Units, HVAC
Lines 6 + 7 + 8 + 9   = _____ VA

11. **List "Fastened-in-Place" Appliances *in addition* to Electric Ranges, Air Conditioners, Clothes Dryers, Space Heaters**

Appliance	VA Load
Water Heater	_____ VA
Dishwasher	_____ VA
Garage Door Opener	_____ VA
Food Waste Disposer	_____ VA
_____	_____ VA
_____	_____ VA
_____	_____ VA
_____	_____ VA
_____	_____ VA
Total	_____ VA

12. **Apply 75% Demand Factor *(Section 220-17)* if four or more "Fastened-in-Place" Appliances. If less than four, figure at 100%. Do not include electric ranges, clothes dryers, or HVAC equipment.**

    Line 11 _____ × 0.75   = _____ VA

13. **Total Computed Load (Lighting, Small Appliance, Laundry, Ranges, Dryers, HVAC, "Fastened-in-Place" Appliances)**

    Line 10 _____ + Line 12 _____   = _____ VA

14. **Add 25% of Largest Motor *(Section 220-14)***   = _____ VA

15. **Total Line 13 + Line 14**   = _____ VA

16. **Minimum Ampacity for Ungrounded Service-Entrance Conductors**

    $$\text{Amperes} = \frac{\text{Line 15}}{240} = \frac{\rule{2cm}{0.4pt}}{240}$$ = _____ amperes

17. **Minimum Ampacity for Neutral Service-Entrance Conductor, *Sections 220-22, Note 3* to *Table 310-16*. Do Not Include Straight 240-Volt Loads.**

    a. Line 6 _____   = _____ VA

    b. Line 7 _____ × 0.70   = _____ VA

    c. Line 8 _____ × 0.70   = _____ VA

    d. Line 11 (include only 120-volt loads)

_____	_____ VA
_____	_____ VA
_____	_____ VA
_____	_____ VA
_____	_____ VA
Total	_____ VA

e. Line d Total @ 75% Demand Factor if four or more, otherwise use 100%.

Line d Total _____ × 0.75   = _____ VA

f. Add 25% of largest 120-volt motor

_____ × 0.25   = _____ VA

g. Total a + b + c + e + f   = _____ VA

$$\text{Amperes} = \frac{\text{Line g}}{240} = \frac{}{240}$$ = _____ amperes

18. Ungrounded Conductor Size (copper) *(Note 3, Table 310-16)*   _____

Note: *Note 3*, *Table 310-16* may be used only for 120/240-volt, 3-wire, single-phase service-entrance, service lateral, and feeder conductors that serve as the main power feeder to a dwelling unit.

19. Neutral Conductor Size (copper) *(Section 220-22)*   _____

Note: *Note 3* of *Table 310-16* states that the neutral conductor is permitted to be smaller than the ungrounded "hot" conductors provided the requirements of *Sections 215-2*, *220-22*, and *230-22* are met.

20. Grounding Electrode Conductor Size (copper) *(Table 250-94)*   _____

21. Conduit Size   _____

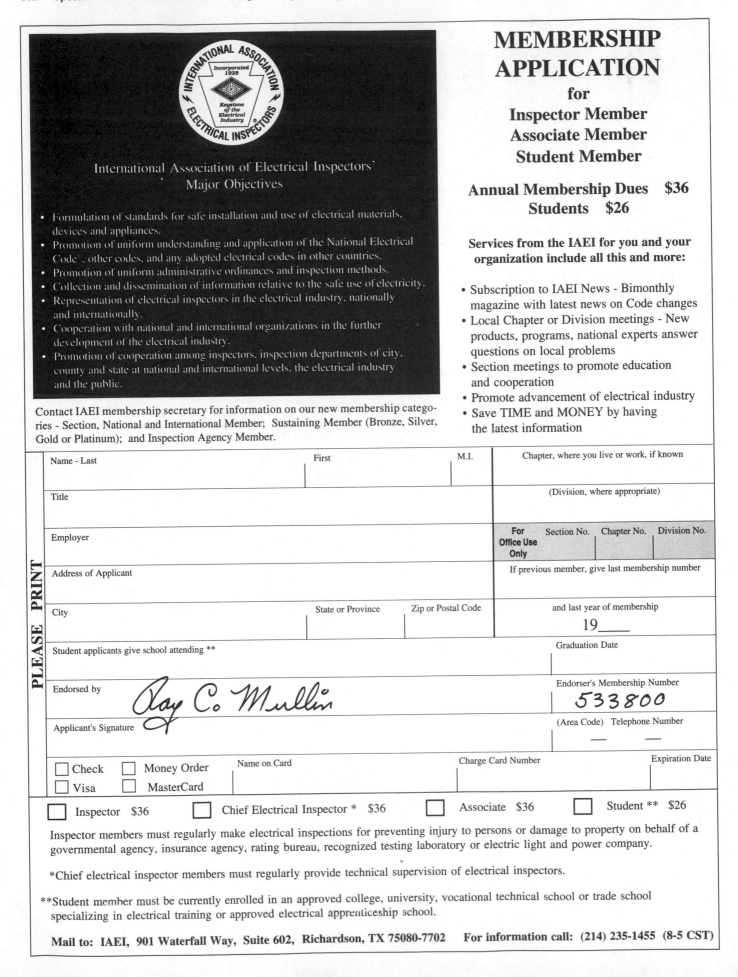

MEMBERSHIP APPLICATION

for
Inspector Member
Associate Member
Student Member

Annual Membership Dues $36
Students $26

Services from the IAEI for you and your organization include all this and more:

- Subscription to IAEI News - Bimonthly magazine with latest news on Code changes
- Local Chapter or Division meetings - New products, programs, national experts answer questions on local problems
- Section meetings to promote education and cooperation
- Promote advancement of electrical industry
- Save TIME and MONEY by having the latest information

International Association of Electrical Inspectors'
Major Objectives

- Formulation of standards for safe installation and use of electrical materials, devices and appliances.
- Promotion of uniform understanding and application of the National Electrical Code, other codes, and any adopted electrical codes in other countries.
- Promotion of uniform administrative ordinances and inspection methods.
- Collection and dissemination of information relative to the safe use of electricity.
- Representation of electrical inspectors in the electrical industry, nationally and internationally.
- Cooperation with national and international organizations in the further development of the electrical industry.
- Promotion of cooperation among inspectors, inspection departments of city, county and state at national and international levels, the electrical industry and the public.

Contact IAEI membership secretary for information on our new membership categories - Section, National and International Member; Sustaining Member (Bronze, Silver, Gold or Platinum); and Inspection Agency Member.

PLEASE PRINT

Name - Last	First	M.I.	Chapter, where you live or work, if known
Title			(Division, where appropriate)
Employer			**For Office Use Only** — Section No. / Chapter No. / Division No.
Address of Applicant			If previous member, give last membership number
City	State or Province	Zip or Postal Code	and last year of membership 19___
Student applicants give school attending **			Graduation Date
Endorsed by *Ray C. Mullin*			Endorser's Membership Number 533800
Applicant's Signature			(Area Code) Telephone Number ___ ___

☐ Check   ☐ Money Order   ☐ Visa   ☐ MasterCard
Name on Card — Charge Card Number — Expiration Date

☐ Inspector $36   ☐ Chief Electrical Inspector * $36   ☐ Associate $36   ☐ Student ** $26

Inspector members must regularly make electrical inspections for preventing injury to persons or damage to property on behalf of a governmental agency, insurance agency, rating bureau, recognized testing laboratory or electric light and power company.

*Chief electrical inspector members must regularly provide technical supervision of electrical inspectors.

**Student member must be currently enrolled in an approved college, university, vocational technical school or trade school specializing in electrical training or approved electrical apprenticeship school.

Mail to: IAEI, 901 Waterfall Way, Suite 602, Richardson, TX 75080-7702   For information call: (214) 235-1455 (8-5 CST)

CODE INDEX

INDEX

NOTE: **Boldfaced** numbers indicate non-text material.